Results and Problems in Cell Differentiation

Volume 78

Results and Problems in Cell Differentiation is an up-to-date book series that presents and explores selected questions of cell and developmental biology. Each volume focuses on a single, well-defined topic. Reviews address basic questions and phenomena, but also provide concise information on the most recent advances. Together, the volumes provide a valuable overview of this exciting and dynamically expanding field. This series is indexed in PubMed.

Malgorzata Kloc • Jarek Wosik
Editors

Mechanotransduction and Mechanosensing

Roles and Effects on Biological Systems under Physiological and Pathological Conditions

Springer

Editors
Malgorzata Kloc
Houston Methodist Research Institute and
The Houston Methodist Hospital
Houston, TX, USA

Jarek Wosik
Electrical and Computer Engineering
Department and Texas Center for
Superconductivity
University of Houston
Houston, TX, USA

ISSN 0080-1844 ISSN 1861-0412 (electronic)
Results and Problems in Cell Differentiation
ISBN 978-3-032-19508-1 ISBN 978-3-032-19509-8 (eBook)
https://doi.org/10.1007/978-3-032-19509-8

This Springer imprint is published by the registered company Springer Nature Switzerland AG
The registered company address is: Gewerbestrasse 11, 6330 Cham, Switzerland

Preface

All organisms on our planet are constantly exposed to mechanical forces generated by animate and inanimate environments. Thus, an organism's survival depends on how well it responds and adjusts to these forces. Besides mechanical forces from the environment, all living beings generate internal (intraorganellar, intracellular, or intra-organ) mechanical forces that must be sensed and reacted to. In response to the combination of external and internal mechanical forces, all organisms developed an extremely sophisticated and complex system of mechanosensors and mechanotransduction pathways that convert mechanical force cues into structural, biochemical, and functional information and adaptations.

This volume presents three interconnected parts exploring mechanoreception and mechanotransduction at the molecular, cellular, tissue, and organ levels. Part I describes mechanoreceptors, mechanosensors, channels, and organelles enabling cells to detect and respond to mechanical forces generated by external, intracellular, and intra-organ triggers. It also explains how mathematical and computational models of mechanosensing complement experiments and generate testable hypotheses. The last chapter of Part I presents the digital image correlation toolkit (StrainMapperJ) that can be used for the exploration and quantification of deforming tissue mechanics.

Part II examines how mechanotransduction pathways and intracellular and extracellular mechanical forces act at higher, systemic levels. The first chapter introduces the concept of network phenomena, in which simpler constituents cooperate to transmit and respond to mechanical force and stress at a higher organizational level. It also reviews the impact of emergent intra- and extracellular network models on studies of disease pathogenesis and progression, drug effects, and treatments. The following chapters describe how mechanical forces affect cellular membranes, drive gene expression, remodel the cell nucleus, regulate interactions with the extracellular matrix, control mitotic and meiotic cell divisions, and affect the timing and fidelity of chromosome segregation.

Part III examines tissue- and organ-specific features of mechanotransduction in skeletal, musculoskeletal, and neural systems. It reviews how a mechanosensitive Runt-related transcription factor 2, along with its epigenetic regulation, coordinates osteoblast differentiation and bone remodeling. It also discusses potential therapeutic approaches targeting mechanosensing components to promote regeneration and wound healing.

This book, by covering organellar, molecular, and cellular aspects of mechanosensing and mechanotransduction, offers an up-to-date overview of the mechanisms and challenges and aims to inspire the development of novel mechano-targeted therapies for many human diseases.

Houston, TX, USA

Malgorzata Kloc
Jarek Wosik

Contents

Part I
Mechanoperception Molecules and Organelles, Mechanosensing Models and Mechanics Quantification Tools

Chapter 1
Mechanosensitive TMEM63 Ion Channels: Structure, Gating Mechanism, and Emerging Physiological and Pathological Roles

Jin Ou, Yunqing Zhou, Chaoqun Chen, Jincheng Wang, and Wang Zheng

Abstract The TMEM63 family of mechanosensitive ion channels has recently emerged as a distinct group of mammalian force sensors evolutionarily related to plant OSCA channels. Structural and functional studies establish TMEM63s as monomeric, high-threshold mechanotransducers featuring an extended IL2 domain, a hydrophobic latch at the membrane interface, and a lateral lipid-accessible pore that may enable both ion conduction and lipid scrambling. These features define a unique gating mechanism, distinct from other mechanosensitive ion channels. Growing genetic and animal evidence indicates that TMEM63s play diverse and essential roles in the nervous system and other tissues, where altered mechanotransduction can lead to severe neurological and systemic dysfunctions. In this review, we summarize recent progress in the structure, gating mechanisms, and physiological functions of TMEM63 channels, with an emphasis on their force-sensing elements, lipid scrambling activities, and implications in human diseases. We also outline future directions for understanding how TMEM63s convert physical forces into biological signals in health and disease.

1.1 Mechanotransduction

The survival of all organisms depends on their ability to sense and respond to the physical world. Mechanotransduction, the process by which cells convert mechanical forces into biochemical signals, underlies this fundamental capability (Wozniak and Chen 2009). These mechanically driven signals influence gene expression,

---

J. Ou · C. Chen · J. Wang · W. Zheng (✉)
Department of Neuroscience, University of Wisconsin-Madison, Madison, WI, USA
e-mail: wzheng86@wisc.edu

Y. Zhou
Departments of Otolaryngology & Neurology, Boston Children's Hospital and Harvard Medical School, Boston, MA, USA

M. Kloc, J. Wosik (eds.), *Mechanotransduction and Mechanosensing*, Results and Problems in Cell Differentiation 78, https://doi.org/10.1007/978-3-032-19509-8_1

activate signaling pathways, and modulate cell behavior. Over the past decade, accumulating evidence has established that mechanotransduction is implicated in diverse physiological processes, such as hearing, touch, vascular tone regulation, and neuronal growth, while its disruption contributes to various diseases, including cardiovascular disease, cancer metastasis, and neurological disorders (Wozniak and Chen 2009; Douguet and Honore 2019; Kefauver et al. 2020; Cox et al. 2019; Jin et al. 2020), which together emphasize pivotal roles of mechanotransduction in human physiology and disease.

At the molecular level, mechanotransduction is mediated by structural elements and force-sensing proteins. The cytoskeleton, consisting of F-actin filaments, microtubules, and intermediate filaments, serves as the structural framework that maintains cell shape and provides resistance to both external and internal forces (Chuang and Chen 2022). Within this network, focal adhesion complexes link extracellular matrix signals to the intracellular cytoskeletal system through proteins such as integrin, talin, and tensin (Jansen et al. 2017). The G-protein-coupled receptor (GPCR) family contains numerous membrane receptors that detect mechanical stimuli to affect cell behavior (Wilde et al. 2022). The plasma membrane contains mechanosensitive ion channels (MSCs) that control ion movement which affects cellular functions (Ranade et al. 2015). These molecules work together to detect mechanical forces and generate appropriate cellular responses.

MSCs are evolutionarily conserved pore-forming membrane proteins that convert mechanical stimuli into ion fluxes, providing a direct and rapid means of mechanotransduction (Kefauver et al. 2020; Ranade et al. 2015). In bacteria, channels such as MscL and MscS open in response to increased membrane tension during osmotic shock, releasing solutes to prevent cell lysis (Perozo 2006). In mammals, MSCs have diversified to mediate complex processes. Prominent and well-studied examples include PIEZOs and TMCs. PIEZO channels, including PIEZO1 and PIEZO2, were first identified as touch receptors but were subsequently found to mediate many other physiologies, such as blood pressure regulation, vascular development, and interoception (Murthy et al. 2017; Szczot et al. 2021; Xiao 2024). TMCs, especially TMC1 and TMC2, are expressed in sensory hair cells of the inner ear to mediate hearing and balance by detecting sound- or movement-evoked mechanical stimuli (Kawashima et al. 2011; Pan et al. 2018; Pan et al. 2013; Zheng and Holt 2021). TMEM63s, comprising TMEM63A, TMEM63B, and TMEM63C, constitute a newly identified family of MSCs, emerging as critical force-sensing proteins in mammals with characteristics distinct from PIEZOs and TMCs (Hou et al. 2014; Murthy et al. 2018; Zheng et al. 2023). This review focuses on current progress in the molecular mechanisms of TMEM63 gating and physiological and pathological implications.

1.2 Discovery of Mammalian Mechanosensitive TMEM63 Channels

The identification of TMEM63 proteins as MSCs in mammals tracks back to plant biology, where the homologous OSCA proteins in *Arabidopsis thaliana* were first characterized as hyperosmosensitive ion channels in guard cells. In 2014, two groups independently reported OSCA1 as a hyperosmotic stress-activated calcium channel essential for water homeostasis and osmotic stress responses (Hou et al. 2014; Yuan et al. 2014). Four years later, in 2018, two other groups demonstrated that OSCAs form bona fide MSCs, by electrophysiological recordings in the presence of mechanical stimuli, and solved atomic structures of OSCA1.1 and OSCA1.2 with cryogenic-electron microscopy (cryo-EM) (Murthy et al. 2018; Jojoa-Cruz et al. 2018; Liu et al. 2018; Zhang et al. 2018). These structural studies revealed a dimeric assembly of OSCAs and identified a putative ion permeation pathway in each subunit (Jojoa-Cruz et al. 2018; Liu et al. 2018; Zhang et al. 2018). Through genome sequence alignment, Hou et al. identified TMEM63 as the mammalian homologs of OSCAs (Hou et al. 2014), and subsequent functional studies in heterologous HEK293T cells by Murthy et al. revealed membrane stretch-induced activation of TMEM63A and TMEM63B (Murthy et al. 2018). Compared to OSCAs, which display rapid activation kinetics and large single-channel conductance (~100 pS for OSCA1.1 and OSCA1.2; ~30 pS for OSCA3.1), mammalian TMEM63A and TMEM63B exhibit much slower activation kinetics, small macroscopic current amplitudes with undetectable single-channel currents, and high thresholds of activation by force (Murthy et al. 2018; Zheng et al. 2023). These characteristics of TMEM63-mediated currents have been confirmed by many other groups (Zhang et al. 2023; Li et al. 2024; Niloy et al. 2024; Sindoni and Grandl 2025), which collectively established TMEM63 as a new family of MSCs in mammals.

1.3 Structures of TMEM63s

Recent structural studies using cryo-EM have provided substantial insights into the biophysical properties of TMEM63 channels and their molecular gating mechanisms.

1.3.1 Overall Architecture of TMEM63 Channels

All published structures of TMEM63s to date (Zheng et al. 2023, 2025; Zhang et al. 2023; Wu et al. 2024; Han et al. 2024; Miyata et al. 2024; Qin et al. 2023) have revealed a monomeric state, in contrast to the dimeric assembly of OSCAs

(Jojoa-Cruz et al. 2018; Liu et al. 2018; Zhang et al. 2018, 2023; Maity et al. 2019; Shan et al. 2024a; Jojoa-Cruz et al. 2024a, b) and related TMEM16 family of channels/phospholipid scramblases and TMCs (Jan and Jan 2024; Clark et al. 2024; Jeong et al. 2022) (Fig. 1.1a). All these proteins are now grouped into a single superfamily due to their shared overall protein-folding architecture (Pan et al. 2018; Jan and Jan 2024; Ballesteros et al. 2018). The structural similarity suggests that these protein families might have one common ancestor. Each TMEM63 protein contains 11 transmembrane (TM) helices (named TM0 to TM10), with an extracellular N-terminus and an intracellular C-terminus (Fig. 1.1a, b). The intracellular loop 2 (IL2) between TM2 and TM3 constitutes the majority of the cytosolic region and is in proximity to the membrane. TM3 and TM4 traverse the lipid bilayer at a tilted angle rather than perpendicular to the membrane plane. TM3–TM7 form the putative channel pore, while other TMs act as a scaffold to support the pore. Although TMEM63s share the overall architecture with OSCA subunits, they also exhibit distinctive structural features that may account for TMEM63's functional characteristics (see below).

1.3.2 Mammalian TMEM63s Are Monomers

The monomeric state of TMEM63s was first revealed by cryo-EM structural analysis and was further supported by various biochemical and biophysical approaches on heterologously expressed and native TMEM63 proteins (Zheng et al. 2023; Zhang et al. 2023; Qin et al. 2023). Comparative structural analyses suggest that this divergence in oligomerization may originate from the IL2 domain, which mediates the dimerization in OSCAs (Zheng et al. 2023). Domain-swapping experiments show that replacement of OSCA1.2 IL2 with that of human TMEM63A (hTMEM63A) is sufficient to convert OSCA1.2 into a monomer (Zheng et al. 2023). Further mutagenesis investigation indicates that TMEM63s evolutionarily adapt to monomers due to sequence divergence and a 20-amino acid (aa) insertion that disrupts the dimerization interface (Zheng et al. 2023). Besides, compared to OSCAs, the TM10 in hTMEM63A is rotated toward the central cavity, which could further prevent the potential dimeric interface formation (Zhang et al. 2023). Functionally, monomeric OSCA1.2 variants preserve stretch-activated currents but display higher activation thresholds compared with the dimeric OSCA1.2 (Zheng et al. 2023). These findings suggest that oligomerization status is a determinant of the mechanosensitivity of OSCA/TMEM63 channels and TMEM63s may evolve to high-threshold MSCs by disrupting dimerization to fulfill their physiological functions in mammals.

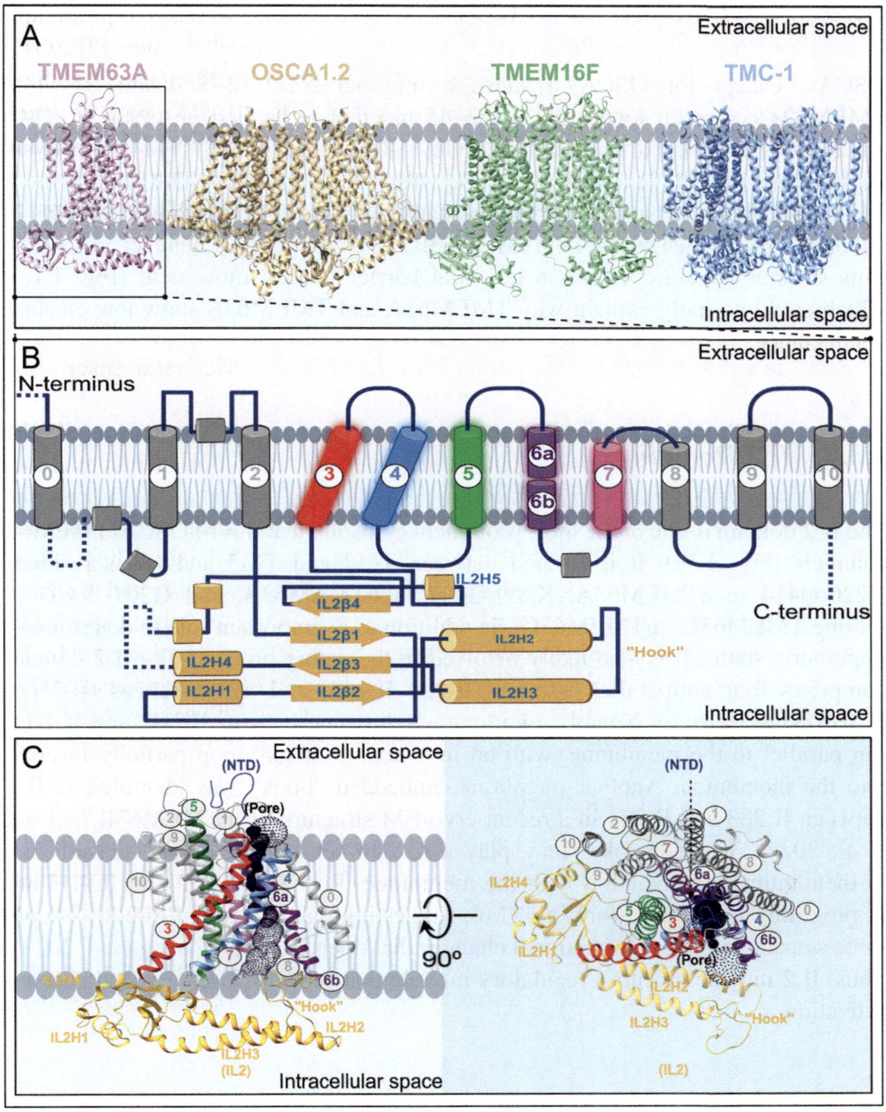

Fig. 1.1 Structure of TMEM63s. (**a**) Cryo-EM structures of monomeric human TMEM63A (PDB: 8EHW), dimeric *Arabidopsis thaliana* OSCA1.2 (PDB: 6MGV), mouse TMEM16F (PDB: 8SUN), and *Caenorhabditis elegans* TMC1 (PDB: 7USX). (**b**) Topological diagram of TMEM63A. The putative pore-lining TM3–TM7 were colored. Dashed lines denote regions unresolved in the density map. (**c**) Side view and top view of TMEM63A structure (PDB: 8EHW) with major structural features indicated. (Created with BioRender.com)

1.3.3 N-Terminal Domain (NTD)

OSCAs, except for OSCA4.1, have a minimal NTD (2–3 residues), while TMEM63s contain a longer NTD (33–45 aa) that is located upstream of TM0 and extends into the extracellular space (Jojoa-Cruz et al. 2018). Cryo-EM analysis suggests an overall flexibility of NTD, with the region near TM0 being more rigid and forming a small hairpin that caps the pore (i.e., residues S30-V45 in hTMEM63A and P19-L36 in hTMEM63B) (Fig. 1.1b, c) (Zheng et al. 2023). This structural feature creates a potential barrier for ion movement (Fig. 1.1c), which could partially explain why TMEM63A and TMEM63B show low channel conductance.

1.3.4 IL2 Domain

The IL2 domain is one of the most prominent cytosolic features of OSCA/TMEM63 channels (Fig. 1.1c). It is located between TM2 and TM3 and spans residues Y220-Q414 in hTMEM63A, K229-G427 in hTMEM63B, and G207-W405 in murine TMEM63C (mTMEM63C). In addition to its important role in determining oligomeric status, it is also likely involved in the gating process. The IL2 domain comprises four antiparallel β-sheets (IL2β1–IL2β4) and five α-helices (IL2H1–IL2H5) (Fig. 1.1b, c). Notably, two long amphipathic helices, IL2H2 and IL2H3, run parallel to the membrane, with an intervening "hook" loop partially inserted into the membrane. Another membrane-embedded "hook" was identified in IL2 between IL2β3 and IL2β4 in a recent cryo-EM structure of mTMEM63B (Miyata et al. 2024). These "hooks" may play a role in sensing membrane tension due to their intimate association with the membrane. Besides, IL2H2 and IL2H3 are in proximity to the pore-forming TM6, suggesting that IL2 might directly couple force sensing to the conformational changes that open the pore (Zheng et al. 2023). Thus, IL2 may be a central regulatory module that governs force-induced channel activation.

1.3.5 Ion-Conducting Pore

TM3–TM7 in TMEM63s form the putative ion-conducting pore that contains a highly extended narrow section known as the "neck" (Zheng et al. 2023, 2025; Zhang et al. 2023; Wu et al. 2024; Han et al. 2024; Miyata et al. 2024; Qin et al. 2023) (Fig. 1.1b, c). This "neck" section is ~12 Å long in OSCA1.2, while it remarkably extends over 50 Å in TMEM63s (Zheng et al. 2023) (Fig. 1.1c). It is plausible that this elongated "neck" contributes to the low single-channel conductance in TMEM63s.

1.4 Activation Mechanisms of TMEM63s

1.4.1 TMEM63s Are High-Threshold, Low Conductance MSCs

TMEM63s are activated by membrane stretch induced by negative pressures applied in electrophysiological recording pipettes in cell-attached or excised configuration (Murthy et al. 2018; Zheng et al. 2023; Zhang et al. 2023; Li et al. 2024; Niloy et al. 2024; Sindoni and Grandl 2025). They are also reported to respond to hypoosmotic and hyperosmotic stress in calcium imaging assays (Qin et al. 2023; Du et al. 2020; Yang et al. 2024; Zou et al. 2025). Prominent features of TMEM63 channels include their high activation threshold and low channel conductance. When heterologously expressed in HEK293T cells, the PIEZO1 channel shows a half-maximal activation pressure (P_{50}) of about −30 mmHg and maximal current amplitude (I_{max}) above 100 pA (Murthy et al. 2018; Coste et al. 2010, 2015), with single-channel conductance around 30 pS (Saotome et al. 2018). However, mammalian TMEM63 channels were reported to have P_{50} values between −50 and −130 mmHg and I_{max} below 40 pA in different studies (Table 1.1), with single-channel currents unresolvable in electrophysiological recordings. Through nonstationary noise analysis, *Drosophila* TMEM63 (DmTMEM63) was estimated to have a single-channel conductance of only 1.9 pS (Li et al. 2024). These unique features suggest that TMEM63s may have distinct gating mechanisms from PIEZOs.

1.4.2 Force-From-Lipid Model

Two principal models have been proposed to explain how MSCs detect and respond to mechanical stimuli: the force-from-filament and force-from-lipid mechanisms (Cox et al. 2019). In the force-from-filament model, channels are mechanically coupled to intracellular or extracellular structures, such as the cytoskeleton, extracellular matrix, or accessory proteins, that transmit force to the channel. A well-characterized example is NOMPC, a *Drosophila* TRP channel whose large N-terminal ankyrin repeat domain forms a spring-like tether connected to microtubules, allowing cytoskeletal displacements to directly gate the channel (Yan et al. 2013; Zhang et al. 2015; Jin et al. 2017). In contrast, the force-from-lipid model involves the channel itself directly sensing changes in the membrane tension, thickness, or curvature, which lead to protein conformational changes and pore opening. This mechanism is exemplified by bacterial MscL and MscS channels, which can be activated in purified lipid bilayers without cytoskeleton or accessory partners (Reddy et al. 2019; Perozo et al. 2002; Zhang et al. 2021).

The force-from-lipid mechanism appears to be widely applicable to mammalian MSCs. Structural and imaging studies of PIEZO channels have provided compelling evidence that these channels use membrane curvature as their force-sensing mecha-

Table 1.1 Biophysical properties of TMEM63-mediated currents

TMEM63s	Expression cell/system	Imax (pA) (−80 mV)	Activation tau (ms)	Inactivation tau (ms)	P_{50} (-mmHg)	References
hTMEM63A	HEK-P1KO	12.4 ± 4.4	188 ± 18	427 ± 94	59.7 ± 3.2	Murthy et al. (2018)
	HEK-P1KO	19.2 ± 3.7	196 ± 27	147 ± 31 ms	64.0 ± 2.1	Niloy et al. (2024)
	HEK293T	33.5 ± 7.5	–	–	123.3 ± 11.2	Zheng et al. (2023)
	Proteoliposome	18.7 ± 3.1	–	–	142.6 ± 9.1	Zheng et al. (2023)
hTMEM63B	HEK293T	17.4 ± 2.3	–	–	134.2 ± 11.4	Zheng et al. (2023)
	Proteoliposome	16.3 ± 1.8	–	–	143.9 ± 7.3	Zheng et al. (2023)
mTMEM63A	HEK-P1KO	18.75 ± 3.0	126 ± 19	323 ± 30	60 ± 5.4	Murthy et al. (2018)
mTMEM63B	HEK-P1KO	17.97 ± 3.9	245 ± 38	323 ± 30	66 ± 7.8	Murthy et al. (2018)
mTMEM63C	HEK-P1KO	4.73 ± 0.8	–	–	–	Murthy et al. (2018)
DmTMEM63	HEK-P1KO	14.2 ± 4.5	22.6 ± 4.3	154 ± 34	90 ± 2	Murthy et al. (2018)

Note: *P1KO* PIEZO1 knockout, – not determined; All values are presented as mean ± SEM

nism (Yang et al. 2022; Vaisey et al. 2022; Mulhall et al. 2023). In the resting state, PIEZOs assemble into a large trimeric structure that forms a bowl shape, which naturally bends the surrounding membrane. The mechanical force is believed to cause the curvature to become flat, which results in conformational changes and pore opening. Compared to PIEZOs, TMEM63s form much smaller monomeric channels (Zheng et al. 2023, 2025; Zhang et al. 2023; Wu et al. 2024; Han et al. 2024; Miyata et al. 2024; Qin et al. 2023), which are unlikely to induce large-scale membrane curvature in the resting state. Thus, TMEM63s are probably gated by force, with a mechanism distinct from that of PIEZOs. Functional studies have shown that purified TMEM63 or OSCA proteins are sufficient to be activated in a reconstituted lipid bilayer (Murthy et al. 2018; Zheng et al. 2023; Zhang et al. 2023), strongly suggesting a force-from-lipid mechanism for OSCA/TMEM63s.

1.4.3 Open-State Structures Captured by Cryo-EM and Scramblase Activity

Initial structures of either TMEM63s or OSCAs resolved by cryo-EM are largely in the closed state. A major obstacle in defining the gating mechanism is the difficulty of capturing an activated state, due to the lack of effective approaches to apply force to the proteins during structural studies. Recent cryo-EM studies have explored alternative strategies to capture open conformations of both OSCAs and TMEM63s, providing important insights into how the pore opens. For OSCA1.1 and OSCA1.2, structures in open conformation were obtained by introducing a mutation (F516A in OSCA1.1) (Shan et al. 2024a) and by reconstituting purified proteins (OSCA1.2) into small-diameter liposomes, mimicking membrane curvature and stretch (Han et al. 2024). These strategies revealed an outward "blooming" movement of pore-forming TM3–TM6 and expansion of the pore diameter from ≤1 Å to ~5 Å at the narrowest segment. Moreover, the open structure of OSCA1.2 in liposomes revealed a "proteo-lipidic" pore architecture where lipid molecules form part of the permeation pathway wall. For TMEM63s, two open-conformation structures have also been resolved. The hTMEM63A V53M gain-of-function mutant (corresponding to the disease-causing mutation V44M in hTMEM63B) was captured in a markedly dilated pore conformation relative to the closed structure (Zheng et al. 2025). In parallel, an antibody-bound mTMEM63B structure captured an open-like conformation in which a Fab was developed to bind the extracellular loops (Miyata et al. 2024). These open structures of TMEM63s revealed conformational changes largely like those observed in OSCAs. Overall, the current open-conformation structures, achieved through mutagenesis, liposome reconstitution, and Fab-mediated conformational trapping, reveal a unifying scheme across the OSCA/TMEM63s for the pore opening.

Remarkably, the open structures of OSCA/TMEM63s (Zheng et al. 2025; Han et al. 2024; Miyata et al. 2024; Shan et al. 2024a) also reveal a wide and laterally exposed pore architecture that provides structural clues to their potential lipid scramblase activity. Lipid scramblases are membrane proteins that facilitate the bidirectional, energy-independent movement of phospholipids between the inner and outer leaflets of the lipid bilayer (Jan and Jan 2024), thereby disrupting the normally asymmetric distribution of phospholipids, in particular phosphatidylserine (PS), which is typically confined to the inner leaflet. The externalization of PS triggers a range of biological effects. During apoptosis, exposed PS serves as an "eat me" signal that facilitates recognition and clearance by phagocytes (Krahling et al. 1999; Callahan et al. 2000). In other settings, transient PS exposure participates in various physiological processes, including cell-cell fusion, blood coagulation, angiogenesis, and vesicle fusion (Bevers and Williamson 2010; Yang et al. 2012; Zhang et al. 2020; Shan et al. 2024b). The laterally exposed pore in OSCA/TMEM63s forms a hydrophilic groove that creates a potential solvent-accessible pathway for lipid headgroups to cross the membrane (Fig. 1.2). Indeed, lipid-like densities lining this laterally exposed pore were observed in the cryo-EM structures of OSCA1.2 (Han et al. 2024) and TMEM63A V53M mutant (Zheng et al. 2025), reminiscent of the scrambling grooves seen in the TMEM16 family of ion channel/scramblase (Jan and Jan 2024; Kalienkova et al. 2019). Functionally, constitutive scramblase activities of OSCA/TMEM63s were first reported when pore-lining residues were mutated (Lowry et al. 2024). More importantly, many disease-causing mutations in TMEM63B were found to be gain-of-function mutations that evoke constitutive scramblase activities (Zheng et al. 2025; Miyata et al. 2024, 2025) (Table 1.2), highlighting the clinical relevance. Miyata et al. also showed that wild-type (WT) TMEM63B proteins can function as scramblases responding to membrane curvature (Miyata et al. 2024). In addition, Lin et al. reported that lipid scrambling in TMEM63A and TMEM63B can be activated by mechanical forces in the membrane (Lin et al. 2025). Collectively, these findings raise the possibility that TMEM63s may form a family of mechanically activated ion channels and scramblases.

1.4.4 "Hydrophobic Latch" as a Potential Force-Sensing Element

Since the IL2 domain in OSCA/TMEM63s is positioned in proximity to the membrane and potentially interacts with the membrane through amphipathic IL2H2/H3 and the intervening "hook" (Fig. 1.1c), it was proposed to serve as the force-sensing domain. Recent cryo-EM analysis of human TMEM63A and TMEM63B has identified a distinctive "hydrophobic latch" that offers an alternative or complementary mechanism for force sensation (Zheng et al. 2025). This latch is formed by a conserved cluster of hydrophobic residues spanning TM0, TM4, and TM6, located just beneath the outer leaflet of the membrane (Fig. 1.2). In the closed conformation, these residues, including the V53 in hTMEM63A and the V44 in hTMEM63B, are tightly packed to anchor TM4 against adjacent TM6 and TM0 around the gate

Table 1.2 Scramblase activities detected in WT or mutant TMEM63s

TMEM63s	Expression cell/system	Mutation/WT	Scramblase activity	Methods	References
hTMEM63A	HEK-16FKO	V53M, V53M/T469N	Constitutive	Annexin V	Zheng et al. (2025)
	HEK-P1KO/GUV	WT	Force activated	Annexin V/NBD-PC and Rhod-PE	Lin et al. (2025)
mTMEM63A	HEK-16FKO	W472K, S475K	Constitutive	Annexin V	Lowry et al. (2024)
	HEK-P1&16FKO	W472K, W472G	Constitutive	Annexin V	Lin et al. (2025)
hTMEM63B	Ba/F3	WT	Force activated	Annexin V/NBD-PC or -SM	Miyata et al. (2025)
	Ba/F3	V44M, D459E, V463I, I475del, T481N, G580, G580S, R660T, F697L	Constitutive	Annexin V/NBD-PC or -SM	Miyata et al. (2025)
	HEK-16FKO	V44M, V44M/T481N	Constitutive	Annexin V	Zheng et al. (2025)
mTMEM63B	Ba/F3	WT	Force activated	NBD-PC or -SM	Miyata et al. (2024)
	Ba/F3	V44M	Force activated	NBD-PC or -SM	Miyata et al. (2024)

Note: *16FKO* TMEM16F knockout, *P1KO* PIEZO1 knockout, *GUV* giant unilamellar vesicle

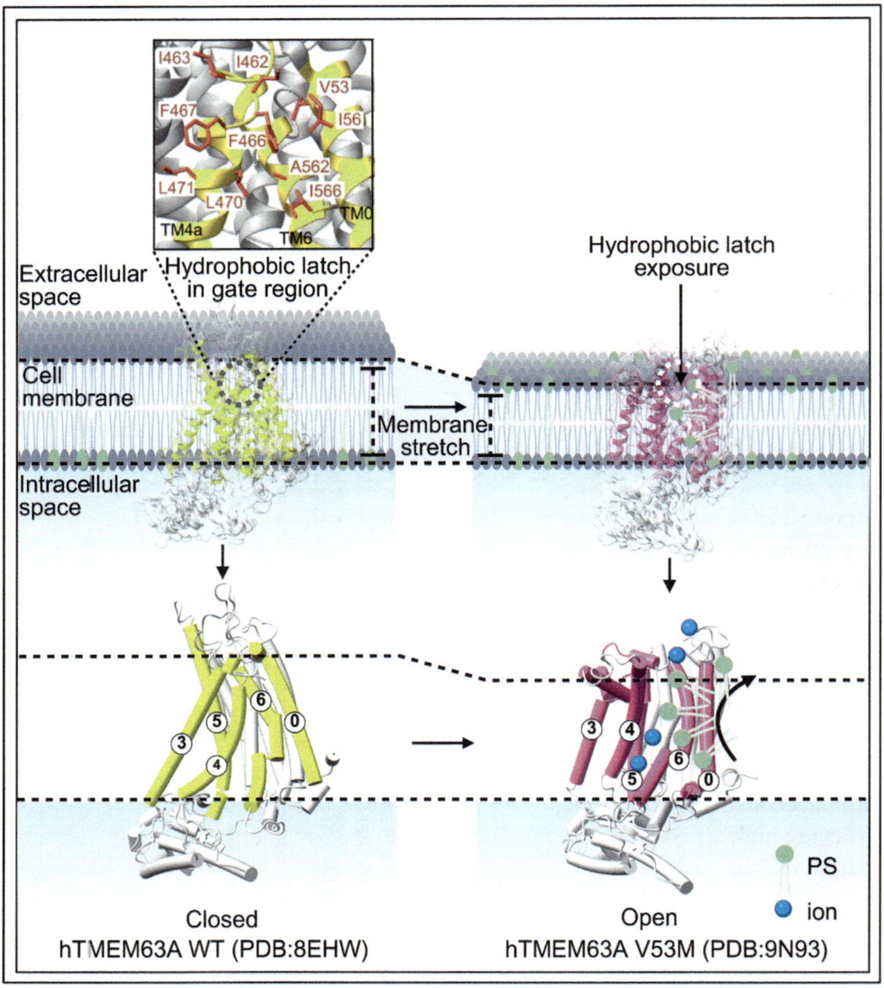

Fig. 1.2 Hypothetic gating mechanism for TMEM63 activation by membrane stretch. In the absence of membrane stretch, the hydrophobic latch in TMEM63s is buried underneath the surface of the cell membrane and maintains the protein in the closed conformation. In the presence of stretch, the membrane becomes thinner, which exposes the hydrophobic latch to extracellular space, leading to a hydrophobic mismatch. This would then cause rearrangement of TM3–TM5, leading to pore opening. (Created with BioRender.com)

region (Fig. 1.2), thereby maintaining a closed pore. Disruption of this network, e.g., via disease-associated mutation hTMEM63B V44M or corresponding hTMEM63A V53M, was hypothesized to destabilize the hydrophobic interactions, allowing rearrangement of TM3–TM5, which opens the lateral lipid translocation cleft and triggers constitutive lipid scramblase activity even in the absence of mechanical stimuli (Zheng et al. 2025). Because the hydrophobic latch is near the membrane-water interface, it likely senses changes in membrane thickness or tension: when stretch induces membrane thinning to expose the hydrophobic latch, the resulting hydrophobic mismatch destabilizes the latch, initiating conformational transitions that open the pore (Fig. 1.2) (Zheng et al. 2025). Thus, the hydrophobic latch potentially acts as a force-sensing gatekeeper, linking membrane deformation to structural rearrangements and mechanical activation in TMEM63 channels.

1.5 Clinical Implications and Physiological Roles of TMEM63s

The in vivo roles of TMEM63s have been increasingly elucidated through converging evidence from clinical genetics and animal studies. Below, we will summarize the current progress in this aspect.

1.5.1 Clinical Implications of TMEM63s

1.5.1.1 TMEM63A and Hypomyelinating Leukodystrophy 19 (HLD19)

Heterozygous variants in TMEM63A, including G168E, I462N, G553V, G553D, Y559H, and G567S, have been linked to infantile transient HLD19 (OMIM #618688) (Fig. 1.3) (Fukumura et al. 2022; Tonduti et al. 2021; Yan et al. 2019, 2021; Yoneno et al. 2024), highlighting a role of TMEM63A in brain myelination and development. This rare disorder typically affects infants in the first months of life with nystagmus, hypotonia, and motor delay, and magnetic resonance imaging (MRI) shows diffuse hypomyelination that improves substantially through childhood and often approaches normalization by early school age (Yan et al. 2019; Yoneno et al. 2024). These variants fail to induce stretch-activated currents when expressed in HEK293T cells, indicating loss of MSC function (Yan et al. 2019; Halford et al. 2025). Single-cell RNA sequencing data suggests that TMEM63A expression is enriched in oligodendrocytes (Fig. 1.4), consistent with its roles in myelination. More recently, Halford et al. reported that in mice, global or oligodendrocyte-specific knockout of TMEM63A induces transient hypomyelination, and in zebrafish, disruption of TMEM63A via CRISPR reagent injection of one-cell zygotes results in a transient spinal cord myelination (Halford et al. 2025), which further supports TMEM63A as an evolutionarily conserved regulator of myelination.

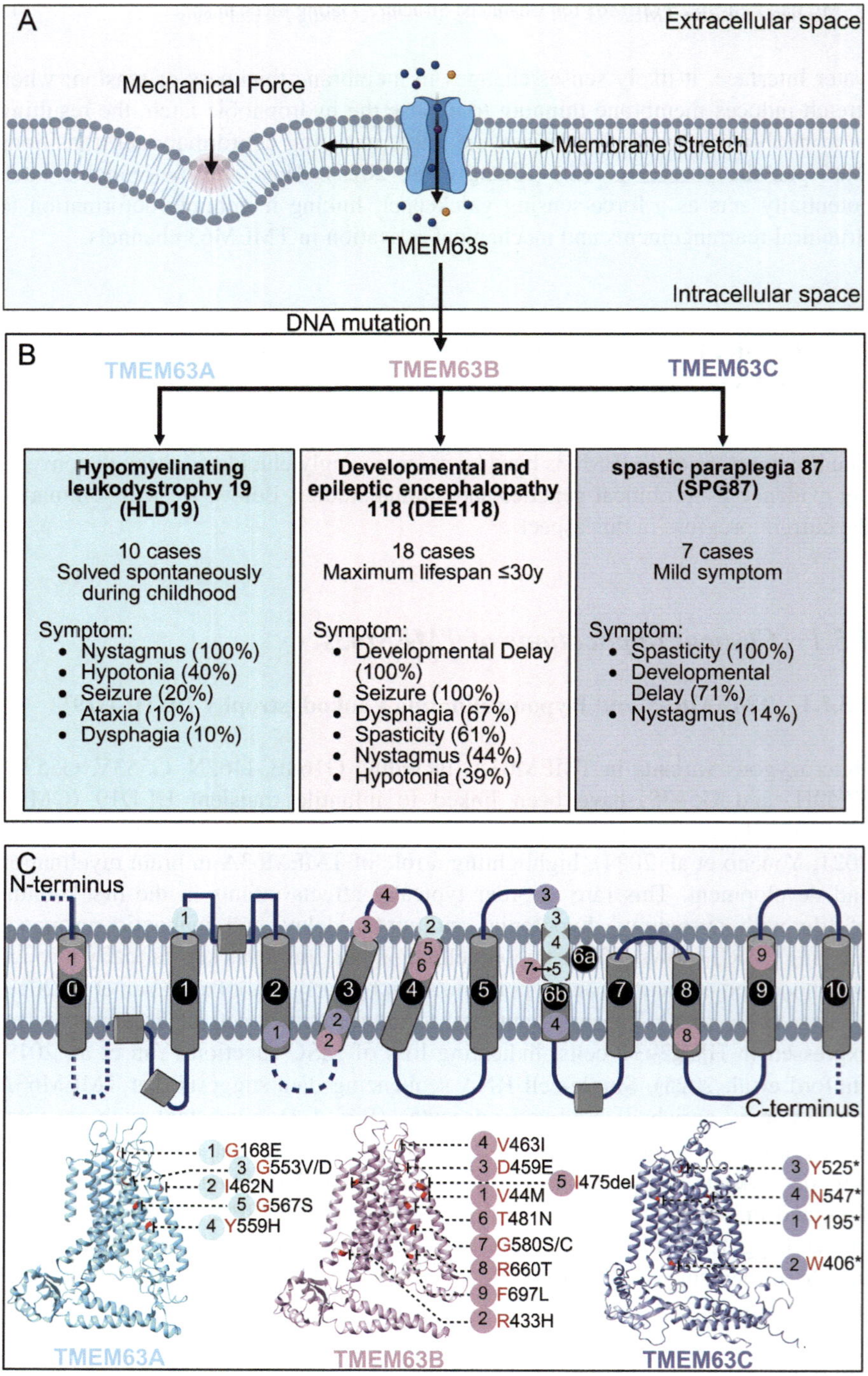

Fig. 1.3 Neurological disorders caused by genetic mutations in TMEM63s. (**a**) Schematic illustration showing TMEM63s function as MSCs activated by membrane stretch. (**b**) Genetic diseases caused by mutations in TMEM63A, TMEM63B, and TMEM63C, respectively. (**c**) Summary of reported mutations in TMEM63s, with the mutation sites illustrated. Mutations in TMEM63C cause either protein premature stop or frameshift and are indicated by "*". (Created with BioRender.com)

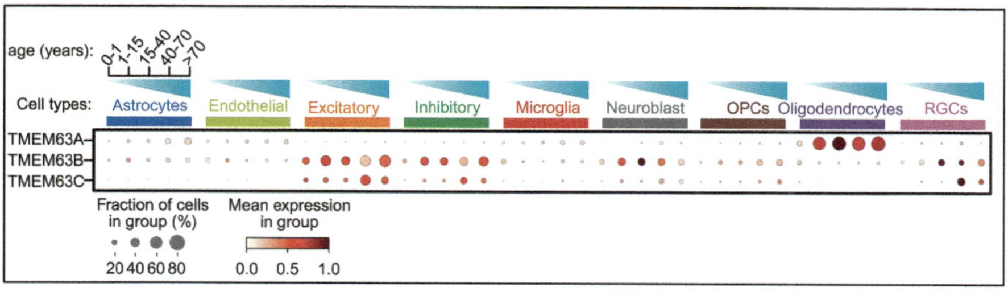

Fig. 1.4 Single-cell RNA expression level of TMEM63s in cell types of human brain across different ages. Data were retrieved from Zenodo (https://zenodo.org/records/10939707), a Brain Transcriptome Single-cell Atlas. *OPCs* oligodendrocyte progenitor cells, *RGCs* radial glial cells. (Created with BioRender.com)

1.5.1.2 TMEM63B and Developmental and Epileptic Encephalopathy 118 (DEE118)

De novo heterozygous variants in TMEM63B have recently been associated with DEE118 (OMIM #621250) (Fig. 1.3) (Sasaki et al. 2024; Vetro et al. 2023), a severe neurodevelopmental disorder characterized by early-onset seizures, global developmental delay, intellectual disability, motor impairment, and feeding or swallowing difficulties. This finding underscores the critical role of TMEM63B in brain development. In contrast to TMEM63A, TMEM63B is predominantly expressed in neuronal cells, including both excitatory and inhibitory neurons, according to single-cell RNA sequencing data (Fig. 1.4), but its functions in these cells remain largely unknown. Initial electrophysiological recordings reported that six of these mutations (V44M, R433H, T481N, G580S, R660T, F697L) evoke leak currents in the absence of mechanical force, suggesting gain-of-function effects (Vetro et al. 2023). However, subsequent structural and functional studies have shown that the V44M mutation has little effect on either leak currents or mechanically activated currents but instead evokes constitutive lipid scramblase activity (Zheng et al. 2025). Other groups have also reported scramblase activities of TMEM63B V44M and other disease-associated mutants (Table 1.2) (Miyata et al. 2024, 2025), suggesting that this might be a common mechanism for the DEE118.

1.5.1.3 TMEM63C and Spastic Paraplegia 87 (SPG87)

TMEM63C has been implicated in hereditary SPG87 (OMIM #619966) (Fig. 1.3) (Tabara et al. 2022), a neurodegenerative disorder characterized by progressive lower-limb spasticity, motor dysfunction, and mild intellectual disability. Unlike heterozygous missense mutations identified in TMEM63A and TMEM63B, SPG87 arises from biallelic loss-of-function variants in TMEM63C, including nonsense and frameshift mutations that abolish protein expression (Tabara et al. 2022). TMEM63C is likely present in brain neurons (Fig. 1.4), where its functions remain

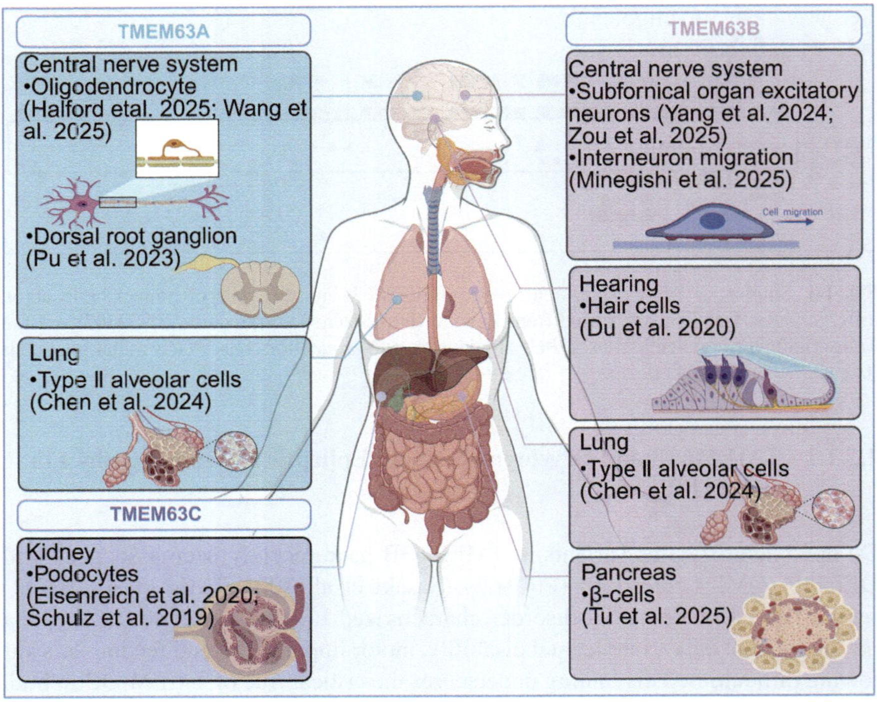

Fig. 1.5 Schematic summary of expression/function of TMEM63s in specific cell types. (Created with BioRender.com)

to be elucidated. Cellular studies have reported that loss of TMEM63C disrupts ER-mitochondria tethering and impairs mitochondria morphology (Tabara et al. 2022), which may lead to defective energy homeostasis and axonal maintenance.

1.5.2 Physiological Roles of TMEM63s

Recent animal studies, mainly using mouse models, have started to reveal a wide range of physiological functions of TMEM63s in multiple cell types/tissues (summarized in Fig. 1.5).

1.5.2.1 TMEM63A

Consistent with TMEM63A's clinical implication in HLD19, loss-of-function studies in mice and zebrafish have confirmed that in the central nervous system, TMEM63A is essential for developmental myelination, presumably acting cell autonomously in oligodendrocytes to regulate differentiation of precursor cells and initiation of myelin wrapping (Halford et al. 2025; Wang et al. 2025). In the peripheral nervous system, TMEM63A was reported to be expressed in non-peptidergic nociceptors in the dorsal root ganglia, mediating mechanical allodynia and chronic post-amputation pain in mice (Pu et al. 2023). Beyond the nervous system, TMEM63A, together with TMEM63B, was shown to act as a mechanosensor in lung alveolar type II epithelial cells, where they couple mechanical stretch to surfactant secretion, ensuring proper lung inflation and gas exchange at birth (Chen et al. 2024). TMEM63A was also found to localize to the lysosomal membrane (Li et al. 2024), where it is presumably involved in regulating organelle morphology and lysosomal degradative capacity under osmotic and mechanical stress.

1.5.2.2 TMEM63B

In the brain, TMEM63B was shown to be highly expressed in subfornical organ (SFO) neurons, where it is proposed to detect membrane stretch during hyperosmotic stress and drive thirst behavior; deletion of TMEM63B in mice suppresses osmotic activation of SFO neurons and water-seeking behavior (Yang et al. 2024; Zou et al. 2025). These studies suggest an important role of TMEM63B in thirst sensation. Besides, recent work identified TMEM63B as a critical mechanosensor for neuronal migration during brain development (Minegishi et al. 2025). Using live imaging and force probes, Minegishi et al. showed that the extension of the leading process in migrating neurons increases membrane tension, activating TMEM63B to trigger calcium influx, which then stimulates myosin II-dependent contractility to drive directional movement. Consistently, loss of TMEM63B in 3D cultures leads to abolishment of the tension-induced calcium transients and disruption of neuronal migration. In the inner ear, TMEM63B is present in the basal membrane of outer hair cells, where it was hypothesized to mediate calcium influx and volume regulation in response to osmotic stress (Du et al. 2020). Knockout of TMEM63B in mice causes outer hair cell degeneration and hearing loss (Du et al. 2020). In the pancreas, TMEM63B was reported to be highly expressed in β-cells, in which it is activated osmotically to mediate calcium influxes to facilitate insulin secretion (Tu et al. 2025).

1.5.2.3 TMEM63C

Existing studies suggest that TMEM63C is potentially involved in kidney podocyte survival (Eisenreich et al. 2020; Schulz et al. 2019). In hypertensive rats and zebrafish, loss of TMEM63C correlates with podocyte injury and albuminuria (Schulz et al. 2019). In cultured podocytes, TMEM63C expression is induced by angiotensin II via NF-κB, and its knockdown increases apoptosis and reduces Akt signaling (Eisenreich et al. 2020), indicating a pro-survival function. TMEM63C thus appears to support podocyte integrity, though its mechanosensory mechanism remains unclear.

1.6 Perspective

In recent years, we have witnessed remarkable advances in our understanding of the TMEM63 family of mechanosensitive ion channels. Structural studies using cryo-EM have uncovered the monomeric architecture of TMEM63s and identified key structural features, including the IL2 domain, a hydrophobic latch, and a lateral lipid-accessible pore, which indicates distinct force-sensing and gating mechanisms. These discoveries, together with clinical and animal studies, have increasingly revealed the versatile mechanosensing functions of TMEM63s.

Despite these significant insights, many critical questions remain unanswered. It is still unclear how changes in membrane tension or thickness trigger conformational transitions at the hydrophobic latch or IL2 domain to induce pore opening. Once activated, how could ions and lipids pass through the pore simultaneously? How do the channel activity and scramblase activity of TMEM63 proteins contribute to their physiological functions in vivo? The diverse functions of TMEM63A, TMEM63B, and TMEM63C across different cell types and tissues raise important questions regarding the molecular basis underlying their distinct force sensitivity and specificity. Although TMEM63 mutations have been linked to various neurological disorders, the cellular mechanisms translating TMEM63 dysfunction into pathology remain to be elucidated.

Looking forward, it will be necessary to combine structural, biophysical, and physiological approaches to comprehensively dissect how TMEM63s detect and transduce mechanical stimuli, from atomic resolution to animal level. High-resolution structures under membrane tension will be instrumental to help understand force-induced activation mechanism. Live imaging of mechanical signaling in primary isolated cells and native tissues will be invaluable to connect molecular mechanisms to physiological functions in vivo. Given the emerging links between TMEM63 dysfunction and human neurological disorders, selectively modulating their ion-conducting or lipid-scrambling activities may offer new therapeutic opportunities. TMEM63s thus represent a new frontier in mechanobiology, offering profound insights into how mechanical forces shape cellular physiology and human health.

Acknowledgments This work was supported by UW-Madison start-up funds (to W.Z.).

Author Contributions J.O. wrote original draft and generated figures. Y.Z., C.C., and J.W. reviewed and edited the draft. W.Z. provided conceptualization and funding and revised the manuscript.

Competing Interests All authors declare no competing interests.

References

Ballesteros A, Fenollar-Ferrer C, Swartz KJ (2018) Structural relationship between the putative hair cell mechanotransduction channel TMC1 and TMEM16 proteins. elife 7. https://doi.org/10.7554/eLife.38433

Bevers EM, Williamson PL (2010) Phospholipid scramblase: an update. FEBS Lett 584:2724–2730. https://doi.org/10.1016/j.febslet.2010.03.020

Callahan MK, Williamson P, Schlegel RA (2000) Surface expression of phosphatidylserine on macrophages is required for phagocytosis of apoptotic thymocytes. Cell Death Differ 7:645–653. https://doi.org/10.1038/sj.cdd.4400690

Chen GL, Li JY, Chen X, Liu JW, Zhang Q, Liu JY, Wen J, Wang N, Lei M, Wei JP et al (2024) Mechanosensitive channels TMEM63A and TMEM63B mediate lung inflation-induced surfactant secretion. J Clin Invest 134. https://doi.org/10.1172/JCI174508

Chuang YC, Chen CC (2022) Force from filaments: the role of the cytoskeleton and extracellular matrix in the gating of mechanosensitive channels. Front Cell Dev Biol 10:886048. https://doi.org/10.3389/fcell.2022.886048

Clark S, Jeong H, Posert R, Goehring A, Gouaux E (2024) The structure of the Caenorhabditis elegans TMC-2 complex suggests roles of lipid-mediated subunit contacts in mechanosensory transduction. Proc Natl Acad Sci USA 121:e2314096121. https://doi.org/10.1073/pnas.2314096121

Coste B, Mathur J, Schmidt M, Earley TJ, Ranade S, Petrus MJ, Dubin AE, Patapoutian A (2010) Piezo1 and Piezo2 are essential components of distinct mechanically activated cation channels. Science 330:55–60. https://doi.org/10.1126/science.1193270

Coste B, Murthy SE, Mathur J, Schmidt M, Mechioukhi Y, Delmas P, Patapoutian A (2015) Piezo1 ion channel pore properties are dictated by C-terminal region. Nat Commun 6:7223. https://doi.org/10.1038/ncomms8223

Cox CD, Bavi N, Martinac B (2019) Biophysical principles of ion-channel-mediated mechanosensory transduction. Cell Rep 29:1–12. https://doi.org/10.1016/j.celrep.2019.08.075

Douguet D, Honore E (2019) Mammalian mechanoelectrical transduction: structure and function of force-gated ion channels. Cell 179:340–354. https://doi.org/10.1016/j.cell.2019.08.049

Du H, Ye C, Wu D, Zang YY, Zhang L, Chen C, He XY, Yang JJ, Hu P, Xu Z et al (2020) The cation channel TMEM63B is an osmosensor required for hearing. Cell Rep 31:107596. https://doi.org/10.1016/j.celrep.2020.107596

Eisenreich A, Orphal M, Bohme K, Kreutz R (2020) Tmem63c is a potential pro-survival factor in angiotensin II-treated human podocytes. Life Sci 258:118175. https://doi.org/10.1016/j.lfs.2020.118175

Fukumura S, Hiraide T, Yamamoto A, Tsuchida K, Aoto K, Nakashima M, Saitsu H (2022) A novel de novo TMEM63A variant in a patient with severe hypomyelination and global developmental delay. Brain Dev 44:178–183. https://doi.org/10.1016/j.braindev.2021.09.006

Halford J, Senatore AJ, Berryman S, Munoz A, Semidey D, Doan RA, Coombs AM, Noimany B, Emberley K, Emery B et al (2025) TMEM63A, associated with hypomyelinating leukodystrophies, is an evolutionarily conserved regulator of myelination. Proc Natl Acad Sci USA 122:e2507354122. https://doi.org/10.1073/pnas.2507354122

Han Y, Zhou Z, Jin R, Dai F, Ge Y, Ju X, Ma X, He S, Yuan L, Wang Y et al (2024) Mechanical activation opens a lipid-lined pore in OSCA ion channels. Nature 628:910–918. https://doi.org/10.1038/s41586-024-07256-9

Hou C, Tian W, Kleist T, He K, Garcia V, Bai F, Hao Y, Luan S, Li L (2014) DUF221 proteins are a family of osmosensitive calcium-permeable cation channels conserved across eukaryotes. Cell Res 24:632–635. https://doi.org/10.1038/cr.2014.14

Jan LY, Jan YN (2024) Wide-ranging cellular functions of ion channels and lipid scramblases in the structurally related TMC, TMEM16 and TMEM63 families. Nat Struct Mol Biol. https://doi.org/10.1038/s41594-024-01444-x

Jansen KA, Atherton P, Ballestrem C (2017) Mechanotransduction at the cell-matrix interface. Semin Cell Dev Biol 71:75–83. https://doi.org/10.1016/j.semcdb.2017.07.027

Jeong H, Clark S, Goehring A, Dehghani-Ghahnaviyeh S, Rasouli A, Tajkhorshid E, Gouaux E (2022) Structures of the TMC-1 complex illuminate mechanosensory transduction. Nature 610:796–803. https://doi.org/10.1038/s41586-022-05314-8

Jin P, Bulkley D, Guo Y, Zhang W, Guo Z, Huynh W, Wu S, Meltzer S, Cheng T, Jan LY et al (2017) Electron cryo-microscopy structure of the mechanotransduction channel NOMPC. Nature 547:118–122. https://doi.org/10.1038/nature22981

Jin P, Jan LY, Jan YN (2020) Mechanosensitive ion channels: structural features relevant to mechanotransduction mechanisms. Annu Rev Neurosci 43:207–229. https://doi.org/10.1146/annurev-neuro-070918-050509

Jojoa-Cruz S, Saotome K, Murthy SE, Tsui CCA, Sansom MS, Patapoutian A, Ward AB (2018) Cryo-EM structure of the mechanically activated ion channel OSCA1.2. elife 7. https://doi.org/10.7554/eLife.41845

Jojoa-Cruz S, Burendei B, Lee WH, Ward AB (2024a) Structure of mechanically activated ion channel OSCA2.3 reveals mobile elements in the transmembrane domain. Structure 32:157–167. https://doi.org/10.1016/j.str.2023.11.009

Jojoa-Cruz S, Dubin AE, Lee WH, Ward AB (2024b) Structure-guided mutagenesis of OSCAs reveals differential activation to mechanical stimuli. elife 12. https://doi.org/10.7554/eLife.93147

Kalienkova V, Clerico Mosina V, Bryner L, Oostergetel GT, Dutzler R, Paulino C (2019) Stepwise activation mechanism of the scramblase nhTMEM16 revealed by cryo-EM. elife 8. https://doi.org/10.7554/eLife.44364

Kawashima Y, Geleoc GS, Kurima K, Labay V, Lelli A, Asai Y, Makishima T, Wu DK, Della Santina CC, Holt JR, Griffith AJ (2011) Mechanotransduction in mouse inner ear hair cells requires transmembrane channel-like genes. J Clin Invest 121:4796–4809. https://doi.org/10.1172/JCI60405

Kefauver JM, Ward AB, Patapoutian A (2020) Discoveries in structure and physiology of mechanically activated ion channels. Nature 587:567–576. https://doi.org/10.1038/s41586-020-2933-1

Krahling S, Callahan MK, Williamson P, Schlegel RA (1999) Exposure of phosphatidylserine is a general feature in the phagocytosis of apoptotic lymphocytes by macrophages. Cell Death Differ 6:183–189. https://doi.org/10.1038/sj.cdd.4400473

Li K, Guo Y, Wang Y, Zhu R, Chen W, Cheng T, Zhang X, Jia Y, Liu T, Zhang W et al (2024) Drosophila TMEM63 and mouse TMEM63A are lysosomal mechanosensory ion channels. Nat Cell Biol 26:393–403. https://doi.org/10.1038/s41556-024-01353-7

Lin Y, Zhou Z, Han Y, Cheng D, Wang H, Ju AL, Zhang Y, Cox CD, Corry B (2025) TMEM63 proteins act as mechanically activated cholesterol modulated lipid scramblases contributing to membrane mechano-resilience. bioRxiv, 2025.2007.2026.664997. https://doi.org/10.1101/2025.07.26.664997

Liu X, Wang J, Sun L (2018) Structure of the hyperosmolality-gated calcium-permeable channel OSCA1.2. Nat Commun 9:5060. https://doi.org/10.1038/s41467-018-07564-5

Lowry AJ, Liang P, Song M, Wan Y, Pei ZM, Yang H, Zhang Y (2024) TMEM16 and OSCA/TMEM63 proteins share a conserved potential to permeate ions and phospholipids. elife 13. https://doi.org/10.7554/eLife.96957

Maity K, Heumann JM, McGrath AP, Kopcho NJ, Hsu PK, Lee CW, Mapes JH, Garza D, Krishnan S, Morgan GP et al (2019) Cryo-EM structure of OSCA1.2 from Oryza sativa elucidates the mechanical basis of potential membrane hyperosmolality gating. Proc Natl Acad Sci USA 116:14309–14318. https://doi.org/10.1073/pnas.1900774116

Minegishi T, Hasebe H, Aoyama T, Naruse K, Takahashi Y, Inagaki N (2025) Mechanical signaling through membrane tension induces somal translocation during neuronal migration. EMBO J 44:767–780. https://doi.org/10.1038/s44318-024-00326-8

Miyata Y, Takahashi K, Lee Y, Sultan CS, Kuribayashi R, Takahashi M, Hata K, Bamba T, Izumi Y, Liu K et al (2024) Membrane structure-responsive lipid scrambling by TMEM63B to control plasma membrane lipid distribution. Nat Struct Mol Biol. https://doi.org/10.1038/s41594-024-01411-6

Miyata Y, Nishimura M, Nagata A, Jing X, Sultan CS, Kuribayashi R, Takahashi K, Lee Y, Nishizawa T, Segawa K (2025) Membrane structure-responsive lipid scramblase activity of the TMEM63/OSCA family. FEBS Lett 599:656–666. https://doi.org/10.1002/1873-3468.15084

Mulhall EM, Gharpure A, Lee RM, Dubin AE, Aaron JS, Marshall KL, Spencer KR, Reiche MA, Henderson SC, Chew TL, Patapoutian A (2023) Direct observation of the conformational states of PIEZO1. Nature 620:1117–1125. https://doi.org/10.1038/s41586-023-06427-4

Murthy SE, Dubin AE, Patapoutian A (2017) Piezos thrive under pressure: mechanically activated ion channels in health and disease. Nat Rev Mol Cell Biol 18:771–783. https://doi.org/10.1038/nrm.2017.92

Murthy SE, Dubin AE, Whitwam T, Jojoa-Cruz S, Cahalan SM, Mousavi SAR, Ward AB, Patapoutian A (2018) OSCA/TMEM63 are an evolutionarily conserved family of mechanically activated ion channels. elife 7. https://doi.org/10.7554/eLife.41844

Niloy SI, Strege PR, Hannan EC, Cowan LM, Linsenmeier F, Friedrich O, Farrugia G, Beyder A (2024) Stretch response of the mechano-gated channel TMEM63A in membrane patches and single cells. Am J Physiol Cell Physiol 326:C622–C631. https://doi.org/10.1152/ajpcell.00583.2023

Pan B, Geleoc GS, Asai Y, Horwitz GC, Kurima K, Ishikawa K, Kawashima Y, Griffith AJ, Holt JR (2013) TMC1 and TMC2 are components of the mechanotransduction channel in hair cells of the mammalian inner ear. Neuron 79:504–515. https://doi.org/10.1016/j.neuron.2013.06.019

Pan B, Akyuz N, Liu XP, Asai Y, Nist-Lund C, Kurima K, Derfler BH, Gyorgy B, Limapichat W, Walujkar S et al (2018) TMC1 forms the pore of mechanosensory transduction channels in vertebrate inner ear hair cells. Neuron 99:736–753. https://doi.org/10.1016/j.neuron.2018.07.033

Perozo E (2006) Gating prokaryotic mechanosensitive channels. Nat Rev Mol Cell Biol 7:109–119. https://doi.org/10.1038/nrm1833

Perozo E, Cortes DM, Sompornpisut P, Kloda A, Martinac B (2002) Open channel structure of MscL and the gating mechanism of mechanosensitive channels. Nature 418:942–948. https://doi.org/10.1038/nature00992

Pu S, Wu Y, Tong F, Du WJ, Liu S, Yang H, Zhang C, Zhou B, Chen Z, Zhou X et al (2023) Mechanosensitive ion channel TMEM63A gangs up with local macrophages to modulate chronic post-amputation pain. Neurosci Bull 39:177–193. https://doi.org/10.1007/s12264-022-00910-0

Qin Y, Yu D, Wu D, Dong J, Li WT, Ye C, Cheung KC, Zhang Y, Xu Y, Wang Y et al (2023) Cryo-EM structure of TMEM63C suggests it functions as a monomer. Nat Commun 14:7265. https://doi.org/10.1038/s41467-023-42956-2

Ranade SS, Syeda R, Patapoutian A (2015) Mechanically activated ion channels. Neuron 87:1162–1179. https://doi.org/10.1016/j.neuron.2015.08.032

Reddy B, Bavi N, Lu A, Park Y, Perozo E (2019) Molecular basis of force-from-lipids gating in the mechanosensitive channel MscS. elife 8. https://doi.org/10.7554/eLife.50486

Saotome K, Murthy SE, Kefauver JM, Whitwam T, Patapoutian A, Ward AB (2018) Structure of the mechanically activated ion channel Piezo1. Nature 554:481–486. https://doi.org/10.1038/nature25453

Sasaki K, Nakashima M, Fujii Y, Itamura S, Saitsu H, Kato M (2024) A new case of developmental and epileptic encephalopathy and macrocytic anemia with stretched-activated ion channel TMEM63B variant. Brain Dev Case Rep 2:100043. https://doi.org/10.1016/j.bdcasr.2024.100043

Schulz A, Muller NV, van de Lest NA, Eisenreich A, Schmidbauer M, Barysenka A, Purfurst B, Sporbert A, Lorenzen T, Meyer AM et al (2019) Analysis of the genomic architecture of a complex trait locus in hypertensive rat models links Tmem63c to kidney damage. elife 8. https://doi.org/10.7554/eLife.42068

Shan Y, Zhang M, Chen M, Guo X, Li Y, Zhang M, Pei D (2024a) Activation mechanisms of dimeric mechanosensitive OSCA/TMEM63 channels. Nat Commun 15:7504. https://doi.org/10.1038/s41467-024-51800-0

Shan KZ, Le T, Liang P, Dong P, Lowry AJ, Kremmyda P, Claesson-Welsh L, Yang H (2024b) TMEM16F scramblase regulates angiogenesis via endothelial intracellular signaling. J Cell Sci 137. https://doi.org/10.1242/jcs.261566
Sindoni M, Grandl J (2025) A closed-loop system for millisecond readout and control of membrane tension. Biophys J. https://doi.org/10.1016/j.bpj.2025.03.025
Szczot M, Nickolls AR, Lam RM, Chesler AT (2021) The form and function of PIEZO2. Annu Rev Biochem 90:507–534. https://doi.org/10.1146/annurev-biochem-081720-023244
Tabara LC, Al-Salmi F, Maroofian R, Al-Futaisi AM, Al-Murshedi F, Kennedy J, Day JO, Courtin T, Al-Khayat A, Galedari H et al (2022) TMEM63C mutations cause mitochondrial morphology defects and underlie hereditary spastic paraplegia. Brain 145:3095–3107. https://doi.org/10.1093/brain/awac123
Tonduti D, Mura E, Masnada S, Bertini E, Aiello C, Zini D, Parmeggiani L, Cantalupo G, Talenti G, Veggiotti P et al (2021) Spinal cord involvement and paroxysmal events in "infantile onset transient Hypomyelination" due to TMEM63A mutation. J Hum Genet 66:1035–1037. https://doi.org/10.1038/s10038-021-00921-1
Tu JJ, Ye C, Teng XY, Zang YY, Sun XY, Chen S, Chen J, Shi YS (2025) Osmosensor TMEM63B facilitates insulin secretion in pancreatic beta-cells. Sci China Life Sci 68:1714–1726. https://doi.org/10.1007/s11427-024-2833-3
Vaisey G, Banerjee P, North AJ, Haselwandter CA, MacKinnon R (2022) Piezo1 as a force-through-membrane sensor in red blood cells. elife 11. https://doi.org/10.7554/eLife.82621
Vetro A, Pelorosso C, Balestrini S, Masi A, Hambleton S, Argilli E, Conti V, Giubbolini S, Barrick R, Bergant G et al (2023) Stretch-activated ion channel TMEM63B associates with developmental and epileptic encephalopathies and progressive neurodegeneration. Am J Hum Genet. https://doi.org/10.1016/j.ajhg.2023.06.008
Wang YY, Wu D, Zhan Y, Li F, Zang YY, Teng XY, Zhang L, Duan GF, Wang H, Xu R et al (2025) Cation channel TMEM63A autonomously facilitates oligodendrocyte differentiation at an early stage. Neurosci Bull 41:615–632. https://doi.org/10.1007/s12264-024-01338-4
Wilde C, Mitgau J, Suchy T, Schoneberg T, Liebscher I (2022) Translating the force-mechano-sensing GPCRs. Am J Physiol Cell Physiol 322:C1047–C1060. https://doi.org/10.1152/ajpcell.00465.2021
Wozniak MA, Chen CS (2009) Mechanotransduction in development: a growing role for contractility. Nat Rev Mol Cell Biol 10:34–43. https://doi.org/10.1038/nrm2592
Wu X, Shang T, Lu X, Luo D, Yang D (2024) A monomeric structure of human TMEM63A protein. Proteins 92:750–756. https://doi.org/10.1002/prot.26660
Xiao B (2024) Mechanisms of mechanotransduction and physiological roles of PIEZO channels. Nat Rev Mol Cell Biol 25:886–903. https://doi.org/10.1038/s41580-024-00773-5
Yan Z, Zhang W, He Y, Gorczyca D, Xiang Y, Cheng LE, Meltzer S, Jan LY, Jan YN (2013) Drosophila NOMPC is a mechanotransduction channel subunit for gentle-touch sensation. Nature 493:221–225. https://doi.org/10.1038/nature11685
Yan H, Helman G, Murthy SE, Ji H, Crawford J, Kubisiak T, Bent SJ, Xiao J, Taft RJ, Coombs A et al (2019) Heterozygous variants in the mechanosensitive ion channel TMEM63A result in transient hypomyelination during infancy. Am J Hum Genet 105:996–1004. https://doi.org/10.1016/j.ajhg.2019.09.011
Yan H, Ji H, Kubisiak T, Wu Y, Xiao J, Gu Q, Yang Y, Xie H, Ji T, Gao K et al (2021) Genetic analysis of 20 patients with hypomyelinating leukodystrophy by trio-based whole-exome sequencing. J Hum Genet 66:761–768. https://doi.org/10.1038/s10038-020-00896-5
Yang H, Kim A, David T, Palmer D, Jin T, Tien J, Huang F, Cheng T, Coughlin SR, Jan YN, Jan LY (2012) TMEM16F forms a Ca2+-activated cation channel required for lipid scrambling in platelets during blood coagulation. Cell 151:111–122. https://doi.org/10.1016/j.cell.2012.07.036
Yang X, Lin C, Chen X, Li S, Li X, Xiao B (2022) Structure deformation and curvature sensing of PIEZO1 in lipid membranes. Nature 604:377–383. https://doi.org/10.1038/s41586-022-04574-8

Yang G, Jia M, Li G, Zang YY, Chen YY, Wang YY, Zhan SY, Peng SX, Wan G, Li W et al (2024) TMEM63B channel is the osmosensor required for thirst drive of interoceptive neurons. Cell Discov 10:1. https://doi.org/10.1038/s41421-023-00628-x

Yoneno S, Yamamoto K, Tabata K, Shimizu-Motohashi Y, Tomita A, Hayashi T, Maki H, Sato N, Inoue K, Saitsu H, Komaki H (2024) A novel heterozygous TMEM63A variant in a familial case with early onset nystagmus, severe hypomyelination, and a favorable adult prognosis. J Hum Genet 69:607–611. https://doi.org/10.1038/s10038-024-01268-z

Yuan F, Yang H, Xue Y, Kong D, Ye R, Li C, Zhang J, Theprungsirikul L, Shrift T, Krichilsky B et al (2014) OSCA1 mediates osmotic-stress-evoked Ca2+ increases vital for osmosensing in Arabidopsis. Nature 514:367–371. https://doi.org/10.1038/nature13593

Zhang W, Cheng LE, Kittelmann M, Li J, Petkovic M, Cheng T, Jin P, Guo Z, Gopfert MC, Jan LY, Jan YN (2015) Ankyrin repeats convey force to gate the NOMPC mechanotransduction channel. Cell 162:1391–1403. https://doi.org/10.1016/j.cell.2015.08.024

Zhang M, Wang D, Kang Y, Wu JX, Yao F, Pan C, Yan Z, Song C, Chen L (2018) Structure of the mechanosensitive OSCA channels. Nat Struct Mol Biol 25:850–858. https://doi.org/10.1038/s41594-018-0117-6

Zhang Y, Le T, Grabau R, Mohseni Z, Kim H, Natale DR, Feng L, Pan H, Yang H (2020) TMEM16F phospholipid scramblase mediates trophoblast fusion and placental development. Sci Adv 6:eaba0310. https://doi.org/10.1126/sciadv.aba0310

Zhang Y, Daday C, Gu RX, Cox CD, Martinac B, de Groot BL, Walz T (2021) Visualization of the mechanosensitive ion channel MscS under membrane tension. Nature 590:509–514. https://doi.org/10.1038/s41586-021-03196-w

Zhang M, Shan Y, Cox CD, Pei D (2023) A mechanical-coupling mechanism in OSCA/TMEM63 channel mechanosensitivity. Nat Commun 14:3943. https://doi.org/10.1038/s41467-023-39688-8

Zheng W, Holt JR (2021) The mechanosensory transduction machinery in inner ear hair cells. Annu Rev Biophys 50:31–51. https://doi.org/10.1146/annurev-biophys-062420-081842

Zheng W, Rawson S, Shen Z, Tamilselvan E, Smith HE, Halford J, Shen C, Murthy SE, Ulbrich MH, Sotomayor M et al (2023) TMEM63 proteins function as monomeric high-threshold mechanosensitive ion channels. Neuron 111:3195–3210. https://doi.org/10.1016/j.neuron.2023.07.006

Zheng W, Lowry AJ, Smith HE, Xie J, Rawson S, Wang C, Ou J, Sotomayor M, Fu TM, Yang H, Holt JR (2025) Structural and functional basis of mechanosensitive TMEM63 channelopathies. Neuron 113:2474–2489. https://doi.org/10.1016/j.neuron.2025.05.009

Zou W, Deng S, Chen X, Ruan J, Wang H, Zhan W, Wang J, Liu Z, Yan Z (2025) TMEM63B functions as a mammalian hyperosmolar sensor for thirst. Neuron 113:1430–1445. https://doi.org/10.1016/j.neuron.2025.02.012

Chapter 2
How Touch Triggers Mechanotransduction in Cutaneous Mechanoreceptors

Alexandre Bergheau, Ombeline Parbaud, Gabriel Cazorla, Youcef Ben-Khalifa, and Hassan Zahouani

Abstract Understanding the mechanotransduction mechanisms underlying the tactile dimension of skin aging remains an underexplored aspect of dermatological and neurosciences. This investigation introduces a novel mechanosensory responsiveness index, tactile acuity index (TAI), derived from a time-frequency analysis of friction-induced vibrations recorded during controlled touch interactions utilizing the wearable tactile acquisition device (WTAD). By decomposing tactile signals across frequency bands specific to Merkel, Meissner, Ruffini, and Pacinian mechanoreceptors, TAI quantifies both the amplitude and spectral organization of vibrational energy transmitted through the skin.

In a first experimental phase, TAI exhibited a clear monotonic decline across seven grades of paper textures, confirming that smoother surfaces produce weaker and spectrally narrower mechanical stimulation, thereby reducing mechanoreceptor recruitment. The strongest correlations were observed for Ruffini ($R^2 = 0.89$) and Pacinian ($R^2 = 0.95$) receptors, reflecting their sensitivity to macroscopic strain and high-frequency vibration. In an ensuing aging study involving 40 women aged 20–65, TAI revealed a pronounced reduction in mechanosensory activity across all receptor types, with the largest declines in deep mechanotransductive pathways. The associated time-frequency spectrograms further highlighted shorter, lower-amplitude bursts and restricted spectral bandwidths in older skin, indicating a progressive simplification of tactile encoding with age.

Together, these findings validate TAI as a quantitative biomarker of tactile vitality, capable of distinguishing between textural and biological sources of sensory decline. By providing receptor-specific insights into mechanotransduction efficiency, this approach establishes a novel framework for diagnosing, monitoring, and potentially restoring tactile function through targeted cosmetic or therapeutic interventions.

A. Bergheau (✉) · O. Parbaud · H. Zahouani
Univ Lyon, Ecole Centrale de Lyon, CNRS, ENTPE, LTDS, UMR5513, Ecully, France

G. Cazorla · Y. Ben-Khalifa
Innovation Research & Development, Chanel Fragrance & Beauty, Pantin, France

M. Kloc, J. Wosik (eds.), *Mechanotransduction and Mechanosensing*, Results and Problems in Cell Differentiation 78, https://doi.org/10.1007/978-3-032-19509-8_2

2.1 Introduction

The sense of touch originates from mechanotransduction, a biophysical process through which mechanical forces applied to the skin are converted into electrical activity in peripheral sensory neurons. This conversion is enabled by mechanically gated ion channels whose opening is triggered by membrane deformation induced by stretch, pressure, shear, or vibration. The resulting neural impulses form the primary biological code that initiates tactile perception, making mechanical-to-neural conversion the fundamental gateway between external mechanical stimuli and sensory interpretation.

Beyond its fundamental biophysical role, mechanotransduction also shapes tactile acuity by determining how effectively mechanical events—such as friction, localized deformation, or surface-induced vibrations—are transmitted through the skin layers to activate cutaneous mechanoreceptors (Shyti et al. 2025). This process sustains tactile acuity, enabling the discrimination of textures, the detection of skin deformation, and rapid protective responses to mechanical stimuli. While cutaneous aging has been widely studied through visible manifestations such as wrinkles, pigmentation, and loss of elasticity, the mechanical-neural dimension of aging skin remains comparatively underexplored (Amaied et al. 2015). Yet declines in tactile acuity with age—including higher vibration thresholds, poorer texture discrimination, and reduced perception of skin stretch—have direct implications for hand function, haptics, and sensory well-being (Burns et al. 2010).

Cutaneous mechanoreception is mediated by distinct receptor populations: Merkel cells (SA I), specialized for sustained pressure and fine texture discrimination; Meissner corpuscles (RA I), responsive to low-frequency flutter and light motion; Ruffini endings (SA II), detecting skin stretch and providing proprioceptive feedback; and Pacinian corpuscles (RA II), tuned to high-frequency vibration and deep pressure (Derler et al. 2007). Each receptor class contributes unique spectral-temporal signatures to the tactile afferent signal. However, differentiating their respective decline across the lifespan has remained challenging because most available methods rely on subjective reporting or coarse psychophysical thresholds (Derler et al. 2009).

Wearable tactile acquisition device (WTAD), a high-precision tactile sensing platform, enables objective measurement of mechanotransduction during natural human touch (Shyti et al. 2025; Orr et al. 2006). Instead of imposing artificial forces, the device records the mechanical events that occur when a user touches, slides, or presses on the skin or on a material. During this interaction, it captures minute variations in friction, pressure, and vibration at high temporal and spectral resolution. This transforms the inherently subjective act of touching into a quantitative, reproducible metrology of mechanoreceptor function.

Furthermore, a core strength of this approach lies in its time-frequency signal analysis. Instead of describing tactile responses only in the time domain, we decompose them into joint temporal-spectral representations employing methods such as the short-time Fourier transform analysis (Abdouni et al. 2018). This enables the detection of subtle receptor-specific patterns:

- Attenuation of high-frequency components (>300 Hz) suggesting Pacinian decline.
- Damping of low-frequency responses (<50 Hz) implicating Meissner or Merkel dysfunction.

By systematically mapping these features across age groups or anatomical sites, WTAD makes it possible to construct a data-driven atlas of mechanotransductive aging. Such quantitative phenotyping distinguishes normal interindividual variability from early mechanosensory decline and provides mechanistic insight that may guide personalized interventions—from haptic training and sensory rehabilitation to targeted skincare strategies aimed at preserving tactile function.

This work positions WTAD not only as a device for surface softness or material feel but also as a mechanotransduction analysis platform capable of dissecting the functional state of cutaneous mechanoreceptors with great precision (Shyti et al. 2025; Blake et al. 1997).

2.2 Material and Methods

2.2.1 Introduction

Human skin functions not only as a physical barrier against environmental stress but also as a complex sensory structure due to the presence of specialized mechanoreceptors distributed throughout the epidermis and dermis. These receptors—Merkel cells (SA I), Meissner corpuscles (RA I), Ruffini endings (SA II), and Pacinian corpuscles (RA II)—convert mechanical stimulation into neural signals through a process known as mechanotransduction (Dunn and Olmedo 2016). In this mechanism, deformation of the skin activates mechanosensitive ion channels located in the afferent nerve endings, leading to neuronal depolarization and transmission of tactile information to the central nervous system (Orr et al. 2006).

Each type of mechanoreceptor is sensitive to specific mechanical inputs: Merkel cells detect sustained pressure and fine surface details, Meissner corpuscles respond to low-frequency vibrations, Ruffini endings encode skin stretch, and Pacinian corpuscles are highly responsive to high-frequency vibrations and deep mechanical perturbations. Hinged upon the literature, these receptors can be characterized according to vibration frequency sensitivity ranges (Table 2.1): Merkel (0.3–3 Hz) and Meissner (3–50 Hz) receptors are situated near the epidermis-dermis junction, whereas Ruffini (50–300 Hz) and Pacinian (300–1000 Hz) receptors are located deeper within the dermis and hypodermis. Figure 2.1 illustrates this anatomical organization in relation to their respective frequency activation bands.

Table 2.1 Frequency range response of each mechanoreceptor

Quantized parameters in the Merkel cell frequency band	Quantized parameters in the Meisner cell frequency band	Quantized parameters in the Ruffini cell frequency band	Quantized parameters in the Pacini cell frequency band
	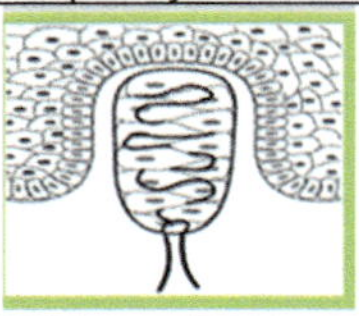	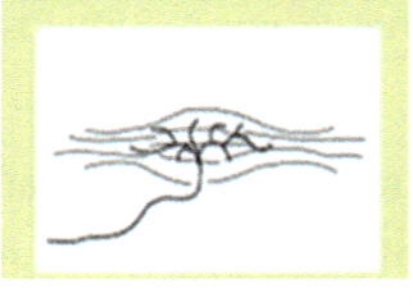	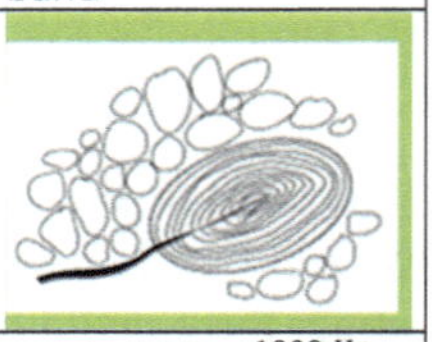
$\begin{bmatrix} La(dB) \\ f_c \to \lambda(f_c) \end{bmatrix}_{0;3Hz}^{3Hz}$	$\begin{bmatrix} La(dB) \\ f_c \to \lambda(f_c) \end{bmatrix}_{3Hz}^{50Hz}$	$\begin{bmatrix} La(dB) \\ f_c \to \lambda(f_c) \end{bmatrix}_{50Hz}^{400\ Hz}$	$\begin{bmatrix} La(dB) \\ f_c \to \lambda(f_c) \end{bmatrix}_{400\ Hz}^{1000\ Hz}$

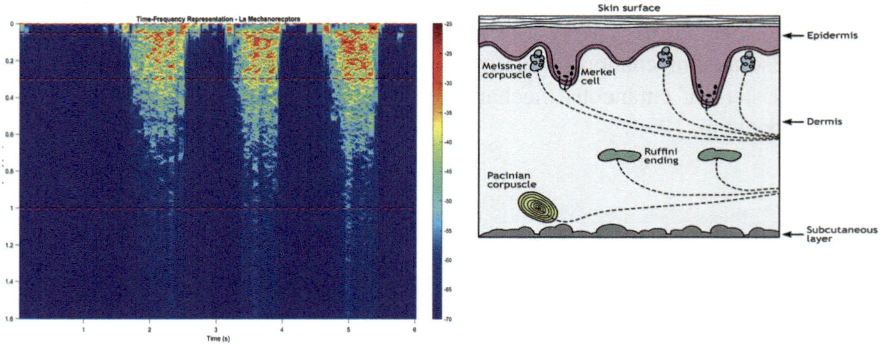

Fig. 2.1 Time frequency vs mechanoreceptors

As the skin ages, structural and biochemical alterations progressively affect its mechanical and sensory functions. Collagen degradation, reduced hydration, and flattening of the epidermal-dermal junction disrupt the extracellular environment necessary for optimal mechanoreceptor activity. In parallel, age-related changes in peripheral nerve conduction and mechanosensitive ion channel expression may impair the fidelity of mechanotransduction (Blake et al. 1997). This progressive sensory decline reduces tactile acuity and affects proprioception, motor control, and skin homeostasis. Therefore, investigating the functional integrity of mechanotransduction pathways is essential for understanding the physiological impact of skin aging, especially since tactile function is both a marker and a regulator of skin health.

2.3 Wearable Tactile Acquisition Device Technology

This study makes use of a recently developed wearable device combined with a dedicated signal processing approach designed to characterize tactile interactions with

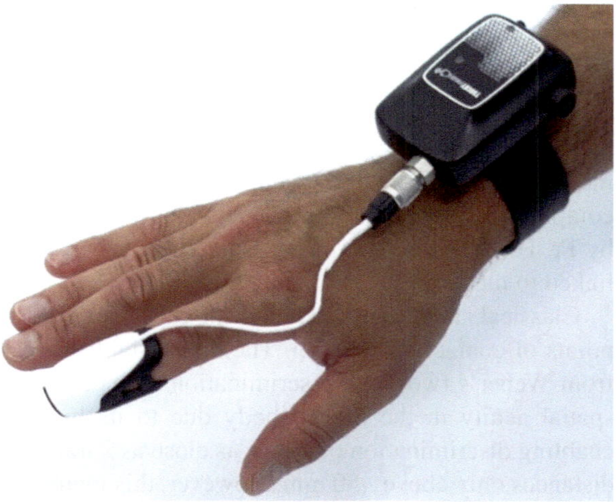

Fig. 2.2 Wearable tactile acquisition device (WTAD)

textured surfaces through friction-induced vibrations: WTAD (Fig. 2.2) (Shyti et al. 2025). When a surface is explored with the fingertip, mechanical contact generates micro-vibrations that are captured and processed by the WTAD. These friction-induced vibrations are known to play a crucial role in fine texture perception, making them a reliable indicator for evaluating surface discrimination abilities. The purpose of this method is to provide a consistent and objective framework for analyzing these recorded signals and identifying perceptible differences during sliding contact between the skin and a material surface.

2.3.1 *Key Parameters*

The primary quantitative descriptor extracted from the recorded signals is the vibratory level La (expressed in decibels, dB) (Abdouni et al. 2018), calculated from the acceleration signal using the following expression:

$$\mathrm{La} = 20\log\left(\frac{A_{\mathrm{RMS}}}{A_{\mathrm{ref}}}\right)$$

This parameter reflects the intensity of vibration produced during tactile interaction. Higher surface roughness typically generates larger acceleration amplitudes and therefore higher La values (Fujimoto et al. 2003). In addition to the global vibratory level, La is also computed within specific frequency bands corresponding to mechanoreceptor sensitivity ranges: Merkel (0–3 Hz), Meissner (3–50 Hz), Ruffini (50–300 Hz), and Pacinian (300–1000 Hz). These band-specific calculations allow selective evaluation of the contribution of each mechanoreceptor type to tactile perception, in accordance with Eq. (2.1). In this expression, A_{RMS} represents the root mean square of the measured acceleration (m/s^2), while A_{ref} denotes the minimum detectable acceleration of the sensor, equal to 10^{-6}m/s^2 and does not depend on the frequency.

To increase the ability to reveal subtle differences within the dataset, an additional parameter is derived from the acquired acceleration signal. For each mechanoreceptor-related frequency band, a wavelength variable, denoted λ, is computed as a function of frequency. This parameter, developed based on prior work by Pr. H. Zahouani, can be interpreted as a dynamic spatial sensitivity descriptor linked to mechanoreceptor activation.

Classical tactile acuity refers to the capacity to distinguish two spatially separated points of contact on the skin. The most established assessment protocol originates from Weber's two-point discrimination method. The fingertip exhibits the highest spatial acuity in the human body due to its high density of mechanoreceptors, enabling discrimination of points as close as 2 mm apart, whereas the back resolves distances only above ~50 mm. However, this metric remains static in nature.

Unlike static measures, dynamic tactile acuity—defined here as the ability of mechanoreceptors to resolve surface-related mechanical cues during active touch—remains largely unexplored. This dynamic dimension is likely influenced by surface micro-geometry, spatial patterns, and material mechanical properties. Assessing this behavior provides insight into how different mechanoreceptor populations encode surface information and which spatial scales of roughness they are most sensitive to. In this pilot work, vibrations recorded at the finger-surface interface are used to estimate the characteristic spatial wavelengths perceptible by each mechanoreceptor group.

The approach relies on wave dispersion principles in layered biological media, where propagation wavelength depends on both the wave speed and its frequency:

$$\lambda\,(f, t) = \frac{v(t)}{f}$$

where v is the propagation velocity (m/s) and f is frequency (Hz).

To estimate these wavelengths from experimental data, the recorded acceleration signal $\gamma(t)$, defined as the time derivative of velocity, is used:

$$\gamma(t) = \frac{dv(t)}{\text{dt}}$$

The Fourier transform (TF) of this signal is then expressed as

$$\text{TF}\,[\gamma(t)] = \text{TF}\left[\frac{dv(t)}{\text{dt}}\right]$$

Leveraging Fourier derivative properties yields

$$\mathrm{TF}\left[\frac{dv(t)}{\mathrm{dt}}\right] = j * 2\pi f * \mathrm{TF}\left[v(t)\right] \tag{2.1}$$

Since velocity in the frequency domain can be written as a function of wavelength and frequency, TF[v(t)]=$\lambda(f, t) * f$, substitution gives.

$$\mathrm{TF}\left[\gamma(t)\right] = \mathrm{TF}\left[\frac{dv}{\mathrm{dt}}\right] = \mathrm{j} * 2\pi\mathrm{f} * TF\left[v(t)\right]$$

$$= \mathrm{j} * 2\pi\mathrm{f} * \lambda\,(f, t) * f = j * 2\pi f^2 * \lambda\,(f, t)$$

Taking the modulus of the complex expression results in

$$\|\mathrm{TF}\left[\gamma(t)\right]\| = 2\pi f^2 * \lambda\,(f, t)$$

Finally, solving for $\lambda(f, t)$ leads to the wavelength estimation used in this study:

$$\lambda\,(f, t) = \frac{\|\mathrm{TF}\left[\gamma(t)\right]\|}{2\pi f^2}$$

To further characterize tactile perception and monitor its evolution with aging, we finally introduce a quantitative criterion of tactile acuity (TAI), defined as

$$\mathrm{TAI} = La(f) * \lambda\,(f, t) * f^2$$

where *La* represents the vibratory level across the acquired touch signal within the mechanoreceptor band frequency range (captured by WTAD) in dB and f is the frequency associated with that wavelength in Hz. Therefore, TAI is in dBm/s^2. This composite metric integrates the vibratory amplitude of the touch interaction with its spectral-spatial properties, providing a single, scalable parameter to track mechanotransductive performance across individuals or over time. By consolidating the energy and frequency content of tactile signals into a single value, TAI allows the detection of subtle sensory decline that may not be evident through conventional threshold testing.

2.4 Time-Frequency and Tactile Acuity Index (TAI) Analysis Validations

In this first experimental phase, cutaneous mechanosensory responses were objectively quantified during controlled horizontal sliding contact between the index finger and seven grades of commercial hand paper towels exhibiting progressive variation in microstructural roughness (Lederman 1983; Zahouani et al. 2013;

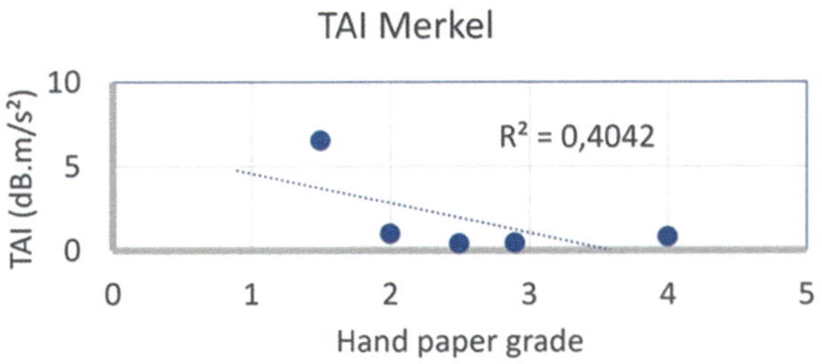

Fig. 2.3 Tactile acuity index of Merkel cells of the first experimental phase

Delhaye et al. 2012). Using the WTAD, friction-induced vibrations produced at the skin-surface interface were recorded with high temporal precision. Each hand paper grade measurement involve three strokes repeated three times and then averaged. These micro-vibrations arise from the intermittent stick-slip transitions that occur as the fingertip traverses microscopic surface asperities, and they represent the primary mechanical stimulus that activates cutaneous mechanoreceptors (Thieulin et al. 2017; Bruce 1980). To assess tactile sensitivity, a newfangled mechanosensory responsiveness index, referred to as TAI, was defined and calculated across mechanoreceptor-specific frequency bands. The TAI revealed a clear monotonic decrease across paper grades, as shown in Figs. 2.3, 2.4, 2.5, and 2.6, indicating that as surfaces became smoother, they generated weaker mechanical stimulation and thus reduced mechanoreceptor recruitment. This trend was most pronounced for Ruffini (Fig. 2.5) ($R^2 = 0.8356$) and Pacinian (Fig. 2.6) ($R^2 = 0.7949$) receptors, which are known to be sensitive to macroscopic deformation and high-frequency dynamic excitation, respectively. Their marked decline with increasing grade demonstrates that tactile feedback from smoother surfaces lacks the transient energy fluctuations required to trigger subsurface mechanotransduction. Merkel ($R^2 = 0.4042$) and Meissner ($R^2 = 0.5584$) (Figs. 2.3 and 2.4) receptors also exhibited decreasing activity across grades but with weaker correlations, suggesting that superficial tactile encoding is progressively dampened as fine surface irregularities are lost, though not to the same extent as deeper vibration-based sensory input (Scheibert et al. 2009). Moreover, the "R" in R^2 refers to the correlation coefficient, a statistical measure that describes the strength and direction of the linear relationship between two variables. Its value ranges from -1 to $+1$: a value close to $+1$ indicates a strong positive linear association, a value close to -1 indicates a strong negative association, and a value near 0 suggests no linear relationship. R^2 is simply this correlation coefficient squared, representing the proportion of the variance in the data that is explained by the model.

To gain mechanistic insight into the temporal organization of tactile encoding, time-frequency analysis (Figs. 2.7, 2.8, 2.9, 2.10, 2.11, 2.12, and 2.13) was applied to the acquired vibration signals. Unlike static metrics, the spectrograms provide a dynamic map of mechanoreceptor engagement by decomposing tactile responses across time and frequency. The roughest paper sample (0p9 in Fig. 2.7) produced a rich spectrotemporal signature, characterized by intense broadband components spanning low to high frequencies. These reflect robust mechanical interactions

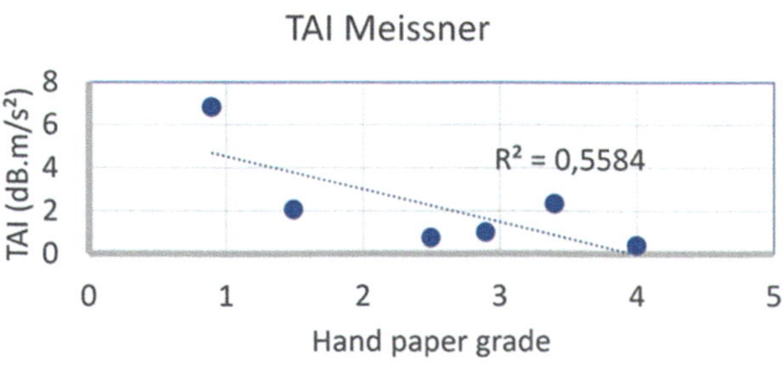

Fig. 2.4 Tactile acuity index of Meissner corpuscle of the first experimental phase

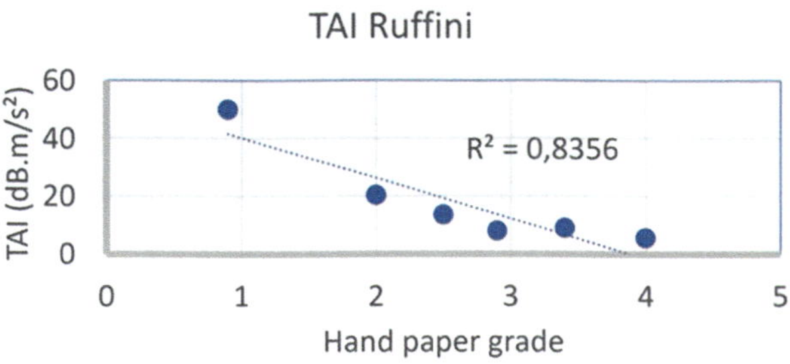

Fig. 2.5 Tactile acuity index of Ruffini corpuscle of the first experimental phase

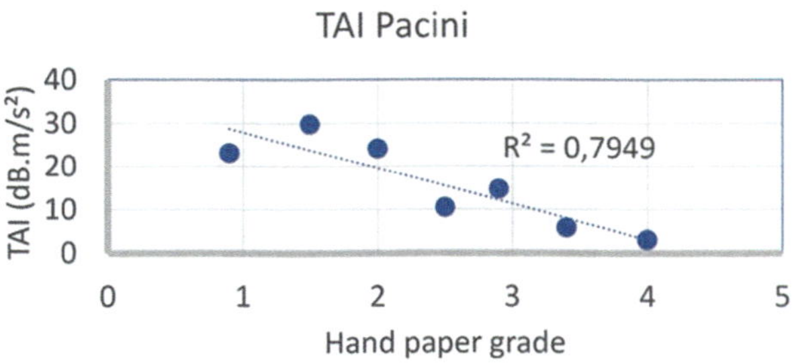

Fig. 2.6 Tactile acuity index of Pacini corpuscle of the first experimental phase

involving both superficial and deep skin layers, manifested by pronounced bursts in the Meissner band (3–50 Hz) and strong sustained energy in the Ruffini and Pacinian bands above 50 Hz. The presence of recurrent high-frequency spikes indicates the existence of repetitive micro-slip events due to high surface roughness, which mechanically stimulate the fingertip via rapidly adapting receptors. Sample 2p (Fig. 2.9) preserved this multimodal vibrational structure but exhibited slightly lower spectral energy, suggesting a reduction in mechanical instabilities and a partial decrease in subdermal strain transmission. This trend became more evident in sample 2p5 (Fig. 2.10), where spectrograms showed localized mid-frequency activity and diminished contributions above 300 Hz. The attenuation of Pacinian activation beyond this point marks a turning threshold, beyond which deep mechanoreceptive channels become progressively disengaged.

For samples with intermediate textures, such as 3p4 in Fig. 2.12, the spectrogram revealed a significant reorganization of spectral energy. High-frequency activity was no longer sustained, indicating reduced rapid indentation forces and lower vibrational penetration into deep tissue layers. Instead, moderate vibrational energy

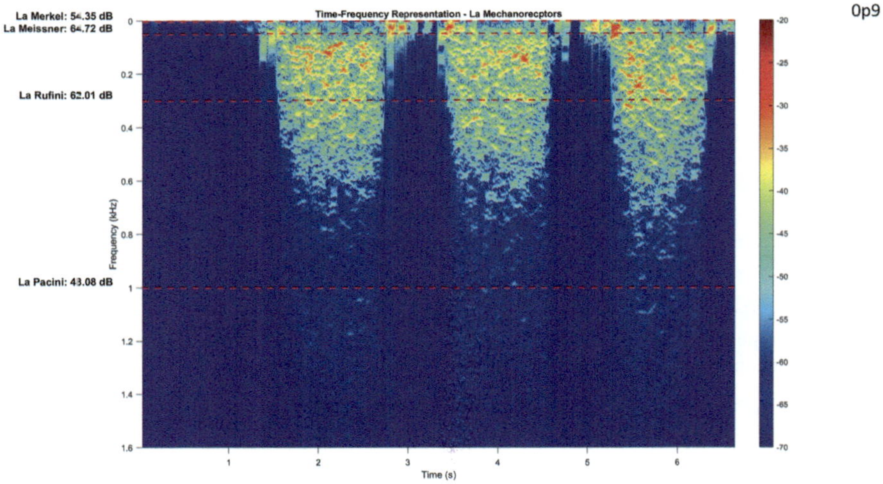

Fig. 2.7 Time-frequency response of hand paper grade 0p9 of the first experimental phase

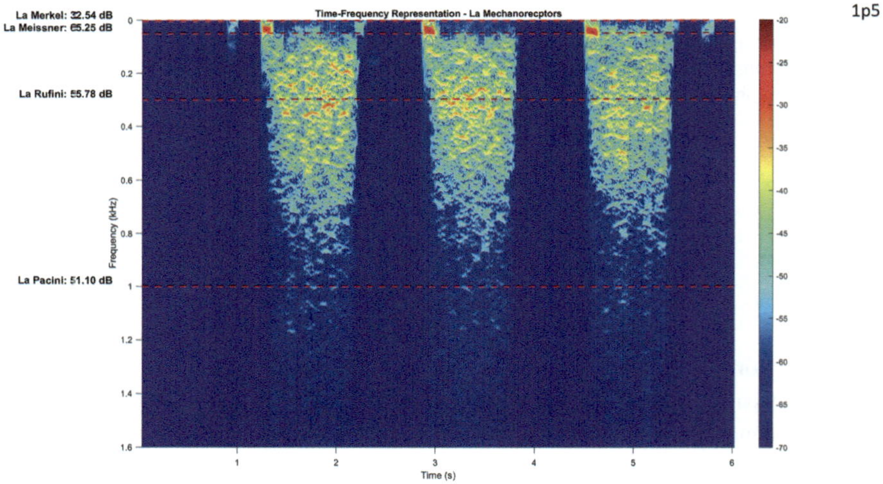

Fig. 2.8 Time-frequency response of hand paper grade 1p5 of the first experimental phase

persisted within the Meissner and Ruffini ranges, suggesting that tactile feedback was increasingly dominated by shear deformation of the skin rather than surface-induced impulse loading. In smoother samples such as 0p9 and 2p, the spectrograms displayed a narrow spectral bandwidth with fast signal decay, indicating that these surfaces failed to generate prolonged dynamic events or variability in frequency content (Cascio and Sathian 2001). These responses were largely confined to low-frequency fluctuations consistent with superficial Merkel-Meissner activation, demonstrating a sensory shift from a multimodal tactile response to one dominated

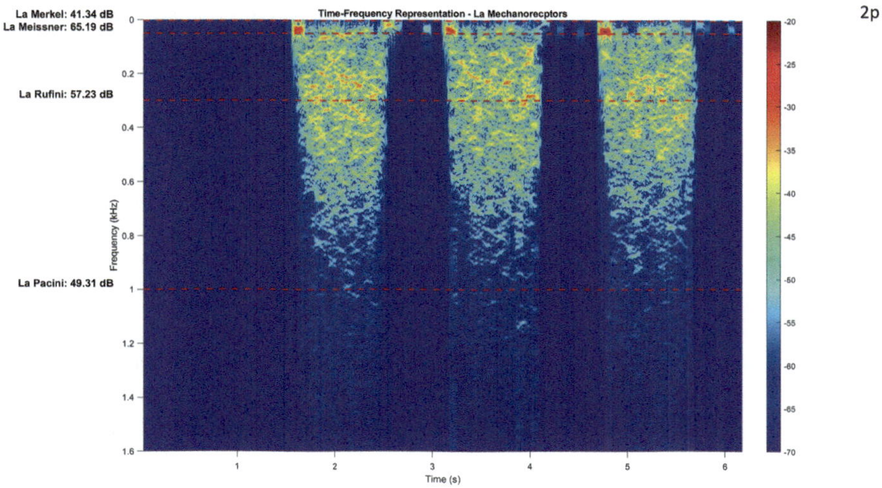

Fig. 2.9 Time-frequency response of hand paper grade 2p of the first experimental phase

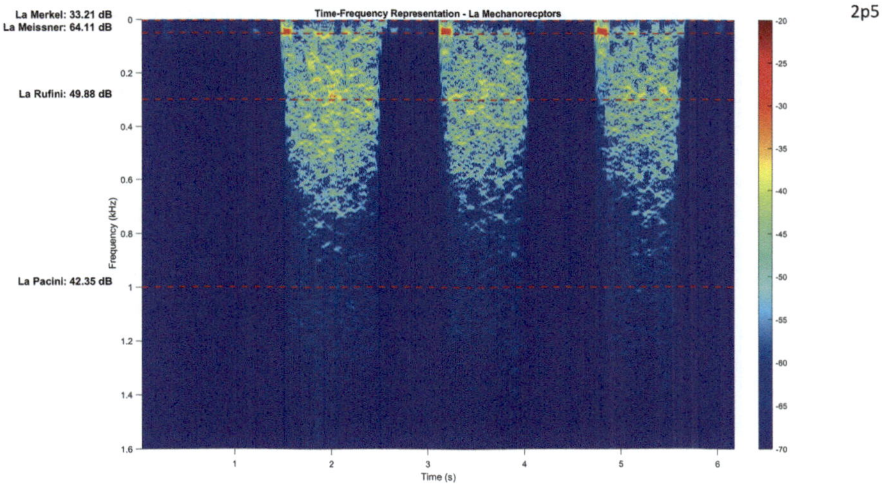

Fig. 2.10 Time-frequency response of hand paper grade 2p5 of the first experimental phase

by low-intensity cues. Finally, the smoothest grade (4p in Fig. 2.13) produced minimal mechanosensory activation, with spectrograms showing merely faint, discontinuous traces of vibrational energy. The absence of mid and high-frequency signatures confirms that nearly uniform surfaces generate minimal contact variability, leading to near-silent tactile encoding with low perceptual richness.

These findings highlight that tactile encoding is not simply a function of average surface roughness but is contingent upon the preservation of frictional heterogeneity along the contact path. The combination of TAI analysis and time-frequency decom-

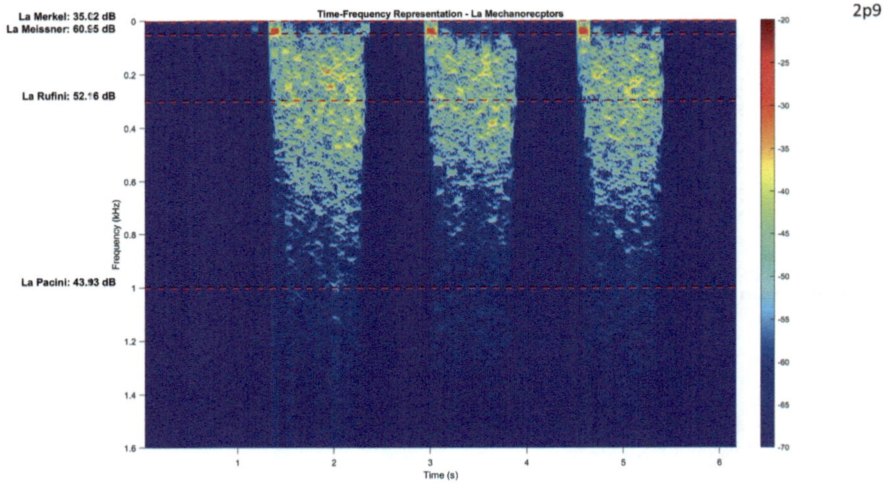

Fig. 2.11 Time-frequency response of hand paper grade 2p9 of the first experimental phase

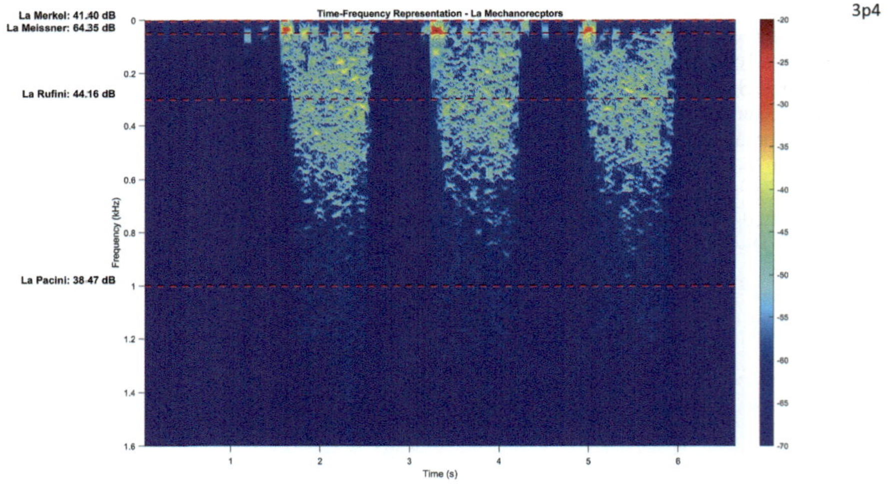

Fig. 2.12 Time-frequency response of hand paper grade 3p4 of the first experimental phase

position reveals that the tactile system uses a hierarchical encoding strategy: coarse textures engage full-spectrum mechanoreceptor recruitment involving both rapidly and slowly adapting pathways, while smoother textures reduce tactile complexity and confine perception to superficial and low-frequency cues. By demonstrating a direct and quantifiable relationship between vibrational richness and mechanoreceptor activation, this investigation provides a mechanical explanation for why coarse materials feel more perceptually informative than smooth ones. In addition, the spectral evolution observed across grades confirms that touch is inherently dynamic

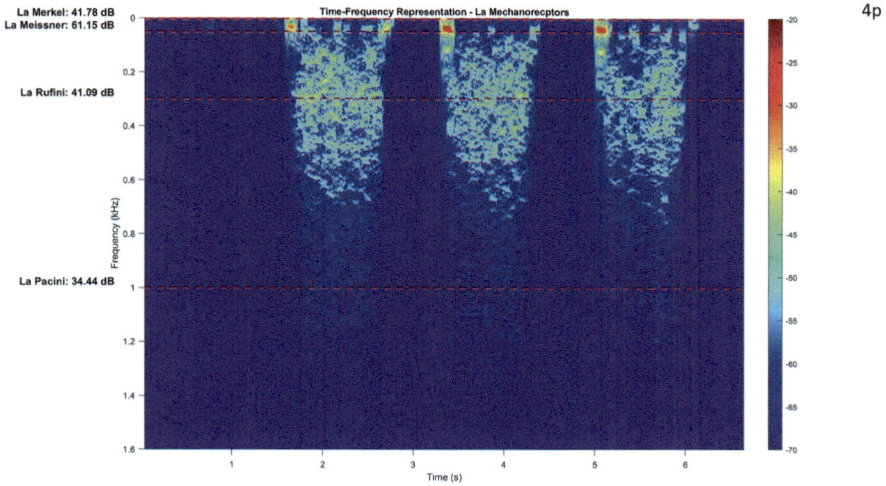

Fig. 2.13 Time-frequency response of hand paper grade 4p of the first experimental phase

and that a single time-domain measure cannot capture the structural complexity of tactile perception. The present results validate the use of WTAD as a high-resolution tactile evaluation platform and establish the TAI combined with time-frequency analysis as a powerful framework for quantifying material tactility.

In this second experimental phase, cutaneous mechanosensory responses were quantitatively assessed during controlled vertical sliding contact between the index finger and thumb, employing five types of commercial hand paper towels with progressively varying microstructural roughness. The experimental protocol was identical to that of the previous study.

The TAI showed a consistent monotonic decline across paper towel grades, as illustrated in Figs. 2.14, 2.15, 2.16, and 2.17, indicating that smoother surfaces produced weaker mechanical stimulation and consequently reduced activation of cutaneous mechanoreceptors. This trend was most evident for Ruffini ($R^2 = 0.89$ in Fig. 2.16) and Pacinian ($R^2 = 0.95$ in Fig. 2.17) receptors, which are primarily responsive to macroscopic strain and high-frequency dynamic stimuli, respectively. Their pronounced decline with increasing smoothness confirms that tactile feedback from fine, even surfaces lack the transient mechanical fluctuations necessary to elicit strong subsurface mechanotransduction. Merkel ($R^2 = 0.77$ in Fig. 2.14) and Meissner ($R^2 = 0.499$ in Fig. 2.15) receptors also displayed a decreasing trend, though with less consistency, suggesting that superficial tactile encoding becomes progressively dampened as micro texture irregularities diminish, albeit to a lesser extent than deeper vibration-sensitive pathways.

The time-frequency analyses reinforce these quantitative findings by revealing the dynamic evolution of tactile signal energy within receptor-specific bands during each sliding event. For the roughest paper (sample 0p4 in Fig. 2.18), the spectrogram exhibits dense, high-energy regions extending from low to high frequencies, with

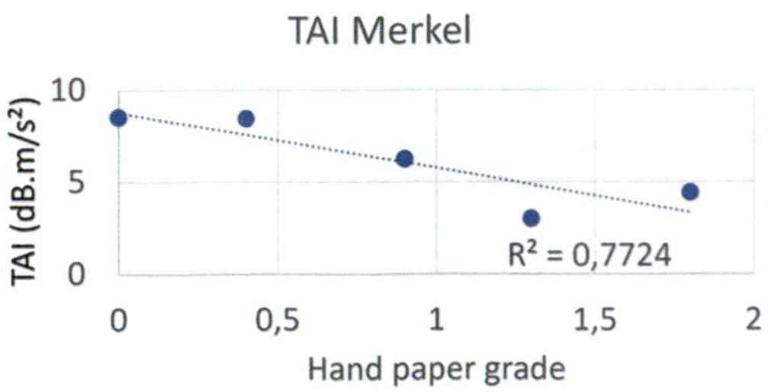

Fig. 2.14 Tactile acuity index of Merkel cells of the second experimental phase

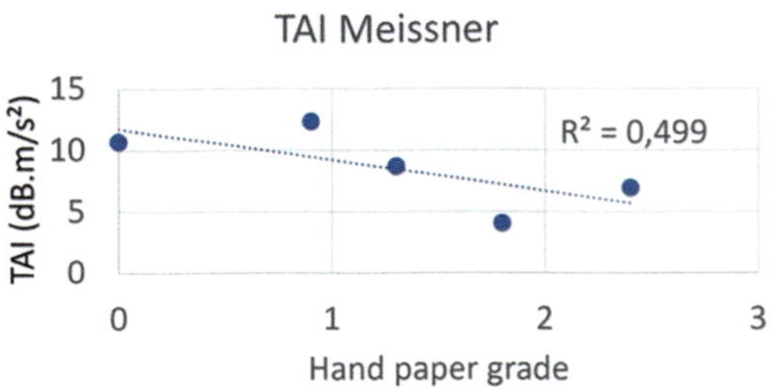

Fig. 2.15 Tactile acuity index of Meissner corpuscle of the second experimental phase

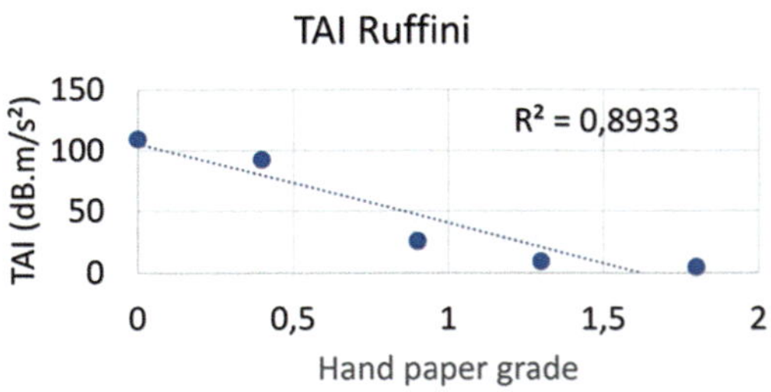

Fig. 2.16 Tactile acuity index of Ruffini corpuscle of the second experimental phase

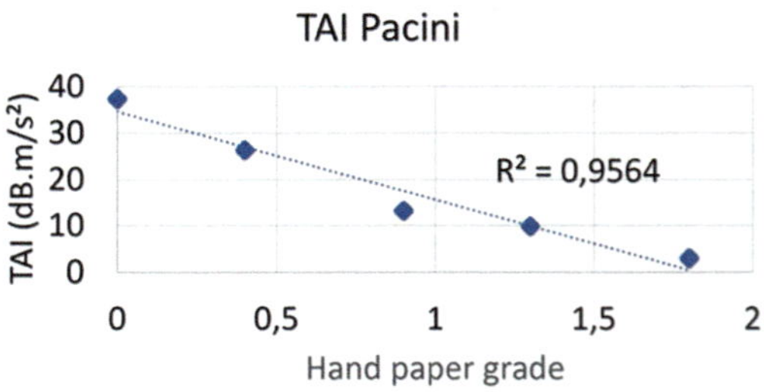

Fig. 2.17 Tactile acuity index of Pacini corpuscle of the second experimental phase

La Meissner = 63.13 dB, La Ruffini = 69.60 dB, and La Pacinian = 56.67 dB. The wide spectral spread and strong persistence of vibrational energy indicate that the coarse texture generated substantial mechanical excitation across all mechanoreceptor types. Frequent high-frequency bursts above 300 Hz correspond to transient micro-slip phenomena, producing deep tissue deformation and pronounced activation of Pacinian corpuscles. The consistent energy plateau in the 50–300 Hz range reveals sustained strain in the dermal layer, strongly engaging Ruffini endings.

For sample 0p9 (Fig. 2.19), a reduction in spectral intensity and bandwidth is evident, with La Ruffini decreasing to 62.78 dB and La Pacinian to 47.63 dB.

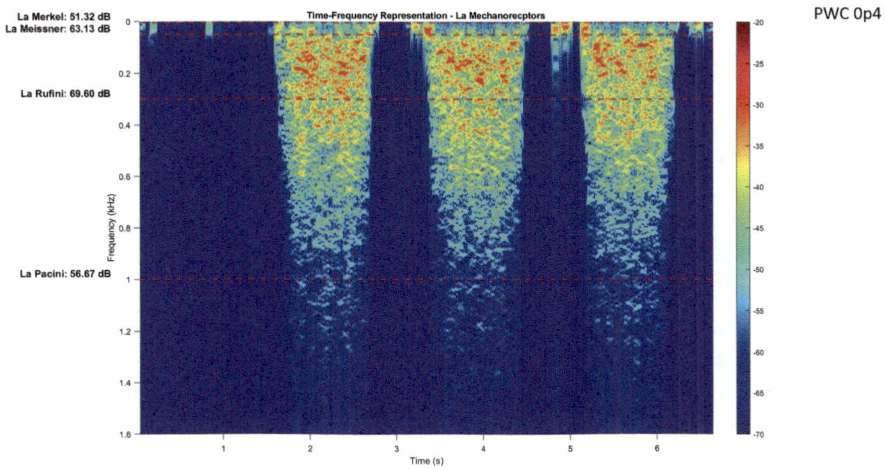

Fig. 2.18 Time-frequency response of hand paper grade 0p4 of the second experimental phase

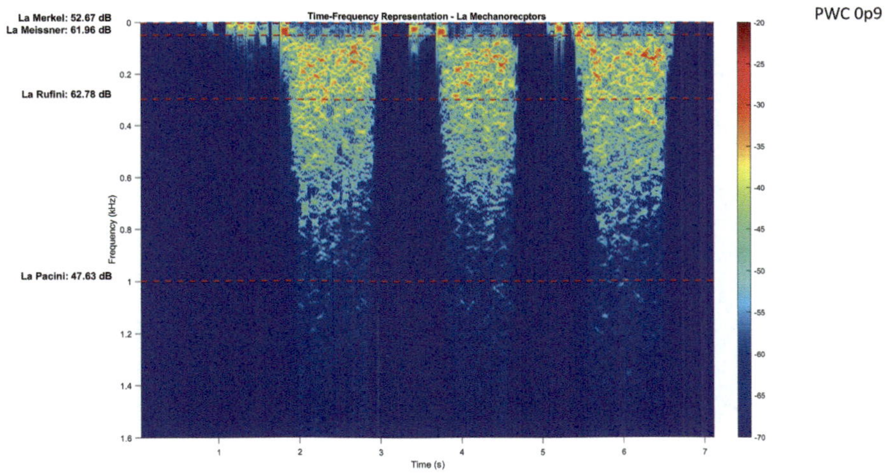

Fig. 2.19 Time-frequency response of hand paper grade 0p9 of the second experimental phase

Although surface asperities still induce intermittent high-frequency bursts, these are shorter and less dense, reflecting fewer frictional instabilities and diminished subsurface vibration transmission. The time-frequency map retains Meissner activity below 50 Hz, confirming that superficial receptors continue to encode fine temporal changes in contact, while deep receptor involvement starts to wane. In sample 1p3 (Fig. 2.20), the overall spectral energy further declines, and the color scale shows a dominance of mid-frequency tones (Meissner and lower Ruffini domains), indicating that the tactile signal is increasingly shaped by surface-level sliding rather

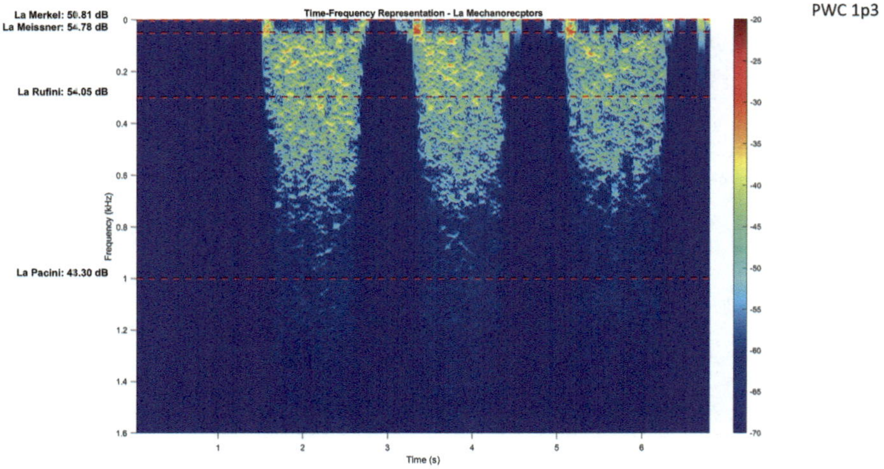

Fig. 2.20 Time-frequency response of hand paper grade 1p3 of the second experimental phase

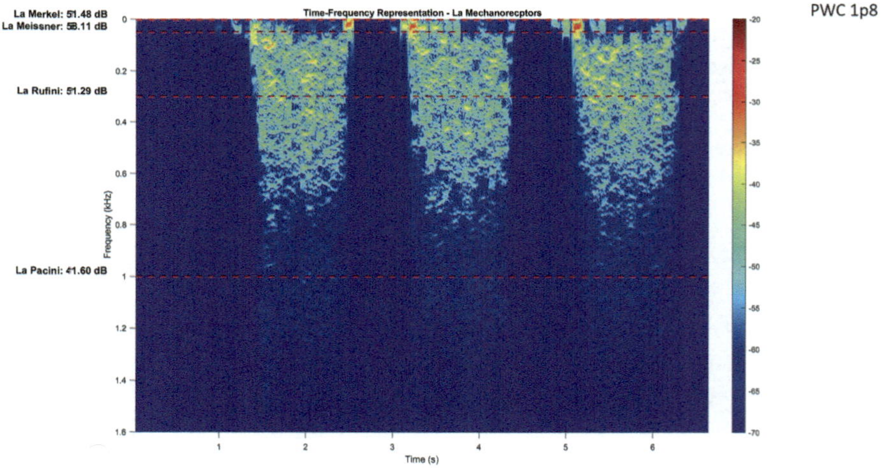

Fig. 2.21 Time-frequency response of hand paper grade 1p8 of the second experimental phase

than subsurface strain. Pacinian activity above 300 Hz becomes sparse, confirming limited excitation of deep, rapidly adapting receptors.

Sample 1p8 in Fig. 2.21 illustrates the transition toward smoother tactile behavior, with La Ruffini = 51.29 dB and La Pacinian = 41.60 dB. The spectrogram depicts shorter bursts with lower amplitude and reduced frequency dispersion, pointing to a dominance of quasi-static skin deformation and attenuated dynamic feedback. The tactile event appears less temporally rich, with energy localized near the beginning of each sliding phase and quickly dissipating thereafter. Finally, the smoothest grade (2p4 in Fig. 2.22) exhibits minimal vibrational amplitude and

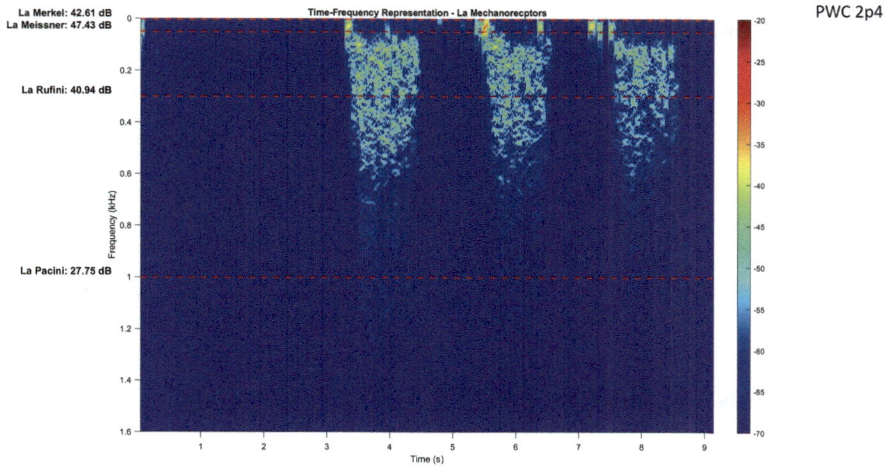

Fig. 2.22 Time-frequency response of hand paper grade 2p4 of the second experimental phase

spectral density. Energy is largely confined to low frequencies, with La Pacinian dropping to 27.75 dB. The absence of pronounced spectral features confirms that uniform, low-roughness materials produce almost no transient excitation capable of triggering deep mechanoreceptor pathways. The tactile signal is characterized by weak, short-lived Meissner activity and negligible Ruffini or Pacinian engagement, corresponding to a perceptually soft, low-contrast surface feel.

Altogether, these observations demonstrate that the vertical sliding configuration elicits a distinct but consistent mechanosensory pattern compared to horizontal motion. While both configurations reveal a monotonic decline in receptor activation with smoother surfaces, the vertical geometry emphasizes the effect of compressive load and normal deformation on Ruffini and Pacinian activation. The gradual spectral narrowing observed across grades reflects the loss of mechanical heterogeneity and a transition from complex, broadband tactile cues to simple, low-frequency oscillations. The time-frequency representation thus provides not only a validation of the TAI but also a multidimensional insight into how tactile encoding evolves with contact mechanics. It illustrates that texture discrimination depends on the coexistence of transient and sustained spectral components: the former drive the perception of vibrational richness, while the latter encode surface continuity. As the spectral diversity collapses with increasing smoothness, tactile perception becomes dominated by low-intensity superficial feedback, resulting in reduced textural differentiation. These findings confirm that WTAD's analytical framework can capture the full temporal-spectral fingerprint of touch, establishing a sensitive methodology for studying fine variations in texture-induced mechanoreceptor activation and their relation to perceived tactile quality.

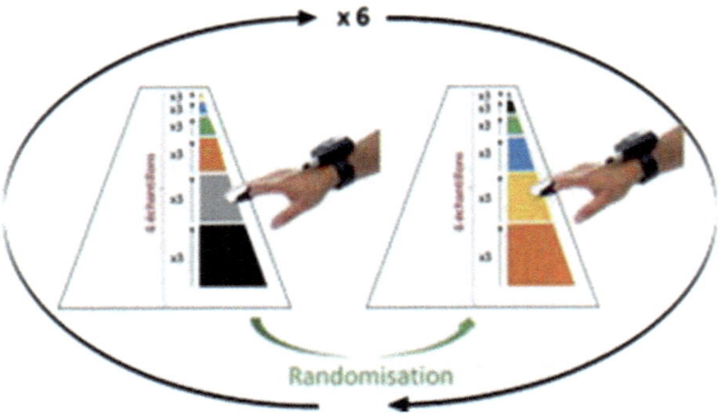

Fig. 2.23 Diagram of the test portion of the protocol

2.5 Results Implementation of a Study

The study was conducted with 40 right-handed female volunteers aged 20–65 (divided into "young" and "older" groups) with diverse Caucasian and Asian skin types (phototypes 1–3). Over 13 days, sessions were scheduled for four participants per day (two in the morning and two in the afternoon), with each session lasting 2 h. Recruitment was external, and strict inclusion/exclusion criteria were applied, including an evaluation of manual ability to ensure consistent touch sensitivity.

The protocol consisted of several phases: an initial 15-min acclimatization to laboratory conditions and WTAD fitting and a 10-min blind texture presentation where volunteers rated various tactile characteristics. This was followed by 20 min of WTAD training using standardized gestures. During the subsequent 20-min test phase, volunteers touched 6 randomly (Fig. 2.23) ordered samples repeatedly (totaling 108 measurements). The session concluded with a 6-min valence questionnaire to assess texture pleasantness and a final 20-min phase for equipment removal and cleaning.

2.6 Mechanotransduction and Activation of Mechanoreceptors Upon Contact with Textures

The difference in mechanoreceptor activation according to texture was tested.

The first graph (Fig. 2.24) shows the intensity of stimulation of each mechanoreceptor for each texture touched.

Overall, all mechanoreceptors are activated and seem to follow the same activation pattern regardless of the texture, though some differences are observable. Indeed, for the VDL plaques, the order of stimulation remains unchanged and is very similar in terms of intensity (Meissner > Pacini > Merkel > Ruffini). However, this is not the case for the other samples. For the fabric, Ruffini appears to be much more stimulated than for the other textures. For the velvet and synthetic skin, however, Pacini seems to be less stimulated.

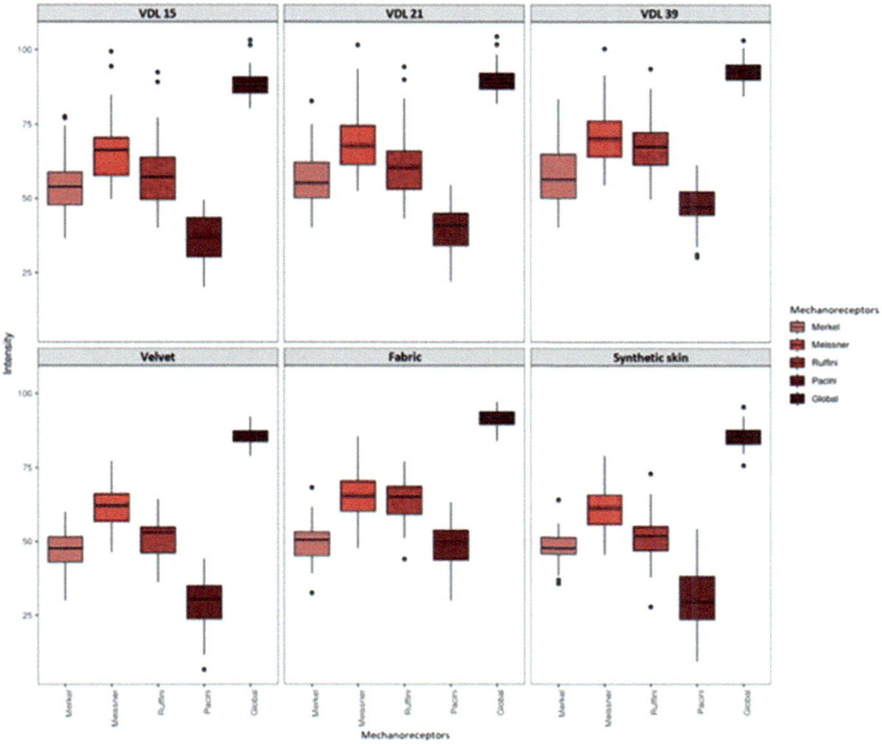

Fig. 2.24 Mechanoreceptor activation for the different textures

Next, boxplots were drawn representing mechanoreceptor activation, separated by type of texture (Fig. 2.25).

The letters above the boxplots represent "significance groups." All boxplots sharing the same letter belong to the same group, and within a group, there is no significant difference. If two boxplots do not share any letter, they are significantly different.

For example, for Merkel, E3 and E2 both belong to group A, so they are not significantly different from each other. However, both differ from E1, which has the letter B. Sample E6 has letters C and D, so it is simultaneously not different from any boxplots labeled C and any labeled D (here, not significantly different from E5 and E4).

E1, which corresponds to VDL plaque 15, thus differs from the others in terms of Merkel, Ruffini, and Pacini activation. VDL plaque 21 differs in terms of Meissner, Ruffini, and Pacini activation. VDL plaque 39 differs in terms of Meissner and Ruffini activation, and the fabric stands out for its Ruffini activation.

As for synthetic skin and velvet, each mechanoreceptor's activation refers to a different texture.

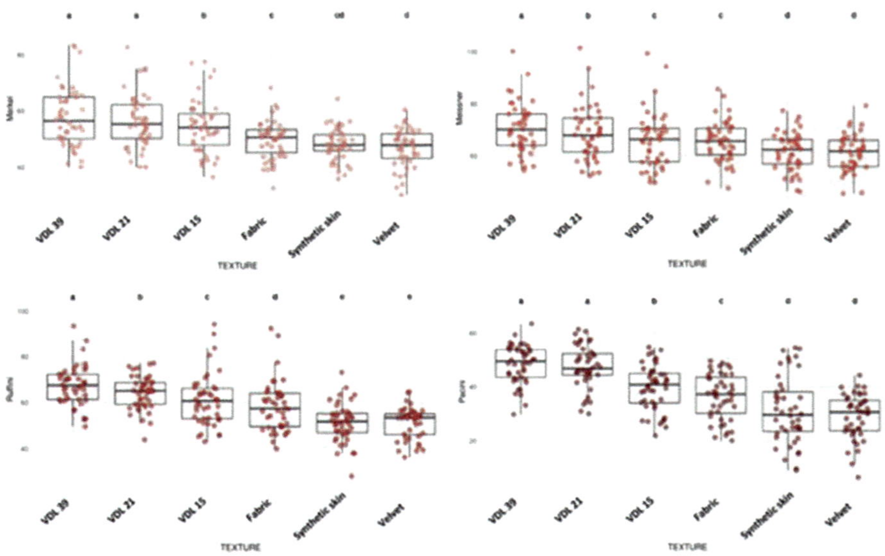

Fig. 2.25 Mechanoreceptor activation by different textures

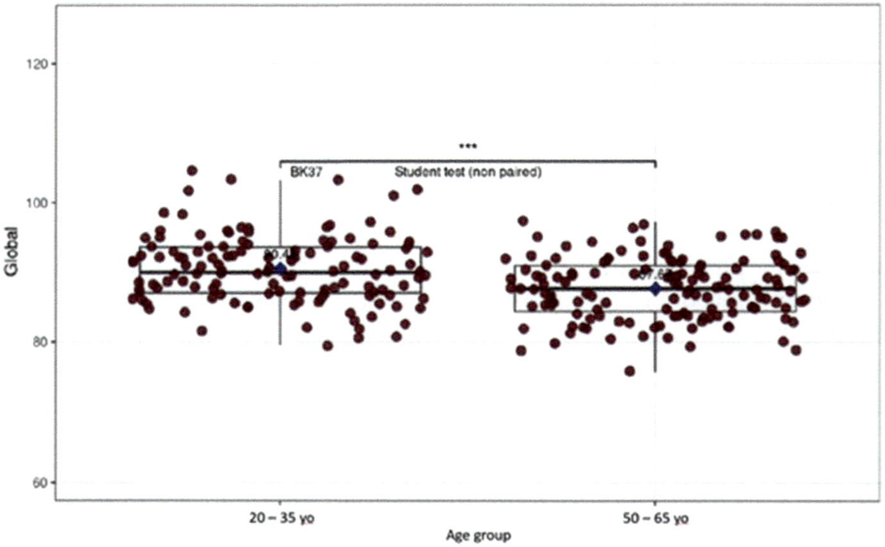

Fig. 2.26 Overall acoustic vibration for the two age groups

2.6.1 Influence of Age on Mechanoreceptors

Below in Fig. 2.26, the boxplot of the overall acoustic vibration level ("La," measured in dB) provides a global visualization of the results obtained.

Corpuscles and Cells	All textures	VDL 15	VDL 21	VDL 39	Velvet	Fabric	Synthetic skin
Merkel	Significant difference	Significant difference	Tendency (pValue = 0.062)	Significant difference	Non Significant difference	Non Significant difference	Non Significant difference
Meissner	Significant difference	Tendency (pValue = 0.051)	Significant difference	Significant difference	Significant difference	Significant difference	Tendency (pValue = 0.054)
Ruffini	Significant difference	Significant difference	Significant difference	Significant difference	Significant difference	Significant difference	Tendency (pValue = 0.072)
Pacini	Significant difference	Tendency (pValue = 0.071)	Tendency (pValue = 0.066)	Tendency (pValue = 0.1)	Significant difference	Non Significant difference	Non Significant difference
Global	Significant difference	Significant difference	Significant difference	Significant difference	Significant difference	Significant difference	Non Significant difference

Fig. 2.27 Summary of the influence of age on mechanoreceptor activation

A significant age effect on the activation of all mechanoreceptors is indeed present. The older group on average depicts weaker mechanoreceptor activation compared to the younger group.

The following table (Fig. 2.27) summarizes the results obtained from the boxplots for each condition and the statistical tests.

Overall, there is a significant difference in mechanoreceptor activation. By texture, nuances appear. The VDL plates are more discriminating in terms of age. For velvet and fabric, there is no significant difference for Merkel cells, but there is a marked difference for the other mechanoreceptors. The tested synthetic skins illustrate no significant differences between the young and older groups, although a trend appears for the Meissner and Ruffini mechanoreceptors.

The comparative analysis of the TAI (Figs. 2.28, 2.29, 2.30, 2.31, 2.32, 2.33, 2.34, and 2.35) between young and aged groups reveals a marked attenuation of mechanosensory responsiveness across all receptor types. In younger subjects, the Cr profiles exhibit strong, multimodal activation for coarse or rigid textures (VDL plates 1–3), with particularly high responses for Ruffini and Pacinian receptors. These receptors, associated respectively with dermal stretch and high-frequency vibration, demonstrate elevated TAI values, indicating robust subsurface mechanotransduction and rich vibrational encoding. Merkel and Meissner receptors, responsible for fine texture and low-frequency flutter perception, also display distinct activation peaks within this range but show lower amplitude overall, consistent with their role in encoding superficial tactile cues. As the explored surfaces become smoother and more compliant (4–6), TAI values for all receptor types decrease sharply, reflecting the loss of frictional heterogeneity and the consequent reduction in mechanical energy available for receptor activation.

In contrast, older skin exhibits a global compression of the TAI dynamic range. The amplitude of TAI declines drastically—by approximately 80% across receptor types—and the differences between textures become less pronounced. Merkel and Meissner receptors show diminished sensitivity to fine and low-frequency cues, suggesting a decline in superficial mechanotransduction efficiency. The Ruffini and

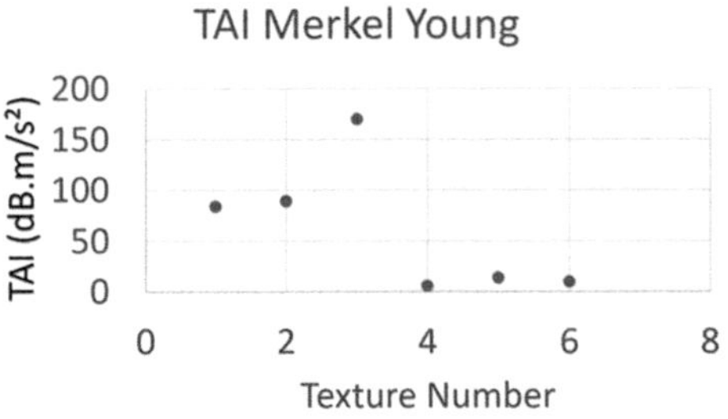

Fig. 2.28 Tactile acuity index of Merkel cells of the textures for the young age group

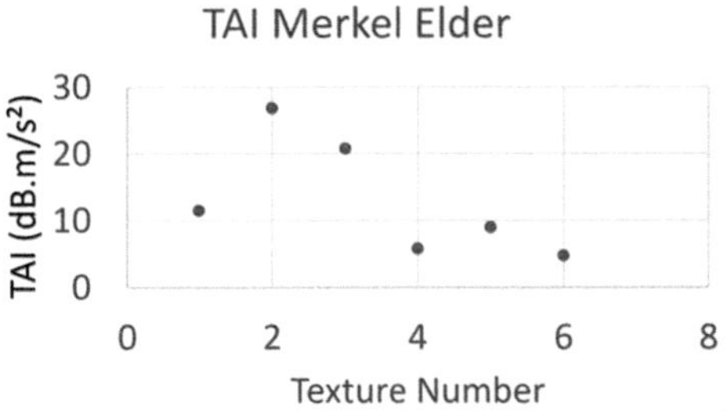

Fig. 2.29 Tactile acuity index of Merkel cells of the textures for the elder age group

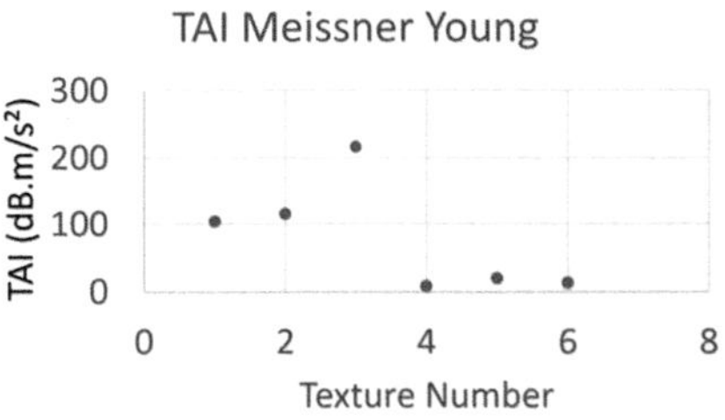

Fig. 2.30 Tactile acuity index of Meissner corpuscle of the textures for the young age group

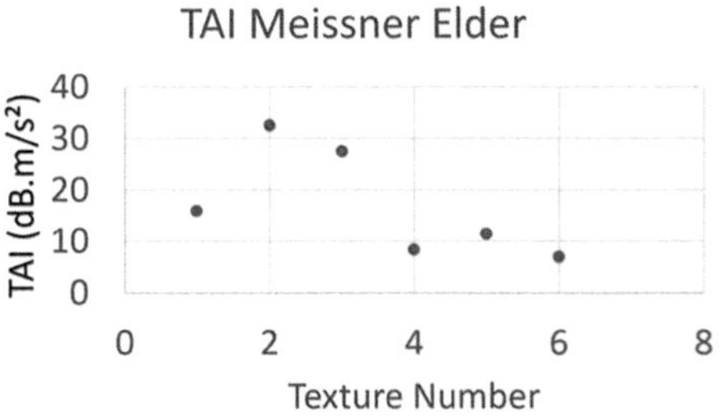

Fig. 2.31 Tactile acuity index of Meissner corpuscle of the textures for the elder age group

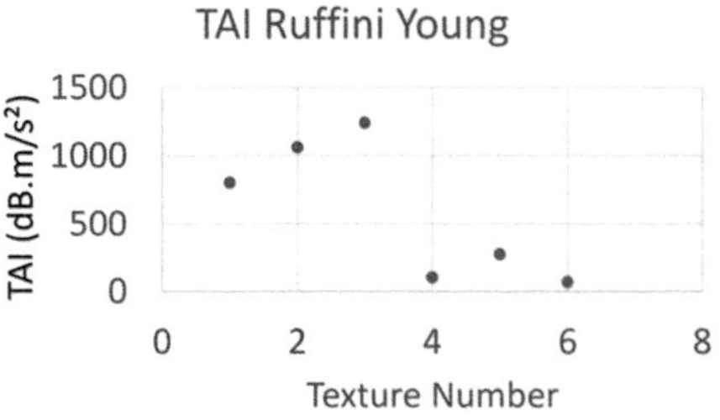

Fig. 2.32 Tactile acuity index of Ruffini corpuscle of the textures for the young age group

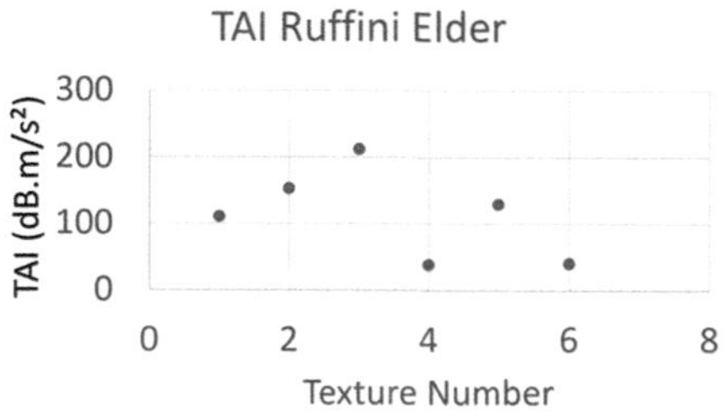

Fig. 2.33 Tactile acuity index of Ruffini corpuscle of the textures for the elder age group

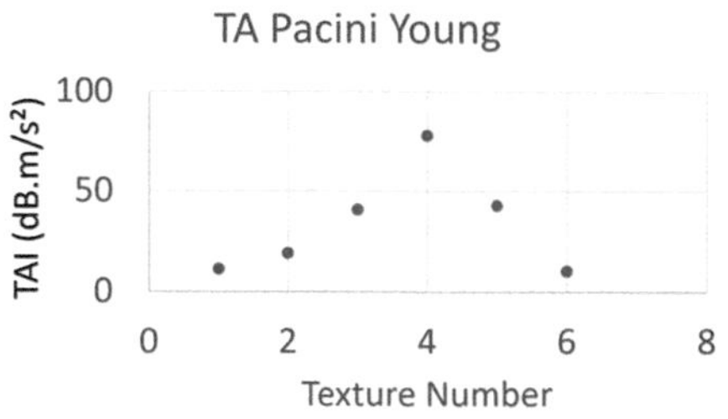

Fig. 2.34 Tactile acuity index of Pacini corpuscle of the textures for the young age group

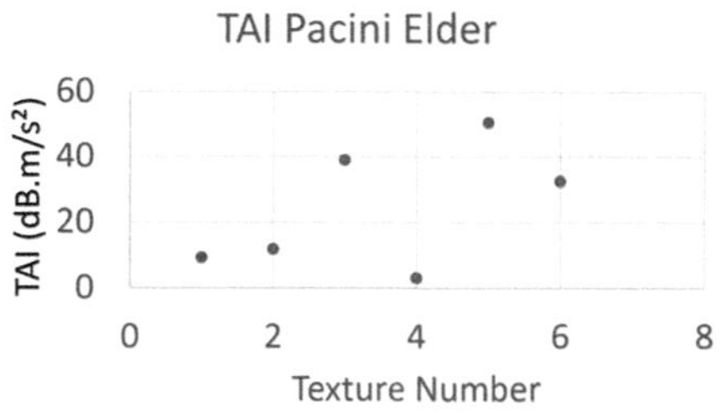

Fig. 2.35 Tactile acuity index of Pacini corpuscle of the textures for the elder age group

Pacinian receptors, which govern deeper stretch and vibration perception, display the most substantial reductions, confirming that age-related structural degradation of the dermal matrix and reduced elasticity strongly impair the propagation of vibrational energy into deeper layers. The resulting flattening of the TAI curves across textures demonstrates that older skin generates weaker and less differentiated tactile feedback, leading to a perceptual homogenization of surface feel.

Overall, these findings establish TAI as a sensitive quantitative marker of tactile aging. By capturing both the intensity and frequency structure of mechanoreceptor activation, the TAI provides a unified metric capable of distinguishing texture-driven mechanosensory variation from biologically driven sensory decline. The pronounced reductions observed in Ruffini and Pacinian pathways highlight that tactile aging predominantly originates from deeper mechanotransductive impairment rather than from superficial receptor desensitization alone. Consequently, the TAI offers a mechanistic framework to monitor age-related tactile loss and to evaluate the effectiveness of receptor-targeted interventions aimed at restoring vibrational responsiveness and tactile acuity.

2.7 Discussion

The present study introduces a new mechanosensory responsiveness index, TAI, as a quantitative descriptor of tactile acuity, and demonstrates its potential to characterize both material-dependent and age-related variations in skin mechanotransduction. Through a combination of controlled tactile interactions and time-frequency decomposition, TAI captures the amplitude and spectral composition of vibrational energy transmitted through the skin, directly reflecting the recruitment dynamics of the four main classes of cutaneous mechanoreceptors.

Across both experimental phases, TAI exhibited a monotonic decline with surface smoothness, validating that coarse textures elicit broad-spectrum mechanical excitation engaging both superficial and deep receptors, while smooth materials produce narrowband, low-energy signals insufficient to sustain robust activation. The strong correlations observed for Ruffini ($R^2 = 0.89$) and Pacinian ($R^2 = 0.95$) receptors emphasize that these deeper mechanosensory pathways are particularly sensitive to the loss of transient high-frequency fluctuations. Conversely, the more moderate correlations for Merkel ($R^2 = 0.77$) and Meissner ($R^2 = 0.499$) receptors reflect the persistence of low-frequency tactile encoding even when textural irregularities diminish, consistent with their superficial localization and lower activation thresholds.

When extended to biological aging, the TAI revealed a global attenuation of mechanosensory performance. In older skin, the amplitude of TAI decreased by nearly 80% across all receptor classes, accompanied by a loss of differentiation between textures. This flattening of the TAI profiles underscores a fundamental simplification of tactile encoding with age, in which the skin's ability to generate and transmit complex vibrational cues is compromised. The decline was most pronounced in the Ruffini and Pacinian bands, confirming that age-related changes in dermal elasticity and subcutaneous coupling primarily impair deep mechanotransductive processes. The reduced responsiveness of Merkel and Meissner receptors, though less severe, further indicates a gradual dampening of superficial tactile sensitivity. Together, these findings provide a mechanistic explanation for the observed decrease in texture discrimination, pressure sensitivity, and vibrational acuity in aged skin.

The integration of TAI with time-frequency analysis offers a multidimensional understanding of tactile decline: while the spectrograms visualize the temporal dispersion and amplitude decay of mechanoreceptor activation, TAI condenses these dynamics into a single, scalable metric suitable for interindividual and longitudinal comparisons. This dual approach bridges the gap between biophysical signal analysis and perceptual function, enabling the quantification of tactile aging with unprecedented precision. It also establishes a framework for evaluating the impact of cosmetic treatments, massage protocols, or neurosensory training aimed at restoring receptor activity.

2.8 Conclusion

To begin with, this study demonstrates that the TAI, combined with time-frequency analysis of tactile vibrations, provides a powerful and objective framework for quantifying the mechanosensory dimension of skin aging. By linking vibrational energy patterns to specific mechanoreceptor pathways, the approach captures both the structural degradation of the skin and the functional decline in receptor responsiveness that accompany aging. The results highlight that tactile deterioration is not limited to surface-level desensitization but involves a deeper loss of mechanotransductive efficiency, particularly within Ruffini and Pacinian channels.

TAI thus emerges as a novel biomarker of tactile vitality, capable of differentiating young and aged skin, mapping receptor-specific vulnerabilities, and assessing the efficacy of receptor-targeted interventions. Beyond its diagnostic value, this metric paves the way for a novel generation of sensor-based skincare research that evaluates not merely how the skin appears but also how it perceives, responds, and adapts to touch.

Conflict of Interest This study was partially funded by CHANEL PB. Gabriel Cazorla and Youcef Ben-Khalifa are employed by CHANEL PB and contributed to the experimental design and interpretation of skincare-related results. The authors declare no other competing interests.

Ethic Statement All procedures involving human participants were approved by the institutional ethics review board: Società Scientifica Italiana per le Indagini Cliniche Non Farmacologiche [Prot. FR0000908/23 Versione N.02 del 06/11/2023]. Participants gave written informed consent prior to participation. The study complied with the principles of the Declaration of Helsinki.

References

Abdouni A, Vargiolu R, Zahouani H (2018) Impact of finger biophysical properties on touch gestures and tactile perception: aging and gender effects. Sci Rep 8(1):12605. https://doi.org/10.1038/s41598-018-30677-2

Amaied E, Vargiolu R, Bergheau JM, Zahouani H (2015) Aging effect on tactile perception: experimental and modelling studies. Wear 332–333:715–724. https://doi.org/10.1016/j.wear.2015.02.030

Blake DT, Hsiao SS, Johnson KO (1997) Neural coding mechanisms in tactile pattern recognition: the relative contributions of slowly and rapidly adapting mechanoreceptors to perceived roughness. J Neurosci 17(19):7480–7489. https://doi.org/10.1523/JNEUROSCI.17-19-07480.1997

Bruce MF (1980) The relation of tactile thresholds to histology in the fingers of elderly people. J Neurol Neurosurg Psychiatry 43(8):730–734

Burns T, Breathnach SM, Cox N, Griffiths C (2010) Rook's textbook of dermatology, 8th edn. Wiley, Chichester

Cascio CJ, Sathian K (2001) Temporal cues contribute to tactile perception of roughness. J Neurosci 21(14):5289–5296. https://doi.org/10.1523/JNEUROSCI.21-14-05289.2001

Delhaye B, Hayward V, Lefèvre P, Thonnard J-L (2012) Texture-induced vibrations in the forearm during tactile exploration. Front Behav Neurosci 6. https://doi.org/10.3389/fnbeh.2012.00037

Derler S, Schrade U, Gerhardt L-C (2007) Tribology of human skin and mechanical skin equivalents in contact with textiles. Wear 263(7–12):1112–1116. https://doi.org/10.1016/j.wear.2006.11.031

Derler S, Gerhardt L-C, Lenz A, Bertaux E, Hadad M (2009) Friction of human skin against smooth and rough glass as a function of the contact pressure. Tribol Int 42(11–12):1565–1574. https://doi.org/10.1016/j.triboint.2008.11.009

Dunn SL, Olmedo ML (2016) Mechanotransduction: relevance to physical therapist practice – understanding our ability to affect genetic expression through mechanical forces. Phys Ther 96(5):712–721. https://doi.org/10.2522/ptj.20150073

Fujimoto I, Yamada Y, Morizono T, Umetani Y, Maeno T (2003) Development of artificial finger skin to detect incipient slip for realization of static friction sensation. In: Proceedings of IEEE international conference on multisensor fusion and integration for intelligent systems, MFI2003. IEEE, Tokyo, pp 15–20. https://doi.org/10.1109/MFI-2003.2003.1232571

Lederman SJ (1983) Tactual roughness perception: spatial and temporal determinants. Can J Psychol Rev Can Psychol 37(4):498–511. https://doi.org/10.1037/h0080750

Orr AW, Helmke BP, Blackman BR, Schwartz MA (2006) Mechanisms of Mechanotransduction. Dev Cell 10(1):11–20. https://doi.org/10.1016/j.devcel.2005.12.006

Scheibert J, Leurent S, Prevost A, Debrégeas G (2009) The role of fingerprints in the coding of tactile information probed with a biomimetic sensor. Science 323(5920):1503–1506

Shyti R, Nays P, Vargiolu R, Zahouani H (2025) An innovative wearable device for sensing mechanoreceptor activation during touch. Sci Rep 15(1):9986. https://doi.org/10.1038/s41598-025-92379-w

Thieulin C, Pailler-Mattei C, Vargiolu R, Lancelot S, Zahouani H (2017) Study of the tactile perception of bathroom tissues: comparison between the sensory evaluation by a handfeel panel and a tribo-acoustic artificial finger. Colloids Surf B Biointerfaces 150:417–425. https://doi.org/10.1016/j.colsurfb.2016.11.006

Zahouani H, Mezghani S, Vargiolu R, Hoc T, El Mansori M (2013) Effect of roughness on vibration of human finger during a friction test. Wear 301(1–2):343–352. https://doi.org/10.1016/j.wear.2012.11.028

Chapter 3
PIEZO1 in Immune Cells: From Force to Function

Dipanwita Ghosh and Dipyaman Ganguly

Abstract As a professional mechanosensor, PIEZO1 converts mechanical stimuli—such as substrate stiffness, fluid shear stress, and membrane tension—into intracellular calcium influx, which in turn regulates a wide array of immune processes. The chapter details its significance across both innate and adaptive immunity, including T cell activation and migration, macrophage polarization, dendritic cell activation, natural killer cell cytotoxicity, neutrophil extracellular trap (NET) formation, and B cell antigen discrimination and class-switching to IgA. PIEZO1 emerges as a central regulator that fine-tunes immune responses by integrating signals from the cellular microenvironment, influencing inflammation, pathogen clearance, and metabolic reprogramming. Despite these advances, key questions remain regarding its role in chronic disease, autoimmunity, and cancer immunity. Targeting PIEZO1 presents a promising therapeutic strategy for conditions driven by mechanical stress and immune dysfunction, such as fibrosis, atherosclerosis, and solid tumours, potentially inaugurating a new era of mechano-immunotherapy.

3.1 Introduction

Cellular life is tightly regulated by both intracellular and extracellular signals (Zhang and Wang 2012). The mechanical environment surrounding the cell, including extracellular matrix (ECM) and other cells, plays a crucial role in regulating diverse cellular functions. The term mechanotransduction can be defined as a cellular process in which cells sense and respond to mechanical stimuli, such as stretching, fluid shear stress, and cyclic force, arising from the extracellular

D. Ghosh · D. Ganguly (✉)
Department of Biology, Trivedi School of Biosciences, Ashoka University, Sonipat, Haryana, India

Datla Human Immunome Lab, Trivedi School of Biosciences, Ashoka University, Sonipat, Haryana, India
e-mail: dipyaman.ganguly@ashoka.edu.in

M. Kloc, J. Wosik (eds.), *Mechanotransduction and Mechanosensing*, Results and Problems in Cell Differentiation 78, https://doi.org/10.1007/978-3-032-19509-8_3

environment. In the context of immunology, biophysical cues from cytoskeleton or external forces from the ECM can be sensed and responded to by cell as they develop, activate, differentiate, and proliferate. Leukocyte extravasation (Schaefer and Hordijk 2015), macrophage polarization (Mei et al. 2024; Adams et al. 2019), T cell activation (Liu et al. 2018; Hope et al. 2022; Jeffreys et al. 2024), T cell migration (Liu et al. 2024; Rossy et al. 2018), and NK cell cytotoxicity (Barda-Saad et al. 2020) are just a few of the immune cell processes that are regulated by these biophysical cues. The key mechanotransduction channels known to function in the immune system, such as PIEZO1 (Solis et al. 2019), TRP (Yang et al. 2024), and TREK1 channels (Immanuel et al. 2024), convert mechanical signals to biochemical signals, playing an important role in immune mechanotransduction. Recent studies find that mechanotransduction plays a crucial role in immune-related illnesses, which are linked to immune dysfunction, caused by the loss or disruption of signalling (Yang et al. 2024). Almost all cells in the mammalian immune system, ranging from innate to adaptive immunity, express the PIEZO1 non-selective cation channel, as a primary mechanosensor and communicate with different signalling pathways through a second messenger, calcium ion (Ca^{2+}). Mechanotransduction by PIEZO1 is important for immune cell migration, trafficking, antigen recognition, synapse formation, cytokine production, and immune metabolism—discussed in this review.

3.2 Discovery of PIEZO Proteins

Mechanotransduction plays diverse physiological function in living organisms, encompassing embryonic development (Ranade et al. 2014), nociception (Kim et al. 2012), tactile sensation (Maksimovic et al. 2014), proprioception (Woo et al. 2015), auditory processing (Lee et al. 2024), vascular tone regulation, urinary system regulation (Zhang et al. 2025), pulmonary growth and function (Xiong et al. 2022), the maintenance of bone and muscle integrity (Wang et al. 2022), as well as tumour progression and metastasis (Agrawal et al. 2025). Although the physiological importance of mechanotransduction has long been recognized, the field lacked a clear understanding of the specific molecular sensor that directly responds to the mechanical stimuli in the mammalian system. In 2010, Coste et al. identified a novel class of mammalian mechanosensitive, pore-forming, non-selective cation channels, PIEZO1 and PIEZO2, within the neuro2A mouse neuroblastoma cell line (Coste et al. 2010, 2015). PIEZO1 is predominantly expressed in the bladder, colon, kidney, lung, and skin (Holt et al. 2021). While PIEZO2 is also detected in the bladder (Marshall et al. 2020), colon, and lung, it shows comparatively lower expression in the kidney and skin. Notably, PIEZO2 displays robust expression in dorsal root ganglion (DRG) sensory neurones, implicating it as a key molecular mediator in somatosensory mechanotransduction (Woo et al. 2014, 2015; Servin-Vences et al. 2023; Villarino et al. 2023).

3.3 PIEZO1 Is an Evolutionarily Conserved Protein with a Highly Polymorphic Nature

Mammalian PIEZO1/2 proteins are exceptionally large mechanosensitive channels, comprising over 2500 amino acids (~ 900kD molecular weight) with 24–36 predicted transmembrane segments, and interestingly, these proteins share no recognizable sequence homology with any known class of ion channel (Coste et al. 2010, 2015; Xiao 2024). In humans, the PIEZO1 gene (DHS/FAM38A/LMPH3/LMPHM6) is located on chromosome 16 (region 16q24.3) and contains 51 exons. Its murine counterpart, referred to as 9630020 g22, Fam3, Fam38a, Pie, or mLIAA0233, consists of 53 exons and resides on chromosome 8 (region 8 E1), according to data from the NCBI database. The human PIEZO1 gene shows a high degree of polymorphism, with numerous mutations implicated in various pathological conditions (Murthy et al. 2017). These mutations are generally categorized into gain-of-function (GOF) and loss-of-function (LOF) types. Disorders like hereditary xerocytosis and dehydrated hereditary stomatocytosis, which involve abnormal red blood cell hydration and membrane stability, are examples of GOF mutants (Albuisson et al. 2013). A representative condition associated with LOF variants is autosomal recessive congenital lymphatic dysplasia, a disorder characterized by defective lymphatic development and function (Lukacs et al. 2015).

3.4 Structure of PIEZO1

Structurally, mammalian PIEZO1 homotrimer adopts a distinct three-bladed propeller architecture, with a cap (also called C-terminal extracellular region, CED) domain located extracellularly at the top of the central helix (Ge et al. 2015). PIEZO1 is predicted to have an unprecedented 38-TM topology, consisting of approximately 36 transmembrane helices with the outer (OH) and inner helices (IH). These helices are organized into nine transmembrane helical units (THUs), each composed of four TM segments that collectively form a blade-like structure crucial for mechanotransduction (Fig. 3.1). Each THU is connected by intracellular loops, while the first two TM helices of every THU extend into the extracellular loop essential for proper channel assembly, expression, and mechanical gating (Xiao 2024; Coste et al. 2015). The blades are structurally coupled with the pore domain via a helical beam approximately 9 nm in length (corresponding to residues 1300–1365) (Xiao 2024). This beam lies centrally within the protein, linking TM28 of THU7 to TM29 of THU8, and acts as a pivotal element in transmitting conformational changes across the channel. An anchor domain, located between the OH-IH and the C- terminal domain (CTD), connects TM36 of THU (with TM37 of the OH-cap-IH pore module, producing a swapped configuration of a cap domain (Zhao et al. 2018). Together, these, blade-beam-CTD-anchor-IH/OH, form a lever-

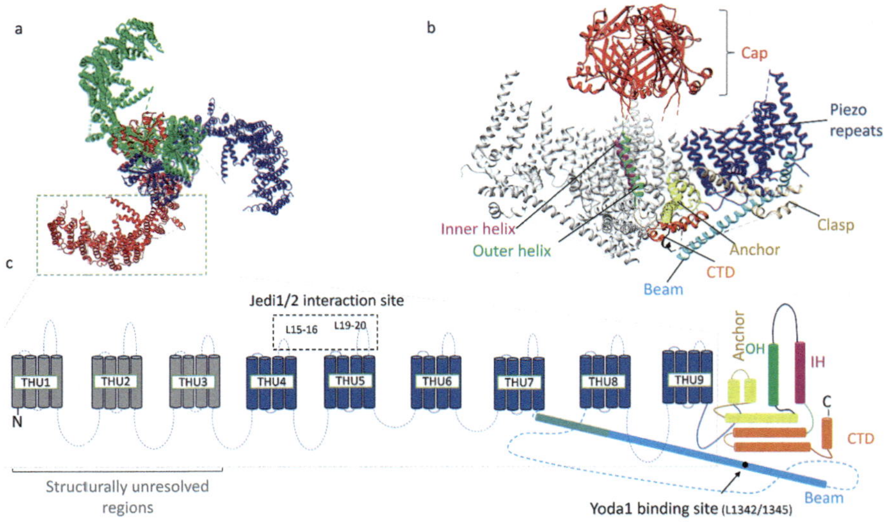

Fig. 3.1 Structural architecture and the domain arrangements of the mechanosensitive ion channel PIEZO1. Top (**a**) and side (**b**) view of mPIEZO1 cartoon representation with each subunit and domains colour coded. (**c**) Schematic 38-TM topology model of a Piezo1, colour coded to match the cartoon mode in (**b**)

like mechanotransduction axis that is important for activation of PIEZO1 (Zhao et al. 2018; Li et al. 2017). Beam domain is followed by a series of critical gating modules, including central plug, latch, lateral plug, and, a yet uncharacterized, clasp domain.

3.4.1 Pore-Forming Region of PIEZO1 Plays a Central Role in Ion Conductance

Multi-gating properties of PIEZO1 depend on three lateral portals, extracellular cap gate, transmembrane gate, and intracellular lateral plug gate, each serving as a potential ion exit route (Saotome et al. 2018). Additionally, the entry of the cation through the channel pore happens either by moving vertically through the central cavity aligned with the threefold axis with the cap domain or by entering laterally through the openings formed by the flexible linkers that connect the cap to the inner and outer helices (Saotome et al. 2018). Under resting membrane tension, PIEZO1 adopts a flattened configuration, thought to represent its inactive state, where both cap and lateral plug gates remain closed (Xiao 2024; Shan et al. 2025). Conversely, during mechanical stimulation, conformational rearrangements in the blade, beam, and cap domains are predicted to open all three gating elements, facilitating ion permeation through the channel's central and lateral pathways (Y. Wang and Xiao

2018; Xiao 2024). The C-terminal region forms the ion-conducting pore modules, composed of the outer helix (OH), inner helix (IH), C-terminal domain (CTD), and cap (Coste et al. 2015). Therefore region OH-CED-IH-CTD functions as a pore-forming module (Ge et al. 2015; Guo and MacKinnon 2017; Saotome et al. 2018; Zhao et al. 2016).

3.4.2 PIEZO1 Integrates Membrane Tension and Cytoskeletal Dynamics to Mediate Mechanotransduction

PIEZO1 channels exhibit intrinsic sensitivity to asymmetric lipid bilayer curvature, allowing them to respond dynamically to variations in local membrane architecture, consistent with the FFL model (Syeda et al. 2016; Lewis and Grandl 2015). Beyond lipid-driven curvature, PIEZO1 channels are also biochemically and functionally linked to the actin cytoskeleton through a mechanotransduction complex composed of E-cadherin, β-catenin, and vinculin (Cox et al. 2016). Collectively, these findings suggest that PIEZO channels integrate both force-from-lipid (FFL) and force-from-filament (FFF) mechanism. Such dual sensitivity enables them to operate as highly adaptable mechanotransducers, capable of detecting not only local changes in membrane curvature and tension but also mechanical signals propagated over long distances within or between cells (Jiang et al. 2021).

3.4.3 Fast Activation and Adaptive Inactivation Kinetics of Mammalian PIEZO1 Channel

Functional studies have demonstrated that PIEZO channels are both necessary and sufficient to generate mechanically activated (MA) currents across multiple cell types (Coste et al. 2010). Additionally, mechanical perturbation applied directly to the lipid bilayer is adequate to trigger the fast activation of PIEZO channels, highlighting their intrinsic capability to function as a molecular force sensor and transducer (Syeda et al. 2016). PIEZO1 also exhibits distinct single-channel conductance, sub-conductance, and rapid inactivation dynamics (Coste et al. 2010; Lewis and Grandl 2020; Wu et al. 2017; Zheng et al. 2019; Ullah et al. 2024). Notably after mechanical activation, PIEZO1 undergoes substantial inactivation despite of sustained mechanical stimuli, describing the presence of at least two (or more) kinetically distinct inactivation state and a four stage model of channel gating (closed, open, and two inactivation stages) (Bae et al. 2013b; Lewis et al. 2017; Ullah et al. 2024; Wijerathne et al. 2023). The inactivation kinetics differ between PIEZO1 and PIEZO2, with PIEZO1 displaying slower inactivation, reflecting their specialized physiological role across tissues (Moroni et al. 2018; Coste et al. 2010; Wu et al. 2017). Both human and mouse PIEZO1 (hPIEZO1 and mPIEZO1) ion channels display conductance characteristics determined largely by their pore structures (Coste et al. 2015; Bae et al. 2013b). It has been shown that hPIEZO1 exhibits more rapid inactivation kinetics than its murine counterparts under identical voltage conditions (Coste et al. 2010; Bae et al. 2013a. Systematic examination of PIEZO1 variants carrying disease-associated mutations revealed a consistent trend—each of the six mutations implicated in dehydrated hereditary stomatocytosis (DHS) shows similar prolonged channel inactivation (Albuisson et al. 2013; Zarychanski et al. 2012; Bae et al. 2013a). In addition, MDFIC/MDFI, a Myo-D interacting protein, functions as an accessory unit of both hPIEZO1 and mPIEZO1 mechanosensitive ion channel (Zhou et al. 2023; Shan et al. 2025). MDFIC, in particular, modulates the response of hPIEZO1 to varying mechanical stimuli by altering the pore module through its interactions with the acyl chain of pore lipids, forming a stable hPIEZO1-milti-lipidated MDFIC-pore lipid complex (Shan et al. 2025). Pore-associated lipids play a crucial role in mediating the rapid inactivation of the PIEZO1 channel (Fig. 3.2). The hydrophobic acyl chains of these lipids penetrate the hydrophobic pore region, effectively sealing it (Shan et al. 2025; Romero et al. 2019). This structural arrangement is supported by electrophysiological studies showing that the hydrophobic residue lining the pore with hydrophilic counterparts significantly slows inactivation in both PIEZO1 and

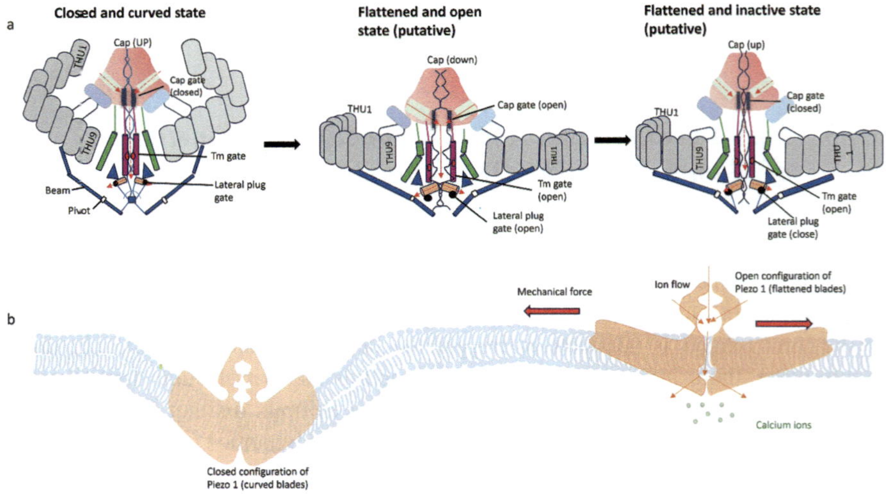

Fig. 3.2 Schematic representation of the dynamic gating model and activation of PIEZO1 channels. Closed state (**a**) PIEZO1 adopts a curved conformation with all gates (cap, Tm, and lateral plug) closed blocking the ion flow (dashed red arrow). Under the membrane tension, the channel flattens, and the cap moves downwards, opening all the gates and permitting io movements (red solid arrows). During inactivation the cap returns to the up position, closing the cap gate while the blades remain flattened; (**b**) increased membrane tension drives PIEZO1 to change the conformation from a curved, bowl-shaped closed state to a flattened open state, enabling ion influx (indicated by red arrow)

PIEZO2, highlighting the importance of hydrophobic pore lipid interaction in gating kinetics (Zheng et al. 2019; Shan et al. 2025).

3.5 Chemical Modulators of PIEZO1

3.5.1 Positive Modulators

To check nonmechanical activation of the PIEZO1 channel, Syeda et al. discovered a synthetic agonist, Yoda1, active on both human and mouse PIEZO channels. Experiments in purified lipid bilayer confirmed that Yoda1 directly activates PIEZO1 without additional cellular components. This compound stabilizes the open state of the channel leading to the slower inactivation of PIEZO1 channels (Syeda et al. 2015). Yoda1 functions as a molecular wedge that separates the upper region of repeats A from repeats B, enabling the distal segment of the arm to extend more under membrane tension (Botello-Smith et al. 2019).

A new class of PIEZO1 chemical activator, Jedi1/2, binds to the extracellular blade region of PIEZO1 and stimulates the channel (Wang et al. 2018). Both Jedi1/2 and Yoda1 activate PIEZO1 independently by distinct molecular mechanisms. Jedi1/2, hydrophilic in nature, thought to interact with distal extracellular loops of the channel while Yoda1, which is hydrophobic, rapidly diffuses (within 2 ns) and accesses its binding pockets from the intracellular sides (Wang et al. 2018; Botello-Smith et al. 2019). This observation highlights diverse chemical activation pathways for PIEZO1, providing deeper insights into its structural and functional complexities.

3.5.2 Negative Regulators

GsMTx4, a lysine-rich peptide purified from spider venom, targets the mechanosensitive cation channels, including PIEZO1, and inhibits their activity. The binding of GsMTx4 is thought to decrease the inter-blading spacing within the PIEZO1 channel and inhibit it by stabilizing a more compact, closed channel formation. However, activity of GsMTx4 is not exclusive to PIEZO1; it also affects other mechanosensitive channels, such as PIEZO2, TRPC5, TRPC6, voltage-gated sodium channels, and potassium channels. This lack of specificity is likely linked to its mode of action, which involves interaction with membrane lipids that modify lateral membrane tension (Gnanasambandam et al. 2017).

Dooku1, a Yoda1 analogue, formed by substituting the pyrazin-2-yl thiadiazol group in Yoda1 with a pyrrole 2-yl oxadiazole moiety, competitively inhibit Yoda1-activated PIEZO1 channel (Evans et al. 2018). Ruthenium red (RR) is an inorganic polycationic dye also known to inhibit PIEZO1 by binding at negatively charged residues located on the intracellular side of PIEZO1 (Coste et al. 2012). Similarly GD3+, a trivalent lanthanide ion, also suppresses PIEZO1-dependent mechanosensitive currents (Coste et al. 2010). Apart from these, some saturated and polyunsaturated fatty acid (PUFA) can also inhibit PIEZO1 (Romero et al. 2019).

3.6 PIEZO1 in Innate Immune Cells

The significance of innate immune cells in sensing and responding to the physical properties of their microenvironment is increasingly being recognized. Innate immune cells, including macrophages, neutrophils, dendritic cells, NK cells, and eosinophils, can translate mechanical forces into biochemical signals that regulate their functional behaviour (Pageon et al. 2018; Orsini et al. 2021). Almost all cells in the innate immune system express the PIEZO1 proteins (Solis et al. 2019). In the following section, we discuss the role of PIEZO1 in modulating the function of distinct populations of innate immune cells.

3.6.1 PIEZO1 in Macrophages

Macrophages are functionally diverse immune cells that play an essential role in metabolic regulation, tissue homeostasis, pathogen defence, and response to injuries. Their functional diversity arises from the capacity to adopt distinct polarization states, secrete a wide spectrum of cytokines, and respond rapidly and flexibly to cues from the surrounding microenvironment (Smith et al. 2017; Sica and Mantovani 2012; Jain et al. 2019). In addition to soluble mediators, biophysical cues such as matrix stiffness, mechanical forces, and matrix architecture are also critical regulators of macrophage function (Xiong et al. 2024; Blakney et al. 2012).

3.6.1.1 PIEZO1 Expression Enhances the Inflammatory Response and Reduces the Wound Healing Response in Macrophages

The activation of PIEZO1 in macrophages enhances pro-inflammatory pathways triggered by interferon-γ (IFNγ) and lipopolysaccharide (LPS), while concurrently inhibiting the wound healing responses induced by interleukin-4 (IL4) and interleukin-13 (IL13) (Atcha et al. 2021b). PIEZO1 has also been shown to regulate macrophage function by facilitating calcium entry in response to substrate rigidity. This response is particularly enhanced on a stiffer substrate, thereby amplifying inflammatory response (Atcha et al. 2021a). Nuclear factor kappa beta (NFkB) and signal transducer and activator of transcription 6 (STAT6) are essential transcription factors for inflammation and wound healing pathway, respectively. Both are regulated in a Ca^{2+}-dependent manner, where elevated intracellular Ca^{2+}

levels promote NFkB activation but suppress STAT6 signalling on a stiff surface. Altogether, these evidences suggest that PIEZO1, through its role in modulating intracellular Ca^{2+} flux, serves as a critical regulator of the transcriptional balance between inflammatory and healing responses in macrophages (Atcha et al. 2021a).

3.6.1.2 The Crosstalk Between CD11b and PIEZO1 Modulates Mechanotransduction in Macrophages

Macrophage activation is sustained by a positive feedback loop in which PIEZO1-mediated signalling and actin remodelling mutually enhance one another (Atcha et al. 2021a). Various research indicates that PIEZO1, integrin, and actin cytoskeleton engage in functional crosstalk, influencing cellular mechanotransduction (Nourse and Pathak 2017; McHugh et al. 2010). In macrophages, integrins and stretch-activated ion channels are pivotal regulators of cellular processes such as migration, activation, and mechanosensing (Hsieh et al. 2019). Among these, CD11b (αM integrin) is the most abundant and well-established marker for macrophage differentiation (Misharin et al. 2013). It plays a crucial role in modulating inflammatory responses. Mechanical stress, along with soluble stimuli, regulates macrophage inflammatory functions. IFN-γ/LPS stimulation, along with cyclic stretch, upregulates CD11b, while concomitantly decreasing PIEZO1 expression, leading to decreased inflammatory response. Prolonged mechanical stretch downregulates PIEZO1 expression. Consequently, this reduction of PIEZO1-driven inflammation promotes CD11b-mediated suppression of inflammatory responses. These findings suggest that mechanical stretch dynamically regulates CD11b expression, thereby influencing the macrophage response to soluble inflammatory cues. A reciprocal relationship exists between PIEZO1 and CD11b. Knockdown of either receptor eliminated the changes in the inflammatory response induced by stretching, strongly suggesting that a functional crosstalk exists between integrins and ion channels (Atcha et al. 2021b).

3.6.1.3 PIEZO1 Amplifies TLR4 Signalling to Enhance Macrophage Responses During Bacterial Infection

Toll-like receptors (TLRs) on phagocytes play an important role in host defence by recognizing pathogens, triggering innate immune signalling, and shaping antigen-specific adaptive immunity (Kawai and Akira 2010). In macrophages, PIEZO1 modulates the TLR4-driven innate immune response to microbial invasion (Geng et al. 2021). Upon stimulation by lipopolysaccharide (LPS) or bacterial pathogens, TLR4 engages PIEZO1 to form functional complexes that facilitate F actin cytoskeleton remodelling, enhanced phagocytosis, and mitochondria phagosome interactions necessary for reactive oxygen species (ROS) generation and effective bacterial clearance (West et al. 2011; Kawai and Akira 2010; Salauddin et al. 2025; Geng et al. 2021). This coordination allows TLR4 to recruit the PIEZO1, thereby activating the CaMKII-Mst1/2 signalling axis, which ensures macrophages' effector functions essential for host defence (Chen 2019; Geng et al. 2015; Geng et al. 2021). As a mechanosensitive ion channel, PIEZO1 mediates calcium influx into the cytoplasm, which in turn activates Ca^{2+}/calmodulin-dependent protein kinase II (CaMKII), leading to phosphorylation of the stress-activated serine/threonine protein kinase involved in the Hippo signalling pathway, MST1. MST1/2 kinases further promote phagosome mitochondrion coupling and stimulate mitochondrial ROS (mROS) production by coordinating Rac1-driven cytoskeletal remodelling, crucial for phagocytosis (Geng et al. 2015, 2021).

3.6.1.4 Activation of PIEZO1 Drives a Metabolic Shift Towards Aerobic Glycolysis in Macrophages

Beyond its roles in the cytoskeleton and signalling regulation, PIEZO1 also affects macrophage metabolism (Leng et al. 2022). Deficiency of PIEZO1 reduced glycolytic activity in macrophages, leading to diminished production of proinflammatory cytokines (Viola et al. 2019). This effect is mediated through PIEZO1's regulation of Ca^{2+}/calmodulin-dependent protein kinase II (CaMKII) and stabilizes the hypoxia-inducible factor 1a (HIF-1a) (Kelly and O'Neill 2015), both of which are critical for the metabolic priming of TLR signalling. So PIEZO1-mediated mechanosensation promotes a metabolic switch towards aerobic glycolysis, thereby enhancing TLR-driven inflammatory responses and ensuring effective immune activation (Leng et al. 2022).

3.6.1.5 Shear Stress Triggers NLRP3 Inflammasome Activation in LPS-Primed Macrophages Through PIEZO1 Mechanosensor

Shear stress acts as a mechanical cue that can prime macrophages for NLRP3 inflammasome activation, playing a crucial role in the regulation of inflammatory response against various pathogens and cellular damage (Fish et al. 2025; Fish and Kulkarni 2024). It has been shown that a higher magnitude shear stress promoted inflammasome activation through PIEZO1-IKKb-NFkB-NLRP3 signalling axis in immortalized bone marrow-derived macrophages (iBMDMs) (Fish and Kulkarni 2024), as evidenced by an increase of ASC speck [apoptosis-associated speck-like protein containing a caspase recruitment domain (CARD)] formation and IL-1B secretion, prominent marker for inflammasome activation (Bauernfeind et al. 2009).

3.6.2 PIEZO1 in Neutrophils

Neutrophils are the most abundant granulocyte and an essential part of the innate immune system (Herro and Grimes 2024). Generated in the bone marrow, they are released into the bloodstream to conduct immune surveillance within the tissues (Hidalgo et al. 2019; Mayadas et al. 2014; Teng et al. 2017). Upon encountering pathogens, neutrophils rapidly eliminate them by multiple mechanisms, including phagocytosis (Lee et al. 2003), degranulation (Lacy 2006), oxidative burst (Roos et al. 2003), and neutrophil extracellular trap formation (Papayannopoulos 2018; Singhal and Kumar 2022). Beyond their well-known antimicrobial functions, they are also important for wound healing (Yang et al. 2025) and fibrotic diseases (Wu and Yang 2024). Furthermore, the phenotype and activity of neutrophils are dynamically modulated by signals derived from their local environment (Abaricia et al. 2021).

3.6.2.1 Substrate Stiffness Promotes NET Formation by Activating Focal Adhesion Kinase

Substrate stiffness alone can serve as a mechanical cue that induces neutrophil extracellular trap (NET) formation through integrin-dependent signalling (Khanmohammadi et al. 2024; Abaricia et al. 2021). Neutrophils exposed to a higher-stiffness environment show increased NETosis and elevated secretion of pro-inflammatory cytokines and chemokines (Abaricia et al. 2021). Stiffness-induced NET formation was potentiated by increasing the availability of integrin binding domain and abolished by inhibiting focal adhesion kinase (FAK), a downstream mediator of integrin signalling (Abaricia et al. 2021).

3.6.2.2 Shear Stress Induced PIEZO1-Mediated NET Formation

Neutrophils, like other leukocytes, are continuously exposed to haemodynamic forces such as shear stress, which regulates their adhesion and migration towards sites of inflammation. PIEZO1, a key mechanosensitive channel, gates directly in response to mechanical stress (Lai et al. 2022; Nourse and Pathak 2017; Beech and Kalli 2019). A hallmark of neutrophil function is the release of DNA into the extracellular environment, forming neutrophil extracellular traps (NETs). This weblike chromatin networks coated with cytosolic and granule proteins trap and neutralize invading pathogens. Shear stress has been shown to directly induce NETosis and enhance neutrophil responsiveness to additional NET-inducing stimuli, including lipopolysaccharide (LPS) and ATP, but does not affect the phorbol-12-myristate-13-acetate (PMA)-induced NETosis (Khanmohammadi et al. 2024). Activation of PIEZO1 by shear stress triggers calcium influx, leading to calpain activation. Calpain activity facilitates chromatin decondensation and NETosis while modulating actin remodelling, required for the early NETosis. Altogether, PIEZO1 activation through shear stress not only triggers NET formation but also enhances neutrophil responsiveness to NET-inducing stimuli (Baratchi et al. 2024).

3.6.2.3 PIEZO1 Plays a Crucial Role in NETosis and Macrophage Differentiation During Antiviral Infection

PIEZO1 has been a key regulator of neutrophil extracellular trap formation and M1 macrophage polarization during influenza virus infection. During virus infection, PIEZO1 regulates ROS production, cytokine release, and NET formation. NET-derived DNA activates TLR9 and cGAS-STING pathway, promoting M1 macrophage differentiation. PIEZO1 knockdown in neutrophils diminishes NET-associated DNA release, which in turn suppresses the downstream TLR9 and cGAS-STING signalling (Wang et al. 2025a; Apel et al. 2021) and induces a phenotypic switch in macrophages from M1 to M2. In influenza infection, PIEZO1 integrates magnesium flux through the PIEZO1-SIRT2-HIFα axis to drive NETosis and modulate macrophage polarization, highlighting its potential as a therapeutic target in viral inflammation (Wang et al. 2025b).

3.6.3 Dendritic Cell and PIEZO1

Dendritic cells (DCs) represent central regulators of immune surveillance, integrating signals from both the innate and adaptive arms of the immune system. As the most potent antigen-presenting cells, DCs are subjected to a spectrum of mechanical stimuli arising from the extracellular matrix (ECM) and stromal architecture. Among those, substrate stiffness has emerged as a critical biophysical cue that governs DC biology (Mennens et al. 2017). During tissue surveillance, DCs are continuously exposed to fluid shear stress (FSS), which is a mechanical force generated by blood flow that immune cells encounter while circulating (Dombroski et al. 2023). Immune cells, including T cells (Hope et al. 2022), macrophages (Fish et al. 2025), and neutrophils (Zhu et al. 2024), respond to shear stress by enhancing their proliferation and cytokine secretion. In the case of dendritic cells (DCs), studies have shown that applying cyclic strain leads to the upregulation of MHC class II and costimulatory molecules (Lewis et al. 2013), although research on DC activation through shear stress remains limited. Importantly, FSS has been found to activate DCs while only minimally affecting their viability, making it a potentially safe strategy for ex vivo immune activation (Dombroski et al. 2023). Comparative analysis between stimulated and unstimulated DCs reveals distinct metabolic and morphological changes, with FSS inducing phosphorylation of NFkB and c-Fos (Dombroski et al. 2023). The elevated expression of these transcription factors points to the signalling pathways underlying FSS-mediated DC activation. Since NFkB and c-Fos are crucial regulators of DC maturation and cytokine production required for T cell priming (Yoshida et al. 2012), these findings suggest that the observed activation is likely mediated by mechanosensitive ion channels such as PIEZO1.

Recent studies have begun to uncover the mechanisms by which dendritic cells (DCs) sense and translate extracellular mechanical stiffness into functional outcomes, particularly in shaping their inflammatory capacity. Central to this process is the Hippo pathway effector Taz, along with Ca^{2+}-dependent channels including PIEZO1, which mediate dendritic cell metabolism and function in response to the mechanical tension (Chakraborty et al. 2021). Together with its paralog YAP, TAZ is also a well-established mechanosensitive regulator in various non-immune cell types (Chakraborty et al. 2021).

Plasmacytoid dendritic cells (pDCs) constitute a specialized subset of DCs distinguished by their potent capacity to produce type 1 interferons (IFNs) in response to viral and self-derived nucleic acids recognized through Toll-like receptor 9 (TLR9) (Gilliet et al. 2008; Kerkmann et al. 2003). A Recent non-peer-reviewed article identified the mechanosensitive ion channel PIEZO1 as a crucial regulator of ligand-dependent IFN signalling in human primary pDCs (Chinnathambi et al. 2012). Upon engagement with synthetic oligodeoxynucleotides (ODNs) such as CpG-A, which tend to self-assemble into large aggregates, membrane deformation during endocytic uptake induces mechanical stress that activates PIEZO1 (Pattanayak et al. 2025; Guiducci et al. 2006; Kerkmann et al. 2003). This activation promotes localization of Ca^{2+} influx and actin cytoskeletal remodelling, facilitating the formation of an F-actin-rich domain that retains CpG-A within early endosomes (Pattanayak et al. 2025). The prolonged endosomal residence of these complexes supports sustained IRF7 nuclear translocation and robust type I IFN secretion (Honda et al. 2005a, b). Collectively, the observation positions PIEZO1-dependent mechanosensing at the plasma membrane as a pivotal control point in nucleic-driven immune activation (Pattanayak et al. 2025), suggesting that the physical characteristics of nucleic acid ligands can dictate distinct functional outcome in antiviral (Lund et al. 2003) and autoimmune context (Chaudhary et al. 2025; Ganguly et al. 2013).

3.6.4 Other Innate Immune Cells

Eosinophils are highly versatile leukocytes that participate in many immune and inflammatory processes (Hogan et al. 2008). Traditionally recognized for their role in parasitic infections, allergic reactions, and certain bacterial and viral diseases, eosinophils have emerged as important contributors to several homeostatic processes (Sp et al. 2008; Raap and Wardlaw 2008). They participate in tumour surveillance (Gatault et al. 2012; Prizment et al. 2016), tissue remodelling and fibrosis (Aceves 2014), metabolic regulation, and immune modulation (Kanda et al. 2021; de Oliveira et al. 2023). Upon activation, they migrate from the bloodstream to sites of inflammation, where they influence immune activity through multiple mechanisms (Constantine and Klion 2022).

3.6.4.1 PIEZO1 in Eosinophils Regulates Both the Immune System and the Sensory Nervous System

Eosinophils possess mechanosensitive properties, allowing them to sense and respond to physical cues such as fluid shear stress in tissues or the vasculature (Son et al. 2021). A recent study demonstrated that Yoda1 (Botello-Smith et al. 2019), a selective activator of the mechanosensitive ion channel PIEZO1, induces calcium influx in the AML 14.3D10 eosinophil cell line. This calcium signalling cascade regulates eosinophil function, particularly the release of cytokines that contribute to inflammatory and pathological responses (Son et al. 2021; Hwang et al. 2024). Upon activation, AML14.3D10 cells secrete a diverse set of cytokines into their extracellular environment, which, in turn, enhances the activity of TRPV1 and ATP receptors, key mediators involved in the transmission of pain and itch sensations detected by dorsal root ganglion (DRG) neurones (Hwang et al. 2024). Furthermore, the subsequent analysis revealed that PIEZO1 activation not only governs calcium influx but also regulates the expression and secretion of both pro-inflammatory and anti-inflammatory cytokines within AML14.3D 10 cells (Hwang et al. 2024). These observations suggest a functional crosstalk between eosinophils and DRG neurones, in which PIEZO1 serves as a critical molecular mediator linking eosinophil mechanotransduction to immune and neuroinflammatory processes.

3.6.4.2 PIEZO1 in Natural Killer (NK) Cells

NK cells are a central component of the innate immune system, recognized for their ability to identify and eliminate aberrant cells such as tumour cells (Wang et al. 2023; Meza Guzman et al. 2020) and pathogen-infected cells (Björkström et al. 2022). Beyond their direct cytotoxicity against malignant cells, NK cells contribute significantly to antiviral defence, particularly against cytoplasmic viruses such as herpesviruses (Münz and Chijioke 2017) and functional adenoviruses. The activation and effector function of NK cells are tightly regulated by interacting with MHC class 1 molecules on their surface, and the lack of MHC-specific engagement results in target cell lysis (Moretta et al. 2002).

In recent years, accumulating evidence has underscored the influence of biophysical cues, especially the mechanical properties of the microenvironment, on NK cell activity (Mordechay et al. 2021). Among all biophysical cues, surface stiffness has emerged as an important factor modulating NK cell cytotoxicity (Mordechay et al. 2021). Numerous studies have demonstrated that cancer cells are generally softer than their normal counterparts (Alibert et al. 2017). NK cell-mediated killing is strongly influenced by the mechanical properties of tumour cells (Mordechay et al. 2021). Specifically, the time required to induce apoptosis or necrosis in tumour cells is significantly extended when the target cells are mechanically softened, compared with their stiffer counterparts (Yanamandra et al. 2024; Mordechay et al. 2021). In human NK cells, mechanosensing of target cell stiffness is predominantly governed by the ion channel PIEZO1 (Yanamandra et al. 2024). Disruption of PIEZO1 function effectively eliminates the stiffness-dependent responsiveness of NK cells, demonstrating its essential role in mechanotransduction (Yanamandra et al. 2024). Mechanistically, the reduced cortical tension of softened tumour cells hampers perforin-mediated pore formation and subsequent perforin-dependent lysis, leading to a lengthened cytotoxic process (Basu et al. 2016). Previous studies have also shown that engagement with a new target cell can accelerate NK cell detachment from previously engaged ones (Netter et al. 2017), while unsuccessful killing events are often associated with extended contact duration (Jenkins et al. 2015; Anft et al. 2020). Consistent with these studies, NK cells detach from stiffened tumour cells much sooner after initiating apoptosis than they do from softer cells. This suggests that alterations in tumour cell stiffness not only affect cytolytic efficiency but also regulate the timing of NK target cell disengagement. Collectively, these results emphasize that PIEZO1-dependent mechanosensing is pivotal for orchestrating NK cell cytotoxicity (Yanamandra et al. 2024). Targeting this pathway could therefore provide a promising strategy to enhance NK cell efficacy, particularly in the treatment of solid tumours, where altered mechanical properties in the tumour microenvironment often impair immune-mediated clearance.

Although PIEZO1 has been implicated in regulating cytotoxicity, migration, and inflammatory signalling in specific immune subsets (Fig. 3.3), the mechanism of PIEZO1 activation remains unclear. Further research is therefore necessary to elucidate the full spectrum of PIEZO1-mediated mechanotransduction within innate immunity and to determine whether its modulation could serve as a universal strategy for enhancing immune defence and immunotherapy effectiveness.

3.7 Adaptive Immune System and PIEZO1

The adaptive immune system (AIS) attracts attention of researchers due to its remarkable specificity and diversity (Flajnik and Kasahara 2010). The adaptive immune system functions like a natural learning network, using signals, feedback, and early-life training (e.g. maternal antibodies IgG, self-recognition, microbiome) to shape immune responses, and this process is called natural machine learning (NML) (Cohen 2025). For functional purposes, cells of the adaptive immune system, T and B lymphocytes, possess the ability to recognize and kill a wide range of pathogens while maintaining a high degree of specificity. Another important phenomenon of AIS is its capacity for immunological memory, which underpins vaccination and enables the body to mount a defence against a familiar threat (Lam et al. 2024).

In the dynamic environment of adaptive immunity, lymphocytes must traverse and adapt to a variety of physiological environments (Meng et al. 2020), which critically shape lymphocytes' survival, proliferation (Lebon et al. 2024), activation (Hope et al. 2022), and migration (Liu et al. 2024). Recent research has illuminated the central role of the mechanosensitive ion channels, particularly PIEZO1, in translating mechanical stimuli into biochemical signals the regulate immune cell

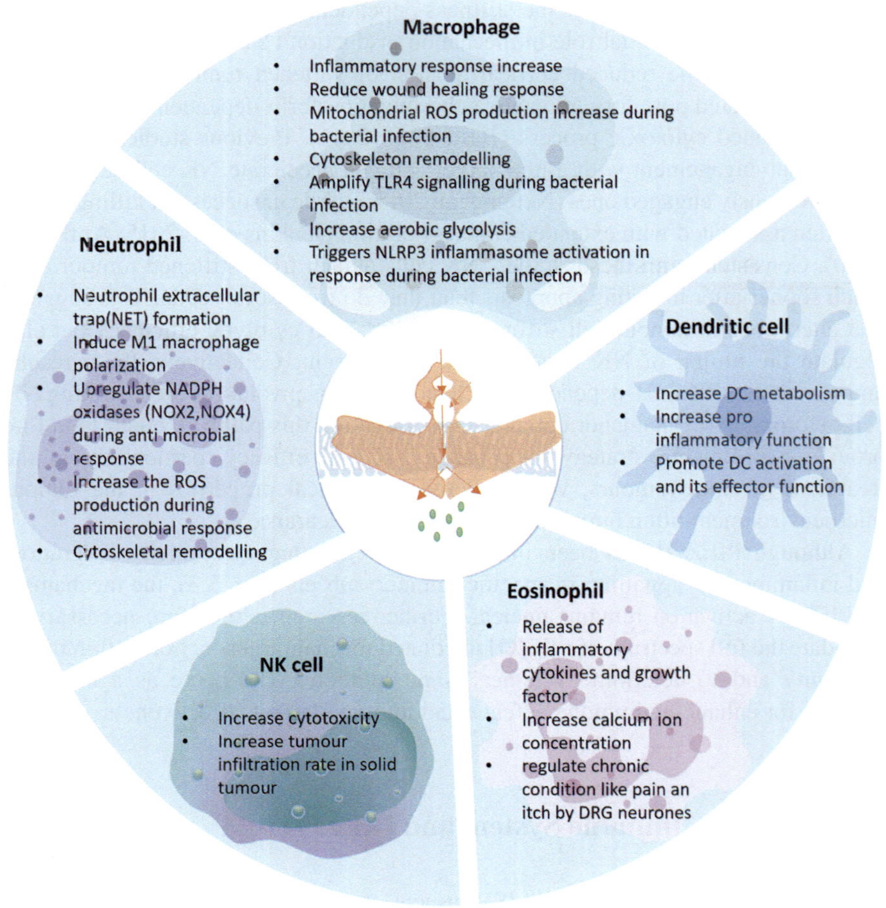

Fig. 3.3 PIEZO1 regulates diverse functional responses across innate immune cell population. The figure depicts the central role of the mechanosensitive PIEZO1 channel across innate immune cell populations, including macrophages, neutrophils, dendritic cells, NK cells, and eosinophils by facilitating the influx of cations, specifically $Ca2^{+}$

behaviour. The next section focuses on how PIEZO1-mediated mechanotransduction orchestrates the responses of key adaptive immune cells, including T and B lymphocytes (Fig. 3.4). Of note here, it was in the context of human T lymphocyte function that a role of PIEZO1 in immune regulation was first described (Liu et al. 2018; Liu and Ganguly 2019).

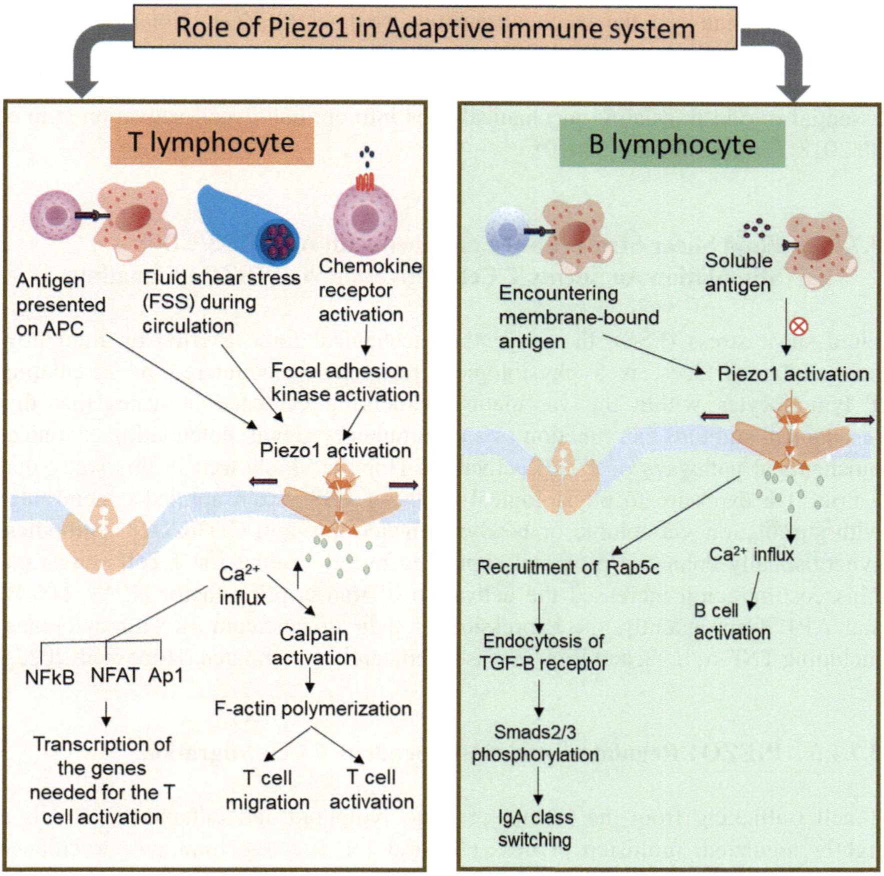

Fig. 3.4 Mechanotransduction by Piezo1 regulates B and T lymphocyte function. Following antigen encounter, PIEZO1 drives the T cell activation and migration and different gene expression which further promotes T cell activation. In B lymphocytes, Piezo1 is activated upon encountering membrane-bound antigen and plays an important role in IgA class switching as well as B cell activation

3.7.1 PIEZO1 and T Lymphocytes

3.7.1.1 PIEZO1 Enables Mechanical Sensing During TCR Engagement

Recent studies highlight that mechanical force plays a pivotal role in modulating every stage of T cell biology, including T cell activation, migration, antigen recognition, and other effector functions. The activation of T cells depends on antigen recognition by T cell receptor (TCR) engagement with antigen presenting cells (APCs). Liu et al. demonstrated that TCR crosslinking by antibody-coated beads, as opposed to soluble antibodies, generates a superior activation response, which critically depends on the mechanosensitive channel PIEZO1. Notably, PIEZO1 deficiency in $CD4^+$ and $CD8^+$ T cells significantly impair hallmark activation events such as ZAP70 phosphorylation and CD69 expression, highlighting its essential role in translating mechanical cues into optimal T cell activation (Liu et al. 2018; Liu and Ganguly 2019).

3.7.1.2 Fluid Shear Stress (FSS) Combined with Anti-CD3/CD28 Stimulation Augments T Cell Activation via PIEZO1 Signalling

Fluid shear stress (FSS), the tangential mechanical force exerted by fluid flow on cellular surfaces, is a physiological parameter encountered by circulating T lymphocytes within the vasculature. Emerging evidence indicates that this mechanical stimulus can function as a costimulatory signal, potentiating canonical biochemical pathways of T cell activation. Hope et al. showed in this study that a brief 1-h exposure to physiological levels of FSS, when applied concurrently with stimulation via soluble or bead-conjugated anti-anti-CD3/CD28 antibodies, synergistically enhances proximal signalling events critical for T cell activation. This costimulation increased the activation of transcription factor NFAT, NFκB, and AP1. Consequently, the expression of their downstream effector cytokines, including TNF-α, IL-2, and IFN-γ, was significantly upregulated (Hope et al. 2022).

3.7.1.3 PIEZO1 Regulates Integrin-Dependent T Cell Migration

T cell trafficking from the vasculature into lymphoid and inflamed tissues is a tightly regulated, multistep process essential for effective immune surveillance and response. T cell migration depends on chemokine gradients, integrin-mediated adhesion, and dynamic actomyosin activity. Upon chemokine receptor engagements, focal adhesion kinase (FAK) is phosphorylated, initiating the recruitment of the leading edge. These adhesions generate localized increases in membrane tension by tethering the extracellular matrix to the intracellular cytoskeleton. PIEZO1 is recruited to this tension-enriched region and facilitates T cell migration. In chemokine-stimulated T cells, PIEZO1 and F-actin exhibit a distinct spatial polarization: PIEZO1 is accumulated at the leading edge, while F-actin is concentrated posteriorly. At the leading edge, PIEZO1 activity promotes the accumulation and activation of the integrin LFA-1, which in turn facilitates robust F-actin polymerization necessary for forward propulsion. In summary, Fak-mediated membrane tension activates PIEZO1, which in turn promotes integrin-mediated adhesion and actin dynamics for directed T cell migration (Liu et al. 2024).

3.7.2 *Piezo1 and B Lymphocytes*

3.7.2.1 PIEZO1 Mediates B Cell Activation Through Membrane-Bound Antigen

B lymphocytes, a central component of the adaptive immune system, can recognize both soluble and membrane-bound antigens in in vitro and in vivo engagement with the B cell receptor (Batista and Harwood 2009). The signalling pathway of soluble antigen and membrane-bound antigens is different, but the underlying mechanism remains elusive. Kwak et al. identified the mechanosensitive ion channel PIEZO1 as a key mediator in this selectivity. PIEZO1-mediated mechanosensing at the immunological synapses provides a parallel calcium signalling axis that enables B cells to specifically discriminate between membrane-bound and soluble antigens, thereby enhancing the selectivity of the adaptive immune response (Kwak et al. 2019, 2023).

3.7.2.2 PIEZO1 as a Critical Mediator in IgA Class Switching

PIEZO1 has been shown to selectively modulate transforming growth factor-beta 1 (TGF-β1) signalling to enhance immunoglobulin A (IgA) class switching in mouse B cells primarily upregulating Rab5c expression. Rab5c is a GTPase that facilitates TGF-β1 receptor internalization via early endosome, thereby enabling the Smad2/3 (a transcription factor) phosphorylation required for IgA class switching. PIEZO1-mediated TGF-β1-induced IgA class switching appears to be independent of calcium signalling. The selectivity of PIEZO1 suggests that it acts as a fine-tuning modulator rather than a primary driver of the overall immune response (Jung et al. 2025).

3.8 Conclusion

PIEZO1 has emerged as a central mechanosensor in immune system. Activation of PIEZO1 plays a diverse role in different immune cells including macrophage polarization, dendritic cell activation, T cell activation, and many more. This had created a new research avenue for exploring the role of physical cues in immunocellular regulation, apart from the conventionally explored biochemical cues. However, several critical research gaps remain like the molecular mechanisms linking PIEZO1 activation with discrete downstream cell biological events as well as the role of PIEZO1 in chronic stimulation, tumour immunity, autoimmunity, and in general in terms of its influence in immune heterogeneity among healthy individuals. Future direction in research on PIEZO1 should also envisage therapeutic targeting of PIEZO1 utilizing different strategies ranging from small molecules, monoclonal antibodies, as well as external transduction of physical cues. These efforts will have the potential to initiate a new era of mechano-therapeutics that integrate biophysics with immunology.

References

Abaricia JO, Shah AH, Olivares-Navarrete R (2021) Substrate stiffness induces neutrophil extracellular trap (NET) formation through focal adhesion kinase activation. Biomaterials 271:120715. https://doi.org/10.1016/j.biomaterials.2021.120715

Aceves SS (2014) Remodeling and fibrosis in chronic eosinophil inflammation. Digestive Diseases (Basel, Switzerland) 32:15–21. https://doi.org/10.1159/000357004

Adams S, Wuescher LM, Worth R, Yildirim-Ayan E (2019) Mechano-immunomodulation: mechanoresponsive changes in macrophage activity and polarization. Ann Biomed Eng 47(11):2213–2231. https://doi.org/10.1007/s10439-019-02302-4

Agrawal A, Javanmardi Y, Watson SA, Serwinski B, Djordjevic B, Li W, Aref AR, Jenkins RW, Moeendarbary E (2025) Mechanical signatures in cancer metastasis. NPJ Biol Phys Mech 2(1):3. https://doi.org/10.1038/s44341-024-00007-x

Albuisson J, Murthy SE, Bandell M, Coste B, Louis-dit-Picard H, Mathur J, Fénéant-Thibault M, Tertian G, De Jaureguiberry J-P, Syfuss P-Y, Cahalan S, Garçon L, Toutain F, Simon Rohrlich P, Delaunay J, Picard V, Jeunemaitre X, Patapoutian A (2013) Dehydrated hereditary stomatocytosis linked to gain-of-function mutations in mechanically activated PIEZO1 ion channels. Nat Commun 4(1):1884. https://doi.org/10.1038/ncomms2899

Alibert C, Goud B, Manneville J-B (2017) Are cancer cells really softer than normal cells? Biol Cell 109(5):167–189. https://doi.org/10.1111/boc.201600078

Anft M, Netter P, Urlaub D, Prager I, Schaffner S, Watzl C (2020) NK cell detachment from target cells is regulated by successful cytotoxicity and influences cytokine production. Cell Mol Immunol 17(4):347–355. https://doi.org/10.1038/s41423-019-0277-2

Apel F, Andreeva L, Knackstedt LS, Streeck R, Frese CK, Goosmann C, Hopfner K-P, Zychlinsky A (2021) The cytosolic DNA sensor cGAS recognizes neutrophil extracellular traps. Sci Signal 14(673):eaax7942. https://doi.org/10.1126/scisignal.aax7942

Atcha H, Jairaman A, Holt JR, Meli VS, Nagalla RR, Veerasubramanian PK, Brumm KT, Lim HE, Othy S, Cahalan MD, Pathak MM, Liu WF (2021a) Mechanically activated ion channel PIEZO1 modulates macrophage polarization and stiffness sensing. Nat Commun 12(1):3256. https://doi.org/10.1038/s41467-021-23482-5

Atcha H, Meli VS, Davis CT, Brumm KT, Anis S, Chin J, Jiang K, Pathak MM, Liu WF (2021b) Crosstalk between CD11b and PIEZO1 mediates macrophage responses to mechanical cues. Front Immunol 12:689397. https://doi.org/10.3389/fimmu.2021.689397

Bae C, Gnanasambandam R, Nicolai C, Sachs F, Gottlieb PA (2013a) Xerocytosis is caused by mutations that alter the kinetics of the mechanosensitive channel PIEZO1. Proc Natl Acad Sci USA 110(12):E1162–E1168. https://doi.org/10.1073/pnas.1219777110

Bae C, Gottlieb PA, Sachs F (2013b) Human PIEZO1: removing inactivation. Biophys J 105(4):880–886. https://doi.org/10.1016/j.bpj.2013.07.019

Baratchi S, Danish H, Chheang C, Zhou Y, Huang A, Lai A, Khanmohammadi M, Quinn KM, Khoshmanesh K, Peter K (2024) PIEZO1 expression in neutrophils regulates shear-induced NETosis. Nat Commun 15(1):7023. https://doi.org/10.1038/s41467-024-51211-1

Barda-Saad M, Ben-Shmuel A, Biber G, Raiff A, Sabag B, Matalon O (2020) Targeting intracellular check points to unleash natural killer cell cytotoxicity. J Immunol 204(1_Supplement):69.8. https://doi.org/10.4049/jimmunol.204.Supp.69.8

Basu R, Whitlock BM, Husson J et al (2016) Cytotoxic T cells use mechanical force to potentiate target cell killing. Cell 165(1):100–110. https://doi.org/10.1016/j.cell.2016.01.021

Batista FD, Harwood NE (2009) The who, how and where of antigen presentation to B cells. Nat Rev Immunol 9(1):15–27. https://doi.org/10.1038/nri2454
Bauernfeind FG, Horvath G, Stutz A, Alnemri ES, MacDonald K, Speert D, Fernandes-Alnemri T, Wu J, Monks BG, Fitzgerald KA, Hornung V, Latz E (2009) Cutting edge: NF-kappaB activating pattern recognition and cytokine receptors license NLRP3 inflammasome activation by regulating NLRP3 expression. J Immunol 183(2):787–791. https://doi.org/10.4049/jimmunol.0901363
Beech DJ, Kalli AC (2019) Force sensing by PIEZO channels in cardiovascular health and disease. Arterioscler Thromb Vasc Biol 39(11):2228–2239. https://doi.org/10.1161/ATVBAHA.119.313348
Björkström NK, Strunz B, Ljunggren H-G (2022) Natural killer cells in antiviral immunity. Nat Rev Immunol 22(2):112–123. https://doi.org/10.1038/s41577-021-00558-3
Blakney AK, Swartzlander MD, Bryant SJ (2012) The effects of substrate stiffness on the in vitro activation of macrophages and in vivo host response to poly(ethylene glycol)-based hydrogels. J Biomed Mater Res A 100(6):1375–1386. https://doi.org/10.1002/jbm.a.34104
Botello-Smith WM, Jiang W, Zhang H, Ozkan AD, Lin Y-C, Pham CN, Lacroix JJ, Luo Y (2019) A mechanism for the activation of the mechanosensitive PIEZO1 channel by the small molecule Yoda1. Nat Commun 10(1):4503. https://doi.org/10.1038/s41467-019-12501-1
Chakraborty M, Chu K, Shrestha A, Revelo XS, Zhang X, Gold MJ, Khan S, Lee M, Huang C, Akbari M, Barrow F, Chan YT, Lei H, Kotoulas NK, Jovel J, Pastrello C, Kotlyar M, Goh C, Michelakis E et al (2021) Mechanical stiffness controls dendritic cell metabolism and function. Cell Rep 34(2):108609. https://doi.org/10.1016/j.celrep.2020.108609
Chaudhary V, Mishra B, Ah Kioon MD, Du Y, Ivashkiv LB, Crow MK, Barrat FJ (2025) Mechanosensing regulates pDC activation in the skin through NRF2 activation. J Exp Med 222(3):e20240852. https://doi.org/10.1084/jem.20240852
Chen L (2019) Non-canonical hippo signaling regulates immune responses. Adv Immunol 144:87–119. https://doi.org/10.1016/bs.ai.2019.07.001
Chinnathambi S, Chen S, Ganesan S, Hanagata N (2012) Binding mode of CpG oligodeoxynucleotides to nanoparticles regulates bifurcated cytokine induction via toll-like receptor 9. Sci Rep 2:534. https://doi.org/10.1038/srep00534
Cohen I (2025) Natural machine learning (NML) shapes the adaptive immune system (No. 2025050563). Preprints. https://doi.org/10.20944/preprints202505.0563.v1
Constantine GM, Klion AD (2022) Recent advances in understanding the role of eosinophils. Faculty Rev 11:26. https://doi.org/10.12703/r/11-26
Coste B, Mathur J, Schmidt M, Earley TJ, Ranade S, Petrus MJ, Dubin AE, Patapoutian A (2010) PIEZO1 and PIEZO2 are essential components of distinct mechanically activated cation channels. Science 330(6000):55–60. https://doi.org/10.1126/science.1193270
Coste B, Xiao B, Santos JS, Syeda R, Grandl J, Spencer KS, Kim SE, Schmidt M, Mathur J, Dubin AE, Montal M, Patapoutian A (2012) PIEZOs are pore-forming subunits of mechanically activated channels. Nature 483(7388):176–181. https://doi.org/10.1038/nature10812
Coste B, Murthy SE, Mathur J, Schmidt M, Mechioukhi Y, Delmas P, Patapoutian A (2015) PIEZO1 ion channel pore properties are dictated by C-terminal region. Nat Commun 6(1):7223. https://doi.org/10.1038/ncomms8223
Cox CD, Bae C, Ziegler L, Hartley S, Nikolova-Krstevski V, Rohde PR, Ng C-A, Sachs F, Gottlieb PA, Martinac B (2016) Removal of the mechanoprotective influence of the cytoskeleton reveals PIEZO1 is gated by bilayer tension. Nat Commun 7(1):10366. https://doi.org/10.1038/ncomms10366
de Oliveira MC, Silveira ALM, de Oliveira ACC, Lana JP, Costa KA, Vieira ÉLM, Pinho V, Teixeira MM, Merabtene F, Marcelin G, Clément K, Ferreira AVM (2023) Eosinophils protect from metabolic alterations triggered by obesity. Metabolism 146. https://doi.org/10.1016/j.metabol.2023.155613
Dombroski JA, Rowland SJ, Fabiano AR, Knoblauch SV, Hope JM, King MR (2023) Fluid shear stress enhances dendritic cell activation. Immunobiology 228(6):152744. https://doi.org/10.1016/j.imbio.2023.152744

Evans EL, Cuthbertson K, Endesh N, Rode B, Blythe NM, Hyman AJ, Hall SJ, Gaunt HJ, Ludlow MJ, Foster R, Beech DJ (2018) Yoda1 analogue (Dooku1) which antagonizes Yoda1-evoked activation of PIEZO1 and aortic relaxation. Br J Pharmacol 175(10):1744–1759. https://doi.org/10.1111/bph.14188

Fish A, Kulkarni A (2024) Flow-induced shear stress primes NLRP3 inflammasome activation in macrophages via PIEZO1. ACS Appl Mater Interfaces 16(4):4505–4518. https://doi.org/10.1021/acsami.3c18645

Fish A, Forster J, Malik V, Kulkarni A (2025) Shear-stress initiates signal two of NLRP3 inflammasome activation in LPS-primed macrophages through PIEZO1. ACS Appl Mater Interfaces 17(5):7363–7376. https://doi.org/10.1021/acsami.4c18845

Flajnik MF, Kasahara M (2010) Origin and evolution of the adaptive immune system: genetic events and selective pressures. Nat Rev Genet 11(1):47–59. https://doi.org/10.1038/nrg2703

Ganguly D, Haak S, Sisirak V, Reizis B (2013) The role of dendritic cells in autoimmunity. Nat Rev Immunol 13(8):566–577. https://doi.org/10.1038/nri3477

Gatault S, Legrand F, Delbeke M, Loiseau S, Capron M (2012) Involvement of eosinophils in the anti-tumour response. Cancer Immunol Immunother 61(9):1527–1534. https://doi.org/10.1007/s00262-012-1288-3

Ge J, Li W, Zhao Q, Li N, Chen M, Zhi P, Li R, Gao N, Xiao B, Yang M (2015) Architecture of the mammalian mechanosensitive PIEZO1 channel. Nature 527(7576):64–69. https://doi.org/10.1038/nature15247

Geng J, Sun X, Wang P, Zhang S, Wang X, Wu H, Hong L, Xie C, Li X, Zhao H, Liu Q, Jiang M, Chen Q, Zhang J, Li Y, Song S, Wang H-R, Zhou R, Johnson RL et al (2015) Kinases Mst1 and Mst2 positively regulate phagocytic induction of reactive oxygen species and bactericidal activity. Nat Immunol 16(11):1142–1152. https://doi.org/10.1038/ni.3268

Geng J, Shi Y, Zhang J, Yang B, Wang P, Yuan W, Zhao H, Li J, Qin F, Hong L, Xie C, Deng X, Sun Y, Wu C, Chen L, Zhou D (2021) TLR4 signalling via PIEZO1 engages and enhances the macrophage mediated host response during bacterial infection. Nat Commun 12(1):3519. https://doi.org/10.1038/s41467-021-23683-y

Gilliet M, Cao W, Liu Y-J (2008) Plasmacytoid dendritic cells: sensing nucleic acids in viral infection and autoimmune diseases. Nat Rev Immunol 8(8):594–606. https://doi.org/10.1038/nri2358

Gnanasambandam R, Ghatak C, Yasmann A, Nishizawa K, Sachs F, Ladokhin AS, Sukharev SI, Suchyna TM (2017) GsMTx4: mechanism of inhibiting mechanosensitive ion channels. Biophys J 112(1):31–45. https://doi.org/10.1016/j.bpj.2016.11.013

Guiducci C, Ott G, Chan JH, Damon E, Calacsan C, Matray T, Lee K-D, Coffman RL, Barrat FJ (2006) Properties regulating the nature of the plasmacytoid dendritic cell response to toll-like receptor 9 activation. J Exp Med 203(8):1999–2008. https://doi.org/10.1084/jem.20060401

Guo YR, MacKinnon R (2017) Structure-based membrane dome mechanism for PIEZO mechanosensitivity. elife 6:e33660. https://doi.org/10.7554/eLife.33660

Harwood NE, Batista FD (2010) Early events in B cell activation. Annu Rev Immunol 28(1):185–210

Herro R, Grimes HL (2024) The diverse roles of neutrophils from protection to pathogenesis. Nat Immunol 25(12):2209–2219. https://doi.org/10.1038/s41590-024-02006-5

Hidalgo A, Chilvers ER, Summers C, Koenderman L (2019) The neutrophil life cycle. Trends Immunol 40(7):584–597. https://doi.org/10.1016/j.it.2019.04.013

Hogan SP, Rosenberg HF, Moqbel R, Phipps S, Foster PS, Lacy P, Kay AB, Rothenberg ME (2008) Eosinophils: biological properties and role in health and disease. Clin Exp Allergy 38(5):709–750. https://doi.org/10.1111/j.1365-2222.2008.02958.x

Holt JR, Zeng W-Z, Evans EL, Woo S-H, Ma S, Abuwarda H, Loud M, Patapoutian A, Pathak MM (2021) Spatiotemporal dynamics of PIEZO1 localization controls keratinocyte migration during wound healing. elife 10:e65415. https://doi.org/10.7554/eLife.65415

Honda K, Ohba Y, Yanai H, Negishi H, Mizutani T, Takaoka A, Taya C, Taniguchi T (2005a) Spatiotemporal regulation of MyD88-IRF-7 signalling for robust type-I interferon induction. Nature 434(7036):1035–1040. https://doi.org/10.1038/nature03547

Honda K, Yanai H, Negishi H, Asagiri M, Sato M, Mizutani T, Shimada N, Ohba Y, Takaoka A, Yoshida N, Taniguchi T (2005b) IRF-7 is the master regulator of type-I interferon-dependent immune responses. Nature 434(7034):772–777. https://doi.org/10.1038/nature03464

Hope JM, Dombroski JA, Pereles RS, Lopez-Cavestany M, Greenlee JD, Schwager SC, Reinhart-King CA, King MR (2022) Fluid shear stress enhances T cell activation through PIEZO1. BMC Biol 20(1):61. https://doi.org/10.1186/s12915-022-01266-7

Hsieh JY, Keating MT, Smith TD, Meli VS, Botvinick EL, Liu WF (2019) Matrix crosslinking enhances macrophage adhesion, migration, and inflammatory activation. APL Bioengineering 3(1):016103. https://doi.org/10.1063/1.5067301

Hwang S-M, Song J-M, Choi JJ, Jung Y, Park C-K, Kim YH (2024) Functional role of PIEZO1 in the human eosinophil cell line AML14.3D10: implications for the immune and sensory nervous systems. Biomolecules 14(9):1157. https://doi.org/10.3390/biom14091157

Immanuel CN, Teng B, Dong BE, Gordon EM, Luellen C, Lopez B, Harding J, Cormier SA, Fitzpatrick EA, Schwingshackl A, Waters CM (2024) Two-pore potassium channel TREK-1 (K2P2.1) regulates NLRP3 inflammasome activity in macrophages. Am J Physiol Lung Cell Mol Physiol 326(3):L367–L376. https://doi.org/10.1152/ajplung.00313.2023

Jain N, Moeller J, Vogel V (2019) Mechanobiology of macrophages: how physical factors Coregulate macrophage plasticity and phagocytosis. Annu Rev Biomed Eng 21:267–297. https://doi.org/10.1146/annurev-bioeng-062117-121224

Jeffreys N, Brockman JM, Zhai Y, Ingber DE, Mooney DJ (2024) Mechanical forces amplify TCR mechanotransduction in T cell activation and function. Appl Phys Rev 11(1):011304. https://doi.org/10.1063/5.0166848

Jenkins MR, Rudd-Schmidt JA, Lopez JA, Ramsbottom KM, Mannering SI, Andrews DM, Voskoboinik I, Trapani JA (2015) Failed CTL/NK cell killing and cytokine hypersecretion are directly linked through prolonged synapse time. J Exp Med 212(3):307–317. https://doi.org/10.1084/jem.20140964

Jiang Y, Yang X, Jiang J, Xiao B (2021) Structural designs and mechanogating mechanisms of the mechanosensitive PIEZO channels. Trends Biochem Sci 46(6):472–488. https://doi.org/10.1016/j.tibs.2021.01.008

Jung Y, Han Y, Kang J, Yu S-L, Park S-R (2025) PIEZO1 selectively enhances TGF-β1-induced IgA class switching by B cells. Cell Mol Life Sci 82(1):243. https://doi.org/10.1007/s00018-025-05789-4

Kanda A, Yun Y, Bui DV, Nguyen LM, Kobayashi Y, Suzuki K, Mitani A, Sawada S, Hamada S, Asako M, Iwai H (2021) The multiple functions and subpopulations of eosinophils in tissues under steady-state and pathological conditions. Allergol Int 70(1):9–18. https://doi.org/10.1016/j.alit.2020.11.001

Kawai T, Akira S (2010) The role of pattern-recognition receptors in innate immunity: update on toll-like receptors. Nat Immunol 11(5):373–384. https://doi.org/10.1038/ni.1863

Kelly B, O'Neill LAJ (2015) Metabolic reprogramming in macrophages and dendritic cells in innate immunity. Cell Res 25(7):771–784. https://doi.org/10.1038/cr.2015.68

Kerkmann M, Rothenfusser S, Hornung V, Towarowski A, Wagner M, Sarris A, Giese T, Endres S, Hartmann G (2003) Activation with CpG-A and CpG-B oligonucleotides reveals two distinct regulatory pathways of type I IFN synthesis in human plasmacytoid dendritic cells. J Immunol 170(9):4465–4474. https://doi.org/10.4049/jimmunol.170.9.4465

Khanmohammadi M, Danish H, Sekar NC, Suarez SA, Chheang C, Peter K, Khoshmanesh K, Baratchi S (2024) Cyclic stretch enhances neutrophil extracellular trap formation. BMC Biol 22:209. https://doi.org/10.1186/s12915-024-02009-6

Kim SE, Coste B, Chadha A, Cook B, Patapoutian A (2012) The role of drosophila PIEZO in mechanical nociception. Nature 483(7388):209–212. https://doi.org/10.1038/nature10801

Knaus UG, Clark RA, Nauseef WM, Dinauer MC, Leto TL, Jaquet V (2024) Challenging the role of a NOX4-PIEZO1 axis in neutrophil bactericidal function. Immunity 57(8):1716–1718. https://doi.org/10.1016/j.immuni.2024.07.012

Kwak K, Akkaya M, Pierce SK (2019) B cell signaling in context. Nat Immunol 20(8):963–969. https://doi.org/10.1038/s41590-019-0427-9

Kwak K, Sohn H, George R, Torgbor C, Manzella-Lapeira J, Brzostowski J, Pierce SK (2023) B cell responses to membrane-presented antigens require the function of the mechanosensitive cation channel PIEZO1. Sci Signal 16(804):eabq5096. https://doi.org/10.1126/scisignal.abq5096

Lacy P (2006) Mechanisms of degranulation in neutrophils. Allergy Asthma Clin Immunol 2(3):98–108. https://doi.org/10.1186/1710-1492-2-3-98

Lai A, Thurgood P, Cox CD, Chheang C, Peter K, Jaworowski A, Khoshmanesh K, Baratchi S (2022) PIEZO1 response to shear stress is controlled by the components of the extracellular matrix. ACS Appl Mater Interfaces 14(36):40559–40568. https://doi.org/10.1021/acsami.2c09169

Lam N, Lee Y, Farber DL (2024) A guide to adaptive immune memory. Nat Rev Immunol 24(11):810–829. https://doi.org/10.1038/s41577-024-01040-6

Lee WL, Harrison RE, Grinstein S (2003) Phagocytosis by neutrophils. Microbes Infect 5(14):1299–1306. https://doi.org/10.1016/j.micinf.2003.09.014

Lee JH, Perez-Flores MC, Park S, Kim HJ, Chen Y, Kang M, Kersigo J, Choi J, Thai PN, Woltz RL, Perez-Flores DC, Perkins G, Sihn C-R, Trinh P, Zhang X-D, Sirish P, Dong Y, Feng WW, Pessah IN et al (2024) The PIEZO channel is a mechano-sensitive complex component in the mammalian inner ear hair cell. Nat Commun 15(1):526. https://doi.org/10.1038/s41467-023-44230-x

Leng S, Zhang X, Wang S, Qin J, Liu Q, Liu A, Sheng Z, Feng Q, Hu X, Peng J (2022) Ion channel PIEZO1 activation promotes aerobic glycolysis in macrophages. Front Immunol 13:976482. https://doi.org/10.3389/fimmu.2022.976482

Lewis AH, Grandl J (2015) Mechanical sensitivity of PIEZO1 ion channels can be tuned by cellular membrane tension. Elife 4:e12088. https://doi.org/10.7554/eLife.12088

Lewis AH, Grandl J (2020) Inactivation kinetics and mechanical gating of PIEZO1 ion channels depend on subdomains within the cap. Cell Rep 30(3):870–880.e2. https://doi.org/10.1016/j.celrep.2019.12.040

Lewis JS, Dolgova N, Chancellor TJ, Acharya AP, Karpiak JV, Lele TP, Keselowsky BG (2013) Dendritic cell activation is influenced by cyclic mechanical strain when cultured on adhesive substrates. Biomaterials 34(36):9063–9070. https://doi.org/10.1016/j.biomaterials.2013.08.021

Lewis AH, Cui AF, McDonald MF, Grandl J (2017) Transduction of repetitive mechanical stimuli by PIEZO1 and PIEZO2 ion channels. Cell Rep 19(12):2572–2585. https://doi.org/10.1016/j.celrep.2017.05.079

Li W, Gao N, Yang M (2017) The structural basis for sensing by the PIEZO1 protein. Curr Top Membr 79:135–158. https://doi.org/10.1016/bs.ctm.2016.10.001

Liu CSC, Raychaudhuri D, Paul B, Chakrabarty Y, Ghosh AR, Rahaman O, Talukdar A, Ganguly D (2018) Cutting edge: PIEZO1 Mechanosensors optimize human T cell activation. J Immunol 200(4):1255–1260. https://doi.org/10.4049/jimmunol.1701118

Liu CSC, Ganguly D. Mechanical Cues for T Cell Activation: Role of Piezo1 Mechanosensors (2019) Crit Rev Immunol. 39(1):15–38. https://doi.org/10.1615/CritRevImmunol.2019029595

Liu CSC, Mandal T, Biswas P, Hoque MA, Bandopadhyay P, Sinha BP, Sarif J, D'Rozario R, Sinha DK, Sinha B, Ganguly D (2024) PIEZO1 mechanosensing regulates integrin-dependent chemotactic migration in human T cells. Elife 12. https://doi.org/10.7554/eLife.91903.2

Lukacs V, Mathur J, Mao R, Bayrak-Toydemir P, Procter M, Cahalan SM, Kim HJ, Bandell M, Longo N, Day RW, Stevenson DA, Patapoutian A, Krock BL (2015) Impaired PIEZO1 function in patients with a novel autosomal recessive congenital lymphatic dysplasia. Nat Commun 6(1):8329. https://doi.org/10.1038/ncomms9329

Lund J, Sato A, Akira S, Medzhitov R, Iwasaki A (2003) Toll-like receptor 9-mediated recognition of herpes simplex virus-2 by plasmacytoid dendritic cells. J Exp Med 198(3):513–520. https://doi.org/10.1084/jem.20030162

Maksimovic S, Nakatani M, Baba Y, Nelson AM, Marshall KL, Wellnitz SA, Firozi P, Woo S-H, Ranade S, Patapoutian A, Lumpkin EA (2014) Epidermal Merkel cells are mechanosensory cells that tune mammalian touch receptors. Nature 509(7502):617–621. https://doi.org/10.1038/nature13250

Marshall KL, Saade D, Ghitani N, Coombs AM, Szczot M, Keller J, Ogata T, Daou I, Stowers LT, Bönnemann CG, Chesler AT, Patapoutian A (2020) PIEZO2 in sensory neurons and urothelial cells coordinates urination. Nature 588(7837):290–295. https://doi.org/10.1038/s41586-020-2830-7

Mayadas TN, Cullere X, Lowell CA (2014) The multifaceted functions of neutrophils. Annu Rev Pathol 9:181–218. https://doi.org/10.1146/annurev-pathol-020712-164023

McHugh BJ, Buttery R, Lad Y, Banks S, Haslett C, Sethi T (2010) Integrin activation by Fam38A uses a novel mechanism of R-Ras targeting to the endoplasmic reticulum. J Cell Sci 123(1):51–61. https://doi.org/10.1242/jcs.056424

Mei F, Guo Y, Wang Y, Zhou Y, Heng BC, Xie M, Huang X, Zhang S, Ding S, Liu F, Deng X, Chen L, Yang C (2024) Matrix stiffness regulates macrophage polarisation via the PIEZO1-YAP signalling axis. Cell Prolif 57(8):e13640. https://doi.org/10.1111/cpr.13640

Meng KP, Majedi FS, Thauland TJ, Butte MJ (2020) Mechanosensing through YAP controls T cell activation and metabolism. J Exp Med 217(8):e20200053. https://doi.org/10.1084/jem.20200053

Mennens SFB, Bolomini-Vittori M, Weiden J, Joosten B, Cambi A, van den Dries K (2017) Substrate stiffness influences phenotype and function of human antigen-presenting dendritic cells. Sci Rep 7(1):17511. https://doi.org/10.1038/s41598-017-17787-z

Meza Guzman LG, Keating N, Nicholson SE (2020) Natural killer cells: tumour surveillance and signaling. Cancer 12(4):952. https://doi.org/10.3390/cancers12040952

Misharin AV, Morales-Nebreda L, Mutlu GM, Budinger GRS, Perlman H (2013) Flow cytometric analysis of macrophages and dendritic cell subsets in the mouse lung. Am J Respir Cell Mol Biol 49(4):503–510. https://doi.org/10.1165/rcmb.2013-0086MA

Mordechay L, Le Saux G, Edri A, Hadad U, Porgador A, Schvartzman M (2021) Mechanical regulation of the cytotoxic activity of natural killer cells. ACS Biomater Sci Eng 7(1):122–132. https://doi.org/10.1021/acsbiomaterials.0c01121

Moretta L, Biassoni R, Bottino C, Cantoni C, Pende D, Mingari MC, Moretta A (2002) Human NK cells and their receptors. Microbes Infect 4(15):1539–1544. https://doi.org/10.1016/S1286-4579(02)00037-0

Moroni M, Servin-Vences MR, Fleischer R, Sánchez-Carranza O, Lewin GR (2018) Voltage gating of mechanosensitive PIEZO channels. Nat Commun 9(1):1096. https://doi.org/10.1038/s41467-018-03502-7

Münz C, Chijioke O (2017) Natural killer cells in herpesvirus infections. F1000Res 6:F1000. https://doi.org/10.12688/f1000research.11197.1

Murthy SE, Dubin AE, Patapoutian A (2017) PIEZOs thrive under pressure: mechanically activated ion channels in health and disease. Nat Rev Mol Cell Biol 18(12):771–783. https://doi.org/10.1038/nrm.2017.92

Netter P, Anft M, Watzl C (2017) Termination of the activating NK cell immunological synapse is an active and regulated process. J Immunol 199(7):2528–2535. https://doi.org/10.4049/jimmunol.1700394

Nourse JL, Pathak MM (2017) How cells channel their stress: interplay between PIEZO1 and the cytoskeleton. Semin Cell Dev Biol 71:3–12. https://doi.org/10.1016/j.semcdb.2017.06.018

Orsini EM, Perelas A, Southern BD, Grove LM, Olman MA, Scheraga RG (2021) Stretching the function of innate immune cells. Front Immunol 12:767319. https://doi.org/10.3389/fimmu.2021.767319

Pageon SV, Govendir MA, Kempe D, Biro M (2018) Mechanoimmunology: molecular-scale forces govern immune cell functions. Mol Biol Cell 29(16):1919–1926. https://doi.org/10.1091/mbc.E18-02-0120

Papayannopoulos V (2018) Neutrophil extracellular traps in immunity and disease. Nat Rev Immunol 18(2):134–147. https://doi.org/10.1038/nri.2017.105

Pattanayak S et al (2025) Spatio-temporal regulation of ligand trafficking and TLR9 activation involves PIEZO1 mechanosensing in human plasmacytoid dendritic cells. bioRxiv, 2025.07.31.667789. https://doi.org/10.1101/2025.07.31.667789

Prizment AE, Vierkant RA, Smyrk TC, Tillmans LS, Lee JJ, Sriramarao P, Nelson HH, Lynch CF, Thibodeau SN, Church TR, Cerhan JR, Anderson KE, Limburg PJ (2016) Tumour eosinophil infiltration and improved survival of colorectal cancer patients: Iowa women's health study. Mod Pathol 29(5):516–527. https://doi.org/10.1038/modpathol.2016.42

Raap U, Wardlaw AJ (2008) A new paradigm of eosinophil granulocytes: neuroimmune interactions. Exp Dermatol 17(9):731–738. https://doi.org/10.1111/j.1600-0625.2008.00741.x

Ranade SS, Qiu Z, Woo S-H, Hur SS, Murthy SE, Cahalan SM, Xu J, Mathur J, Bandell M, Coste B, Li Y-SJ, Chien S, Patapoutian A (2014) PIEZO1, a mechanically activated ion channel, is required for vascular development in mice. Proc Natl Acad Sci 111(28):10347–10352. https://doi.org/10.1073/pnas.1409233111

Romero LO, Massey AE, Mata-Daboin AD, Sierra-Valdez FJ, Chauhan SC, Cordero-Morales JF, Vásquez V (2019) Dietary fatty acids fine-tune PIEZO1 mechanical response. Nat Commun 10(1):1200. https://doi.org/10.1038/s41467-019-09055-7

Roos D, van Bruggen R, Meischl C (2003) Oxidative killing of microbes by neutrophils. Microbes Infect 5(14):1307–1315. https://doi.org/10.1016/j.micinf.2003.09.009

Rossy J, Laufer JM, Legler DF (2018) Role of Mechanotransduction and tension in T cell function. Front Immunol 9. https://doi.org/10.3389/fimmu.2018.02638

Salauddin M, Bhattacharyya D, Samanta I, Saha S, Xue M, Hossain MG, Zheng C (2025) Role of TLRs as signaling cascades to combat infectious diseases: a review. Cell Mol Life Sci 82(1):122. https://doi.org/10.1007/s00018-025-05631-x

Saotome K, Murthy SE, Kefauver JM, Whitwam T, Patapoutian A, Ward AB (2018) Structure of the mechanically activated ion channel PIEZO1. Nature 554(7693):481–486. https://doi.org/10.1038/nature25453

Schaefer A, Hordijk PL (2015) Cell-stiffness-induced mechanosignaling – a key driver of leukocyte transendothelial migration. J Cell Sci 128(13):2221–2230. https://doi.org/10.1242/jcs.163055

Servin-Vences MR, Lam RM, Koolen A, Wang Y, Saade DN, Loud M, Kacmaz H, Frausto S, Zhang Y, Beyder A, Marshall KL, Bönnemann CG, Chesler AT, Patapoutian A (2023) PIEZO2 in somatosensory neurons controls gastrointestinal transit. Cell 186(16):3386–3399. https://doi.org/10.1016/j.cell.2023.07.006

Shan Y, Guo X, Zhang M, Chen M, Li Y, Zhang M, Pei D (2025) Structure of human PIEZO1 and its slow-inactivating channelopathy mutants. elife 13:RP101923. https://doi.org/10.7554/eLife.101923

Sica A, Mantovani A (2012) Macrophage plasticity and polarization: in vivo veritas. J Clin Invest 122(3):787–795. https://doi.org/10.1172/JCI59643

Singhal A, Kumar S (2022) Neutrophil and remnant clearance in immunity and inflammation. Immunology 165(1):22–43. https://doi.org/10.1111/imm.13423

Smith TD, Nagalla RR, Chen EY, Liu WF (2017) Harnessing macrophage plasticity for tissue regeneration. Adv Drug Deliv Rev 114:193–205. https://doi.org/10.1016/j.addr.2017.04.012

Solis AG, Bielecki P, Steach HR, Sharma L, Harman CCD, Yun S, Palm NW, de Zoete MR, Warnock JN et al (2019) Mechanosensation of cyclical force by PIEZO1 is essential for innate immunity. Nature 573(7772):69–74. https://doi.org/10.1038/s41586-019-1485-8

Son K, Hussain A, Sehmi R, Janssen L (2021) The cycling of intracellular calcium released in response to fluid shear stress is critical for migration-associated actin reorganization in eosinophils. Cells 10(1):157. https://doi.org/10.3390/cells10010157

Sp H, Hf R et al (2008) Eosinophils: biological properties and role in health and disease. Clin Exp Allergy 38(5). https://doi.org/10.1111/j.1365-2222.2008.02958.x

Syeda R, Xu J, Dubin AE, Coste B, Mathur J, Huynh T, Matzen J, Lao J, Tully DC, Engels IH, Petrassi HM, Schumacher AM, Montal M, Bandell M, Patapoutian A (2015) Chemical activation of the mechanotransduction channel PIEZO1. elife 4:e07369. https://doi.org/10.7554/eLife.07369

Syeda R, Florendo MN, Cox CD, Kefauver JM, Santos JS, Martinac B, Patapoutian A (2016) PIEZO1 channels are inherently mechanosensitive. Cell Rep 17(7):1739–1746. https://doi.org/10.1016/j.celrep.2016.10.033

Teng T-S, Ji A, Ji X-Y, Li Y-Z (2017) Neutrophils and immunity: from bactericidal action to being conquered. J Immunol Res 2017:9671604. https://doi.org/10.1155/2017/9671604

Ullah G, Nosyreva ED, Thompson D, Cuello VA, Cuello LG, Syeda R (2024) Analysis of pressure-activated PIEZO1 open and subconductance states at a single channel level. J Biol Chem 300(4):107156. https://doi.org/10.1016/j.jbc.2024.107156

Villarino NW, Hamed YMF, Ghosh B, Dubin AE, Lewis AH, Odem MA, Loud MC, Wang Y, Servin-Vences MR, Patapoutian A, Marshall KL (2023) Labeling PIEZO2 activity in the peripheral nervous system. Neuron 111(16):2488–2501. https://doi.org/10.1016/j.neuron.2023.05.015

Viola A, Munari F, Sánchez-Rodríguez R, Scolaro T, Castegna A (2019) The metabolic signature of macrophage responses. Front Immunol 10:1462. https://doi.org/10.3389/fimmu.2019.01462

Wang Y, Xiao B (2018) The mechanosensitive PIEZO1 channel: structural features and molecular bases underlying its ion permeation and mechanotransduction. J Physiol 596(6):969–978. https://doi.org/10.1113/JP274404

Wang Y, Chi S, Guo H, Li G, Wang L, Zhao Q, Rao Y, Zu L, He W, Xiao B (2018) A lever-like transduction pathway for long-distance chemical- and mechano-gating of the mechanosensitive PIEZO1 channel. Nat Commun 9:1300. https://doi.org/10.1038/s41467-018-03570-9

Wang L, You X, Zhang L, Zhang C, Zou W (2022) Mechanical regulation of bone remodeling. Bone Res 10(1):16. https://doi.org/10.1038/s41413-022-00190-4

Wang L, Chen Z, Liu G, Pan Y (2023) Functional crosstalk and regulation of natural killer cells in tumour microenvironment: significance and potential therapeutic strategies. Genes & Diseases 10(3):990–1004. https://doi.org/10.1016/j.gendis.2022.07.009

Wang Y, Chu T, Meng C, Bian Y, Li J (2025a) PIEZO1-specific deletion in macrophage protects the progression of chronic inflammatory bowel disease in mice. Cell Mol Gastroenterol Hepatol 19(7):101495. https://doi.org/10.1016/j.jcmgh.2025.101495

Wang Y, Yang Q, Dong Y, Wang L, Zhang Z, Niu R, Wang Y, Bi Y, Liu G (2025b) PIEZO1-directed neutrophil extracellular traps regulate macrophage differentiation during influenza virus infection. Cell Death Dis 16(1):60. https://doi.org/10.1038/s41419-025-07395-5

West AP, Brodsky IE, Rahner C, Woo DK, Erdjument-Bromage H, Tempst P, Walsh MC, Choi Y, Shadel GS, Ghosh S (2011) TLR signalling augments macrophage bactericidal activity through mitochondrial ROS. Nature 472(7344):476–480. https://doi.org/10.1038/nature09973

Wijerathne TD, Ozkan AD, Lacroix JJ (2023) Microscopic mechanism of PIEZO1 activation by pressure-induced membrane stretch. J Gen Physiol 155(5):e202213260. https://doi.org/10.1085/jgp.202213260

Woo S-H, Ranade S, Weyer AD, Dubin AE, Baba Y, Qiu Z, Petrus M, Miyamoto T, Reddy K, Lumpkin EA, Stucky CL, Patapoutian A (2014) PIEZO2 is required for Merkel-cell mechanotransduction. Nature 509(7502):622–626. https://doi.org/10.1038/nature13251

Woo S-H, Lukacs V, De Nooij JC, Zaytseva D, Criddle CR, Francisco A, Jessell TM, Wilkinson KA, Patapoutian A (2015) PIEZO2 is the principal mechanotransduction channel for proprioception. Nat Neurosci 18(12):1756–1762

Wu X, Yang Y (2024) Neutrophil extracellular traps (NETs) and fibrotic diseases. Int Immunopharmacol 133:112085. https://doi.org/10.1016/j.intimp.2024.112085

Wu J, Young M, Lewis AH, Martfeld AN, Kalmeta B, Grandl J (2017) Inactivation of mechanically activated PIEZO1 ion channels is determined by the C-terminal extracellular domain and the inner pore helix. Cell Rep 21(9):2357–2366. https://doi.org/10.1016/j.celrep.2017.10.120

Xiao B (2024) Mechanisms of mechanotransduction and physiological roles of PIEZO channels. Nat Rev Mol Cell Biol 25(11):886–903. https://doi.org/10.1038/s41580-024-00773-5

Xiong H, Yang J, Guo J, Ma A, Wang B, Kang Y (2022) Mechanosensitive PIEZO channels mediate the physiological and pathophysiological changes in the respiratory system. Respir Res 23:196. https://doi.org/10.1186/s12931-022-02122-6

Xiong J, Xiao R, Zhao J, Zhao Q, Luo M, Li F, Zhang W, Wu M (2024) Matrix stiffness affects tumour-associated macrophage functional polarization and its potential in tumour therapy. J Transl Med 22(1):85. https://doi.org/10.1186/s12967-023-04810-3

Yanamandra AK, Zhang J, Montalvo G, Zhou X, Biedenweg D, Zhao R, Sharma S, Hoth M, Lautenschläger F, Otto O, Del Campo A, Qu B (2024) PIEZO1-mediated mechanosensing governs NK-cell killing efficiency and infiltration in three-dimensional matrices. Eur J Immunol 54(3):e2350693. https://doi.org/10.1002/eji.202350693

Yang C, Xie R, Cao T, Zhang Y, Wang X, Xu Y, Fan Q, Ye F (2024) Mechanical communication and function regulation of immune cells. Fundam Res. https://doi.org/10.1016/j.fmre.2024.04.008

Yang L, Shi F, Cao F, Wang L, She J, He B, Xu X, Kong L, Cai B (2025) Neutrophils in tissue injury and repair: molecular mechanisms and therapeutic targets. MedComm 6(5):e70184. https://doi.org/10.1002/mco2.70184

Yoshida R, Suzuki M, Sakaguchi R, Hasegawa E, Kimura A, Shichita T, Sekiya T, Shiraishi H, Shimoda K, Yoshimura A (2012) Forced expression of stabilized c-Fos in dendritic cells reduces cytokine production and immune responses *in vivo*. Biochem Biophys Res Commun 423(2):247–252. https://doi.org/10.1016/j.bbrc.2012.05.097

Zarychanski R, Schulz VP, Houston BL, Maksimova Y, Houston DS, Smith B, Rinehart J, Gallagher PG (2012) Mutations in the mechanotransduction protein PIEZO1 are associated with hereditary xerocytosis. Blood 120(9):1908–1915. https://doi.org/10.1182/blood-2012-04-422253

Zhang Y, Wang H (2012) Integrin signalling and function in immune cells. Immunology 135(4):268–275. https://doi.org/10.1111/j.1365-2567.2011.03549.x

Zhang L, Liu Q, Yang X, Su C, Ding H, Hu J, Han W, Wu J, Zhang M, Zuo L, Mei Q (2025) Mechanosensitive Ion Channel PIEZO1 as a key regulator of intestinal fibrosis in Crohn's disease. Inflamm Bowel Dis 31(7):1772–1788. https://doi.org/10.1093/ibd/izaf041

Zhao Q, Wu K, Geng J, Chi S, Wang Y, Zhi P, Zhang M, Xiao B (2016) Ion permeation and mechanotransduction mechanisms of mechanosensitive PIEZO channels. Neuron 89(6):1248–1263. https://doi.org/10.1016/j.neuron.2016.01.046

Zhao Q, Zhou H, Chi S, Wang Y, Wang J, Geng J, Wu K, Liu W, Zhang T, Dong M-Q, Wang J, Li X, Xiao B (2018) Structure and mechanogating mechanism of the PIEZO1 channel. Nature 554(7693):487–492. https://doi.org/10.1038/nature25743

Zheng W, Gracheva EO, Bagriantsev SN (2019) A hydrophobic gate in the inner pore helix is the major determinant of inactivation in mechanosensitive PIEZO channels. elife 8:e44003. https://doi.org/10.7554/eLife.44003

Zhou Z, Ma X, Lin Y, Cheng D, Bavi N, Secker GA, Li JV, Janbandhu V, Sutton DL, Scott HS, Yao M, Harvey RP, Harvey NL, Corry B, Zhang Y, Cox CD (2023) MyoD-family inhibitor proteins act as auxiliary subunits of PIEZO channels. Science 381(6659):799–804. https://doi.org/10.1126/science.adh8190

Zhu Y, Wang T, Yang Y, Wang Z, Chen X, Wang L, Niu R, Sun Z, Zhang C, Luo Y, Hu Y, Gu W (2024) Low shear stress exacerbates atherosclerosis by inducing the generation of neutrophil extracellular traps via PIEZO1-mediated mechanosensation. Atherosclerosis 391. https://doi.org/10.1016/j.atherosclerosis.2024.117473

Chapter 4
Podosomes and Mechanical Force

Malgorzata Kloc, Souhail A. Thabet, Samruddhi Ranmale, and Jarek Wosik

Abstract Podosomes are specialized, highly dynamic adhesion structures protruding from the membrane of animal cells. Their complex ultrastructure enables interactions with the microenvironment and extracellular matrix and the sensory and mechano-effector functions. Although historically attributed to monocytes, macrophages, and dendric cells, podosomes also occur in osteoclasts, endothelial cells, muscle, neural, cancer, and other cell types. This review describes podosome structure and podosomes' role in mechanotransduction and beyond.

4.1 Introduction

Podosomes, often equated with invadopodia, are the micron-sized actin-rich extensions of the cell ventral membrane. They regulate cell adhesion to the substrate, extracellular matrix (ECM) degradation, and cell movement (Schachtner et al. 2013). Podosome consists of the actin filament-rich center surrounded by adhesion proteins such as vinculin. Podosomes were first observed in 1980 in Rous sarcoma virus (RSV)-transformed chicken embryo fibroblasts. The podosome name referring to "cellular feet" structures was coined in 1985. In 1989, after discovering podosome activity in extracellular matrix degradation, they were renamed invadopodia (Chen 1989; David-Pfeuty and Singer 1980; Murphy and Courtneidge 2011; Murthi et al. 2025; Tarone et al. 1985). Although the podosome and invadopodium names

M. Kloc (✉) · S. Ranmale
The Houston Methodist Research Institute, Transplant Immunology, Houston, TX, USA

The Houston Methodist Hospital, Department of Surgery, Houston, TX, USA
e-mail: mkloc@houstonmethodist.org

S. A. Thabet
The Houston Methodist Hospital, Department of Surgery, Houston, TX, USA

J. Wosik
Electrical and Computer Engineering Department, University of Houston, Houston, TX, USA

Texas Center for Superconductivity, University of Houston, Houston, TX, USA

M. Kloc, J. Wosik (eds.), *Mechanotransduction and Mechanosensing*, Results and Problems in Cell Differentiation 78, https://doi.org/10.1007/978-3-032-19509-8_4

are often used interchangeably, as both structures coordinate ECM degradation with cell adhesion and motility, some studies indicate that they vary slightly in architecture and function. For example, there are indications that podosomes are less stable, have a faster turnover, and protrude less deeply into ECM, thus having lesser degradative activity than invadopodia (Ayala et al. 2006; Murphy and Courtneidge 2011; Weaver 2008). However, podosomes and invadopodia may be identical structures, and the observed slight architectural and functional variety may be cell type and circumstance specific. Podosomes/invadopodia are indispensable for the functions of healthy cells, organ and tissue homeostasis, and cell migration during embryonic development and play a role in tumorigenesis and many human diseases. Based on nanostructure and function, three types of podosomes have been defined: matrix-degrading podosomes, frustrated phagocytosis podosomes, and phagocytic (phagosome-associated) podosomes. Of note, any strict categorization of podosomes is unrealistic because of their multifunctionality.

4.2 Podosome Ultrastructure, Molecular Components, and Life Cycle

Individual podosome contains over 300 different components (Ezzoukhry et al. 2018). Podosome has a core of branched actin filaments bound to various actin-regulatory proteins controlling actin polymerization, branching, and cross-linking, such as the Wiskott-Aldrich syndrome protein (WASP) (an effector of the RhoGTPase Cdc42 and a substrate of proto-oncogene tyrosine-protein kinases Src family), Arp2/3 actin polymerization complex, cortactin, paxillin, and α-actinin. WASP and Arp2/3 regulate actin nucleation and branching, cortactin, and paxillin enhance and stabilize actin filaments and branching, and α-actinin cross-links actin filaments and stabilizes podosome core (Luxenburg et al. 2012; Schachtner et al. 2013; van den Dries et al. 2019a, b) (Fig. 4.1). Podosome core is highly dynamic, and its actin polymerization/depolymerization cycles are implemented by core regulatory proteins such as coronin (antagonizes cortactin, dissociates Arp2/3, and destabilizes actin branches), tropomyosins (stabilize actin microfilaments and regulate accessibility to actin-associated proteins), caldesmon (regulates podosome assembly/disassembly), and gelsolin (severs and caps actin filaments, affecting their length and dynamics) (Goode et al. 2023; Gunning et al. 2015; van den Dries et al. 2019a, b). Recent studies indicate that spleen tyrosine kinase Syk, through phosphorylation of actin-binding homolog of cortactin, the hematopoietic lineage cell-specific protein 1 (HS1), and a non-receptor tyrosine kinase Pyk2, regulates podosome formation in macrophages. Additionally, Syk clusters the immunoreceptor tyrosine-based activation motif (ITAM)-containing adaptors, which along with integrins assemble the signaling hubs for the initiation and nucleation of podosomes (Ghasempour et al. 2024).

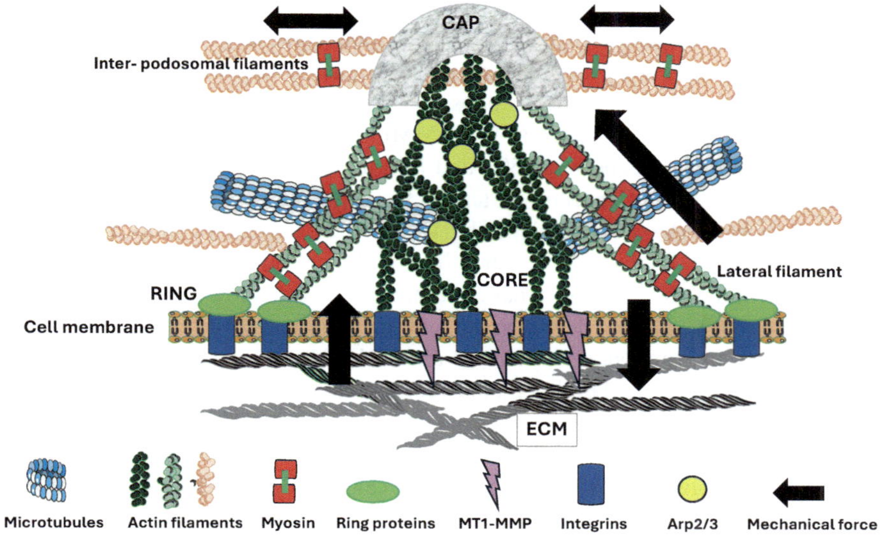

Fig. 4.1 Podosome ultrastructure. The podosome core is composed of branched actin filaments bound to numerous regulatory proteins, such as the Arp2/3 actin polymerization complex (yellow circles). The podosome cap that tops the core is a regulatory module composed of actin-associated proteins that regulate actin and myosin activity and turnover. The podosome core is surrounded by a ring structure that, via adhesion ring proteins (green oval) and integrins (blue cylinders), connects the core to the extracellular matrix (ECM). The lateral actin filaments containing myosin II connect the podosome ring and cap. The inter-podosomal actin filaments integrate individual podosomes into higher-order superstructures. Myosin II (red/green squares) connects podosomes to the microtubules that stabilize podosomes and coordinate matrix degradation function within podosome clusters. Podosome base apposing cell membrane contains transmembrane proteinases such as MT1-MMP (pink jagged arrows), which degrade ECM. The podosome and inter-podosomal actin filaments create and respond to the mechanical force, acting as a mechanosensor and mechano-effector

The podosome cap tops the podosome actin core (Fig. 4.1). Cap is a regulatory module built of actin-associated proteins, such as actin-binding lymphocyte-specific protein 1 (LSP1) and actin-bundling protein fascin, regulating actin and myosin activity and turnover (Linder and Cervero 2020; van den Dries et al. 2019a, b). The podosome ring structure surrounding the core connects the podosome via integrins and adhesion proteins vinculin and talin to the extracellular matrix (Fig. 4.1) (van den Dries et al. 2019a, b; Walde et al. 2014). Recent studies showed that vinculin, by itself, is a mechano-effector and mechanosensitive adaptor that is recruited to adhesion sites by talin and α-catenin. Vinculin acts as a mechanical switch and, in response to mechanical force, undergoes a dynamic unfolding/folding cycle, which allows the recruitment of various binding partners (Liu et al. 2025).

The lateral actin cables containing myosin II connect the podosome ring and cap, while the inter-podosomal actin filaments integrate individual podosomes into higher-order superstructures (Staszowska et al. 2017; van den Dries et al. 2019a, b).

Recent studies with three-dimensional stochastic optical reconstruction microscopy (3D-STORM) showed that myosin II connects podosomes to the microtubules that stabilize podosomes and coordinate matrix degradation function within podosome clusters (Hu et al. 2022) (Fig. 4.1). Podosome base apposing cell membrane contains transmembrane proteinases such as MT1-MMP (Weber et al. 2022) (Fig. 4.1).

Podosome lifespan is between 30 s and 14 min, with the average lifespan ~2 min (Destaing et al. 2003). Consecutive steps of the podosome life cycle from inception to dissolution are described in detail by Weber et al. (2022). In short, podosomes form within zones of low actin contractility, and local inhibition of small GTPase RhoA and its effector ROCK kinase promotes the formation of new podosomes (Chen et al. 2018). During development and maturation, native podosomes acquire all the structural components described above. Fully matured podosomes acquire lytic activity by microtubule-mediated recruitment of vesicles containing MT1-MMP lytic enzymes for extracellular matrix degradation. Podosome dissolution occurs through activation of RhoA-ROCK and myosin II, and degradation of its components involves the WASP pathway and calcium-dependent cysteine protease calpains. Amazingly, after podosome disassembly, the MT1-MMPs remain embedded in the plasma membrane, demarcating the location for seeding a new generation of podosomes (see Sect. 3.3 below) (Weber et al. 2022). Podosomes are often organized into superstructures, such as podosome rosettes, podosome belts, and podosome clusters in which individual podosomes are interconnected by inter-podosomal actin filaments (Figs. 4.1 and 4.2) (Alonso et al. 2019). Studies of fibroblasts showed that podosome rosettes form either de novo or by fission of preexisting rosettes. Authors hypothesize that podosome rosette fission may involve myosin II-dependent contractility regulated by myosin light chain kinase and Rho-associated kinase II (Kuo et al. 2018). Interestingly, the bone-resorbing osteoclasts form a podosome-like superstructure called the sealing zone, which is formed by microtubule-dependent fusion of podosomes and anchors osteoclasts to the mineral substrate. The sealing zone consists of a 4-μm-wide and 4-μm-high ring of actin accompanied by other podosome-typical components (Jurdic et al. 2006).

4.3 Podosome Functions

4.3.1 Cell Adhesion and Movement

Podosome adhesion to extracellular matrix and matrix degradation are closely related to cell motility. Adhesion molecules such as integrins, bound to the podosome ring, attach podosomes to the extracellular matrix, while integrin signaling can induce de novo podosome formation. Matrix degradation by podosomes generates a path for cell migration through the extracellular matrix and basement membranes, while mechanosensing allows podosomes to assess and adjust structurally and functionally to the mechanical properties of the environment and

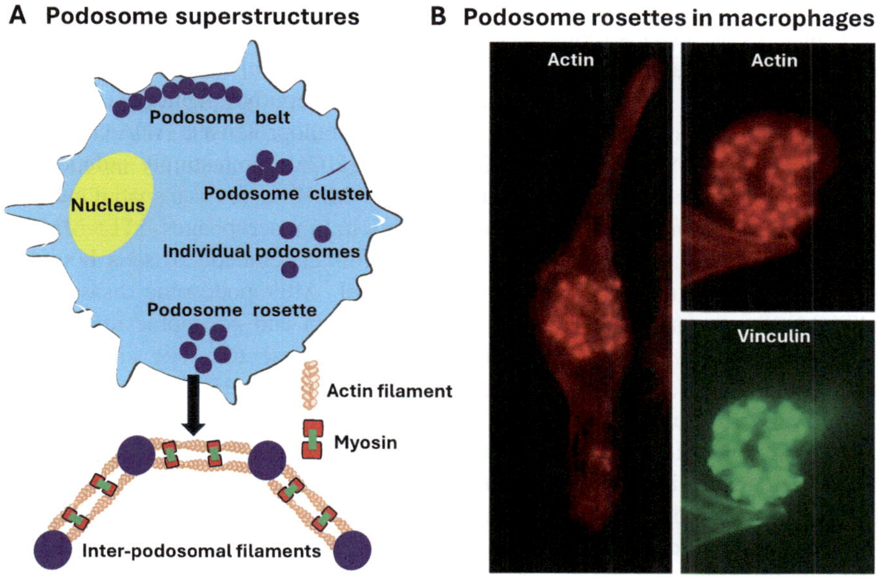

Fig. 4.2 Podosome superstructures. (**a**) Individual podosomes can be arranged into higher-order structures such as podosome clusters, belts, or rosettes. Within the cluster, podosomes are connected by inter-podosomal actin filaments bound to myosin, which coordinate the structure and functions of individual podosomes within the cluster. (**b**) Fluorescent microscope images of mouse peritoneal macrophages stained with rhodamine phalloidin (red) to visualize actin or immunostained with anti-vinculin FITC-conjugated antibody (green) showing podosomes arranged into circular rosette structures

substrate. Adhesion and substrate stiffness reciprocally regulate podosome actin-myosin contractility (Schachtner et al. 2013). The coordination of actin dynamics for adhesion and migration depends on many signaling and structural molecules, such as WASP (Oliveira et al. 2022), actin nucleation-promoting factor cortactin, vimentin, and myosins Myo1e and Myo1f (Cervero et al. 2025; Etienne-Manneville 2025; Huang et al. 2025; Jeannot and Besson 2020). Recent studies of macrophage podosomes using in situ cryo-electron tomography showed that actin polymerization forces at the membrane are not sufficient to generate podosome protrusive force and that the generation of adequate force depends on the bending of the podosome actin network that creates the elastic energy (Jasnin et al. 2022).

4.3.2 *Extracellular Matrix and DNA Degradation*

Matrix-degrading podosomes/invadopodia dock lytic enzyme-containing intracellular vesicles, which fuse with the membrane and release enzymes for extracellular matrix degradation (Cougoule et al. 2018; Linder et al. 2023; Schachtner et al.

2013; Tertrais et al. 2021; Wiesner et al. 2010). Lytic enzymes degrading extracellular matrix components (collagens, fibronectin, vitronectin, laminin, and gelatin) are metalloproteinases (MMPs: MT1-MMP, MMP2, MMP9), serine proteinases (a dual-specificity serine protease seprase and dipeptidyl peptidase 4 DPP4), cathepsins (cathepsin B), and a disintegrin and metalloproteinase ADAM proteins (Edwards et al. 2008; Linder et al. 2023; Zhu et al. 2021). Interestingly, independent of its proteolytic activity, metalloproteinase MT1-MMP serves as a spatial memory device for podosome reformation after dissolution. In macrophages, MT1-MMP is not only in podosomes but also outside as membrane-embedded islets bound to subcortical actin via MT1-MMP cytoplasmic tail. After podosome disassembly, MT1-MMP islets serve as seeds for actin nucleation and podosome reformation (El Azzouzi et al. 2016; Weber et al. 2022). Because the lytic enzymes are active in acidic pH, podosomes/invadopodia contain Na + –H+ exchanger NHE1 (Schelling and Abu Jawdeh 2008). They also contain oxidoreductases that modify extracellular matrix (Quilaqueo-Millaqueo et al. 2024) and DNA-degrading enzyme a glycosylphosphatidylinositol-anchored DNase X (Shiokawa et al. 2007). DNase is recruited to the podosome actin core, and during bacterial infection, the rosettes of macrophage podosomes containing DNase surround bacteria and directly degrade bacterial DNA on the macrophage surface (Pal et al. 2021). Additionally, dendritic cell studies showed that podosomes sample the environment for antigens. The pattern recognition receptors at the podosomes trigger localized endocytic uptake of the antigens to be presented to T cells (Baranov et al. 2014). Such findings indicate that podosomes, in addition to modifying the extracellular matrix, directly participate in innate and adaptive immune defense.

4.3.3 Cell Fusion

Podosomes/invadopodia and podosome-like zipper structures are involved in homotypic and heterotypic cell fusion (Dufrançais et al. 2021; Oikawa et al. 2012; Sens et al. 2010; Sung and Weaver 2011). Studies of body wall muscle development in *Drosophila* showed that the podosomes facilitate adherence and fusion between the fusion-competent myoblasts (FCMs) and muscle founder cells (Sens et al. 2010). Podosomes also dock the fusogenic lipid raft-containing vesicles at the fusion site. During the pre-fusion phase, the actin filament regulator Wiskott-Aldrich syndrome protein (WASP), via Arp2/3 actin polymerization complex, induces podosomes at the surface of FCM, where it contacts the founder cell membrane. The activity of SCAR (WASP-related suppressor of receptor defects) protein in the founder cell induces a thin layer of actin filaments beneath the membrane at the contact site. These two actin structures in fusion partners facilitate the deformation of apposing membranes and initiate fusion pore formation. The next step is actin disassembly and the expansion of the fusion pore (Sens et al. 2010). Studies of osteoclast precursor fusion into multinuclear osteoclasts during osteoclastogenesis and osteoclast fusion with cancer cells showed that the fusion requires the formation, via Tks5/Src pathway, of the circumferential ring of podosomes at the fusion interface (Oikawa et al. 2012). Tks5, a tyrosine kinase substrate with five SH3 domains, is an adaptor protein with an N-terminal phox homology (PX) domain and substrate for proto-oncogene tyrosine-protein kinase Src. Tks5 binds cortactin and other regulators of actin filament assembly, including N-WASP (Kudlik et al. 2020). Circumferential podosomes facilitate the apposition of fusing membranes, formation of the hemifusion stalk, and opening of the fusion pore (Oikawa et al. 2012).

Actin-rich zipper-like structures (ZLSs) have been described in many in vitro systems, but despite their similar appearance, these ZLSs seem to have different origins and functions. Probably, only some are involved in cell fusion per se (Dufrançais et al. 2021). Mouse macrophage model of multinucleated giant cells identified actin-rich zipper-like structures (ZLSs) at the surface of contacting multinuclear giant cells (MGCs) and between the surface of macrophages and implanted biomaterials. Because ZLSs of MGCs form only during long-term IL-4-induced culture and after the cell fusion has already ceased, they are probably not involved in cell fusion per se but create close contact between cells or cells and biomaterials. Actin foci built from small and large actin globules periodically distributed at the membranes of apposing cells give the zipper-like appearance to ZLSs (Fig. 4.3). ZLSs derive from podosomes and disassemble into podosomes. ZLS assembly is sequential and resembles a zipper fastener moving in one direction and "zippering up" plasma membranes of neighboring cells. ZLSs contain components typical for podosomes, such as vinculin, talin, and integrin, and are highly dynamic undergoing assembly/disassembly cycles. Transformation of podosomes into ZLSs occurs when the membranes of neighboring cells are bridged by E-cadherin and nectin-2 junctional proteins, and it requires WASP, Cdc42, and Arp2/3 complex (Balabiyev et al. 2020). ZLSs were also observed in fusing osteoclasts, although their molecular composition differed from macrophage ZLSs. The formation of osteoclast ZLSs seemed independent from podosomes. It results from a continuous retrograde actin flow that pushes membranes of fusing cells into close contact. Osteoclast ZLSs break down and transform into a large actin ring called a podosome belt. However, podosome belts coexisting with ZLSs were also observed (Takito and Nakamura 2012; Takito et al. 2012, 2017). The difference in ZLSs in different cells and the coexistence of podosome belt with ZLSs during cell fusion make it hard to determine which structure is really involved in cell fusion and which is only involved in a cell-cell contact.

4.4 Podosome Role in Mechanosensing and Mechanotransduction

To efficiently function in cell adhesion, migration, and matrix degradation, the podosomes must sense the physical properties of substrate and environment, such

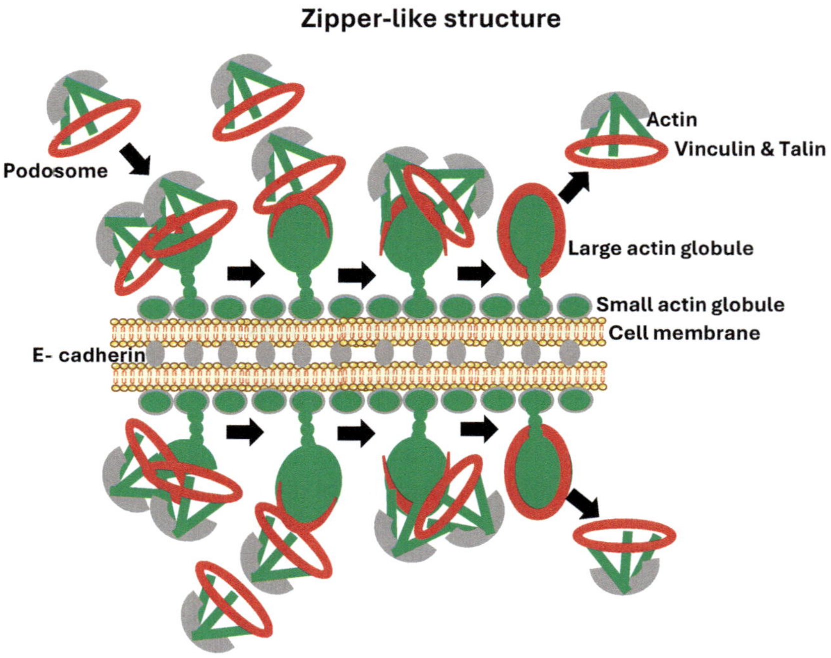

Fig. 4.3 Podosome-dependent formation of the zipper-like structure. The zipper-like structure (ZLS) consists of small and large actin globules periodically distributed at the membranes of apposing cells, giving the zipper-like appearance. Large globules, besides actin, also contain a shell of podosome ring adhesion proteins such as vinculin and talin. ZLSs facilitate cell fusion and derive from podosomes and disassemble into podosomes. ZLS assembly is sequential and resembles a zipper fastener moving in one direction, "zippering up" plasma membranes of neighboring cells. The transformation of podosomes into ZLS occurs when the membranes of neighboring cells are bridged by junctional proteins such as E-cadherin, and it requires WASP, Cdc42, and the Arp2/3 complex

as rigidity, compactness, shape, porosity, and spatial topography, transmit these mechanical cues to the cell interior, and adjust their own ultrastructure, molecular composition, spatial rearrangement, longevity, and functions, accordingly (Linder and Barcelona 2023; Linder et al. 2023; Linder and Wiesner 2016; Parekh and Weaver 2016; van den Dries et al. 2019a, b). Podosomes will not form on soft surfaces below a certain level of rigidity, and as a rule, the stiffer the substrate, the larger, more stable, and more abundant podosomes (Schachtner et al. 2013). Podosomes generate mechanical force, in the range of 10 nN, by the following mechanism. Actin polymerization, dependent on Arp2/3 and myosin II, generates a mechanical force that pushes on the underlying cell membrane and is transmitted to the lateral filaments connected to the podosome cap and podosome ring. Mechanical coupling of the core and lateral filaments generates a counterforce that pulls lateral filaments and adhesive proteins of the ring. Such cyclical pushing and pulling allow podosomes to repeatedly probe the underlying surface and act as mechanosensors

(Fig. 4.1) (Bouissou et al. 2017; Linder and Wiesner 2016). The podosome ring contains mechanosensitive proteins, such as talin. Mechanical stretching of talin exposes binding sites for vinculin that, after binding, stabilizes extended conformation of talin, resulting in generation of protrusion force. Podosome-created mechanical force is also transmitted to the microtubules that transport metalloproteinase-containing vesicles for matrix degradation, thus looping matrix degradation with mechanotransductional properties of podosomes (van den Dries et al. 2019a, b; Weber et al. 2022).

Because individual podosomes within the mesoscale superstructures, such as podosome rosettes, podosome belts, sealing zones, and podosome clusters, are interconnected by lateral actin filaments bound to formin and decorated with myosin IIA, the sensing and responding to mechanical cues are shared between participants to adopt a collective behavior and function as a coordinated mechanosensory entity (Alonso et al. 2019; Hu et al. 2022; Luxenburg et al. 2007; Meddens et al. 2016; van den Dries et al. 2013; Weber et al. 2022). The coordination of dynamics and mechanical functions of individual podosomes within the mesoscale collectives depends on the flow of actin and associated proteins, such as vinculin and talin, powered by actin polymerization and myosin IIA contractility (Gong et al. 2021; Meddens et al. 2016; Weber et al. 2022).

4.5 Phagocytic Podosomes

4.5.1 Phagocytosis and Frustrated Phagocytosis

Phagocytosis is an evolutionarily conserved process in Protozoa and Metazoa. It is the engulfment, internalization, and degradation of pathogens, cell debris, dead cells, nutrients, and other compounds of 0.5 μm diameter and above. Phagocytosis is a multistep process involving (1) recognition of the particle by the appropriate non-opsonic receptors such as scavenger receptors (SRs) and the C-type lectin receptors (CLRs, dectins), which recognize molecular patterns of pathogens and particles, or opsonic receptors, such as Fcγ receptors (FcγRs) recognizing host antibodies and complement receptors (CRs) recognizing complement components; (2) invagination of the cell membrane to form a phagocytic cup (Fig. 4.4); (3) closing the cup and fully enveloping the particle by the cell membrane; and (4) formation of the phagocytic vesicle (phagosome). Subsequently, (5) the phagosome internalizes and fuses with lysosomes, degrading phagosome-enclosed particles. (6) The last phase involves the recycling of membrane and non-membranous components of the phagosome/lysosome complex (Lancaster et al. 2019; Linder and Barcelona 2023).

A variation of phagocytosis is the frustrated phagocytosis (Avila Ponce de León et al. 2021), which involves cell spreading and actin waves that propagate along the membrane. Frustrated phagocytosis occurs when the cells are exposed to an opsonized surface and spread as trying to engulf the nonexistent target without

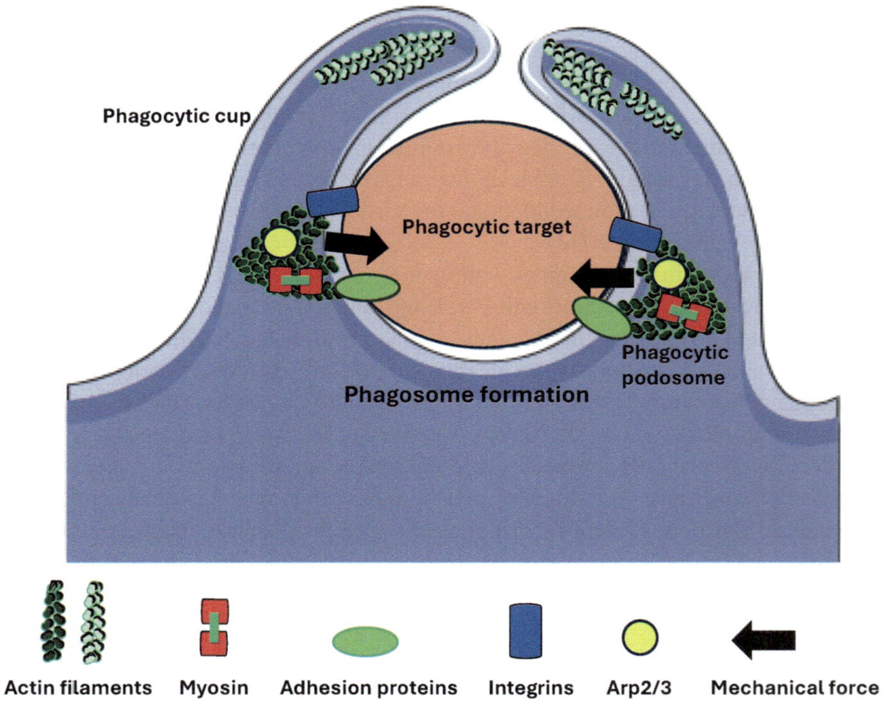

Fig. 4.4 Phagocytic podosomes. All stages of phagocytosis are actin dependent. The diagram shows the actin filament-dependent closure of the phagocytic cup. Phagocytic podosomes contain an actin core and other molecular components of typical podosomes. Phagocytic podosomes exert mechanical pressure on the phagocytic cup membrane, creating a tight seal between the membrane and the engulfed target, and by pushing on the phagosome membrane, they mechanically disrupt the target

the formation of phagosomes (Francis et al. 2022) or when the target is too large for engulfing (such as *Candida albicans* hyphae), causing cell spreading to maximize contact with the target and the formation of incompletely sealed (frustrated) phagosome partially surrounding the target (Maxson et al. 2018). The extent of frustrated phagocytic spreading is limited by the surface area that the cell can generate. Studies showed that human neutrophils spread to ~300% of the resting neutrophil surface, while macrophages can spread to ~600% (Francis et al. 2022). Two main hypotheses, "Brownian zipper" and "protrusive zipper," try to explain the mechanism of phagocytic spreading. Both theories agree that the contact of the phagocyte membrane with the opsonized surface creates a zipper-like adhesive attachment but differ in defining the driving force behind the spreading. In the protrusive zipper hypothesis, the phagocyte cytoskeleton actively creates a protrusive force that expands the cell. In the Brownian zipper hypothesis, the adhesion strength alone is enough to drive the passive spreading of the cell membrane on the target surface (Francis et al. 2022; Heinrich 2015).

Recently, in-depth studies by reflection interference contrast microscopy of human neutrophils phagocytic spreading on surfaces coated with different densities of immunoglobulin G (IgG) antibodies support the protrusive zipper hypothesis by showing that the surface of spreading depends on the density of IgG, thus the strength of adhesion, but the rate of spreading depends on cytoskeleton-generated protrusion force (Francis et al. 2022).

Actin filaments participate in different phases of phagocytosis. In the phagocytic cup, the actin filaments form a dense network underlying the cup membrane and, subsequently, surround the mature phagosome. The amount of actin around the phagosome determines its ability to fuse with a lysosome. The less actin, the higher the propensity for fusion. Additionally, actomyosin-driven, dynamic (undergoing several assembly/disassembly cycles) actin foci (actin flashes) at the phagosome surface contract the phagosome and accelerate the degradation of its content (Poirier et al. 2020; Tertrais et al. 2021). Comprehensive studies of actin flashes during complement receptor (CR)-mediated phagocytosis in macrophages showed that they cyclically assemble/disassemble on the surface of the phagosome within the first-hour post-internalization, and each time they persist for about 3 min. Actin flashes dynamic depends on mechanosensing of the stiffness of phagosome content and recruitment of myosin II, RhoA pathway, and focal adhesion components. This indicates that actomyosin-dependent phagosome contractions physically grind (chew) phagosome content to facilitate subsequent enzymatic digestion (Poirier et al. 2020).

4.5.2 Phagocytic Podosomes

Although the exact role of phagocytic podosomes is not very clear, it seems that by pressing on the phagocytic membrane they create a tight adherence between the membrane and phagocytic target preventing diffusion of signaling molecules. Membrane deformation by the force generated by podosomes enables the phagocytic cup to closely contour the target, whatever its size and shape (Ostrowski et al. 2019) (Fig. 4.4). There is tremendous confusion in the definition and description of phagocytic podosomes. The structures formed at the phagocytic cup and phagosome, consisting of an actin core and associated molecules typical for classic matrix-associated podosomes, have been named phagocytic teeth, phagosome-associated podosomes, and podosome-like structures. We also do not know if the subtle differences in the architecture of these structures are factual or depend on the method of visualization and analysis (Herron et al. 2022). To remedy this nomenclatural chaos, Linder and Barcelona (2023) compared the composition and architecture of the "phagocytic podosomes" described in different cell types and experimental models to bona fide, matrix-associated, podosomes. Phagocytic podosomes have been described in human and murine macrophages, dendritic cells, and macrophage and neutrophil cell lines. The phagocytic targets were complement-opsonized latex beads and zymosan, deformable acrylamide beads, and the opsonized or micropatterned IgG surfaces in the models of frustrated phagocytosis (see Table 1 and references in Linder and Barcelona 2023). In all these cases, phagocytic podosomes had an actin core that was not dome shaped like in classic podosomes but toothlike or hourglass shaped. Peripheral accumulation of vimentin, actin-associated, and adhesion molecules was common between the classic and phagocytic podosomes (Linder and Barcelona 2023).

Like classic matrix-associated podosomes, phagocytic podosomes are protrusive and mechanosensitive. Studies of podosomes (phagocytic teeth) in bone marrow-derived and RAW 264.7 murine macrophages phagocytosing deformable acrylamide beads coated with IgG showed that they pressed on and deformed the beads and sensed bead rigidity increasing in number upon ingestion of softer beads (Barger et al. 2022; Vorselen et al. 2021). It is also possible that phagocytic teeth, like actin flashes (see Sect. 4.1), mechanically disrupt ingested targets by pushing on the phagosome membrane and contracting the phagosome (Vorselen et al. 2021).

4.5.3 Podosomes, Pathogens, and Diseases

Podosome multifunctionality and prevalence among various cell types make them indispensable for the normal functioning of cells, tissues, and organs, proper response to invading pathogens, and overall immune response. Some fungal pathogens, such as *Candida albicans*, try to escape engulfment by host macrophages by forming long, impossible-to-phagocytose hyphae. Macrophages counteract this strategy by folding, with the participation of phagocytic podosomes, and efficiently phagocytosing long hyphae (Bain et al. 2021). *Theileria* and *Toxoplasma* parasites manipulate podosomes and cell adhesion, increasing the motility of infected cells and parasite dissemination within the host (Baumgartner 2011). Similarly, podosome modifications are involved in the spreading of cancer cells and metastasis (Meirson and Gil-Henn 2018). In Alzheimer's disease, the neurofibrillary aggregates of Tau protein deposited in the brain mediate inflammation and neurotoxicity. Degradation of Tau deposits by microglia involves Tau recognition and targeted chemotaxis mediated by podosomes and their component, the purinergic receptor P2Y12. Thus, podosomes are directly involved in clearing Tau deposits (Chinnathambi and Das 2023; Qureshi et al. 2024, 2025). By modulating extracellular matrix, podosomes participate in the development and progression of many diseases. Formation of podosome rosette and increased expression of scaffold protein TKS5, tyrosine kinase substrate with five SH3 domains, essential for podosome formation in lung fibroblasts, promote aberrant deposition of extracellular matrix and idiopathic pulmonary fibrosis (Barbayianni et al. 2023; Kudlik et al. 2020). Podosome mechanosensory functions, such as assessing the stiffness of the extracellular environment that regulates their proteolytic activity and reciprocal interactions with the surrounding, are involved in the development and progression of many cardiovascular diseases, including coronary artery disease, hypertrophic cardiomyopathy, dilated cardiomyopathy, and myocardial infarctions (Sit et al. 2019).

4.5.4 Podosomes in Organ Transplantation

In transplanted organs, podosomes, by degrading extracellular matrix (ECM) and mechanosensing the environment, participate in the infiltration of immune and stromal cells to the graft, its remodeling, and injury (Kardol-Hoefnagel et al. 2020).

In liver transplantation, particularly in marginal liver grafts (grafts that do not meet ideal donor criteria), the oval cells (liver stem cells that can differentiate into either hepatocytes or bile duct cells) develop podosomes that facilitate ECM degradation and migration during their transition toward a mesenchymal phenotype. This enables oval cells to deposit collagen, leading to posttransplant fibrosis (Liu et al. 2017).

In idiopathic pulmonary fibrosis, podosomes are induced by increased expression of the scaffold protein TKS5. Mice with deleted TKS5 have less podosomes and ECM deposition and are resistant to bleomycin-induced pulmonary fibrosis (Barbayianni et al. 2023).

In kidney transplants, macrophage infiltration and proliferation increase during rejection (Bergler et al. 2016; Wyburn et al. 2005). Macrophage migration depends on podosomes and their regulator formin (FMNL1) (Miller and Blystone 2015; Kloc and Ghobrial 2014; Bergler et al. 2016; Wyburn et al. 2005).

Cardiac allograft vasculopathy (CAV) in chronically rejecting heart allografts is caused by the migration of vascular smooth muscle cells (VSMCs) into the intima and intimal hyperplasia. The VSMC podosomes facilitate VSMC. Consequently, podosomes contribute to transplant vasculopathy in chronic rejection of the heart, kidney, liver, and lung (Costello et al. 2013; Kim et al. 2015; Kloc and Ghobrial 2014).

Targeting podosome formation (via their regulatory molecules such as RhoA, Src, TKS5, and FMNL1) represents an emerging strategy to inhibit macrophage migration, thereby potentially limiting allograft inflammation and fibrosis and promoting long-term graft survival (Miller and Blystone 2015; Wyburn et al. 2005; Baruzzi et al. 2015).

Acknowledgments We acknowledge JC Walter Jr.'s support to Houston Methodist Transplantation Center. Some drawings used in the figures were adapted from images at Servier Medical Art (https://smart.servier.com/), licensed under CC BY 4.0 (https://creativecommons.org/licenses/by/4.0/).

References

Alonso F, Spuul P, Daubon T, Kramer I, Génot E (2019) Variations on the theme of podosomes: a matter of context. Biochim Biophys Acta, Mol Cell Res 1866(4):545–553. https://doi.org/10.1016/j.bbamcr.2018.12.009. Epub 2018 Dec 27

Avila Ponce de León MA, Félix B, Othmer HG (2021) A phosphoinositide-based model of actin waves in frustrated phagocytosis. J Theor Biol 527:110764. https://doi.org/10.1016/j.jtbi.2021.110764.Epub

Ayala I, Baldassarre M, Caldieri G, Buccione R (2006) Invadopodia: a guided tour. Eur J Cell Biol 85(3–4):159–164. https://doi.org/10.1016/j.ejcb.2005.09.005. Epub 2005 Oct 11

Bain JM, Alonso MF, Childers DS, Walls CA, Mackenzie K, Pradhan A, Lewis LE, Louw J, Avelar GM, Larcombe DE, Netea MG, Gow NAR, Brown GD, Erwig LP, Brown AJP (2021) Immune cells fold and damage fungal hyphae. Proc Natl Acad Sci USA 118(15):e2020484118. https://doi.org/10.1073/pnas.2020484118. PMID: 33876755; PMCID: PMC8053999

Balabiyev A, Podolnikova NP, Mursalimov A, Lowry D, Newbern JM, Roberson RW, Ugarova TP (2020) Transition of podosomes into zipper-like structures in macrophage-derived multinucleated giant cells. Mol Biol Cell 31(18):2002–2020. https://doi.org/10.1091/mbc.E19-12-0707. Epub 2020 Jun 24. PMID: 32579434; PMCID: PMC7543064

Baranov MV, Ter Beest M, van den Bogaart G (2014) Reaching for far-flung antigen: how solid-core podosomes of dendritic cells transform into protrusive structures. Commun Integr Biol 7(5):e970961I. https://doi.org/10.4161/cib.29084. PMID: 26843902; PMCID: PMC4594491

Barbayianni I, Kanellopoulou P, Fanidis D, Nastos D, Ntouskou ED, Galaris A, Harokopos V, Hatzis P, Tsitoura E, Homer R, Kaminski N, Antoniou KM, Crestani B, Tzouvelekis A, Aidinis V (2023) SRC and TKS5 mediated podosome formation in fibroblasts promotes extracellular matrix invasion and pulmonary fibrosis. Nat Commun 14(1):5882. https://doi.org/10.1038/s41467-023-41614-x. PMID: 37735172; PMCID: PMC10514346

Barger SR, Vorselen D, Gauthier NC, Theriot JA, Krendel M (2022) F-actin organization and target constriction during primary macrophage phagocytosis is balanced by competing activity of myosin-I and myosin-II. Mol Biol Cell 33(14:br24). https://doi.org/10.1091/mbc.E22-06-0210. Epub 2022 Sep 21. PMID: 36129777; PMCID: PMC9727803

Baruzzi A, Remelli S, Lorenzetto E, Sega M, Chignola R, Berton G (2015) Sos1 regulates macrophage podosome assembly and macrophage invasive capacity. J Immunol 195(10):4900–4912. https://doi.org/10.4049/jimmunol.1500579. Epub 2015 Oct 7

Baumgartner M (2011) Enforcing host cell polarity: an apicomplexan parasite strategy towards dissemination. Curr Opin Microbiol 14(4):436–444. https://doi.org/10.1016/j.mib.2011.07.003. Epub 2011 Jul 25

Bergler T, Jung B, Bourier F, Kühne L, Banas MC, Rümmele P, Wurm S, Banas B (2016) Infiltration of macrophages correlates with severity of allograft rejection and outcome in human kidney transplantation. PLoS One 11(6):e0156900. https://doi.org/10.1371/journal.pone.0156900. PMID: 27285579; PMCID: PMC4902310

Bouissou A, Proag A, Bourg N, Pingris K, Cabriel C, Balor S, Mangeat T, Thibault C, Vieu C, Dupuis G, Fort E, Lévêque-Fort S, Maridonneau-Parini I, Poincloux R (2017) Podosome force generation machinery: a local balance between protrusion at the core and traction at the ring. ACS Nano 11(4):4028–4040. https://doi.org/10.1021/acsnano.7b00622. Epub 2017 Apr 4

Cervero P, Barger SR, Verdys P, Herzog R, Paul T, Palmieri M, Poincloux R, Linder S, Krendel M (2025) Myo1e/f at the podosome base regulate podosome dynamics and promote macrophage migration. bioRxiv:2025.04.28.651090. https://doi.org/10.1101/2025.04.28.651090

Chen WT (1989) Proteolytic activity of specialized surface protrusions formed at rosette contact sites of transformed cells. J Exp Zool 251(2):167–185. https://doi.org/10.1002/jez.1402510206

Chen W, Ghobrial RM, Li XC, Kloc M (2018) Inhibition of RhoA and mTORC2/Rictor by Fingolimod (FTY720) induces p21-activated kinase 1, PAK-1 and amplifies podosomes in mouse peritoneal macrophages. Immunobiology 223(11):634–647. https://doi.org/10.1016/j.imbio.2018.07.009. Epub 2018 Jul 7

Chinnathambi S, Das R (2023) Microglia degrade tau oligomers deposit via purinergic P2Y12-associated podosome and filopodia formation and induce chemotaxis. Cell Biosci 13(1):95. https://doi.org/10.1186/s13578-023-01028-0. PMID: 37221563; PMCID: PMC10204346

Costello JP, Mohanakumar T, Nath DS (2013) Mechanisms of chronic cardiac allograft rejection. Tex Heart Inst J 40(4):395–399. PMID: 24082367; PMCID: PMC3783121

Cougoule C, Lastrucci C, Guiet R, Mascarau R, Meunier E, Lugo-Villarino G, Neyrolles O, Poincloux R, Maridonneau-Parini I (2018) Podosomes, but not the maturation status, determine the protease-dependent 3D migration in human dendritic cells. Front Immunol 30(9):846. https://doi.org/10.3389/fimmu.2018.00846. PMID: 29760696; PMCID: PMC5936769

David-Pfeuty T, Singer SJ (1980) Altered distributions of the cytoskeletal proteins vinculin and alpha-actinin in cultured fibroblasts transformed by Rous sarcoma virus. Proc Natl Acad Sci USA 77(11):6687–6691. https://doi.org/10.1073/pnas.77.11.6687. PMID: 6256755; PMCID: PMC350353

Destaing O, Saltel F, Géminard JC, Jurdic P, Bard F (2003) Podosomes display actin turnover and dynamic self-organization in osteoclasts expressing actin-green fluorescent protein. Mol Biol Cell 14(2):407–416. https://doi.org/10.1091/mbc.e02-07-0389. PMID: 12589043; PMCID: PMC149981

Dufrançais O, Mascarau R, Poincloux R, Maridonneau-Parini I, Raynaud-Messina B, Vérollet C (2021) Cellular and molecular actors of myeloid cell fusion: podosomes and tunneling nanotubes call the tune. Cell Mol Life Sci 78(17–18):6087–6104. https://doi.org/10.1007/s00018-021-03875-x. Epub 2021 Jul 23. PMID: 34296319; PMCID: PMC8429379

Edwards DR, Handsley MM, Pennington CJ, The ADAM (2008) Metalloproteinases. Mol Aspects Med 29(5):258–289. https://doi.org/10.1016/j.mam.2008.08.001. Epub 2008 Aug 15. PMID: 18762209; PMCID: PMC7112278

El Azzouzi K, Wiesner C, Linder S (2016) Metalloproteinase MT1-MMP islets act as memory devices for podosome reemergence. J Cell Biol 213(1):109–125. https://doi.org/10.1083/jcb.201510043. PMID: 27069022; PMCID: PMC4828691

Etienne-Manneville S (2025) Comments on 'Vimentin intermediate filaments coordinate actin stress fibers and podosomes to determine the extracellular matrix degradation by macrophages'. J Mol Cell Biol 17:mjaf004. https://doi.org/10.1093/jmcb/mjaf004. Epub ahead of print

Ezzoukhry Z, Henriet E, Cordelières FP et al (2018) Combining laser capture microdissection and proteomics reveals an active translation machinery controlling invadosome formation. Nat Commun 9:2031. https://doi.org/10.1038/s41467-018-04461-9

Francis EA, Xiao H, Teng LH, Heinrich V (2022) Mechanisms of frustrated phagocytic spreading of human neutrophils on antibody-coated surfaces. Biophys J 121(23):4714–4728. https://doi.org/10.1016/j.bpj.2022.10.016. Epub 2022 Oct 14. PMID: 36242516; PMCID: PMC9748254

Ghasempour S, Muise AM, Freeman SA (2024) Podosome nucleation is facilitated by multivalent interactions between Syk and ITAM-containing membrane complexes. J Immunol 213(7):988–997. https://doi.org/10.4049/jimmunol.2400031. PMID: 39140892; PMCID: PMC11404668

Gong Z, van den Dries K, Cambi A, Shenoy VB (2021) Chemo-mechanical diffusion waves orchestrate collective dynamics of immune cell podosomes. bioRxiv:2021.2011.2023.469591

Goode BL, Eskin J, Shekhar S (2023) Mechanisms of actin disassembly and turnover. J Cell Biol 222(12):e202309021. https://doi.org/10.1083/jcb.202309021. Epub 2023 Nov 10. PMID: 37948068; PMCID: PMC10638096

Gunning PW, Hardeman EC, Lappalainen P, Mulvihill DP (2015) Tropomyosin – master regulator of actin filament function in the cytoskeleton. J Cell Sci 128(16):2965–2974. https://doi.org/10.1242/jcs.172502. Epub 2015 Aug 3

Heinrich V (2015) Controlled one-on-one encounters between immune cells and microbes reveal mechanisms of phagocytosis. Biophys J 109(3):469–476. https://doi.org/10.1016/j.bpj.2015.06.042. PMID: 26244729; PMCID: PMC4572503

Herron JC, Hu S, Watanabe T, Nogueira AT, Liu B, Kern ME, Aaron J, Taylor A, Pablo M, Chew TL, Elston TC, Hahn KM (2022) Actin nano-architecture of phagocytic podosomes.

Nat Commun 13(1):4363. https://doi.org/10.1038/s41467-022-32038-0. PMID: 35896550; PMCID: PMC9329332

Hu F, Zhu D, Dong H, Zhang P, Xing F, Li W, Yan R, Zhou J, Xu K, Pan L, Xu J (2022) Super-resolution microscopy reveals nanoscale architecture and regulation of podosome clusters in primary macrophages. iScience 25(12):105514. https://doi.org/10.1016/j.isci.2022.105514. PMID: 36425766; PMCID: PMC9678738

Huang X, Li Z, Huang Y, Zhang Q, Cui Y, Shi X, Jiu Y (2025) Vimentin intermediate filaments coordinate actin stress fibers and podosomes to determine the extracellular matrix degradation by macrophages. Dev Cell. S1534–5807(25):00036-X. https://doi.org/10.1016/j.devcel.2025.01.016. Epub ahead of print

Jasnin M, Hervy J, Balor S, Bouissou A, Proag A, Voituriez R, Schneider J, Mangeat T, Maridonneau-Parini I, Baumeister W, Dmitrieff S, Poincloux R (2022) Elasticity of podosome actin networks produces nanonewton protrusive forces. Nat Commun 13(1):3842. https://doi.org/10.1038/s41467-022-30652-6. PMID: 35789161; PMCID: PMC9253342

Jeannot P, Besson A (2020) Cortactin function in invadopodia. Small GTPases 11(4):256–270. https://doi.org/10.1080/21541248.2017.1405773. Epub 2017 Dec 31. PMID: 29172953; PMCID: PMC7549685

Jurdic P, Saltel F, Chabadel A, Destaing O (2006) Podosome and sealing zone: specificity of the osteoclast model. Eur J Cell Biol 85(3–4):195–202. https://doi.org/10.1016/j.ejcb.2005.09.008. Epub 2005 Oct 24

Kardol-Hoefnagel T, van Logtestijn SALM, Otten HG (2020) A review on the function and regulation of ARHGDIB/RhoGDI2 expression including the hypothetical role of ARHGDIB/RhoGDI2 autoantibodies in kidney transplantation. Transplant Direct 6(5):e548. https://doi.org/10.1097/TXD.0000000000000993. PMID: 32548242; PMCID: PMC7213606

Kim NY, Kohn JC, Huynh J, Carey SP, Mason BN, Vouyouka AG, Reinhart-King CA (2015) Biophysical induction of vascular smooth muscle cell podosomes. PLoS One 10(3):e0119008. https://doi.org/10.1371/journal.pone.0119008. PMID: 25785437; PMCID: PMC4364673

Kloc M, Ghobrial RM (2014) Chronic allograft rejection: a significant hurdle to transplant success. Burns Trauma 2(1):3–10. https://doi.org/10.4103/2321-3868.121646. PMID: 27574640; PMCID: PMC4994504

Kudlik G, Takács T, Radnai L, Kurilla A, Szeder B, Koprivanacz K, Merő BL, Buday L, Vas V (2020) Advances in understanding TKS4 and TKS5: molecular scaffolds regulating cellular processes from Podosome and Invadopodium formation to differentiation and tissue homeostasis. Int J Mol Sci 21(21):8117. https://doi.org/10.3390/ijms21218117. PMID: 33143131; PMCID: PMC7663256

Kuo SL, Chen CL, Pan YR, Chiu WT, Chen HC (2018) Biogenesis of podosome rosettes through fission. Sci Rep 8(1):524. https://doi.org/10.1038/s41598-017-18861-2. PMID: 29323185; PMCID: PMC5765046

Lancaster CE, Ho CY, Hipolito VEB, Botelho RJ, Terebiznik MR (2019) Phagocytosis: what's on the menu? 1. Biochem Cell Biol 97(1):21–29. https://doi.org/10.1139/bcb-2018-0008. Epub 2018 May 23

Linder S, Barcelona B (2023) Get a grip: podosomes as potential players in phagocytosis. Eur J Cell Biol 102(4):151356. https://doi.org/10.1016/j.ejcb.2023.151356. Epub 2023 Aug 22

Linder S, Cervero P (2020) The podosome cap: past, present, perspective. Eur J Cell Biol 99(5):151087. https://doi.org/10.1016/j.ejcb.2020.151087. Epub 2020 May 25

Linder S, Wiesner C (2016) Feel the force: podosomes in mechanosensing. Exp Cell Res 343(1):67–72. https://doi.org/10.1016/j.yexcr.2015.11.026. Epub 2015 Nov 30

Linder S, Cervero P, Eddy R, Condeelis J (2023) Mechanisms and roles of podosomes and invadopodia. Nat Rev Mol Cell Biol 24(2):86–106. https://doi.org/10.1038/s41580-022-00530-6. Epub 2022 Sep 14

Liu Y, Chen W, Wu C, Minze LJ, Kubiak JZ, Li XC, Kloc M, Ghobrial RM (2017) Macrophage/monocyte-specific deletion of Ras homolog gene family member A (RhoA) downregulates fractalkine receptor and inhibits chronic rejection of mouse cardiac allografts.

J Heart Lung Transplant 36(3):340–354. https://doi.org/10.1016/j.healun.2016.08.011. Epub 2016 Aug 20. PMID: 27692539; PMCID: PMC5316307

Liu X, Liu J, Wang Y, Yao M, Baker KB, Klapholz B, Brown NH, Goult BT, Yan J (2025) The mechanical response of vinculin. Sci Adv 11(44):eady6949. https://doi.org/10.1126/sciadv.ady6949

Luxenburg C, Geblinger D, Klein E, Anderson K, Hanein D, Geiger B, Addadi L (2007) The architecture of the adhesive apparatus of cultured osteoclasts: from podosome formation to sealing zone assembly. PLoS One 2(1):e179. https://doi.org/10.1371/journal.pone.0000179. PMID: 17264882; PMCID: PMC1779809

Luxenburg C, Winograd-Katz S, Addadi L, Geiger B (2012) Involvement of actin polymerization in podosome dynamics. J Cell Sci 125(Pt 7):1666–1672. https://doi.org/10.1242/jcs.075903. Epub 2012 Feb 10. PMID: 22328507; PMCID: PMC3346827

Maxson ME, Naj X, O'Meara TR, Plumb JD, Cowen LE, Grinstein S (2018) Integrin-based diffusion barrier separates membrane domains enabling the formation of microbiostatic frustrated phagosomes. elife 19(7):e34798. https://doi.org/10.7554/eLife.34798. PMID: 29553370; PMCID: PMC5897098

Meddens MB, Pandzic E, Slotman JA, Guillet D, Joosten B, Mennens S, Paardekooper LM, Houtsmuller AB, van den Dries K, Wiseman PW, Cambi A (2016) Actomyosin-dependent dynamic spatial patterns of cytoskeletal components drive mesoscale podosome organization. Nat Commun 10(7):13127. https://doi.org/10.1038/ncomms13127. PMID: 27721497; PMCID: PMC5062568

Meirson T, Gil-Henn H (2018) Targeting invadopodia for blocking breast cancer metastasis. Drug Resist Updat 39:1–17. https://doi.org/10.1016/j.drup.2018.05.002. Epub 2018 May 17

Miller MR, Blystone SD (2015) Human macrophages utilize the podosome formin FMNL1 for adhesion and migration. Cellbio (Irvine, Calif) 4(1):1–11. https://doi.org/10.4236/cellbio.2015.41001. Epub 2015 Mar 4. PMID: 26942206; PMCID: PMC4772764

Murphy DA, Courtneidge SA (2011) The 'ins' and 'outs' of podosomes and invadopodia: characteristics, formation and function. Nat Rev Mol Cell Biol 12(7):413–426. https://doi.org/10.1038/nrm3141. PMID: 21697900; PMCID: PMC3423958

Murthi P, Overton E, Brennecke SP, Keogh RJ (2025) Happy feet: the key roles of podosomes and invadopodia in trophoblast invasion at the maternal-fetal interface. Front Endocrinol (Lausanne) 30(16):1576732. https://doi.org/10.3389/fendo.2025.1576732. PMID: 40519515; PMCID: PMC12163045

Oikawa T, Oyama M, Kozuka-Hata H, Uehara S, Udagawa N, Saya H, Matsuo K (2012) Tks5-dependent formation of circumferential podosomes/invadopodia mediates cell-cell fusion. J Cell Biol 197(4):553–568. https://doi.org/10.1083/jcb.201111116. PMID: 22584907; PMCID: PMC3352951

Oliveira MMS, Kung SY, Moreau HD, Maurin M, Record J, Sanséau D, Nylén S, Lennon-Duménil AM, Westerberg LS (2022) The WASp L272P gain-of-function mutation alters dendritic cell coordination of actin dynamics for migration and adhesion. J Leukoc Biol 111(4):793–803. https://doi.org/10.1002/JLB.1AB0821-013RR. Epub 2021 Aug 25

Ostrowski PP, Freeman SA, Fairn G, Grinstein S (2019) Dynamic podosome-like structures in nascent phagosomes are coordinated by phosphoinositides. Dev Cell 50(4):397–410.e3. https://doi.org/10.1016/j.devcel.2019.05.028. Epub 2019 Jun 20

Pal K, Zhao Y, Wang Y, Wang X (2021) Ubiquitous membrane-bound DNase activity in podosomes and invadopodia. J Cell Biol 220(7):e202008079. https://doi.org/10.1083/jcb.202008079. PMID: 33904858; PMCID: PMC8082437

Parekh A, Weaver AM (2016) Regulation of invadopodia by mechanical signaling. Exp Cell Res 343(1):89–95. https://doi.org/10.1016/j.yexcr.2015.10.038. Epub 2015 Nov 4. PMID: 26546985; PMCID: PMC4851576

Poirier MB, Fiorino C, Rajasekar TK, Harrison RE (2020) F-actin flashes on phagosomes mechanically deform contents for efficient digestion in macrophages. J Cell Sci 133(12):jcs239384. https://doi.org/10.1242/jcs.239384

Quilaqueo-Millaqueo N, Brown-Brown DA, Vidal-Vidal JA, Niechi I (2024) NOX proteins and ROS generation: role in invadopodia formation and cancer cell invasion. Biol Res 57(1):98. https://doi.org/10.1186/s40659-024-00577-z. PMID: 39696702; PMCID: PMC11657503

Qureshi T, Desale SE, Chidambaram H, Chinnathambi S (2024) Understanding actin remodeling in neuronal cells through podosomes. Methods Mol Biol 2761:257–266. https://doi.org/10.1007/978-1-0716-3662-6_18

Qureshi T, Chandrashekar M, Ananthanarayana V, Kumarappan M, Rangappa N, Velmurugan G, Chinnathambi S (2025) Nuclear podosomes regulates cellular migration in tau and Alzheimer's disease. Adv Protein Chem Struct Biol 143:411–426. https://doi.org/10.1016/bs.apcsb.2024.09.009. Epub 2024 Oct 2

Schachtner H, Calaminus SD, Thomas SG, Machesky LM (2013) Podosomes in adhesion, migration, mechanosensing and matrix remodeling. Cytoskeleton (Hoboken) 70(10):572–589. https://doi.org/10.1002/cm.21119. Epub 2013 Jul 19

Schelling JR, Abu Jawdeh BG (2008) Regulation of cell survival by Na+/H+ exchanger-1. Am J Physiol Renal Physiol 295(3):F625–F632. https://doi.org/10.1152/ajprenal.90212.2008. Epub 2008 May 14. PMID: 18480176; PMCID: PMC2653110

Sens KL, Zhang S, Jin P, Duan R, Zhang G, Luo F, Parachini L, Chen EH (2010) An invasive podosome-like structure promotes fusion pore formation during myoblast fusion. J Cell Biol 191(5):1013–1027. https://doi.org/10.1083/jcb.201006006. Epub 2010 Nov 22. PMID: 21098115; PMCID: PMC2995175

Shiokawa D, Matsushita T, Shika Y, Shimizu M, Maeda M, Tanuma S (2007) DNase X is a glycosylphosphatidylinositol-anchored membrane enzyme that provides a barrier to endocytosis-mediated transfer of a foreign gene. J Biol Chem 282(23):17132–17140. https://doi.org/10.1074/jbc.M610428200. Epub 2007 Apr 6

Sit B, Gutmann D, Iskratsch T (2019) Costameres, dense plaques and podosomes: the cell matrix adhesions in cardiovascular mechanosensing. J Muscle Res Cell Motil 40(2):197–209. https://doi.org/10.1007/s10974-019-09529-7. Epub 2019 Jun 18. PMID: 31214894; PMCID: PMC6726830

Staszowska AD, Fox-Roberts P, Foxall E, Jones GE, Cox S (2017) Investigation of podosome ring protein arrangement using localization microscopy images. Methods 15(115):9–16. https://doi.org/10.1016/j.ymeth.2016.11.005. Epub 2016 Nov 10

Sung BH, Weaver A (2011) Cell-cell fusion: a new function for invadosomes. Curr Biol 21(3):R121–R123. https://doi.org/10.1016/j.cub.2010.12.024. PMID: 21300277; PMCID: PMC3596881

Takito J, Nakamura M (2012) Precursors linked via the zipper-like structure or the filopodium during the secondary fusion of osteoclasts. Commun Integr Biol 5(5):453–457. https://doi.org/10.4161/cib.20980. PMID: 23181159; PMCID: PMC3502206

Takito J, Nakamura M, Yoda M, Tohmonda T, Uchikawa S, Horiuchi K, Toyama Y, Chiba K (2012) The transient appearance of zipper-like actin superstructures during the fusion of osteoclasts. J Cell Sci 125(Pt 3):662–672. https://doi.org/10.1242/jcs.090886. Epub 2012 Feb 20

Takito J, Otsuka H, Inoue S, Kawashima T, Nakamura M (2017) Symmetrical retrograde actin flow in the actin fusion structure is involved in osteoclast fusion. Biol Open 6(7):1104–1114. https://doi.org/10.1242/bio.025460. PMID: 28711870; PMCID: PMC5550915

Tarone G, Cirillo D, Giancotti FG, Comoglio PM, Marchisio PC (1985) Rous sarcoma virus-transformed fibroblasts adhere primarily at discrete protrusions of the ventral membrane called podosomes. Exp Cell Res 159(1):141–157. https://doi.org/10.1016/s0014-4827(85)80044-6

Tertrais M, Bigot C, Martin E, Poincloux R, Labrousse A, Maridonneau-Parini I (2021) Phagocytosis is coupled to the formation of phagosome-associated podosomes and a transient disruption of podosomes in human macrophages. Eur J Cell Biol 100(4):151161. https://doi.org/10.1016/j.ejcb.2021.151161. Epub 2021 Mar 31

van den Dries K, Meddens MB, de Keijzer S, Shekhar S, Subramaniam V, Figdor CG, Cambi A (2013) Interplay between myosin IIA-mediated contractility and actin network integrity orchestrates podosome composition and oscillations. Nat Commun 4:1412. https://doi.org/10.1038/ncomms2402. PMID: 23361003; PMCID: PMC3562466 van den Dries K, Linder S, Maridonneau-Parini I, Poincloux R (2019a) Probing the mechanical landscape – new insights into podosome architecture and mechanics. J Cell Sci 132(24):jcs236828. https://doi.org/10.1242/jcs.236828

van den Dries K, Nahidiazar L, Slotman JA, Meddens MBM, Pandzic E, Joosten B, Ansems M, Schouwstra J, Meijer A, Steen R, Wijers M, Fransen J, Houtsmuller AB, Wiseman PW, Jalink K, Cambi A (2019b) Modular actin nano-architecture enables podosome protrusion and mechanosensing. Nat Commun 10(1):5171. https://doi.org/10.1038/s41467-019-13123-3. PMID: 31729386; PMCID: PMC6858452

Vorselen D, Barger SR, Wang Y, Cai W, Theriot JA, Gauthier NC, Krendel M (2021) Phagocytic 'teeth' and myosin-II 'jaw' power target constriction during phagocytosis. elife 28(10):e68627. https://doi.org/10.7554/eLife.68627. PMID: 34708690; PMCID: PMC8585483

Walde M, Monypenny J, Heintzmann R, Jones GE, Cox S (2014) Vinculin binding angle in podosomes revealed by high resolution microscopy. PLoS One 9(2):e88251. https://doi.org/10.1371/journal.pone.0088251. PMID: 24523880; PMCID: PMC3921150

Weaver AM (2008) Invadopodia. Curr Biol 18(9):R362–R364. https://doi.org/10.1016/j.cub.2008.02.028

Weber K, Hey S, Cervero P, Linder S (2022) The circle of life: phases of podosome formation, turnover and reemergence. Eur J Cell Biol 101(2):151218. https://doi.org/10.1016/j.ejcb.2022.151218. Epub 2022 Mar 23

Wiesner C, Faix J, Himmel M, Bentzien F, Linder S (2010) KIF5B and KIF3A/KIF3B kinesins drive MT1-MMP surface exposure, CD44 shedding, and extracellular matrix degradation in primary macrophages. Blood 116(9):1559–1569. https://doi.org/10.1182/blood-2009-12-257089. Epub 2010 May 26

Wyburn KR, Jose MD, Wu H, Atkins RC, Chadban SJ (2005) The role of macrophages in allograft rejection. Transplantation 80(12):1641–1647. https://doi.org/10.1097/01.tp.0000173903.26886.20

Zhu H, Wang J, Nie W, Armando I, Han F (2021) ADAMs family in kidney physiology and pathology. EBioMedicine 72:103628. https://doi.org/10.1016/j.ebiom.2021.103628. Epub 2021 Oct 12. PMID: 34653870; PMCID: PMC8517843

Chapter 5
Models of Cellular Mechanosensation

David Dolgitzer and Pablo A. Iglesias

Abstract Over the past decade, theoretical and experimental work in mechanobiology has become increasingly interdependent, with modeling now serving as an essential tool for interpreting experiments and guiding measurements. Cellular mechanosensation, the conversion of mechanical cues into biochemical signals, governs essential processes such as migration, polarization, lineage specification, and tissue homeostasis, and its dysregulation contributes to disease. Quantitative models of mechanosensation have been central to clarifying how cells detect and respond to force across multiple scales, from ion channels and molecular bonds to adhesions, cytoskeletal dynamics, nuclear deformation, and transcriptional regulators such as YAP/TAZ (Yes-associated protein and transcriptional coactivator with PDZ-binding motif). In this chapter, we survey both experimental and theoretical advances in mechanosensation, emphasizing how mathematical and computational models complement experiments, expose limitations, and generate testable hypotheses and conceptual frameworks.

5.1 Introduction

Cellular mechanosensation has come to be recognized as a key regulator of diverse physiological processes, from development and tissue homeostasis to specialized sensory functions and disease. For instance, our sense of hearing relies on hair cells that operate as "biological strain gauges," converting mechanical forces into electrical signals by activating ion channels and generating nerve impulses to the brain (Hudspeth 1989). Beyond highly specialized systems, mechanical cues in the

D. Dolgitzer
Department of Physics and Astronomy, Johns Hopkins University, Baltimore, MD, USA
e-mail: ddolgit1@jhu.edu

P. A. Iglesias (✉)
Department of Electrical & Computer Engineering, Johns Hopkins University, Baltimore, MD, USA
e-mail: pi@jhu.edu

M. Kloc, J. Wosik (eds.), *Mechanotransduction and Mechanosensing*, Results and Problems in Cell Differentiation 78, https://doi.org/10.1007/978-3-032-19509-8_5

cellular environment also profoundly influence cell behavior: substrate rigidity and microenvironmental stiffness can drive cellular polarization, motility, and lineage decisions in stem cells (Pelham and Wang 1997; Engler et al. 2006). Likewise, spatial confinement in micropatterned substrates can switch a cell's response from growth to apoptosis (Chen et al. 1997).

Mechanosensation also plays a significant role in numerous pathological contexts. In the heart, mutations in titin, which contributes to mediating cardiomyocyte stress and stretch signaling, lead to the development of hypertrophic cardiomyopathy (Linke 2008; Neiva-Sousa et al. 2015). In the vascular system, Piezo1 responds to forces such as fluid shear stress, stretch, and extracellular matrix stiffness and is a key contributor to atherosclerosis (Lan et al. 2024). Mutations in genes encoding polycystin, transmembrane proteins that sense fluid flow, are primary contributors to autosomal dominant polycystic kidney disease (ADPKD), a disease that affects about 1:1000 individuals worldwide and can eventually result in dialysis or kidney transplantation (Boletta and Caplan 2025). In persons carrying these mutations, the loss of mechanosensory function necessary for normal tissue morphogenesis leads directly to ADPKD (Nauli et al. 2003). Defects in mechanosensation also contribute to muscular dystrophy: Disruption of the dystrophin-glycoprotein complex (DGC), which links the actin cytoskeleton to the extracellular matrix, weakens muscle fibers and makes them more susceptible to damage from ordinary mechanical forces, ultimately leading to a spectrum of muscular dystrophies (Wilson et al. 2022).

In cancers, aberrant mechanosensation governs nearly every stage of tumor development, from growth and invasion to metastasis, immune evasion, and drug resistance (Discher et al. 2017; Lichtenberg et al. 2023; Pethő et al. 2019; Zanotelli et al. 2021). Abnormal tissue stiffness and changes in the extracellular matrix composition can drive tumor cell proliferation through the activation of mechanosensitive signaling pathways such as YAP/TAZ and integrins (Cai et al. 2021; Dupont et al. 2011; Wada et al. 2011). Mechanical cues from the tumor microenvironment also promote invasion and metastasis by facilitating extracellular matrix remodeling and epithelial-to-mesenchymal transition, enabling cancer cells to migrate and colonize distant organs (Kumar and Weaver 2009). Furthermore, altered mechanical properties can impede immune cell infiltration and modulate immune responses, helping tumors escape immune surveillance (Zhu et al. 2023; Golo et al. 2024). Finally, increased matrix rigidity and dysregulated mechanotransduction contribute to drug resistance by interfering with drug penetration and activating pathways that support cancer cell survival (Kalli et al. 2023). These multifaceted roles of mechanosensation highlight the importance of physical cues in tumor progression and point to promising avenues for therapeutic intervention.

Mechanosensation, by its nature, sits at the intersection of biology and physics. Because mechanical signaling involves complex, multiscale phenomena ranging from molecular interactions to tissue-level forces, computational and theoretical models have become essential tools for unraveling the mechanisms underlying cellular mechanosensation. In the sections that follow, we review major computational approaches that have shaped our current understanding, highlighting how biophysical models complement experimental discovery.

5.2 Force-Dependent Molecular Bonds

In the ideal model, the binding of two molecules is described by constant on- and off- binding rates: k_{on} and k_{off}, respectively. However, many such interactions are known to be affected by mechanical forces, resulting in a mechanosensitive bond. In most cases, applying force to the bond weakens it (k_{off} increases); these are known as *slip* bonds. A *catch* bond shows the opposite behavior: moderate pulling forces stabilize the bond, prolonging its lifetime (k_{off} decreases).

There is ample experimental evidence for both slip and catch bond behavior. These include cell adhesion molecules such as P- and L-selectins that mediate leukocyte rolling as wall shear stress varies (Alon et al. 1995; Finger et al. 1996; Marshall et al. 2003), cadherins that facilitate cells to transduce mechanical forces at cell-cell junctions (Buckley et al. 2014), and integrins that mediate cell-substrate interactions (Kong et al. 2009). A variety of cytoskeletal motors, which convert the energy released by ATP hydrolysis into mechanical movement, also show force dependence (Choi and Zhu 2025). These include dynein (Rao et al. 2019), kinesin (Visscher et al. 1999), and various myosins (Greenberg et al. 2016; Reichl et al. 2008).

In practice, most catch bonds eventually display biphasic behavior, transitioning into slip (Pereverzev et al. 2005), or even triphasic behavior (Barkan and Bruinsma 2024). For example, E-selectin and integrin form slip-catch-slip bonds (Kong et al. 2009; Wayman et al. 2010).

5.2.1 *The Bell Model*

In 1978, George Bell proposed a mathematical model to explain slip bond behavior (Bell 1978). In this framework, mechanical forces alter the rate of bond dissociation exponentially, as follows:

$$k_{\text{off}}(F) = k_{0,\text{slip}} e^{F\delta_{\text{slip}}/k_B T}, \qquad \delta_{\text{slip}} > 0.$$

Here, $k_{\text{off}}(F)$ is the force-dependent off-rate, $k_{0,\text{slip}}$ is the dissociation rate in the absence of force, F is the applied force, $k_B T$ is the thermal energy, and δ_{slip} is a bond-specific parameter, sometimes known as the bond length or distance, that describes how mechanosensitive the bond is. In this model, an increase in force makes the exponential term larger, leading to a faster dissociation rate and a shorter bond lifetime.

Though originally used to predict slip bond behavior, Bell's model can also be used to explain catch bond behavior if it incorporates a negative value in the exponential.

$$k_{\text{off}}(F) = k_{0,\text{catch}} e^{F\delta_{\text{catch}}/k_B T}, \qquad \delta_{\text{catch}} < 0.$$

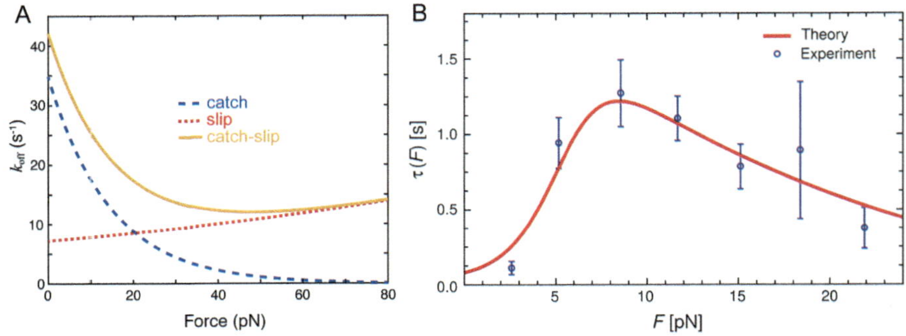

Fig. 5.1 Catch-slip bond behavior. (**a**) Force-dependent dissociation (k_{off}) rates for slip, catch, and catch-slip bond models for L-selectin under constant force (Pereverzev et al. 2005). (**b**) Cadherin-catenin complex/F-actin bond lifetimes. The experimental data comes from Buckley et al. (2014); the model (solid line) and the figure are from Adhikari et al. (2018). Reprinted with permission under the terms of the Creative Commons Attribution License

In the case of a catch-slip bond, the dissociation constant of the bound state at a given force is given by the sum of the two rate constants:

$$k_{\text{off}}(F) = k_{0,\text{slip}} e^{F\delta_{\text{slip}}/k_B T} + k_{0,\text{catch}} e^{F\delta_{\text{catch}}/k_B T}.$$

This formula has been used to fit binding rates for L-selectin and P-selectin under constant forces, with values of δ_{catch} of -0.28 and -2.17 nm, δ_{slip} of 0.034 and 0.51 nm, $k_{0,\text{catch}}$ of 34.9 and 120 s^{-1}, and $k_{0,\text{slip}}$ of 7.2 and 0.25 s^{-1} (Pereverzev et al. 2005) (Fig. 5.1).

Bell's model has proved a powerful tool for incorporating mechanosensitivity into mathematical models. It has the advantage of preserving the traditional form of binding equations, albeit with force-dependent kinetic constants. It should be noted, however, that other possibilities exist. For example, in describing catch bonds, Dembo et al. suggested that the exponential should depend on the *square* of the force (Dembo et al. 1988). Alternatively, Evans and Ritchie posited that the off-rate follows a power law at low forces (Evans and Ritchie 1997).

5.2.2 Molecular Motors Under Load

Force-driven conformational changes in cytoskeletal proteins can affect protein function. Increasing the load on molecular motors increases the duty ratio, meaning the motor spends more time bound to its substrate and extends the duration of its mechanochemical cycle (Sala et al. 2024). For myosin II motors, the size and speed of the power stroke are reduced under load (Reconditi et al. 2004). Human non-muscle myosins IIA and IIB under increased load exhibit load-dependent changes in ADP release kinetics, but not in nucleotide binding (Kovács et al. 2007). These load-dependent adaptations in the mechanochemical behavior of myosin are crucial for modulating its roles in cellular contractility and cargo transport (Hartman and Spudich 2012). Similarly, dynein moving against a load decreases its step size between one-third to one-quarter that of an unloaded motor and produces a force up to 1.1 pN (Mallik et al. 2004).

5.3 The Mechanical Role of Myosin II

Perhaps the protein most closely associated with cellular contractility is myosin II, first identified in muscle cells in the early 1950s and later in non-muscle cells (Huxley and Niedergerke 1954; Clarke and Spudich 1974). When bound to actin in the cell cortex to form actomyosin, myosin II serves as a primary force generator in many eukaryotic cells, driving shape change, motility, and division. Elevated myosin II levels are often found in regions undergoing remodeling or requiring increased stiffness, for example, at the cleavage furrow during cytokinesis and at the rear of a contracting amoeba.

In addition to its role as a contractile motor, experimental evidence implicates myosin II in mechanosensation. Effler et al. (2006) showed that when mitotic *Dictyostelium* cells were subjected to localized stress via micropipette aspiration, myosin II and cortexillin I redistributed from the cleavage furrow to the site of stress (Fig. 5.2a), enabling cells to escape the pipette and complete cytokinesis. Cortexillin I cooperates with myosin II by crosslinking actin filaments, stabilizing the network and allowing myosin II to respond to tension (Luo et al. 2012).

Because its actin binding is load dependent, myosin II was a natural candidate to serve as both sensor and actuator. Experiments with motor and lever-arm mutants showed that lever-arm length controls the cell's ability to detect and respond to load: longer arms enhanced, shorter arms reduced, and myosin II accumulated at stressed sites (Ren et al. 2009) (Fig. 5.3b). Likewise, mutations that reduced motor velocity suppressed myosin II accumulation, supporting its dual roles as force sensor and contractile effector. Myosin II's initial assembly has also been linked to its mechanoresponsiveness (Schiffhauer et al. 2019). Experiments across multiple mammalian cell types and phenotypes compared the initial myosin IIB percent in the cytoskeleton to its accumulation ratio post-stress, finding a biphasic dependency (Fig. 5.2c).

These findings suggest that mechanical stress locks myosin II onto actin, promoting increased bipolar thick filament (BTF) formation (Fig. 5.3a). Further work confirmed that myosin II is not the only protein responding to mechanical load, but it remains central. Luo et al. (2013) combined experiments with a reaction-diffusion models to show that myosin II and α-actinin responded specifically to dilation, accumulating at the tip of the deformed cell (Fig. 5.3b), whereas filamin responded to shear stress, localizing along the neck of the pipette. Moreover, cortexillin I required myosin for its localization, while α-actinin accumulated only in the absence

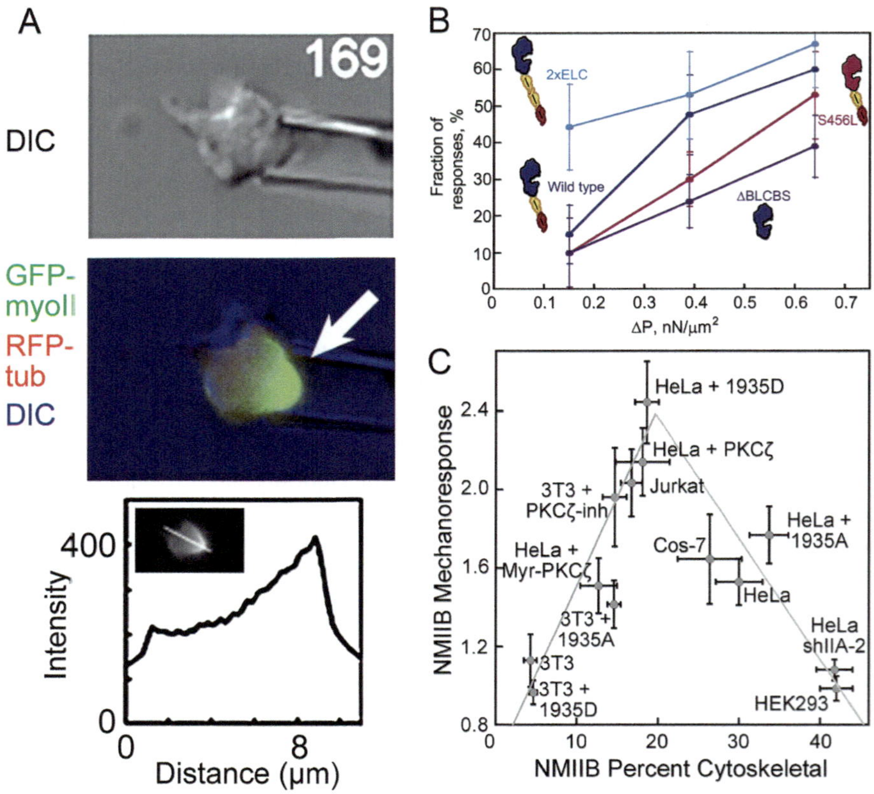

Fig. 5.2 **Myosin II mechanoaccumulation**. (**a**) Myosin II shows higher intensity at the site of stress during micropipette aspiration. Adapted images from Effler et al. (2006), copyright © 2006 Elsevier Ltd. Used by permission. (**b**) Fraction of cells showing myosin II response across lever-arm mutations. At low pressure, 2xELC is more responsive than wild type. At high pressure, *Δ*BLCBS is less responsive than wild type. Image from (Ren et al. 2009), copyright © 2009 Elsevier Ltd. Used by permission. (**c**) myosin IIB response shows a biphasic dependence on its initial cortical percentage across mammalian cell experiments. Image from Schiffhauer et al. (2019), copyright © 2019 Schiffhauer et al. Available under a Creative Commons License (Attribution-Noncommercial-Share Alike 4.0 International license

of myosin, highlighting myosin II's role as the primary mechanoresponsive element even as other proteins contribute to stress-specific pathways.

5.3.1 Higher-Order Complexes: Bipolar Thick Filaments

BTFs increase load-bearing capacity and engage parallel actin structures. From a modeling perspective, a common starting point is a Bell-like myosin-actin binding

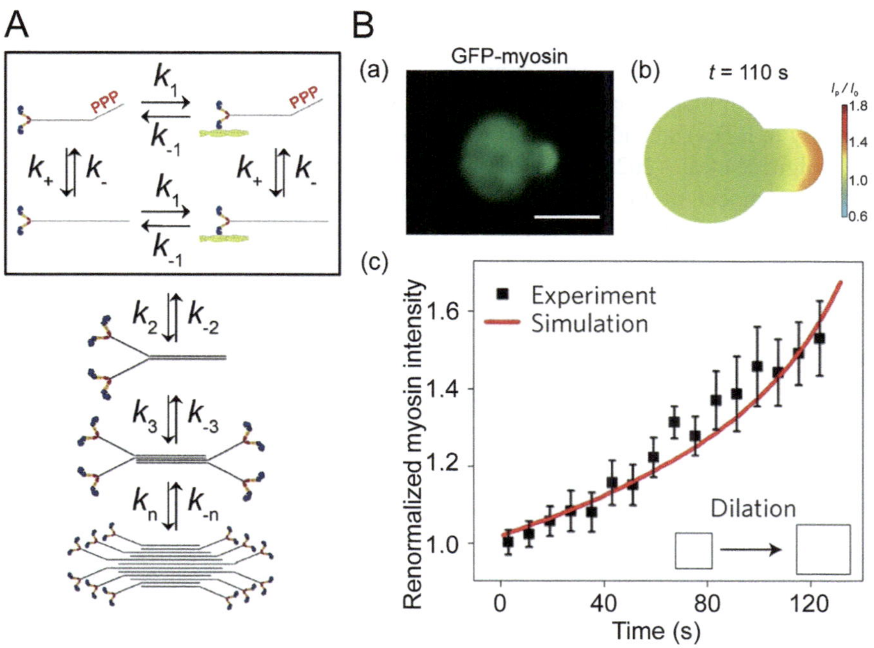

Fig. 5.3 BTF model of myosin II accumulation. (**a**) Bipolar thick filament formation. Myosin II monomers bind to actin and form dimers, tetramers, and BTFs. Phosphorylated myosin II can bind to actin but cannot form BTFs. Image from (Schiffhauer et al. 2019), copyright © 2019 Schiffhauer et al. Available under a Creative Commons License (Attribution-Noncommercial-Share Alike 4.0 International license. (**b**) Myosin II accumulation over time: (**a**) experimental (scale bar is 10 μm); (**b**) simulated myosin II accumulation 110 seconds into aspiration; (**c**) normalized myosin II intensity increases over time in response to dilation stress inside the pipette (right). Adapted images from Luo et al. (2013), copyright © 2013, Springer Nature Limited. Used by permission

reaction in which applied force increases bond lifetime, leading to local myosin enrichment. Initial simulations pointed to a force-sensitive step during the transition between assembly-incompetent and -competent states (Ren et al. 2009).

Building on early models of BTF formation, Luo et al. examined cooperative interactions between myosin II and actin crosslinkers during mechanosensation (Luo et al. 2012). They considered a four-state model in which myosin monomers are either competent or incompetent for thick-filament assembly and either bound or unbound to actin. Conversions between incompetent and competent states are governed by myosin heavy-chain kinase (MHCK) and a phosphatase (Egelhoff et al. 1993). They assumed that BTF assembly is much faster in the presence of abundant actin filaments, so dynamics are dominated by the actin-bound pathway. Mechanosensitivity was implemented by lowering the off-rate in a load-dependent manner via a Bell-type term, thereby accelerating BTF formation.

Schiffhauer et al. (2016) quantified the relative contributions of passive diffusion and active stress-driven recruitment, using a stochastic kinetic framework to show an optimal—so-called Goldilocks—range of actin-binding affinity for maximal mechanoaccumulation: if binding is too weak, interactions are transient; if too strong, mobility is lost. This approach was extended to mammalian isoforms, revealing a similar biphasic relationship between baseline cortical myosin II levels (largely in BTFs) and additional accumulation under load.

Dolgitzer et al. (2025) reproduced these findings using a continuum reaction-diffusion model of myosin II transport to, and assembly within, the cortex in *Dictyostelium*. In the model, myosin II accumulated less when BTF assembly was reduced by phosphorylation and more when assembly was promoted. Across these studies, BTF assembly is treated not as an independent process but as a load-modulated kinetic step nested within cortical recruitment, providing the amplification that converts prolonged force-stabilized binding into the high local contractility required for effective mechanosensation.

5.3.2 Higher-Order Complexes: Contractility Kits

Recent work has clarified how myosin II reaches the cortex before assembly. In addition to locally formed BTFs, *Dictyostelium* cells contain higher-order cytosolic complexes—contractility kits (CKs)—comprising myosin II, the actin crosslinker cortexillin I, and the regulatory proteins IQGAP1 and IQGAP2 (Kee et al. 2012; Kothari et al. 2019) (Fig. 5.4a,b). To understand how these assemblies form, Plaza-Rodríguez et al. (2022) developed a particle-based simulation of CK assembly in SpringSalad, parameterized by experimentally measured affinities (Michalski and Loew 2016). These simulations predicted three CK classes based on IQGAP composition: IQGAP2-containing (mechanoresponsive), IQGAP1-containing (non-mechanoresponsive), and mixed (ambiguous). The first two had been observed previously; the mixed class was subsequently confirmed experimentally. The model also indicated that CKs form primarily in the cytosol before translocating to and binding the membrane.

Dolgitzer and coworkers developed a continuum reaction-diffusion model that combined the BTF accumulation framework with CK data (Dolgitzer et al. 2025). The model showed that CKs are the primary transporters of myosin II and cortexillin I to the cortex, with local transport rates enhanced by external stress. After delivery, CKs disassemble, releasing myosin monomers that form BTFs via the load-modulated assembly described above (Fig. 5.4a). This division of labor, transport via CKs followed by local BTF formation, embodies a common organizational strategy in mechanosensory systems: assemble functional modules in advance and then deploy and mature them on demand. In this way, cells avoid maintaining high-force structures everywhere and instead rely on preassembled intermediates that can be rapidly mobilized when mechanical cues arise. The model also captured the experimentally observed biphasic dependence on initial cortical myosin concentration (Fig. 5.4c).

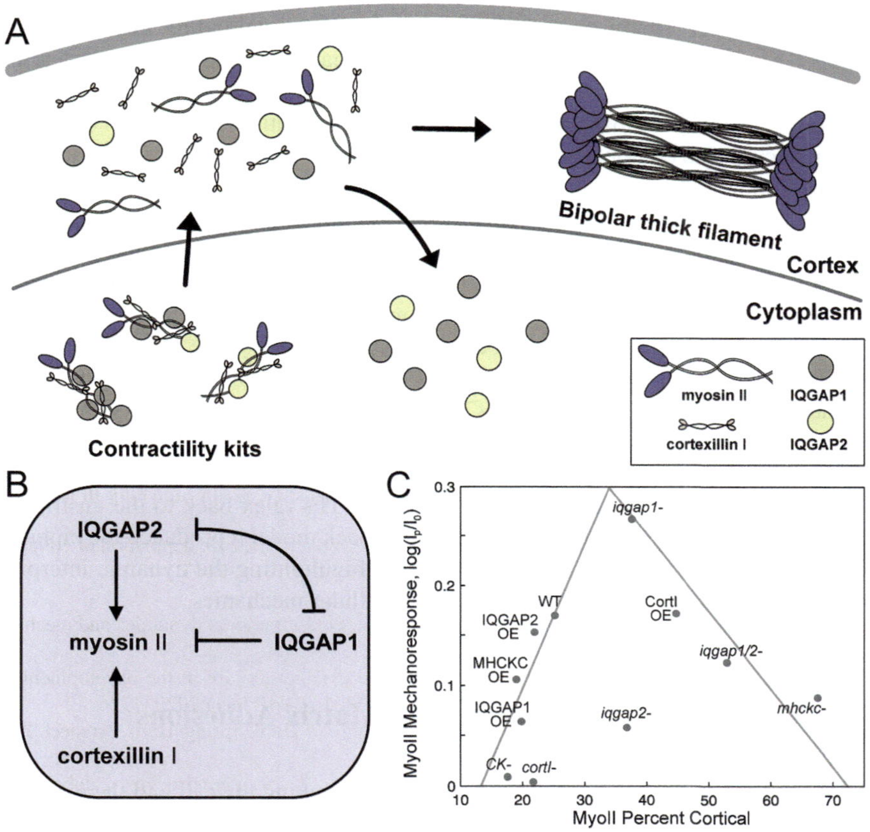

Fig. 5.4 CK model of myosin II accumulation. (**a**) Myosin II, cortexillin I, IQGAP1, and IQGAP2 form CKs in the cytosol and disassemble in the cortex into their monomeric components. Then, myosin II monomers form BTFs, and the IQGAPs diffuse back to the cytosol. Cortexillin I's actin crosslinking is implicit. (**b**) A schematic of the core CK mechanism: myosin II and cortexillin I serve as the primary mechanosensors, while IQGAP2 and IQGAP1 provide positive and negative regulation of mechanoresponsiveness, respectively, and compete for binding the CK complex. (**c**) Myosin II response shows a biphasic dependence on its initial cortical percentage across null and overexpression simulations using a CK-centric model. Adapted image from Dolgitzer et al. (2025), copyright © 2024 Biophysical Society

5.3.3 Feedback Regulation

The models described above focus on protein accumulation at the site of stress, typically imposed by micropipette aspiration. This accumulation is followed by increased contractility that allows the cell to counteract the load. To address this coupling, Mohan et al. (2015) developed a feedback model with several key elements.

The model uses reaction-diffusion equations for BTF assembly and myosin II accumulation, incorporating mechanical stress into the unbinding rate via a Bell model (Luo et al. 2012, 2013). It further assumes that as myosin II accumulates, the applied stress is shared among more bound heads, weakening the Bell effect and creating a negative feedback that slows further accumulation.

The model also computes stresses due to myosin II contractility, surface tension, and aspiration and applies them to a viscoelastic model of cell deformation (Yang et al. 2008). As myosin II accumulates, the resulting contractility counterbalances the aspiration pressure. Simulations showed that this force feedback reduces myosin II accumulation, but alone it could not fully reproduce the biphasic accumulation or the subsequent cell retraction (Mohan et al. 2015).

To capture these features, the model was extended with a two-state description of BTF strain-stiffening and relaxation. BTFs exist in unstrained and strained states. In the absence of external force, BTFs remain unstrained at their steady-state concentration. Immediately after aspiration is applied, the concentration of strained BTFs increases—proportional to the aspiration rate at the cell tip—while that of unstrained BTFs decreases. Over time, strained BTFs relax back to the unstrained state. Simulations of the full biomechanical feedback model reproduced the biphasic myosin II accumulation and cellular retraction, highlighting the dynamic interplay between mechanosensation, contractility, and cellular mechanics.

5.4 Mechanosensing Through Cell-Matrix Adhesions

Cellular sensing of the microenvironment requires linking the cell and the extracellular matrix (ECM) mechanically. In large part, this is achieved through integrin-based focal adhesion sites (Chrzanowska-Wodnicka and Burridge 1996; Kong et al. 2009; Roca-Cusachs et al. 2009); structures that serve as mechanosensitive centers that translate external mechanical cues into intracellular biochemical signals. Focal adhesions strengthen and increase in size proportionally with applied force (Choquet et al. 1997; Balaban et al. 2001), and cells respond to substrate stiffness through changes in focal adhesion assembly (Pelham and Wang 1997). The mechanical coupling is mediated by integrins, which are dynamically linked to the actomyosin cytoskeleton via adaptor proteins such as α-actinin, talin, and vinculin (Kechagia et al. 2019; del et al. 2009).

5.4.1 The Molecular Clutch Framework

The concept of the "molecular clutch" describes the mechanical linkage formed by dynamic associations between ECM-bound integrins and the force-generating actomyosin cytoskeleton (Mitchison and Kirschner 1988). This framework has proven fundamental for understanding mechanosensation at adhesion sites. Integrins exhibit catch-bond behavior, whereby moderate forces strengthen integrin-ligand interactions (Kong et al. 2009). Computational models suggest feedback between mechanical regulation of integrin conformation and binding dynamics: applied force activates integrins, which increases their affinity for ligands and enhances catch-bond behavior (Welf et al. 2012).

Force transmission through the molecular clutch is regulated by the mechanical properties of key adaptor proteins. On stiff substrates, rapid mechanical loading rates induce partial talin unfolding, exposing cryptic vinculin binding sites. Vinculin binding reinforces the clutch by elevating force transmission. Conversely, on soft ECM substrates that deform easily, slow mechanical loading rates are insufficient to induce vinculin-dependent clutch reinforcement, maintaining low force transmission levels.

5.4.2 *The Motor-Clutch Model and Extensions*

Chan and Odde (2008) developed the "clutch-motor" model of force transmission. Molecular clutches and the substrate behave as Hookean springs with constants k_c and k_s, respectively. In the model, molecular clutches stochastically engage the F-actin bundle with characteristic on-and off-rates (Mitchison and Kirschner 1988). Clutches that engage build tension as they are stretched by the retrograde F-actin motion, increasing their off-rate constant, k_{off}, according to Bell's Law:

$$k_{\text{off}} = k_0 e^{F_{\text{clutch}} \delta_b / k_B T}$$

Here, the characteristic bond rupture force is given by $F_b = k_B T / \delta_b$, $F_{\text{clutch}} = k_c(x_i - x_{\text{sub}})$ is the tension within an individual clutch, and x_i and x_{sub} are the positions of the molecular clutch and substrate, respectively.

Tension in the engaged molecular clutches generates a traction force that is balanced by tension and deformation in the compliant substrate. This is accomplished by updating the position of the substrate according to

$$x_{sub} = \frac{k_{\text{clutch}} \sum_{i=1}^{n_{\text{eng}}} x_i}{k_{\text{sub}} + n_{\text{eng}} \times k_{\text{clutch}}}$$

where n_{eng} is the number of engaged molecular clutches. Myosin motors work against this load force, slowing their motor sliding velocity according to a linear force-velocity relation:

$$v_{\text{filament}} = v_u \left(1 - \frac{k_{\text{sub}} x_{\text{sub}}}{F_{\text{stall}}}, \right).$$

where v_u is the unloaded sliding velocity of myosin, $F_{\text{stall}} = n_m F_m$, F_m is the force required to stall the activity of one myosin motor, and n_m is the number of

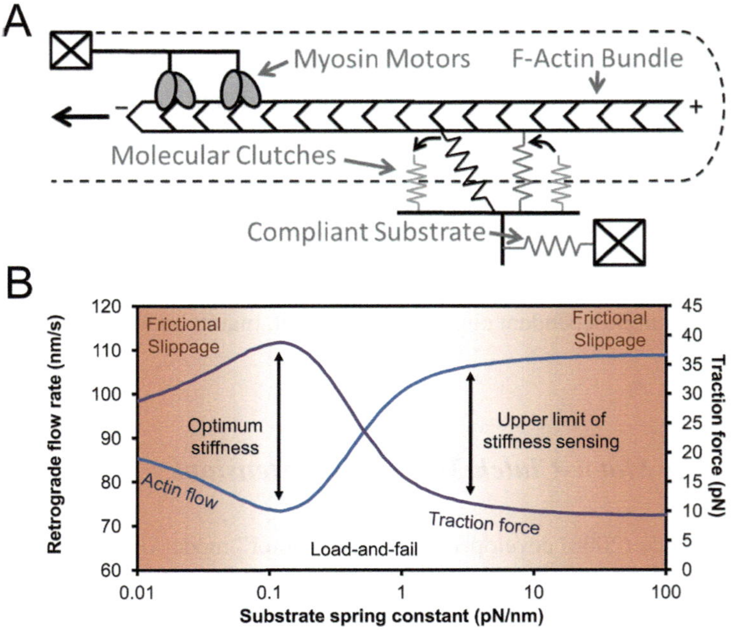

Fig. 5.5 **Motor clutch model.** (**a**) Schematic of the model. Image from Bangasser and Odde (2013), copyright © 2013 Biophysical Society. Used with permission. (**b**) The model from Bangasser et al. predicts varying regimes of stiffness sensing, depending on the substrate spring constant. Two represent regions of frictional slippage, one each on soft and stiff substrates. Between them lies a regime of load-and-fail without frictional slippage. Image from Bangasser et al. (2013), copyright © 2013, Biomedical Engineering Society. Used with permission

motors in the system. Finally, actin polymerizes at the plus end of the F-actin bundle with constant velocity to counter-retrograde flow and maintain the F-actin bundle (Fig. 5.5a).

The model predicts two distinct regions. Whereas linkages show "frictional slippage" due to constant clutch rupture under high mechanical loading rates on stiff surfaces, they display oscillatory "load-and-fail" dynamics, with slower retrograde flow and higher traction forces on soft substrates (Fig. 5.5b). These predictions were tested and confirmed by measuring the nanoscale dynamics of single-growth-cone filopodia in neurons. Subsequently, the model showed that there exists a cell-dependent optimum substrate stiffness (Bangasser et al. 2013).

The basic framework has been expanded over the last fifteen years. For example, Bangasser et al. (2017) incorporated stiffness-dependent behavior and showed how optimal substrate stiffness shifts under different cellular conditions. While the response across the full stiffness range remains biphasic, it could appear monotonic within experimental ranges due to shifting optimal stiffness. This helped explain why fibroblasts show monotonic actin flow-rigidity relations, while neurons exhibit biphasic responses.

Elosegui-Artola et al. (2014) used experimentally measured binding rates between fibronectin and various integrins and included reinforcement of the strengthening and growth of cell-ECM adhesions, commonly observed after force application, by increasing integrin density each time the force in any integrin-FN bond surpassed a threshold before breakage. Subsequent refinements have incorporated mechanochemical aspects, including talin unfolding and vinculin recruitment (Elosegui-Artola et al. 2016). These models revealed talin as a mechanosensitive clutch that exposes vinculin-binding sites under force, with integrin-specific dynamics modifying stiffness-sensing behavior.

Integrating long-range force transmission through cell-cell junctions and actin polymerization at monolayer edges to the motor clutch model predicted symmetric expansion for monolayers attached to a substrate of uniform stiffness. However, under stiffness gradients, the substrate deforms and resists polymerization more at the soft edge, biasing monolayer expansion toward the stiff end (Sunyer et al. 2016). Other extensions considered the interaction of the motor clutches with viscoelastic (Adebowale et al. 2021; Gong et al. 2018; Chaudhuri et al. 2015) or purely viscous (Bennett et al. 2018) substrates. Models incorporating substrate viscoelasticity revealed how stress relaxation influences clutch engagement and cell spreading. Simulations demonstrated that while viscoelasticity did not influence cell spreading on stiff substrates, viscosity serves to stiffen soft substrates on a timescale faster than the clutch off-rate, which enhances cell-ECM adhesion and cell spreading. Extensions to 2D surfaces have also been considered to study cellular migration in confined environments (Prahl et al. 2020), demonstrating that cells migrate toward areas where they can generate maximal traction (Isomursu et al. 2022).

Beyond the molecular clutch framework, other modeling approaches have addressed adhesion mechanosensing at different scales. For example, thermodynamic models coupled mechanical forces and elasticity to adhesion dynamics via force-induced conformational changes (Nicolas et al. 2004; Besser and Safran 2006). These models predict adhesion front and back velocities as functions of applied force and explain anisotropic growth patterns. Integrated multiscale models (Deshpande et al. 2006; Ronan et al. 2014; Cao et al. 2015) couple stress fiber dynamics with focal adhesion behavior, bridging from molecular clutches to whole-cell responses using finite element and thermodynamic frameworks. Nicolas et al. (2004) introduced stochastic models distinguishing between "plaqu" and "cable" adhesion states, while (Bruinsma 2005) applied statistical mechanics to force-induced adhesion growth.

These computational approaches collectively demonstrate that focal adhesion mechanosensing emerges from the interplay of three key factors: molecular-scale force-sensitive protein conformations, dynamic clutch engagement, and collective adhesion assembly. Together, these mechanisms provide a multiscale understanding of how cells sense and respond to their mechanical environment.

5.5 YAP/TAZ Mechanotransduction Pathways

Mechanical stimuli are well-established regulators of gene expression, fundamentally influencing cellular behavior and fate decisions. This mechanically induced transcriptional response requires the transduction of mechanical signals that originate at the cell membrane level to reach the cellular nucleus, where they can modulate gene expression programs. The mechanisms underlying this mechanotransduction process have been the subject of extensive investigation and computational modeling efforts.

Key components of this mechanotransduction machinery are YAP (Yes-associated protein) and TAZ (transcriptional coactivator with PDZ-binding motif), which function as mechanosensitive transcriptional regulators (Dupont et al. 2011; Totaro et al. 2018) by connecting mechanical signals to gene expression levels. YAP/TAZ localization and activity are sensitive to mechanical parameters of the cellular microenvironment, including the extracellular matrix (ECM) stiffness, cell shape and spreading, and cytoskeletal tension (Dupont et al. 2011). On stiff substrates, YAP/TAZ primarily localize to the nucleus and activate target genes. On soft substrates, they remain confined to the cytoplasm, preventing transcriptional activity. F-actin and Rho control this process independently of the canonical Hippo pathway, indicating that multiple mechanotransduction routes regulate YAP/TAZ (Dupont et al. 2011).

YAP/YAZ-mediated mechanosensing has attracted considerable attention in modeling. Models have been used to describe how the mechanosensitivity is initiated, how it is transduced through nuclear import and export, and how it interacts with other signaling pathways. Various modeling frameworks have also been used, including detailed deterministic Ordinary Differential Equation (ODE), stochastic, and spatial PDE models. Here we highlight some of the models; see Jafarinia et al. for a comprehensive review (Jafarinia et al. 2024).

Sun et al. (2016) developed an early computational model integrating three mechanosensitive pathways: Hippo, cytoskeletal tension, and cell-cell adhesion through α-catenin. The model links substrate stiffness to FAK phosphorylation via a Hill function and captures how F-actin sequesters large tumor suppressor (LATS) to promote nuclear YAP/TAZ localization. Key predictions include bistability at intermediate stiffness (explaining population heterogeneity) and different response times for stiffness versus cell-cell contact. The model also revealed competition between F-actin and Hippo pathways in YAP/TAZ regulation, with both pathways required to explain the experimental observations.

Scott et al. (2021) extended on this framework using spatial reaction-diffusion equations solved using finite element methods to capture how cell geometry affects YAP/TAZ localization mechanosensitivity. By incorporating substrate stiffness through its effects on F-actin polymerization and LATS activity, the model exhibited increased nuclear YAP/TAZ for higher surface-to-volume ratios, independent of substrate stiffness. The spatial framework captured subcellular heterogeneity in YAP/TAZ distribution, revealing gradients from the cell periphery to the center that depended on both cell geometry and substrate stiffness, as well as reduced mechanosensitivity in 3D versus 2D environments.

Inspired by both Sun et al. (2016); Scott et al. (2021), Francis and Rangamani (2024) developed a computational model that linked substrate stiffness to circadian clock regulation through the integration of YAP/TAZ and Myocardin-related transcription factor (MRTF) mechanotransduction pathways. They assumed mechanotransduction reaches steady state faster than circadian oscillations, allowing to compute steady-state nuclear concentrations of YAP/TAZ and MRTF algebraically. The model relates stiffer substrates with shortened circadian periods and increased oscillations. Notably, the YAP/TAZ and MRTF pathways contribute synergistically to this mechanical regulation, where disruption of either pathway alone is sufficient to abolish stiffness-dependent changes in circadian dynamics. The model successfully reproduces experimental observations showing that pharmacological inhibition of ROCK, YAP/TAZ, or MRTF eliminates the mechanical modulation of circadian rhythms.

5.6 Mechanosensitive Ion Channels

Many mechanosensors function by altering the conformational state of an associated ion channel in response to external forces. Mechanosensitive ion channels serve as key molecular mediators of mechanosensation, and their gating by mechanical stimuli has been the focus of considerable modeling efforts. Biophysically, two primary models have been proposed to describe how mechanical force gates these channels (Cox et al. 2019). The "force-from-lipids" model posits that force is transmitted through the lipid bilayer itself, which directly leads to channel gating (Cox et al. 2017) (Fig. 5.6a). Examples of the former include mechanosensitive channel of large conductance (MscL) used by bacteria to guard against osmotic shock Perozo et al. (2002; 2002). In this case, mechanosensitivity arises from an iris-like expansion of the channel caused by membrane tension (Betanzos et al. 2002).

In contrast, the "force-from-filament" model suggests that mechanical force is relayed via cytoskeletal or extracellular tethers physically connected to the mechanosensitive channels, which induce conformational changes and channel opening (Delmas and Coste 2013; Gillespie and Müller 2009; Katta et al. 2015) (Fig. 5.6b). For example, this model describes the action of hair cells in the inner ear, whereby the mechanosensitive deflection of stereocilia opens ion channels located on adjacent stereocilia, resulting in an influx of various ions (Hudspeth 1989; Müller 2008).

In these models, channels are embedded in the cell membrane, with protein conformations that change reversibly in response to membrane tension or deformation. In a two-state model in which the inactive channel is closed with probability p_c, or open with probability p_o, the Boltzmann function

$$\frac{p_o}{p_c} = \exp\left(-\Delta G / k_B T\right)$$

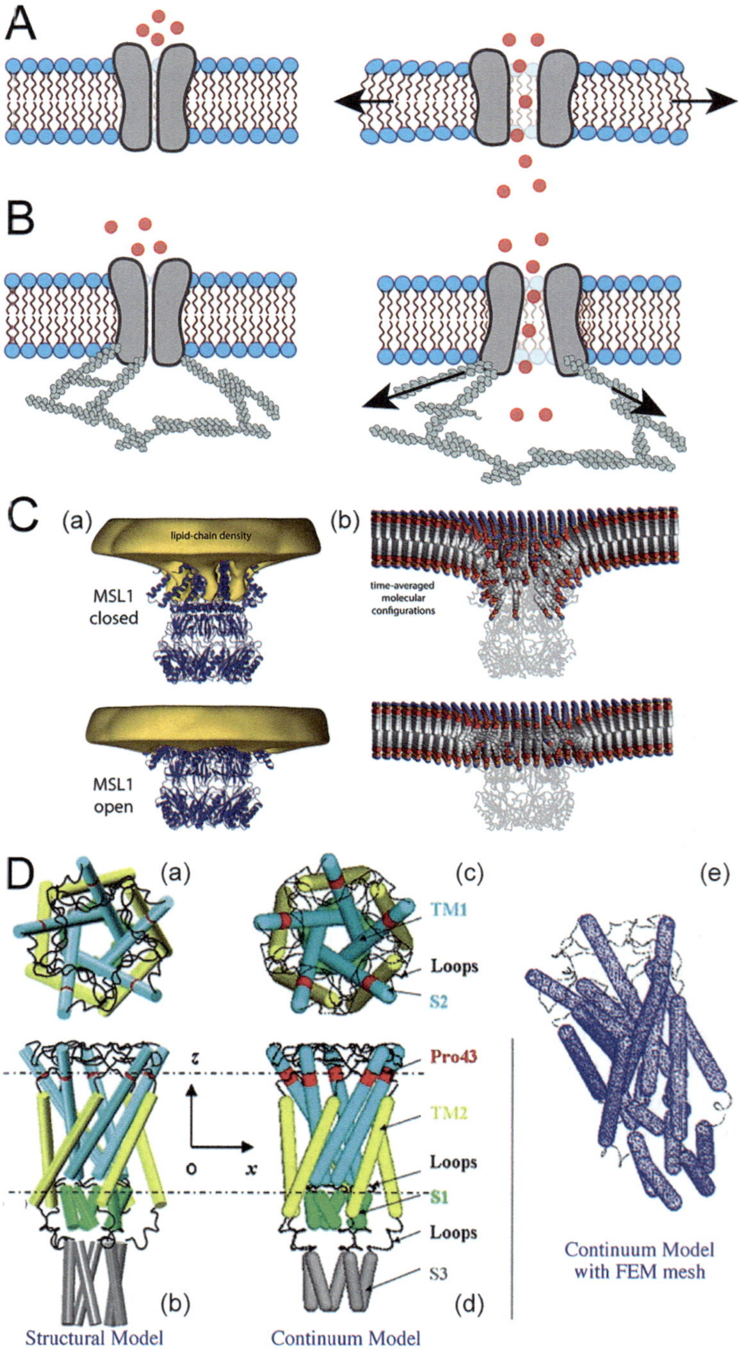

Fig. 5.6 Models of mechanosensitive ion channels. (**a**) "Force-from-lipids" channels open when membrane tension pulls on the channel. (**b**) "Force-from-filament" channels are tethered to cytosk-

relates these probabilities to the difference in free energy (ΔG) between states (Poirier and Iglesias 2007). This change in free energy is generated by the applied mechanical stimulus either through changes induced directly by the lipid bilayer or through mechanical tethering to other cellular components.

Computational approaches to modeling mechanosensitive ion channels have spanned multiple scales and techniques (Nicoletti et al. 2021). At the molecular level, atomistic modeling techniques such as molecular dynamics (MD) simulations use detailed structural data on ion channels in lipid membranes to uncover the mechanisms underlying mechanosensing (Park et al. 2023) (Fig. 5.6c). These simulations are uniquely capable of addressing the kinetics of channel gating and ion conductance in the context of varying mechanical and chemical environments. For example, studies of the TREK-2 channel, a two-pore domain potassium channel, have revealed how distinct "up" (conductive) and "down" (less conductive) conformational states respond to membrane stretch, with stretching favoring transitions toward the up state and increased conductance (Brennecke and de Groot 2018). Such investigations have also uncovered how state-dependent lipid interactions modulate channel behavior, affecting both the inner pore and selectivity filter (Aryal et al. 2017). Similarly, MD simulations of the bacterial MscL ion channel reproduce the reversible transitions of MscL to the open state with measured parameters of lateral expansion and conductivity that match exactly experimental values (Rajeshwar et al. 2021). Despite their strengths, atomistic simulations are computationally intensive and typically limited to very short simulation timescales (much less than 1 ms), restricting their ability to capture slower or collective phenomena.

To overcome some of these limitations, continuum modeling approaches have been developed to study both single channels and channel pools. Within this framework, one approach models the entire channel-membrane system in a continuum fashion, thus bridging atomic and macroscopic length scales while remaining computationally feasible (Tang et al. 2008) (Fig. 5.6d). Another approach is based on detailed kinetic modeling, describing channels as systems with discrete states and transitions whose rates depend on mechanical strain or stress, without resolving the spatial features of the system (Bavi et al. 2016; Pak et al. 2015; Zhu et al. 2019). These models are well-suited for analyzing the aggregate behavior of channels and can incorporate experimental kinetic schemes.

Multiscale modeling further integrates mechanical and electrical phenomena by connecting microscale mechanical dynamics to macroscale tissue behavior and incorporating channel kinetics into neuronal models. Such models use insights from

---

Fig. 5.6 eletal components that exert force to open the channel. (**c**) Molecular dynamics simulations of MSL1 channel in POPC membrane. (**a**) Cryo-EM structures overlaid with lipid density distributions. (b) Time-averaged lipid configurations in the membrane plane. Image from Park et al. (2023), Reprinted with permission under CC0 public domain. (**d**) Bacterial MscL channel models: (**a**), (**b**) top and side views of closed homology model; (**c**), (**d**) top and side views of a continuum model; (**e**) protein mesh with membrane position (dashed line). Image from Tang et al. (2008), Copyright Humana Press Inc. 2008. Reprinted with permission

both atomistic and continuum simulations, enabling the formulation of biophysical or phenomenological coupling laws that link force, channel state, and electrical excitability throughout soft biological tissues. For example, combining normal mode analysis, molecular dynamics, and continuum mechanics, a multiscale model demonstrated that the ankyrin repeat chains of the mechanosensitive NOMPC (No mechanoreceptor potential C) channel act as soft Hookean springs (22 pN/nm spring constant) that, when compressed, generate a rotational torque on the Transient Receptor Potential gating domain through a force-transmission mechanism involving the channel's geometric architecture.

Modeling has also been used to focus on the force-dependent conductance of MS ion channels (Xie et al. 2025). Here, increasing membrane stress shifts the channels from a closed state to either open or inactivated states. The consequent ion flux is described as a function of both chemical and electrical driving forces, and channel kinetics are often integrated into Hodgkin-Huxley-type formalisms to account for membrane potential dynamics. Such models can simulate mechanosensitive channel currents in response to a variety of mechanical stimuli, providing a means for exploring the interplay between mechanical forces and electrical excitability in neurons.

5.7 Nuclear Mechanosensation

Cells integrate physical cues in part through the nucleus, which functions as a mechanosensor converting forces from the microenvironment into changes in gene expression and cell behavior (Kirby and Lammerding 2018; Jones et al. 2022; Lomakin et al. 2020; Venturini et al. 2020). A principal conduit for this mechanotransduction is the linker of nucleoskeleton and cytoskeleton (LINC) complex, which physically spans the nuclear envelope to couple cytoskeletal filaments to the nuclear lamina, thereby transmitting loads to lamins and associated chromatin (Guilluy et al. 2014; Hao and Starr 2019; Janota et al. 2020; Wong et al. 2021).

Classic nuclear micromechanics show that the nucleus is not a simple elastic shell: micropipette aspiration of isolated nuclei, combined with salt-induced swelling, showed that chromatin and lamin B contribute to viscoelasticity in distinct regimes (Dahl et al. 2005). Whereas chromatin bears most of the load in unswollen nuclei, the nuclei become approximately ten times softer after swelling, at which point the lamina bears more of the load. In both cases, deformation follows power-law scaling (stiffer at short times, softer at longer times), indicating a continuous spectrum of reorganization timescales that can tune gene-expression kinetics.

Downstream, forces that reach the nucleus can directly modulate chromatin organization and transcription-factor access, with epigenetic remodeling helping specify gene-response programs to mechanical inputs (Miroshnikova and Wickström 2022). Direct stretching of chromatin fibers can decompact nucleosomal arrays and upregulate transcription, establishing a physical link between load and promoter accessibility (Skinner and Johnson 2017). Complementarily, tension at the nuclear membrane rapidly reduces repressive histone methylation marks, producing a more deformable nucleus that is protected from force-induced damage while shifting the chromatin landscape toward greater accessibility (van Ingen and Kirby 2021). Together, these findings support a view in which cytoskeletal forces, transmitted via LINC and the lamina to chromatin and nuclear bodies, engage multiscale, time-dependent nuclear mechanics and epigenetics to regulate transcriptional specificity.

5.7.1 Computational Models

A broad spectrum of mathematical and computational models has been developed to interrogate nuclear mechanosensation (see, for example, the recent review by Hobson and Stephens (2020)).

Jean et al. (2005) used a finite-element framework to study how mechanical loads are transmitted from adhesions through the cytoskeleton to the nucleus during endothelial cell rounding. Their model, in which the cell is represented as a contractile cytoskeletal continuum coupled to a deformable nucleus, revealed how changes in adhesion geometry and contractility steer stress transfer onto the nucleus and bias its deformation.

Shenoy and colleagues (2016) developed a chemomechanical model in which stress-dependent myosin recruitment drives contractility, polarization, and nuclear strain as functions of matrix stiffness and cell shape. By minimizing the sum of mechanical, chemical, and motor-work terms, they predicted cytoskeletal polarization and a decrease in chemical free energy with increasing stiffness, which in turn predicts nuclear morphologies. Extending this systems view, Alisafaei et al. (2019) built a connected ECM-cytoskeleton-nucleus chemomechanical model with three cytoskeletal elements (myosin-driven tension, compression-bearing microtubule-like elements, and tensile actin) and focal adhesions coupled to a nucleus modeled as a thin elastic shell surrounding an elastic chromatin core. The model clarifies how direct and indirect force paths from the ECM load the nucleus and how adhesion growth and cytoskeletal balance tune nuclear strain.

Stephens et al. (2017) introduced a minimal shell-core description in which an elastic lamin shell geometrically stiffens at large strains while a cross-linked chromatin polymer elevates small-strain stiffness by roughly an order of magnitude. This model captures the nucleus's strain-stiffening transition at 30% deformation, buckling at high strain, and how chromatin unpacking or lamin disruption alter nuclear mechanics.

To interpret atomic force microscopy (AFM) indentation data, Hobson et al. (2020) modeled a spherical indenter pressing an ellipsoidal nucleus with an infinitely thin elastic membrane enclosing an elastic core. The results indicated that the elastic modulus governs stiffness across depths, whereas the membrane stretch modulus dominates only at large indentations, in agreement with their experimental measurements. Lionetti et al. (2020) demonstrated using coarse-grained simulations how cytoskeletal tethers to the lamina control nuclear shape stability. For micropipette aspiration experiments, Vaziri and Mofrad's continuum model (Vaziri and Mofrad 2007) demonstrated that lamina stiffness strongly influences whole-nucleus deformation.

Beyond elastic models, poroelastic models of the nucleus (Lherbette et al. 2017) have captured its permeability and characteristic diffusion times, with fast "stiff" and slow "soft" responses that explain rate-dependent mechanosensation. Finally, adhesion dynamics models (Cao et al. 2015) show how ECM stiffness determines the forces applied to the nucleus, providing mechanical boundary conditions that link extracellular mechanics to nuclear deformation.

5.8 Discussion

The study of cellular mechanotransduction lies at the intersection of physics, engineering, and cell biology, making it inherently interdisciplinary. While the role of chemical signals has long been understood with frameworks standard in cellular biology and biophysics, namely mass-action and enzyme kinetics, ODEs, and network motifs, tools for understanding mechanical stimuli are only now being adapted. Forces are vectorial, spatially distributed, and time-varying; cells are viscoelastic, active materials with anisotropies and complex geometries; and the relevant boundary conditions are often unknown and usually changing in time.

Measurement techniques differ as well: AFM, micropipette aspiration, traction force microscopy, microfluidic shear, and elastic pillar arrays each provides a partial, indirect view of stress and strain, with trade-offs in resolution and throughput. Because of this underlying complexity, computational modeling is essential: it translates heterogeneous mechanical measurements into interpretable quantities (e.g., local membrane tension, focal adhesion stress) and links them to biochemical responses, thus generating quantitative hypotheses and predictions.

Computational models also provide the bridge connecting mechanics to signaling across scales. Mechanochemical models couple membrane tension to mechanosensitive channel gating. At adhesion sites, molecular clutch models reveal how loading-rate-dependent talin unfolding exposes vinculin binding sites, reinforcing integrin-ECM linkages and modulating force transmission. Cooperative interactions between myosin II and actin crosslinkers translate local mechanical inputs into spatial patterns of contractility, providing a mechanochemical route from force to morphology. Finite element and continuum approaches estimate stresses at subcellular structures; ODE/PDE models capture signaling dynamics.

Looking forward, several bottlenecks remain. As with most biological models, parameter identifiability and sensitivity are important considerations for robustness (Villaverde et al. 2016). Unknown and time-varying boundary conditions during live-cell perturbations introduce significant uncertainty. Data-driven approaches and artificial intelligence will likely help address these challenges (Altmeyer et al. 2022;

Abubaker-Sharif et al. 2025; Zhang et al. 2025), particularly for parameter inference and boundary condition estimation; however, model validation is essential to add confidence and avoid overfitting.

Computational models enable virtual experiments, sensitivity and uncertainty analyses, and parameter inference, integrating disparate datasets, from traction maps and biosensor imaging to transcriptomics. As a result, computational modeling is increasingly enabling predictive, mechanistic models, revealing design principles, guiding experimental design, and informing interventions in fibrosis, cancer, cardiovascular disease, and tissue engineering, where mechanical cues are central.

Acknowledgments We thank Doug Robinson (Johns Hopkins Medicine) and his lab members for many years of useful discussions. Research reported in this publication was supported by the National Institute of General Medical Sciences of the National Institutes of Health under award number R01GM149073. During the preparation of this work, the authors used ChatGPT-5 and Claude Opus 4 to review and improve the clarity of the written text. After using this tool, the authors reviewed and edited the content as needed and take full responsibility for the content of the publication.

References

Abubaker-Sharif B, Banerjee T, Devreotes PN, Iglesias PA (2025) Learning stochastic reaction-diffusion models from limited data using spatiotemporal features. bioRxiv p 2024.10.02.616367. https://doi.org/10.1101/2024.10.02.616367

Adebowale K, Gong Z, Hou JC, Wisdom KM, Garbett D, Lee HP, et al (2021) Enhanced substrate stress relaxation promotes filopodia-mediated cell migration. Nat Mater 20(9):1290–1299. https://doi.org/10.1038/s41563-021-00981-w

Adhikari S, Moran J, Weddle C, Hinczewski M (2018) Unraveling the mechanism of the cadherin-catenin-actin catch bond. PLoS Comput Biol 14(8):e1006399. https://doi.org/10.1371/journal.pcbi.1006399

Alisafaei F, Jokhun DS, Shivashankar GV, Shenoy VB (2019) Regulation of nuclear architecture, mechanics, and nucleocytoplasmic shuttling of epigenetic factors by cell geometric constraints. Proc Natl Acad Sci USA 116(27):13200–13209. https://doi.org/10.1073/pnas.1902035116

Alon R, Hammer DA, Springer TA (1995) Lifetime of the P-selectin-carbohydrate bond and its response to tensile force in hydrodynamic flow. Nature. 374(6522):539–42. https://doi.org/10.1038/374539a0

Altmeyer R, Bretschneider T, Janák J, Reiß M (2022) Parameter estimation in an SPDE model for cell repolarization. SIAM/ASA J Uncertainty Quantif 10(1):179–199. https://doi.org/10.1137/20M1373347

Aryal P, Jarerattanachat V, Clausen MV, Schewe M, McClenaghan C, Argent L, et al (2017) Bilayer-mediated structural transitions control mechanosensitivity of the TREK-2 K2P channel. Structure 25(5):708.e2–718.e2. https://doi.org/10.1016/j.str.2017.03.006

Balaban NQ, Schwarz US, Riveline D, Goichberg P, Tzur G, Sabanay I, et al (2001) Force and focal adhesion assembly: a close relationship studied using elastic micropatterned substrates. Nat Cell Biol 3(5):466–72. https://doi.org/10.1038/35074532

Bangasser BL, Odde DJ (2013) Master equation-based analysis of a motor-clutch model for cell traction force. Cell Mol Bioeng 6(4):449–459. https://doi.org/10.1007/s12195-013-0296-5

Bangasser BL, Rosenfeld SS, Odde DJ (2013) Determinants of maximal force transmission in a motor-clutch model of cell traction in a compliant microenvironment. Biophys J 105(3):581–92. https://doi.org/10.1016/j.bpj.2013.06.027

Bangasser BL, Shamsan GA, Chan CE, Opoku KN, Tüzel E, Schlichtmann BW, et al (2017) Shifting the optimal stiffness for cell migration. Nat Commun 8:15313. https://doi.org/10.1038/ncomms15313

Barkan CO, Bruinsma RF (2024) Topology of molecular deformations induces triphasic catch bonding in selectin-ligand bonds. Proc Natl Acad Sci USA 121(6):e2315866121. https://doi.org/10.1073/pnas.2315866121

Bavi O, Vossoughi M, Naghdabadi R, Jamali Y (2016) The combined effect of hydrophobic mismatch and bilayer local bending on the regulation of mechanosensitive ion channels. PLoS One 11(3):e0150578. https://doi.org/10.1371/journal.pone.0150578

Bell GI (1978) Models for the specific adhesion of cells to cells. Science 200(4342):618–27

Bennett M, Cantini M, Reboud J, Cooper JM, Roca-Cusachs P, Salmeron-Sanchez M (2018) Molecular clutch drives cell response to surface viscosity. Proc Natl Acad Sci USA 115(6):1192–1197. https://doi.org/10.1073/pnas.1710653115

Besser A, Safran SA (2006) Force-induced adsorption and anisotropic growth of focal adhesions. Biophys J 90(10):3469–84. https://doi.org/10.1529/biophysj.105.074377

Betanzos M, Chiang CS, Guy HR, Sukharev S (2002) A large iris-like expansion of a mechanosensitive channel protein induced by membrane tension. Nat Struct Biol 9(9):704–10. https://doi.org/10.1038/nsb828

Boletta A, Caplan MJ (2025) Physiologic mechanisms underlying polycystic kidney disease. Physiol Rev. 105(3):1553–1607. https://doi.org/10.1152/physrev.00018.2024

Brennecke JT, de Groot BL (2018) Mechanism of mechanosensitive gating of the TREK-2 potassium channel. Biophys J 114(6):1336–1343. https://doi.org/10.1016/j.bpj.2018.01.030

Bruinsma R (2005) Theory of force regulation by nascent adhesion sites. Biophys J 89(1):87–94. https://doi.org/10.1529/biophysj.104.048280

Buckley CD, Tan J, Anderson KL, Hanein D, Volkmann N, Weis WI, et al (2014) Cell adhesion. The minimal cadherin-catenin complex binds to actin filaments under force. Science 346(6209):1254211. https://doi.org/10.1126/science.1254211

Cai X, Wang KC, Meng Z (2021) Mechanoregulation of YAP and TAZ in cellular homeostasis and disease progression. Front Cell Dev Biol 9:673599. https://doi.org/10.3389/fcell.2021.673599

Cao X, Lin Y, Driscoll TP, Franco-Barraza J, Cukierman E, Mauck RL, et al (2015) A chemomechanical model of matrix and nuclear rigidity regulation of focal adhesion size. Biophys J 109(9):1807–17. https://doi.org/10.1016/j.bpj.2015.08.048

Chan CE, Odde DJ (2008) Traction dynamics of filopodia on compliant substrates. Science 322(5908):1687–91. https://doi.org/10.1126/science.1163595

Chaudhuri O, Gu L, Darnell M, Klumpers D, Bencherif SA, Weaver JC, et al (2015) Substrate stress relaxation regulates cell spreading. Nat Commun 6:6364. https://doi.org/10.1038/ncomms7365

Chen CS, Mrksich M, Huang S, Whitesides GM, Ingber DE (1997) Geometric control of cell life and death. Science 276(5317):1425–8. https://doi.org/10.1126/science.276.5317.1425

Choi HK, Zhu C (2025) Catch bonds in immunology. Annu Rev Immunol 43(1):641–666. https://doi.org/10.1146/annurev-immunol-082423-035904

Choquet D, Felsenfeld DP, Sheetz MP (1997) Extracellular matrix rigidity causes strengthening of integrin-cytoskeleton linkages. Cell 88(1):39–48. https://doi.org/10.1016/S0092-8674(00)81856-5

Chrzanowska-Wodnicka M, Burridge K (1996) Rho-stimulated contractility drives the formation of stress fibers and focal adhesions. J Cell Biol 133(6):1403–15. https://doi.org/10.1083/jcb.133.6.1403

Clarke M, Spudich JA (1974) Biochemical and structural studies of actomyosin-like proteins from non-muscle cells. Isolation and characterization of myosin from amoebae of Dictyostelium discoideum. J Mol Biol 86(2):209–22. https://doi.org/10.1016/0022-2836(74)90013-8

Cox CD, Bavi N, Martinac B (2017) Origin of the force: the force-from-lipids principle applied to piezo channels. Curr Top Membr 79:59–96. https://doi.org/10.1016/bs.ctm.2016.09.001

Cox CD, Bavi N, Martinac B (2019) Biophysical principles of ion-channel-mediated mechanosensory transduction. Cell Rep 29(1):1–12. https://doi.org/10.1016/j.celrep.2019.08.075

Dahl KN, Engler AJ, Pajerowski JD, Discher DE (2005) Power-law rheology of isolated nuclei with deformation mapping of nuclear substructures. Biophys J 89(4):2855–64. https://doi.org/10.1529/biophysj.105.062554

del Rio A, Perez-Jimenez R, Liu R, Roca-Cusachs P, Fernandez JM, Sheetz MP (2009) Stretching single talin rod molecules activates vinculin binding. Science 323(5914):638–41. https://doi.org/10.1126/science.1162912

Delmas P, Coste B (2013) Mechano-gated ion channels in sensory systems. Cell 155(2):278–84. https://doi.org/10.1016/j.cell.2013.09.026

Dembo M, Torney DC, Saxman K, Hammer D (1988) The reaction-limited kinetics of membrane-to-surface adhesion and detachment. Proc R Soc Lond B Biol Sci 234(1274):55–83. https://doi.org/10.1098/rspb.1988.0038

Deshpande VS, McMeeking RM, Evans AG (2006) A bio-chemo-mechanical model for cell contractility. Proc Natl Acad Sci USA 103(38):14015–20. https://doi.org/10.1073/pnas.0605837103

Discher DE, Smith L, Cho S, Colasurdo M, García AJ, Safran S (2017) Matrix mechanosensing: from scaling concepts in 'omics data to mechanisms in the nucleus, regeneration, and cancer. Annu Rev Biophys 46:295–315. https://doi.org/10.1146/annurev-biophys-062215-011206

Dolgitzer D, Plaza-Rodríguez AI, Iglesias MA, Jacob MAC, Todd BA, Robinson DN, et al (2025) A continuum model of mechanosensation based on contractility kit assembly. Biophys J 124(1):62–76. https://doi.org/10.1016/j.bpj.2024.10.020

Dupont S, Morsut L, Aragona M, Enzo E, Giulitti S, Cordenonsi M, et al (2011) Role of YAP/TAZ in mechanotransduction. Nature 474(7350):179–83. https://doi.org/10.1038/nature10137

Effler JC, Kee YS, Berk JM, Tran MN, Iglesias PA, Robinson DN (2006) Mitosis-specific mechanosensing and contractile-protein redistribution control cell shape. Curr Biol. 16(19):1962–7. https://doi.org/10.1016/j.cub.2006.08.027

Egelhoff TT, Lee RJ, Spudich JA (1993) Dictyostelium myosin heavy chain phosphorylation sites regulate myosin filament assembly and localization in vivo. Cell 75(2):363–71

Elosegui-Artola A, Bazellières E, Allen MD, Andreu I, Oria R, Sunyer R, et al (2014) Rigidity sensing and adaptation through regulation of integrin types. Nat Mater 13(6):631–7. https://doi.org/10.1038/nmat3960

Elosegui-Artola A, Oria R, Chen Y, Kosmalska A, Pérez-González C, Castro N, et al (2016) Mechanical regulation of a molecular clutch defines force transmission and transduction in response to matrix rigidity. Nat Cell Biol 18(5):540–8. https://doi.org/10.1038/ncb3336

Engler AJ, Sen S, Sweeney HL, Discher DE (2006) Matrix elasticity directs stem cell lineage specification. Cell 126(4):677–89. https://doi.org/10.1016/j.cell.2006.06.044

Evans E, Ritchie K (1997) Dynamic strength of molecular adhesion bonds. Biophys J 72(4):1541–55. https://doi.org/10.1016/S0006-3495(97)78802-7

Finger EB, Puri KD, Alon R, Lawrence MB, von Andrian UH, Springer TA (1996) Adhesion through L-selectin requires a threshold hydrodynamic shear. Nature 379(6562):266–9. https://doi.org/10.1038/379266a0

Francis EA, Rangamani P (2024) Computational modeling establishes mechanotransduction as a potent modulator of the mammalian circadian clock. J Cell Sci 137(17):jcs261782. https://doi.org/10.1242/jcs.261782

Gillespie PG, Müller U (2009) Mechanotransduction by hair cells: models, molecules, and mechanisms. Cell 139(1):33–44. https://doi.org/10.1016/j.cell.2009.09.010

Golo M, Newman PLH, Kempe D, Biro M (2024) Mechanoimmunology in the solid tumor microenvironment. Biochem Soc Trans 52(3):1489–1502. https://doi.org/10.1042/BST20231427

Gong Z, Szczesny SE, Caliari SR, Charrier EE, Chaudhuri O, Cao X, et al (2018) Matching material and cellular timescales maximizes cell spreading on viscoelastic substrates. Proc Natl Acad Sci USA 115(12):E2686–E2695. https://doi.org/10.1073/pnas.1716620115

Greenberg MJ, Arpağ G, Tüzel E, Ostap EM (2016) A perspective on the role of myosins as mechanosensors. Biophys J 110(12):2568–2576. https://doi.org/10.1016/j.bpj.2016.05.021

Guilluy C, Osborne LD, Van Landeghem L, Sharek L, Superfine R, Garcia-Mata R, et al (2014) Isolated nuclei adapt to force and reveal a mechanotransduction pathway in the nucleus. Nat Cell Biol 16(4):376–81. https://doi.org/10.1038/ncb2927

Hao H, Starr DA (2019) SUN/KASH interactions facilitate force transmission across the nuclear envelope. Nucleus 10(1):73–80. https://doi.org/10.1080/19491034.2019.1595313

Hartman MA, Spudich JA (2012) The myosin superfamily at a glance. J Cell Sci 125(Pt 7):1627–32. https://doi.org/10.1242/jcs.094300

Hobson CM, Stephens AD (2020) Modeling of cell nuclear mechanics: classes, components, and applications. Cells 9(7):1623. https://doi.org/10.3390/cells9071623

Hobson CM, Kern M, O'Brien ET 3rd, Stephens AD, Falvo MR, Superfine R (2020) Correlating nuclear morphology and external force with combined atomic force microscopy and light sheet imaging separates roles of chromatin and lamin A/C in nuclear mechanics. Mol Biol Cell 31(16):1788–1801. https://doi.org/10.1091/mbc.E20-01-0073

Hudspeth AJ (1989) How the ear's works work. Nature 341(6241):397–404. https://doi.org/10.1038/341397a0

Huxley AF, Niedergerke R (1954) Structural changes in muscle during contraction; interference microscopy of living muscle fibres. Nature 173(4412):971–3. https://doi.org/10.1038/173971a0

Isomursu A, Park KY, Hou J, Cheng B, Mathieu M, Shamsan GA, et al (2022) Directed cell migration towards softer environments. Nat Mater 21(9):1081–1090. https://doi.org/10.1038/s41563-022-01294-2

Jafarinia H, Khalilimeybodi A, Barrasa-Fano J, Fraley SI, Rangamani P, Carlier A (2024) Insights gained from computational modeling of YAP/TAZ signaling for cellular mechanotransduction. NPJ Syst Biol Appl 10(1):90. https://doi.org/10.1038/s41540-024-00414-9

Janota CS, Calero-Cuenca FJ, Gomes ER (2020) The role of the cell nucleus in mechanotransduction. Curr Opin Cell Biol 63:204–211. https://doi.org/10.1016/j.ceb.2020.03.001

Jean RP, Chen CS, Spector AA (2005) Finite-element analysis of the adhesion-cytoskeleton-nucleus mechanotransduction pathway during endothelial cell rounding: axisymmetric model. J Biomech Eng 127(4):594–600. https://doi.org/10.1115/1.1933997

Jones ML, Dahl KN, Lele TP, Conway DE, Shenoy V, Ghosh S, et al (2022) The elephant in the cell: Nuclear mechanics and mechanobiology. J Biomech Eng 144(8):080802. https://doi.org/10.1115/1.4053797

Kalli M, Poskus MD, Stylianopoulos T, Zervantonakis IK (2023) Beyond matrix stiffness: targeting force-induced cancer drug resistance. Trends Cancer 9(11):937–954. https://doi.org/10.1016/j.trecan.2023.07.006

Katta S, Krieg M, Goodman MB (2015) Feeling force: physical and physiological principles enabling sensory mechanotransduction. Annu Rev Cell Dev Biol 31:347–71. https://doi.org/10.1146/annurev-cellbio-100913-013426

Kechagia JZ, Ivaska J, Roca-Cusachs P (2019) Integrins as biomechanical sensors of the microenvironment. Nat Rev Mol Cell Biol 20(8):457–473. https://doi.org/10.1038/s41580-019-0134-2

Kee YS, Ren Y, Dorfman D, Iijima M, Firtel R, Iglesias PA, et al (2012) A mechanosensory system governs myosin II accumulation in dividing cells. Mol Biol Cell 23(8):1510–23. https://doi.org/10.1091/mbc.E11-07-0601

Kirby TJ, Lammerding J (2018) Emerging views of the nucleus as a cellular mechanosensor. Nat Cell Biol 20(4):373–381. https://doi.org/10.1038/s41556-018-0038-y

Kong F, García AJ, Mould AP, Humphries MJ, Zhu C (2009) Demonstration of catch bonds between an integrin and its ligand. J Cell Biol 185(7):1275–84. https://doi.org/10.1083/jcb.200810002

Kothari P, Srivastava V, Aggarwal V, Tchernyshyov I, Van Eyk JE, Ha T, et al (2019) Contractility kits promote assembly of the mechanoresponsive cytoskeletal network. J Cell Sci 132(2). https://doi.org/10.1242/jcs.226704

Kovács M, Thirumurugan K, Knight PJ, Sellers JR (2007) Load-dependent mechanism of nonmuscle myosin 2. Proc Natl Acad Sci USA 104(24):9994–9. https://doi.org/10.1073/pnas.0701181104

Kumar S, Weaver VM (2009) Mechanics, malignancy, and metastasis: the force journey of a tumor cell. Cancer Metastasis Rev 28(1-2):113–27. https://doi.org/10.1007/s10555-008-9173-4

Lan Y, Lu J, Zhang S, Jie C, Chen C, Xiao C, et al (2024) Piezo1-mediated mechanotransduction contributes to disturbed flow-Induced atherosclerotic endothelial inflammation. J Am Heart Assoc 13(21):e035558. https://doi.org/10.1161/JAHA.123.035558

Lherbette M, Dos Santos Á, Hari-Gupta Y, Fili N, Toseland CP, Schaap IAT (2017) Atomic force microscopy micro-rheology reveals large structural inhomogeneities in single cell-nuclei. Sci Rep 7(1):8116. https://doi.org/10.1038/s41598-017-08517-6

Lichtenberg JY, Tran S, Hwang PY (2023) Mechanical factors driving cancer progression. Adv Cancer Res 160:61–81. https://doi.org/10.1016/bs.acr.2023.06.002

Linke WA (2008) Sense and stretchability: the role of titin and titin-associated proteins in myocardial stress-sensing and mechanical dysfunction. Cardiovasc Res 77(4):637–48. https://doi.org/10.1016/j.cardiores.2007.03.029

Lionetti MC, Bonfanti S, Fumagalli MR, Budrikis Z, Font-Clos F, Costantini G, et al (2020) Chromatin and cytoskeletal tethering determine nuclear morphology in progerin-expressing cells. Biophys J 118(9):2319–2332. https://doi.org/10.1016/j.bpj.2020.04.001

Lomakin AJ, Cattin CJ, Cuvelier D, Alraies Z, Molina M, Nader GPF, et al (2020) The nucleus acts as a ruler tailoring cell responses to spatial constraints. Science 370(6514):eaba2894. https://doi.org/10.1126/science.aba2894

Luo T, Mohan K, Iglesias PA, Robinson DN (2013) Molecular mechanisms of cellular mechanosensing. Nat Mater 12(11):1064–71. https://doi.org/10.1038/nmat3772

Luo T, Mohan K, Srivastava V, Ren Y, Iglesias PA, Robinson DN (2012) Understanding the cooperative interaction between myosin II and actin cross-linkers mediated by actin filaments during mechanosensation. Biophys J 102(2):238–47. https://doi.org/10.1016/j.bpj.2011.12.020

Luo W, Yu Ch, Lieu ZZ, Allard J, Mogilner A, Sheetz MP, et al (2013) Analysis of the local organization and dynamics of cellular actin networks. J Cell Biol 202(7):1057–73. https://doi.org/10.1083/jcb.201210123

Müller U (2008) Cadherins and mechanotransduction by hair cells. Curr Opin Cell Biol 20(5):557–66. https://doi.org/10.1016/j.ceb.2008.06.004

Mallik R, Carter BC, Lex SA, King SJ, Gross SP (2004) Cytoplasmic dynein functions as a gear in response to load. Nature 427(6975):649–52. https://doi.org/10.1038/nature02293

Marshall BT, Long M, Piper JW, Yago T, McEver RP, Zhu C (2003) Direct observation of catch bonds involving cell-adhesion molecules. Nature 423(6936):190–3. https://doi.org/10.1038/nature01605

Michalski PJ, Loew LM (2016) SpringSaLaD: a spatial, particle-based biochemical simulation platform with excluded volume. Biophys J 110(3):523–529. https://doi.org/10.1016/j.bpj.2015.12.026

Miroshnikova YA, Wickström SA (2022) Mechanical forces in nuclear organization. Cold Spring Harb Perspect Biol 14(1):a039685. https://doi.org/10.1101/cshperspect.a039685

Mitchison T, Kirschner M (1988) Cytoskeletal dynamics and nerve growth. Neuron. 1(9):761–72. https://doi.org/10.1016/0896-6273(88)90124-9

Mohan K, Luo T, Robinson DN, Iglesias PA (2015) Cell shape regulation through mechanosensory feedback control. J R Soc Interface 12(109):20150512. https://doi.org/10.1098/rsif.2015.0512

Nauli SM, Alenghat FJ, Luo Y, Williams E, Vassilev P, Li X, et al (2003) Polycystins 1 and 2 mediate mechanosensation in the primary cilium of kidney cells. Nat Genet. 33(2):129–37. https://doi.org/10.1038/ng1076

Neiva-Sousa M, Almeida-Coelho J, Falcão-Pires I, Leite-Moreira AF (2015) Titin mutations: the fall of Goliath. Heart Fail Rev 20(5):579–88. doi:https://doi.org/10.1007/s10741-015-9495-6

Nicolas A, Geiger B, Safran SA (2004) Cell mechanosensitivity controls the anisotropy of focal adhesions. Proc Natl Acad Sci USA 101(34):12520–5. https://doi.org/10.1073/pnas.0403539101

Nicoletti M, Chiodo L, Loppini A (2021) Biophysics and modeling of mechanotransduction in neurons: a review. Mathematics. 9(4). https://doi.org/10.3390/math9040323

Pak OS, Young YN, Marple GR, Veerapaneni S, Stone HA (2015) Gating of a mechanosensitive channel due to cellular flows. Proc Natl Acad Sci USA 112(32):9822–7. https://doi.org/10.1073/pnas.1512152112

Park YC, Reddy B, Bavi N, Perozo E, Faraldo-Gómez JD (2023) State-specific morphological deformations of the lipid bilayer explain mechanosensitive gating of MscS ion channels. Elife 12:e81445. https://doi.org/10.7554/eLife.81445

Pelham RJ Jr, Wang Yl (1997) Cell locomotion and focal adhesions are regulated by substrate flexibility. Proc Natl Acad Sci USA 94(25):13661–5. https://doi.org/10.1073/pnas.94.25.13661

Pereverzev YV, Prezhdo OV, Forero M, Sokurenko EV, Thomas WE (2005) The two-pathway model for the catch-slip transition in biological adhesion. Biophys J 89(3):1446–54. https://doi.org/10.1529/biophysj.105.062158

Perozo E, Cortes DM, Sompornpisut P, Kloda A, Martinac B (2002) Open channel structure of MscL and the gating mechanism of mechanosensitive channels. Nature. 418(6901):942–8. https://doi.org/10.1038/nature00992

Perozo E, Kloda A, Cortes DM, Martinac B (2002) Physical principles underlying the transduction of bilayer deformation forces during mechanosensitive channel gating. Nat Struct Biol 9(9):696–703. https://doi.org/10.1038/nsb827

Pethő Z, Najder K, Bulk E, Schwab A (2019) Mechanosensitive ion channels push cancer progression. Cell Calcium 80:79–90. https://doi.org/10.1016/j.ceca.2019.03.007

Plaza-Rodríguez AI, Nguyen LTS, Robinson DN, Iglesias PA (2022) Particle-based model of mechanosensory contractility kit assembly. Biophys J 121(23):4600–4614. https://doi.org/10.1016/j.bpj.2022.10.031

Poirier CC, Iglesias PA (2007) An integrative approach to understanding mechanosensation. Brief Bioinform 8(4):258–65. https://doi.org/10.1093/bib/bbm025

Prahl LS, Stanslaski MR, Vargas P, Piel M, Odde DJ (2020) Predicting confined 1D cell migration from parameters calibrated to a 2D motor-clutch model. Biophys J 118(7):1709–1720. https://doi.org/10.1016/j.bpj.2020.01.048

Rajeshwar T R, Anishkin A, Sukharev S, Vanegas JM (2021) Mechanical activation of MscL revealed by a locally distributed tension molecular dynamics approach. Biophys J 120(2):232–242. https://doi.org/10.1016/j.bpj.2020.11.2274

Rao L, Berger F, Nicholas MP, Gennerich A (2019) Molecular mechanism of cytoplasmic dynein tension sensing. Nat Commun 10(1):3332. https://doi.org/10.1038/s41467-019-11231-8

Reconditi M, Linari M, Lucii L, Stewart A, Sun YB, Boesecke P, et al (2004) The myosin motor in muscle generates a smaller and slower working stroke at higher load. Nature 428(6982):578–81. https://doi.org/10.1038/nature02380

Reichl EM, Ren Y, Morphew MK, Delannoy M, Effler JC, Girard KD, et al (2008) Interactions between myosin and actin crosslinkers control cytokinesis contractility dynamics and mechanics. Curr Biol 18(7):471–80. https://doi.org/10.1016/j.cub.2008.02.056

Ren Y, Effler JC, Norstrom M, Luo T, Firtel RA, Iglesias PA, et al (2009) Mechanosensing through cooperative interactions between myosin II and the actin crosslinker cortexillin I. Curr Biol 19(17):1421–8. https://doi.org/10.1016/j.cub.2009.07.018

Roca-Cusachs P, Gauthier NC, Del Rio A, Sheetz MP (2009) Clustering of alpha(5)beta(1) integrins determines adhesion strength whereas alpha(v)beta(3) and talin enable mechanotransduction. Proc Natl Acad Sci USA 106(38):16245–50. https://doi.org/10.1073/pnas.0902818106

Ronan W, Deshpande VS, McMeeking RM, McGarry JP (2014) Cellular contractility and substrate elasticity: a numerical investigation of the actin cytoskeleton and cell adhesion. Biomech Model Mechanobiol 13(2):417–35. https://doi.org/10.1007/s10237-013-0506-z

Sala S, Caillier A, Oakes PW (2024) Principles and regulation of mechanosensing. J Cell Sci 137(18):jcs261338. https://doi.org/10.1242/jcs.261338

Schiffhauer ES, Luo T, Mohan K, Srivastava V, Qian X, Griffis ER, et al (2016) Mechanoaccumulative elements of the mammalian actin cytoskeleton. Curr Biol 26(11):1473–9. https://doi.org/10.1016/j.cub.2016.04.007

Schiffhauer ES, Ren Y, Iglesias VA, Kothari P, Iglesias PA, Robinson DN (2019) Myosin IIB assembly state determines its mechanosensitive dynamics. J Cell Biol 218(3):895–908. https://doi.org/10.1083/jcb.201806058

Scott KE, Fraley SI, Rangamani P (2021) A spatial model of YAP/TAZ signaling reveals how stiffness, dimensionality, and shape contribute to emergent outcomes. Proc Natl Acad Sci USA 118(20):e2021571118. https://doi.org/10.1073/pnas.2021571118

Shenoy VB, Wang H, Wang X (2016) A chemo-mechanical free-energy-based approach to model durotaxis and extracellular stiffness-dependent contraction and polarization of cells. Interface Focus 6(1):20150067. https://doi.org/10.1098/rsfs.2015.0067

Skinner BM, Johnson EEP (2017) Nuclear morphologies: their diversity and functional relevance. Chromosoma 126(2):195–212. https://doi.org/10.1007/s00412-016-0614-5

Stephens AD, Banigan EJ, Adam SA, Goldman RD, Marko JF (2017) Chromatin and lamin a determine two different mechanical response regimes of the cell nucleus. Mol Biol Cell 28(14):1984–1996. https://doi.org/10.1091/mbc.E16-09-0653

Sun M, Spill F, Zaman MH (2016) A computational model of YAP/TAZ mechanosensing. Biophys J 110(11):2540–2550. https://doi.org/10.1016/j.bpj.2016.04.040

Sunyer R, Conte V, Escribano J, Elosegui-Artola A, Labernadie A, Valon L, et al (2016) Collective cell durotaxis emerges from long-range intercellular force transmission. Science 353(6304):1157–61. https://doi.org/10.1126/science.aaf7119

Tang Y, Yoo J, Yethiraj A, Cui Q, Chen X (2008) Mechanosensitive channels: insights from continuum-based simulations. Cell Biochem Biophys 52(1):1–18. https://doi.org/10.1007/s12013-008-9024-5

Totaro A, Panciera T, Piccolo S (2018) YAP/TAZ upstream signals and downstream responses. Nat Cell Biol 20(8):888–899. https://doi.org/10.1038/s41556-018-0142-z

van Ingen MJA, Kirby TJ (2021) LINCing nuclear mechanobiology with skeletal muscle mass and function. Front Cell Dev Biol 9:690577. https://doi.org/10.3389/fcell.2021.690577

Vaziri A, Mofrad MRK (2007) Mechanics and deformation of the nucleus in micropipette aspiration experiment. J Biomech 40(9):2053–62. https://doi.org/10.1016/j.jbiomech.2006.09.023

Venturini V, Pezzano F, Català Castro F, Häkkinen HM, Jiménez-Delgado S, Colomer-Rosell M, et al (2020) The nucleus measures shape changes for cellular proprioception to control dynamic cell behavior. Science 370(6514):eaba2644. https://doi.org/10.1126/science.aba2644

Villaverde AF, Barreiro A, Papachristodoulou A (2016) Structural identifiability of dynamic systems biology models. PLoS Comput Biol 12(10):e1005153. https://doi.org/10.1371/journal.pcbi.1005153

Visscher K, Schnitzer MJ, Block SM (1999) Single kinesin molecules studied with a molecular force clamp. Nature 400(6740):184–9. https://doi.org/10.1038/22146

Wada KI, Itoga K, Okano T, Yonemura S, Sasaki H (2011) Hippo pathway regulation by cell morphology and stress fibers. Development 138(18):3907–14. https://doi.org/10.1242/dev.070987

Wayman AM, Chen W, McEver RP, Zhu C (2010) Triphasic force dependence of E-selectin/ligand dissociation governs cell rolling under flow. Biophys J 99(4):1166–74. https://doi.org/10.1016/j.bpj.2010.05.040

Welf ES, Naik UP, Ogunnaike BA (2012) A spatial model for integrin clustering as a result of feedback between integrin activation and integrin binding. Biophys J 103(6):1379–89. https://doi.org/10.1016/j.bpj.2012.08.021

Wilson DGS, Tinker A, Iskratsch T (2022) The role of the dystrophin glycoprotein complex in muscle cell mechanotransduction. Commun Biol 5(1):1022. https://doi.org/10.1038/s42003-022-03980-y

Wong X, Loo TH, Stewart CL (2021) LINC complex regulation of genome organization and function. Curr Opin Genet Dev 67:130–141. https://doi.org/10.1016/j.gde.2020.12.007

Xie J, Mao F, Yang Y, Jiang H (2025) Mechanical memory of excitable cells induced by mechanosensitive channel inactivation. Biophys J S0006–3495(25), 00374–1. https://doi.org/10.1016/j.bpj.2025.06.014

Yang L, Effler JC, Kutscher BL, Sullivan SE, Robinson DN, Iglesias PA (2008) Modeling cellular deformations using the level set formalism. BMC Syst Biol 2:68. https://doi.org/10.1186/1752-0509-2-68

Zanotelli MR, Zhang J, Reinhart-King CA (2021) Mechanoresponsive metabolism in cancer cell migration and metastasis. Cell Metab 33(7):1307–1321. https://doi.org/10.1016/j.cmet.2021.04.002

Zhang J, Li X, Guo X, You Z, Böttcher L, Mogilner A, et al (2025) Reconstructing noisy gene regulation dynamics using extrinsic-noise-driven neural stochastic differential equations. PLoS Comput Biol 21(9):e1013462. https://doi.org/10.1371/journal.pcbi.1013462

Zhu P, Lu H, Wang M, Chen K, Chen Z, Yang L (2023) Targeted mechanical forces enhance the effects of tumor immunotherapy by regulating immune cells in the tumor microenvironment. Cancer Biol Med 20(1):44–55. https://doi.org/10.20892/j.issn.2095-3941.2022.0491

Zhu L, Cui Q, Xiao H, Liao X, Chen X (2019) Gating and inactivation of mechanosensitive channels of small conductance: a continuum mechanics study. J Mech Behav Biomed Mater 90:502–514. https://doi.org/10.1016/j.jmbbm.2018.10.040

Chapter 6
StrainMapperJ: An Easy-to-Use Digital Image Correlation Toolkit for Exploring and Quantifying the Mechanics of Deforming Tissues

Lance A. Davidson, Sommer Anjum, Jing Yang, and Geneva Masak

Abstract Multicellular tissues are shaped and remodeled during development, organogenesis, wound healing, cancer, and regeneration, as well as during a broad range of diseases. Time-lapse sequences of tissue movements convey a wealth of information about the processes driving their self-assembly and mechanobiology. Here we present StrainMapperJ, a digital image correlation (DIC) tool and supporting macros for the widely used open-source image analysis package ImageJ. StrainMapperJ can produce a diverse set of graphics for mechanical data exploration. At its core, StrainMapperJ uses the ImageJ plugin bUnwarpJ to map deformations in one image to match a second. From these results, StrainMapperJ can generate 2D maps of mechanical strain in the image frame, principal engineering strain and orientation, area strains, synthetic movement trajectories, and vorticity. StrainMapperJ can generate a variety of graphical outputs for quantitative analysis that registers or maps mechanical information directly onto the source image sequence. Additionally, the algorithms used in StrainMapperJ can yield quantitative information from short time-lapse sequences, often only a few minutes in duration, reducing the need for long-term imaging. In addition to these macros, we provide guidance on preparing image datasets, troubleshooting the analysis pipeline, and interpreting StrainMapperJ-produced maps for hypothesis generation and testing.

L. A. Davidson (✉)
Department of Bioengineering, University of Pittsburgh, Pittsburgh, PA, USA

Department of Computational and Systems Biology, School of Medicine, University of Pittsburgh, Pittsburgh, PA, USA
e-mail: lad43@pitt.edu

S. Anjum · J. Yang
Department of Bioengineering, University of Pittsburgh, Pittsburgh, PA, USA

G. Masak
Integrative Systems Biology, School of Medicine, University of Pittsburgh, Pittsburgh, PA, USA

M. Kloc, J. Wosik (eds.), *Mechanotransduction and Mechanosensing*, Results and Problems in Cell Differentiation 78, https://doi.org/10.1007/978-3-032-19509-8_6

6.1 Introduction

Mechanobiology is an emerging field that focuses on the role of environmental mechanical cues and cell-driven forces on biology. In addition to the critical role that mechanics plays in shaping tissues during development and enabling metastases during cancer, mechanical cues can regulate physiology and influence disease progression. Mechanical cues can take many forms. Cues can be driven by both external sources, such as pressure and tension at a distance, and by internal factors generated by the cells that make up the tissue.

The challenge for biological and biomedical researchers seeking to elucidate these roles is in quantifying the magnitude and duration of mechanical cues. Analogous studies on the role of diffusible growth factors are characterized by biophysical, physical-chemical, and genetic parameters such as ligand identity, deposition, binding, transport, and discrete cell signaling pathways. However, methods for quantifying mechanical cues in elastic, viscoelastic, and elastic-plastic materials are not as clear as quantifying flow or delivery of chemical cues. Assessing potential mechanical cues requires the ability to measure deformations acting within living tissues.

Live imaging of cells and tissues undergoing deformation is a key approach for visualizing potential mechanical cues. Time-lapse image sequences contain a rich information on the kinematics of tissue deformation. Data on movement, e.g., kinematics, can be decoded from image sequences acquired with light-sheet and confocal microscopes, epi-illumination compound microscopes, and stereomicroscopes but can also be obtained from other imaging modalities such as optical coherence tomography that provide sufficient contrast.

Common approaches to kinematic analysis involve tracking distinct landmarks or structures, such as cell membranes or nuclei, as the tissue transforms from one geometric configuration to another (Su et al. 2007; Blanchard et al. 2009; Etournay et al. 2016; Tinevez et al. 2017). More generally, these approaches are based on machine vision algorithms that aim to decompose an image into an array of objects. An alternative to classifying and tracking discrete structures is to track the movement of durable patches of texture or nonspecific landmarks in the image. The texture patches can represent random patterns of pigment, extracellular matrix, tissues, or individual cells. Similar to machine vision methods, these patches can be tracked by software (Bookstein 1995; Arganda-Carreras et al. 2008; Wang et al. 2014). Such tracks can yield quantitative spatial and temporal descriptions of tissue kinematics, including deformation, velocities, area changes, and normal or shear strain rates. Since tissue kinematics result from force generation, force transmission, and the mechanical properties of the tissues involved, kinematic data can identify potential mechanical cues and contribute to our understanding of genetic and physical mechanisms underlying the dynamical processes that drive tissue morphogenesis. By identifying potential mechanical cues, kinematic data can reveal how mechanical processes drive cell behavior and cell fate decisions.

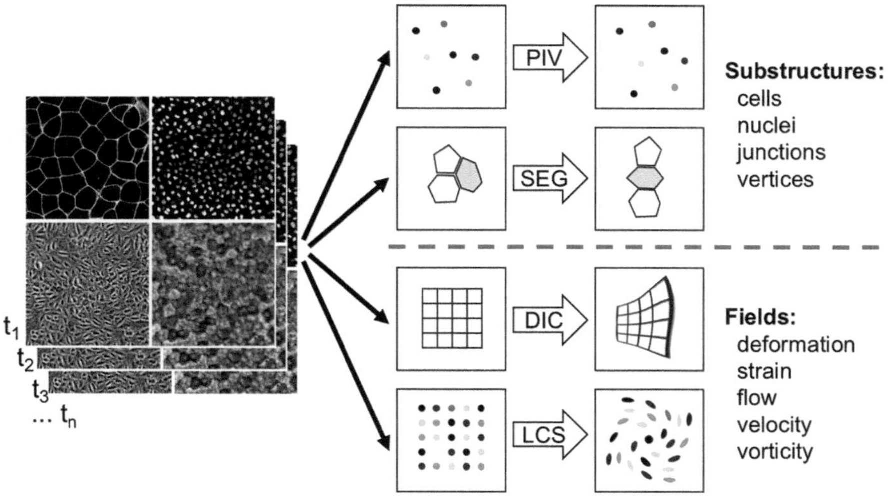

Fig. 6.1 Methods for extracting information about physical processes and potential mechanical cues during tissue morphogenesis. Time-lapse image sequences from a range of visible-light microscopy modalities (confocal, phase, brightfield) can be subjected to machine vision image analysis. Particle image velocimetry (PIV) and segmentation (SEG) analysis can provide information on discrete substructures within tissues. Digital image correlation (DIC) and Lagrangian coherent structure analysis (LCS) can provide information on continuous material kinematics, e.g., "field" wide processes. All of these methods are well suited to study physical processes and generate hypotheses for further analysis of mechanobiological cues

A growing array of machine vision tools is being used to track cells and structures within time lapses (Fig. 6.1). Segmentation (SEG; Aigouy and Prud'homme 2022) and particle image velocimetry (PIV; Petitjean et al. 2010) tools are central to this effort. These methods involve identifying or classifying of unitary structures, such as cells, nuclei, or Golgi, within each frame and can track them as discrete structures over the time-lapse sequence. If a structure can be labeled fluorescently, it can be segmented from the background and tracked. The positions and interplay between multiple structures can be further broken down with foreknowledge of the image scene. For instance, a general scheme of epithelial cell anatomy combined with cell-cell boundaries labeled by plasma membrane-targeted fluorescent proteins can be subdivided into tri-cellular junctions, e.g., vertices, and bicellular junctions and their dynamics quantified through a number of paradigms including force inference (Honda 1983; Ishihara and Sugimura 2012; Weng et al. 2025), nematics (Balasubramaniam et al. 2022; Li et al. 2025), Lagrangian coherent structure (LCS) analysis (e.g., morphoskeletons) (Serra et al. 2020; Serrano Nájera et al. 2025), and tissue tectonics (Blanchard et al. 2009). However, these tools require a steep learning curve and significant dedicated research time to further assess their dynamics and statistical significance. Simpler, easier-to-use tools that rely on continuum features can provide more rapid assessments and highlight where additional resources or more specialized discrete analysis might be needed.

Digital image correlation (DIC) offers an alternative to discrete object-tracking approaches by holistically tracking the motion of textural patterns (Fig. 6.1). Tracking deformation between two images to infer deformation has been applied across a range of fields, from mechanical engineering, where it is known as DIC, to remote sensing of geological motion, where the method is known as satellite radar interferometry (Zamir et al. 2005; Sutton et al. 2009; McCormick and Lord 2010; Blaber et al. 2015; Olufsen et al. 2020; Yang et al. 2022). The use of DIC to track naturally occurring patterns allows for the quantification of motion in the image frame of reference. DIC is descriptive and generally agnostic of constitutive material models, e.g., a priori selection of a solid or fluid representation. Still, the relative motions described by DIC can be output into solid or fluid material frameworks to describe strain behaviors or flow behaviors, respectively.

DIC can generate two-dimensional images that reveal the temporal and spatial patterns of strain, in units of change in length per unit length per time. Strain, or ε, is a unitless description of how materials change in length, e.g., $\varepsilon = (L–L_0)/L_0$. When we consider changes in strain over time (strain rate), we divide the strain by the elapsed time between images. During embryonic morphogenesis in *Xenopus laevis* frog embryos, the strain rates vary over time and space, for instance, tissues that form dorsal structures such as the spinal cord and skeletal muscle undergo 0.15–0.20 strain per hour during convergent extension of late gastrulation and neurulation (Zhou et al. 2015), whereas ventral tissue strain rates peak at 0.15 strain per hour during early tail bud stages (Jackson et al. 2017).

Maps of displacement and deformation generated by DIC can be recast as fluidlike descriptions, since embryonic tissues can also be characterized as if they were highly viscous fluids (Chuai et al. 2023; Serrano Nájera et al. 2025). Displacement maps can be considered velocity fields and further analyzed to yield vorticity maps that reveal whether cells are concentrated or dispersed by the flow.

We have further extended this analysis pipeline with visualization tools. StrainMapperJ (SMJ) provides a set of 2D maps where motion and deformation are encoded as 32-bit floating point, real-valued images where each pixel has an associated real-valued velocity or strain value. Whether solid- or fluidlike analyses are desired, displacements can be averaged over small square domains and visualized with arrow and quiver maps, strains by von Mises circles, and flows by synthetic particle tracking. Synthetic particle tracks can show either step-by-step changes or accumulated start-to-finish changes.

Our group has used strain maps generated by earlier versions of StrainMapper to describe the mechanics of embryonic morphogenesis and explore the role of mechanical cues in developmental programs. These strain maps have been used to describe morphogenesis within intact embryos (Feroze et al. 2015; Jackson et al. 2017) (Fig. 6.2a) as well as morphogenesis of organotypic explants from the animal cap ectoderm (Song et al. 2015; Stepien et al. 2019) (Figure 6.2b). Strain maps have also been used to study more acute responses, such as the mechanical response of epithelia to laser ablation (Arnold et al. 2019) (Fig. 6.2c) and microfluidic stimulation by extracellular ATP (Kim et al. 2014). The method is robust and can even be used to calculate strain rates and evolving stress patterns within force-reporting gels produced by embedded dorsal isolates, organotypic tissues containing all three germ layers that drive convergent extension of the early embryo, and for traction force microscopy of cells undergoing convergent extension (Zhou et al. 2010). Strain maps have been further integrated with models of spreading epithelial sheets within an approximate Bayesian computation analysis (Stepien et al. 2019). These previous applications used a variety of imaging modalities ranging from brightfield stereo microscopy to confocal fluorescence microscopy. Thus, strain maps have broad application in studies of morphogenesis and organogenesis.

The StrainMapperJ toolkit (version 1.0) described here uses DIC of textural features and automated landmark identification to quantify tissue shape changes during morphogenesis. In addition to tracking textural features, we also use landmarks from scale-invariant feature transform (SIFT) and multi-scale oriented patches (MOPS) (Saalfeld et al. 2010). At the core of SMJ lies the elastic registration engine of the ImageJ plugin bUnwarpJ (Arganda-Carreras et al. 2006). bUnwarpJ was initially developed to register sequential images from serial sections, which are often affected by warping and deformation artifacts. Elastic registration relies on minimizing the elastic deformation needed to convert texture patterns and landmarks in one image into the next Our methods use the output from bUnwarpJ's elastic registration for all subsequent calculations. Use of SMJ can save a great deal of time when compared to SEG and PIV approaches.

6.2 Results

The following sections and an accompanying use case describe the operations performed by StrainMapperJ to map tissue movement into graphical products that depict deformation. In brief, we describe image preparation and the use of bUnwarpJ to produce translation maps between two source images. Then we demonstrate how strain maps are calculated in the image frame of reference, as well as principal strains and their orientation. Displacement maps are used to generate velocity maps and vorticity maps. B-spline coefficients are then used to generate synthetic maps of tissue trajectories. From displacement and strain maps, StrainMapperJ generates a variety of visualization plots, including strain ellipsoids, arrow-and-quiver plots, and trajectory maps, using synthetic meshes or particles. The graphical products of StrainMapperJ can be used qualitatively, to generate hypotheses about the mechanical processes that shape tissues and, quantitatively, to measure mechanical features as tissues change shape. We conclude with advice on troubleshooting StrainMapperJ, including methods for validation and identification of spurious results, and guidance on parameters in bUnwarpJ and preprocessing source images to limit misleading output.

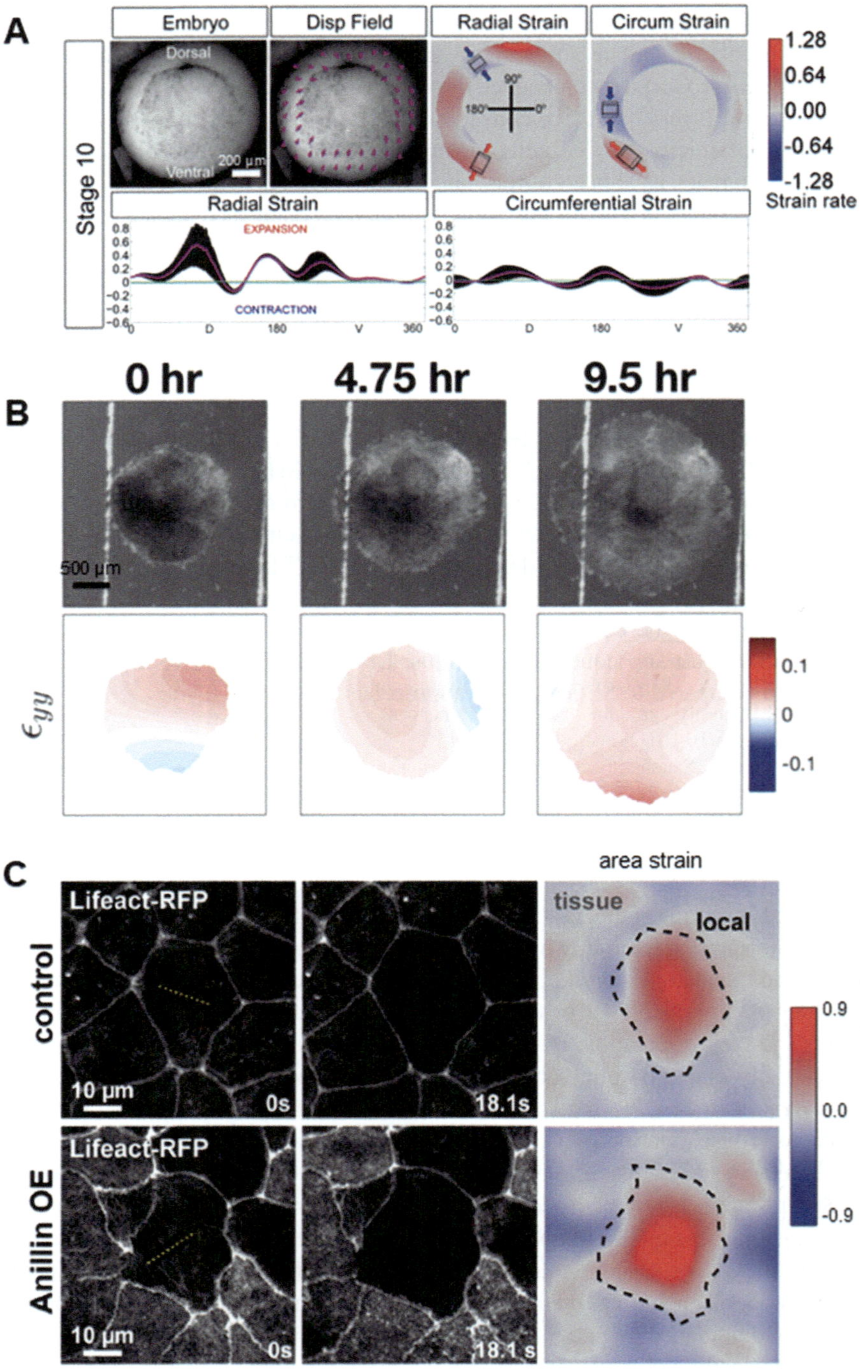

Fig. 6.2 Applications of earlier versions of StrainMapper tools. (**a**) Analysis of the kinematics of blastopore closure in *Xenopus laevis* (Feroze et al. 2015). (**b**) Analysis of the mechanics of spread-

6.2.1 Preparation of Image Stacks or Image Pairs

StrainMapperJ operates on pairs of 2D images in which tissues change shape within the plane. Time-lapse sequences can be collected using either brightfield or confocal microscopy during phases of tissue movement. These sequences can be collected as 4D sets, e.g., time-lapse z-stack sequences or as pairs of 2D or z-stacks images where the tissue sample is subjected to experimental forces. 2D image pairs from z-stacks can be generated from multiple optical sections using maximum or average intensity projection algorithms or via extended depth-of-focus algorithms in ImageJ (Unser 2008; Schroeder et al. 2021). Additionally, we find registration can be improved by cropping the region of interest so that the tissue of interest fills the entire field leaving no background. Cropping can be completed during the pre-processing steps, however, our macro also allows the user to pass a user-defined mask that is used by bUnwarpJ. Once the time-lapse is prepared, the macro "StrainMapperJ" can separate the time-lapse into a set of image pairs. Following this first step, we have also written a macro "StrainMapperJ StackAnalysis," to control subsequent processing and the generation of visual outputs. Macros and a step-by-step tutorial are available through GitHub (see https://github.com/ladavidson). An earlier version of StrainMapper, used in Stepien et al. (2019), can also be found on GitHub (https://github.com/tstepien/strain-mapping).

6.2.2 What Does the StrainMapperJ Macro Do? A Step-by-Step Narrative

StrainMapperJ runs a series of functions that operate on an image pair to produce strain rate information and generate a series of output products. The same functions are called when an SMJ is used for analyzing a time-lapse series. The user is initially prompted with a dialog box (Fig. 6.3a) to identify: the analysis type, specify bUnwarpJ settings, and set grid sizes for graphical elements such as arrows, ellipses, and hexagons in output files. After the dialog choices are completed and the image pair or time-lapse stack is selected, a series of functions is initiated. The functions generate raw data files that are then used for additional visualizations of deformation

Fig. 6.2 spreading epithelial sheets isolated from *Xenopus blastula* stage embryos (Stepien et al. 2019). This work is an example of machine learning used to couple image analysis directly to a mechanical model of epithelial morphogenesis. (**c**) Analysis of the mechanical effects of laser ablation on *Xenopus* epithelial cells that have been perturbed by overexpression of the actin cross-linker anillin (Arnold et al. 2019). This application demonstrates use of SMJ to assess mechanics of subcellular structures such as the medioapical cortex. (**a**) and (**b**) applications were applied to time lapses acquired with brightfield stereomicroscopy, while (**c**) demonstrates usefulness of the method with time-lapse sequences collected with confocal microscopy

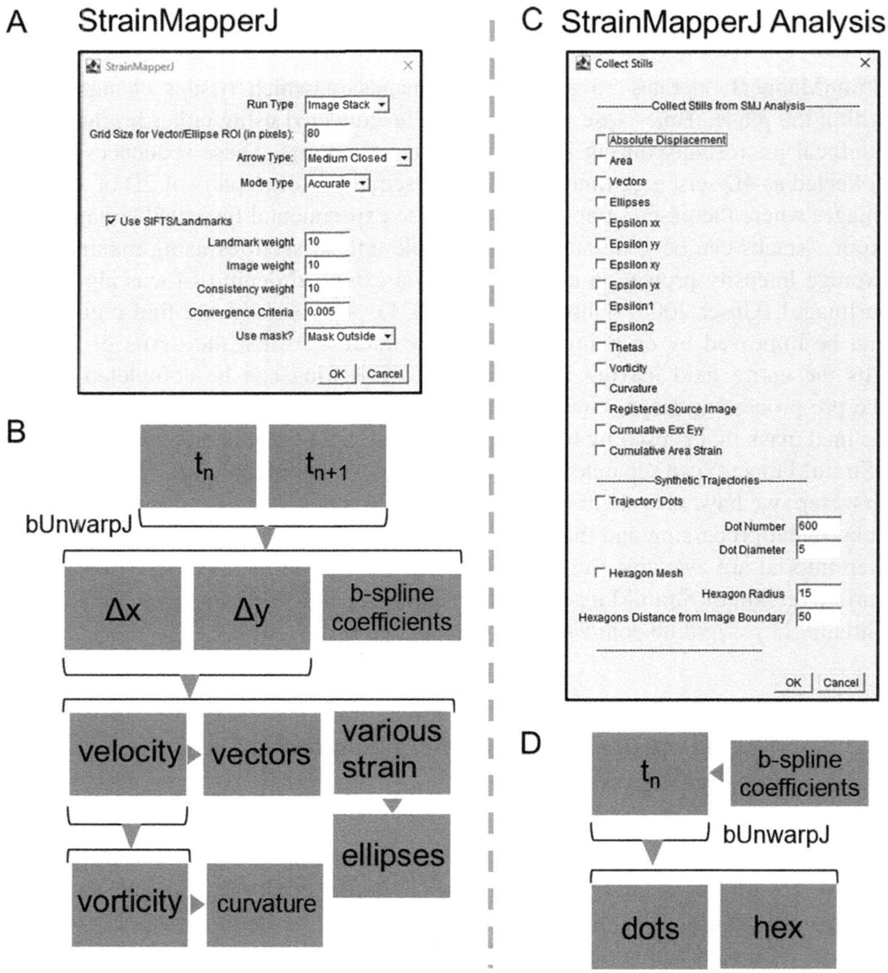

Fig. 6.3 Application and processing steps of StrainMapperJ and postprocessing with StrainMapperJ Analysis macros. (**a**) The dialog box opened by the SMJ macro. User-selectable fields allow for customized visualization products as well as fine-tuning of bUnwarpJ. (**b**) The relationship of various products generated by SML starting with the initial image pair at t_n and t_{n+1}. Displacement maps (Δx and Δy) are generated directly from bUnwarpJ output files. SMJ then uses these maps to calculate velocity maps and vector plots, vorticity and curvature, and strain maps. (**c**) After processing by SMJ, the products from stack analysis are collected by calling the SMJ Analysis macro. The user can select from a range of products including cumulative maps and synthetic trajectory maps using randomized patterns of dots (dots) or regular hexagon arrays (hex). User-selected parameters for the synthetic maps allow customized products for visualization. Graphic products are also generated in overlay format with original images as the background

and strain (Fig. 6.3b). SMJ creates 22 new files within the folder containing the image pair (Table 6.1). The descriptions below present the functions operating in SMJ and the files they generate.

1. **unwarp.** This function runs the plugin "Extract SIFT Correspondences" to automatically identify landmarks (Saalfeld et al. 2010) and then runs the plugin bUnwarpJ that uses elastic registration methods (Arganda-Carreras et al. 2006) to deform the first image into the second. The choice of weights and convergence criteria are discussed in both the ImageJ website for the plugin (https://imagej.net/plugins/bunwarpj/) and the online repository for bUnwarpJ (https://github.com/fiji/bUnwarpJ). Elastic registration of image pairs is one of the most computationally intensive steps in SMJ and is time consuming. At the end of the processing, there will be two new image stacks and both forward and reverse transform files. bUnwarpJ produces two separate solutions for elastic registration: one to transform the source into the target image and the other to transform the target into the source image. The transform files, direct_transf.txt and inverse_transf.txt, contain the basis-spline (b-spline) coefficients that encode the elastic mappings from source to target and target to source, respectively. The transform files are used to generate displacements but also are used later in post-analysis to generate synthetic trajectory and hexagon mesh deformation maps.
2. **trans2raw.** This function is also carried out by bUnwarpJ and converts the b-spline transforms to raw positions using bUnwarpJ. One can consider the raw positions as a series of X and Y values that report the position of a source pixel in the second image ($x + \Delta x, y + \Delta y$). For instance, a pixel at (x, y): (187, 203) in the first image may occupy the position ($x + \Delta x, y + \Delta y$): (183.2, 210.6) in the second image. The raw file (raw.txt) is an ASCII file containing the arrays of new x and y positions, separated by tabs, in raster format.
3. **raw2abs.** This function converts raw positions to displacement vectors. The displacement vector is represented at each pixel by a Δx and Δy value. For the previous example, the raw position (187, 203) in the first image is moved to (183.2, 210.6) in the second image. The displacement vector for that pixel is then $(\Delta x, \Delta y) = (-3.8, 7.6)$. The values of Δx and Δy are stored in two 32-bit real-valued images: deltaX.tif and deltaY.tif.
4. **abs2eps.** This function converts displacement vectors to ε_{xx}, ε_{yy}, ε_{xy}, and ε_{yx} Cauchy strains. Cauchy strains are defined as $\varepsilon = \Delta L/L$, where L is the initial length and ΔL is the change in length. Strain maps are calculated from spatial derivatives of ΔX and ΔY images. Strain values of ε_{xx}, ε_{yy}, ε_{xy}, and ε_{yx} for each pixel are stored as 32-bit real image arrays in epsx.tif, epsy.tif, epsxy.tif, and epsyx.tif files, respectively. These files are further revised to remove values from their edges, e.g., the two pixels where the strain calculations are not defined. These revised files are stored with the prefix "fixed_...." If the research involves time-dependent deformation, the strain values may be divided by the time interval between acquisition of the first and second images to produce an appropriately scaled strain rate (e.g., strain per hour or strain per minute).

Table 6.1 Files generated by StrainMapperJ and StrainMapperJ Analysis ([a])

Filename	Description
AbsDispl.tif	Contains the absolute displacement, $d = (\Delta x^2 + \Delta y^2)^{1/2}$, of the first image to second. Uses deltaX and deltaY images for the calculation
Area.tif	Contains the area stretch or contraction per original area of the first image to the second. Uses epsx, epsy, and epsxy for the calculation
cum_ ....tif [a]	Contains cumulative strains of various types. These are generated from cumulative true strains
deltaX.tif	Contains the displacement of the first image to the second in the x-direction. This is a simple conversion of information generated by bUnwarpJ into a 32-bit real image. This uses values stored in the raw file
deltaY.tif	Similar to deltaX but represents the displacement in the y-direction. Note that the y-coordinate system of ImageJ is unconventional. The origin (0,0) is in the upper left corner of the image
direct_transf.txt	Contains the basis-spline coefficients generated by bUnwarpJ. These coefficients represent the transformation of the first image to the second
ellipses.tif	Contains an image of a grid array of ellipses representing the strain circle
eps1.tif	Contains the first principal strain (eps1) at each pixel. This is calculated from the image frame strains, epsx, epsy, and epsxy
eps2.tif	Contains the second principal strain (eps2) at each pixels. This is calculated from the image frame strains, epsx, epsy, and epsxy
epsx.tif	Contains the strain in the x-direction at each pixel
epsxy.tif	Contains the XY shear strain at each pixel
epsy.tif	Contains the strain in the x-direction at each pixel. This is calculated from the deltaY displacement map
epsyx.tif	Contains the YX shear strain at each pixel
fixed ... .tif	The files in the "fixed...." type are versions of the originals in that pixel values in a two-pixel boundary region are set to 0.0, the displayed values are set to be symmetric around 0.0, and the "phase" LUT is applied. These changes are only recognized in ImageJ. If you wish to preserve the LUT and symmetric display, convert the image to an RGB and save
hexmesh.tif [a]	This product contains a synthetic trajectory map of the motion between the two images. The trajectory shown with a deforming hexagon mesh. Similar to the trajectory product
inverse_transf.txt	Contains the basis-spline coefficients generated by bUnwarpJ. These coefficients represent the transformation of the second image to the first
raw.txt	Contains ascii arrays that represent the position of each pixel from the first position in the second image. This file contains information that is reformatted into the deltaX and deltaY image files
Registered Source Image.tif	This is a stack containing images generated by bUnwarpJ that relate the transformation of the first image to the second
Registered Target Image.tif	This is a stack containing images generated by bUnwarpJ that relate the transformation of the second image to the first
start frame with ellipses.tif	This is a composite image that overlays the principal strain ellipse file over the first image in the image pair
start frame with vectors.tif	This is a composite image that overlays displacement vector file over the first image in the image pair

(continued)

Table 6.1 (continued)

Filename	Description
theta.tif	This image contains the angle of the principal strains eps1 and eps2 at each pixel. This is calculated from the image frame strains, epsx, epsy, and epsxy
vectors.tif	Contains an image of a grid array of vectors representing the displacement vector within a gridded region. Each vector represents the displacement in X and Y (deltaX and deltaY) averaged over a small square subregion. The dimensions of the grid are set within the macro
vorticity.tif	Contains an image showing the spatial gradient of velocity at each pixel. This image is calculated from the displacement image above, which is actually a velocity. This product is 32 bit, and each pixel has the units of displacement/distance. The time dimension is assumed to be the time span between the image pair. The value of this produce should be scaled to indicate per-minute or per-hour velocities
trajectory.tif [a]	This product contains a synthetic trajectory map of the motion between the two images. A synthetic image containing randomly placed randomly colored circles is deformed according to the displacements determined by bUnwarpJ. Multiple images can be maximum projected into a single image that appears similar to time projections of nuclei H2B-labeled time lapses

5. **eps2eps12.** This function converts ε_{xx} and ε_{yy} strains to principal strains. Principal strains ε_1 and ε_2 at each pixel can be found after rotating the XY coordinate system by Θ. At this rotation, the shear strain is identically zero (e.g., $\varepsilon_{xy} = \varepsilon_{yx} = 0$). Strain values of ε_1 and ε_2 and the angle of Θ in degrees for each pixel are stored as 32-bit real image arrays in eps1.tif, eps2.tif, and theta.tif files. Again, principal strain rates can be obtained by dividing the strain by the time interval between the first and second image.
6. **eps2ellipse.** This function converts principal strain maps to a grid of strain ellipses. The function generates a user-defined square grid over the image with the size specified in the function. Within each square of the grid, the mean values of ε_1, ε_2, and the angle Θ are calculated and used to produce a strain ellipse, or strain circle, representing the state of strain at the grid point. The major and minor axes of the strain ellipse are determined by ε_1 and ε_2, respectively, and oriented according to the angle Θ. The size of strain ellipses is normalized to the maximum principal strains within that image. A unit circle is drawn in the upper left corner above the vector grid.
7. **abs2vector.** This function converts ΔX and ΔY displacement maps to a grid of displacement vectors. Like the function eps2ellipse, this function generates a user-defined square grid over the image. Within each square of the grid, the mean values of ΔX and ΔY are calculated and used to produce a vector representing the average direction and magnitude of tissue movement in that grid square. The magnitudes of the vectors are normalized to values within that time step, and a unit vector is drawn on the upper-left corner above the vector grid.

The products of these functions are presented as either pseudocolored fields with the "phase" lookup table for real functions symmetric around zero or with the "fire" lookup table for positive valued functions. Graphical elements of strain ellipses and velocity/displacement arrow and quivers are provided as single images and as composite images overlaid on the first frame of the time lapse.

6.2.3 *Overview of Pixel-by-Pixel Displacement*

Images from time-lapse sequences consist of two-dimensional arrays of intensities where each pixel reports the intensity at that position in the image, e.g., an intensity of 212 at (X, Y) of (352, 667). When a tissue is deformed, that location of that intensity can be displaced to another position, say to (345, 670). The displacement of the pixel in the X-direction would be -7 and $+ 3$ in the Y-direction. The ImageJ plugin bUnwarpJ estimates displacements across the entire field of view using a set of b-splines (for a detailed discussion of elastic registration and validation of bUnwarpJ, see Arganda-Carreras et al. 2006; Arganda-Carreras et al. 2008). bUnwarpJ generates displacement fields as 2D arrays of displacements in the X-direction and in the Y-direction. StrainMapperJ places these arrays into two 2D images for subsequent analysis. Later, we will describe the use of the elastic warping function in bUnwarpJ to generate synthetic maps of tissue trajectories and hexagon meshes.

6.2.4 *Overview of Pixel-by-Pixel Strain*

Strain fields or strain maps are calculated directly from displacement fields. Strain is a unitless measure of how much an object is deformed relative to its initial shape (Fung 1981; Vogel 2003). Strains in the laboratory reference frame, e.g., changes in X in the X- (ε_{xx}) and changes in Y- in the Y-direction (ε_{yy}), and changes of X- in the Y-direction or reciprocal and XY- and YX- shear (ε_{xy} or ε_{yx}), are calculated directly from the X- and Y-displacements (Fig. 6.1d). Strain values can be stored in real-valued images, with negative values ($\varepsilon < 0$) representing compressive strain, where the tissue becomes smaller in that direction, and positive values ($\varepsilon > 0$) representing extensional strain, as tissue becomes larger. Strain in the laboratory reference frame can be recast into principal strains (ε_1 and ε_2) in a local frame of reference that minimizes shear. In 2D, principal strains can be represented as the maximum strain ε_1 (eps1), minimum strain ε_2 (eps2), and the orientation angle, Θ (theta), of maximum strain to the laboratory frame of reference. All three are generated as images by StrainMapperJ.

6.2.5 *Post-Processing: Collecting SMJ-Generated Files from a Stack Analysis into "Strain" Stacks and Synthetic Trajectory Maps*

SMJ analyses of time lapses can generate hundreds of files, so we have written ImageJ macros to collect these files into sequences that parallel the original time-lapse sequence. Like SMJ, this process is automated by a custom macro "SMJ Stack Analysis" that begins with a dialog box where the user can select the output files they want collected into a stack (Fig. 6.3c). Most of these files are simply collected into stacks. Several options require post-processing of SMJ-generated files. The first is the generation of cumulative strain. The problem is that the Cauchy strain from one time step to the next is not additive. Since reference configurations, e.g., L_0 configurations, constantly change over time, we must use true strain to visualize cumulative strain: ($\varepsilon_{\text{“true”}} = \Sigma_{(i=1)}^{n}$ [(ln(1+)] ε_i)) Cumulative ε_{xx} and ε_{yy} and area strains are also generated as images or sequences.

As we began using SMJ, we recognized the need to visualize deformation and flow with a consistent framework. Even as SMJ can automate the processing of time-lapse sequences, we wanted a quick way to evaluate the quality of those analyses. We generated synthetic random dot patterns ("polka dots") and regular hexagonal meshes ("honey combs") and used SMJ-calculated bUnwarpJ b-spline coefficients to deform them at each time step (Fig. 6.3d). Time-lapse sequences of dots and hexagons can be overlaid onto the time-lapse image stacks. The quality of tracking can be quickly evaluated by eye. Overlaying these deformed patterns onto the original image sequence provides a simple visual check. Even small errors in the displacement field produce notable distortions, allowing the quality of tracking to be evaluated by eye.

6.2.6 *Using StrainMapperJ as an Exploratory Tool*

Recently, we found and described a series of complex tissue movements in the *Xenopus* embryo. These movements appear to shape the developing posterior axis during tailbud formation (Masak and Davidson 2025). The tailbud is a 3D tissue composed of multiple cell layers with distinct cell identities. Motion of these tissues in vivo appears to involve spatially localized forces and mechanical constraints. which we evaluated with SMJ, along with additional tools. Similar methods can be applied to explorin complex movements in a simpler 2D system. We turned to apply SMJ to an epithelial sheet spreading model of the *Xenopus* animal cap ectoderm (Stepien et al. 2019) where spreading dynamics was perturbed by placing a nonadhesive obstacle akin to a study with cultured cells (Vedula et al. 2014). The advantages of this model are that tissue motions can be easily recorded with brightfield stereomicroscopy, and we can track motions in the sheet before and after "collision" with the obstacle (Fig. 6.4a, a'). These examples include use of SMJ

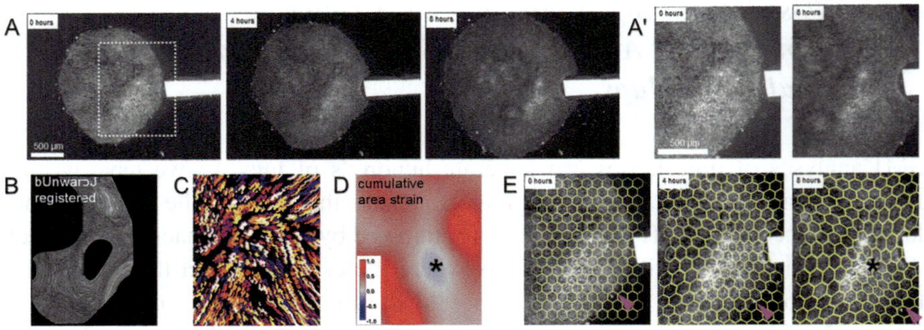

Fig. 6.4 Demonstration of the use of SMJ and SMJ Analysis. (**a**) Preliminary study of "collision" of a spreading epithelial cell sheet with a foreign object (white bar on right). A stereoscope captured the time lapse over 8 h. A') Zoomed in view of the white dashed box in (**a**) showing contact with the obstacle is shown. (**b**) Initial efforts with long interval subsampled time lapse resulted in a catastrophic failure of bUnwarpJ. This failure was avoided by subsampling with a shorter time interval. Initial efforts with bUnwarpJ can be helpful in tuning parameters to allow appropriate elastic registration. (**c**) A time projection of the synthetic dot trajectory reveals spatial complex patterns of spreading as the epithelia contacts and then moves around the obstacle. (**d**) The cumulative strain area map reveals expansion over much of the explant (red). However, one region shows compression (asterisk; blue). Compression is spatially localized to a region nearby but not contacting the obstacle. (**e**) Expansion and compression can also be seen in the synthetic hexagon trajectory map over the full 8 h. The site of compression (asterisk) is clearly seen in the compressed hexagons. "Slippage" of the SMJ deformation is also seen at the margin of the explant as the hexagonal meshwork falls behind the leading edge of the spreading explant (magenta arrowheads)

with brightfield time-lapse sequences of whole embryos and tissue explants but as shown above, SMJ is equally useful for quantifying fluorescence confocal time-lapse sequences.

We applied SMJ to characterize these movements. Initial trials using sparse frames sampled from a high interval time lapse resulted in catastrophic failure of bUnwarpJ (Fig. 6.4b). B-spline solutions did not converge and produced nonsensical deformations showing highly warped tissues. We found that decreasing the time interval between frames prevented this failure and proceeded to characterize tissue displacement using synthetic dot trajectories; maximum intensity projections of the time-lapse sequence of the dot motions, e.g., a t-projection, show how the tissue slows as it encounters the obstacle (Fig. 6.4c). The cumulative area strain map also reveals that this region is compressed (asterisk, Fig. 6.4d). Tissue deformation shown in synthetic hexagon trajectories indicates that tissues spread around the obstacle, but that spreading is limited in a region behind the immediate contact site (asterisk, Fig. 6.4e). Compression is uncommon in spreading epithelial sheets and, in this case, appears 50–200 μm from the contact site, suggesting supracellular forces from collision with the obstacle are transmitted multiple cell diameters and that these forces are integrated with intrinsic force generation by local cells during tissue spreading. We also note that the hex mesh shows "slippage" when compared to tissue motion (arrowheads; Fig. 6.4e). Tissue edges spread and escape the hex mesh as if the mesh were slipping over the tissue's surface. As noted below, slippage may reflect inelastic cell-scale processes that are not adequately represented by the elastic nature of b-spline registration methods. Further analysis with confocal time-lapse microscopy using more discrete methods is needed to describe and identify molecular mechanisms responsible. Rapid analysis of this experimentally simple 2D model with SMJ may shed light on the complex 3D movements that shape the embryo.

StrainMapperJ generates a variety of products; some, like maps of the strain ellipses are more useful for qualitative purposes, while others, such as the area strain maps can be used for quantitative analyses. Consider using tools such as "Analyze- > PlotProfile" or "Analyze- > Measure" to extract quantitative strain features from specific regions of interest. Products from StrainMapperJ should be validated against manually tracked features. For instance, bulk strain can typically be measured manually as the change in distance between two landmark features divided by their original distance. This bulk value can be compared to the average StrainMapperJ-calculated strain over a region spanning the two features.

6.3 Discussion

Strain maps can expose sites where active mechanical processes may contribute to shape change, e.g., where cells generate forces that drive tissue movement. Tissue strain may reflect externally applied forces, e.g., from adjacent tissues, or the sites of internally generated body forces, e.g., produced by tissue-autonomous programs of morphogenesis. Strain maps include both the direction and magnitude of shape change in a tissue and can be used to compare conditions across samples. Compressive strain can reveal the mechanisms underlying force generation. Thus, high-quality strain maps are critical tools for generating hypotheses about the origin of forces that shape tissues and for testing those hypotheses.

StrainMapperJ is one of many tools for digital image correlation. Furthermore, strain can be evaluated through several methods that operate on pairs of images (Boyle et al. 2014). While many of these tools are available from commercial sources, we find that open-source codes are more flexible and have proven to be exceptionally robust. SMJ is coded in ImageJ macro language, a lightweight interpreted language with a broad user base and active user community. SMJ is agnostic to the origin of the images. We have used it with brightfield sequences collected with stereoscopes, compound scopes, and image sequences collected with confocal microscopy. Furthermore, we find SMJ is easily integrated with PIV and other segmentation software that are already available as plugins for ImageJ. SMJ can save a great deal of time and effort compared to SEG- and PIV-based approaches. Surprisingly, we have found that SMJ does not require long-duration sequences and, in fact, can extract information from short duration sequences that have not been previously used to characterize morphogenesis. This last advantage is offered by many DIC-based tools and has not been leveraged in the study of morphogenesis, where long, near-perfect sets of image acquisition are gold standards. The ability to assess the sites of mechanical cues from 10 to 15-min-long time-lapse sequences could be transformative for the field of mechanobiology, enabling significant increases in throughput.

6.3.1 Limitations of SMJ

There are two key assumptions, common to all forms of motion analysis including SMJ: (1) tissues are assumed to be continuous throughout the time-lapse sequence with little loss or creation of new patterns, and (2) tissue movements are assumed to be primarily in the 2D imaging plane of focus. The first of these assumptions that is shared with many tools that map flow and strain is that tissues remain continuous and do not change significantly between sequential images. For instance, temporal discontinuities may occur as cells leave or join the tissue. Additionally, tracked tissues may include material discontinuities such as rigid boundaries or abrupt changes in elastic or viscous material properties.

The second limitation emerges when the surface being tracked is 3D. Out-of-plane motions are ignored by DIC techniques that operate on image sequences produced by maximum-projection or extended-depth-of-focus preprocessing as these reduce 3D surfaces to 2D images. For our applications, minor violations of these assumptions are tolerated, but large violations should be considered and evaluated when interpreting the analysis results.

6.3.2 Trouble-shooting SMJ

Lastly, we find that use of SMJ for each new set of experiments requires some initial trial and error. Commonly, defects in elastic registration are observed in SMJ-generated source and target files. Collected maps of displacement can also reveal catastrophic failure. During the trial-and-error phase, we make use of two tools. First, we generate image pairs and use the plug-ins SIFTS and bUnwarpJ directly to find the best parameter settings. We advise following the advice on the FIJI/Wiki pages for these plug-ins to resolve problems. Changing the time intervals between image pairs or changing the set of weights can greatly improve the performance of bUnwarpJ. Next, we return to the SMJ Analysis macro to track deformation in synthetic hexagon meshes. Catastrophic failure of bUnwarpJ will not produce coherent motion in the hex mesh. Still, even coherent hex mesh tracking motions can reveal "slippage." Slippage occurs when hex mesh displacements lag behind the actual displacements. We never see SMJ displacements running ahead of deforming tissues and suspect that slippage reflects inelastic processes such as cell intercalation, cell extrusion, or proliferation in deforming tissues. Earlier work in our group sought to reduce slippage when tracking epithelial morphogenesis and collective heart progenitor cell migration using a hybrid tracking algorithm ("GridTracker"; (Jackson et al. 2017)) that mixed DIC with PIV techniques.

Our group has been tinkering with StrainMapper for over 15 years. In the short term, several new features could be included for post-analysis, such as normalizing the vector and ellipse products to the stack rather than the individual time step. Overlays such as the hexagon mesh are helpful graphical aids, and we would like to add contour lines of strain, displacement, and vorticity fields. Using Python, we have recently augmented the StrainMapperJ analytic pipeline with tools borrowed from astrophysical analysis (Cuissa and Steiner 2022) to characterize the strength and compactness of tissue flow vortices in the *Xenopus* early tailbud embryo (Masak and Davidson 2025). Porting these codes to ImageJ would extend vorticity analysis to include vortex strength and size. Lastly, future work may use machine learning DIC tools to replace bUnwarpJ (Yang et al. 2022) and reduce the need for user action to select suitable weights and time intervals.

Acknowledgments We thank members and collaborators of the MechMorpho/Davidson Lab for their input on the earlier versions of StrainMapperJ (SMJ). These efforts worked out many of the indexing errors. We would also thank the creators of bUnwarpJ, SIFTS, and the many contributors to ImageJ/FIJI. While many coders may scoff at the use of the ImageJ macro language, it is stable and constantly evolving in a way that is backwards compatible. We also thank the National Institutes of Health for support over the years developing SMJ (R01 HD044750, R37 HD044750 to LAD). Additionally, SA was supported by the Biomechanics in Regenerative Medicine (BiRM) Training Grant from the NIBIB (T32 EB003392).

References

Aigouy B, Prud'homme B (2022) Segmentation and quantitative analysis of epithelial tissues Drosophila: Methods and Protocols. Springer

Arganda-Carreras I, Sorzano COS, Marabini R, Carazo JM, de Solorzano CO, Kybic J (2006) Consistent and elastic registration of histological sections using vector-spline regularization. In: CVAMIA: Computer Vision Approaches to Medical Image Analysis, vol 4241, pp 85–95

Arganda-Carreras I, Sorzano COS, Kybic J, Ortiz-de-Solorzano C (2008) bunwarpj: consistent and elastic registration in imagej, methods and applications. In: Second imageJ user & developer conference, vol 12

Arnold TR, Shawky JH, Stephenson RE, Dinshaw KM, Higashi T, Huq F, Davidson LA, Miller AL (2019) Anillin regulates epithelial cell mechanics by structuring the medial-apical actomyosin network. elife 8:e39065

Balasubramaniam L, Mège R-M, Ladoux B (2022) Active nematics across scales from cytoskeleton organization to tissue morphogenesis. Curr Opin Genet Dev 73:101897

Blaber J, Adair B, Antoniou A (2015) Ncorr: open-source 2D digital image correlation Matlab software. Exp Mech 55(6):1105–1122

Blanchard GB, Kabla AJ, Schultz NL, Butler LC, Sanson B, Gorfinkiel N, Mahadevan L, Adams RJ (2009) Tissue tectonics: morphogenetic strain rates, cell shape change and intercalation. Nat Methods 6(6):458–464

Bookstein FL (1995) How to produce a landmark point: the statistical geometry of incompletely registered images *vision geometry IV*, vol 2573, San Diego

Boyle JJ, Kume M, Wyczalkowski MA, Taber LA, Pless RB, Xia Y, Genin GM, Thomopoulos S (2014) Simple and accurate methods for quantifying deformation, disruption, and development in biological tissues. J R Soc Interface 11(100):20140685
Chuai M, Serrano Nájera G, Serra M, Mahadevan L, Weijer CJ (2023) Reconstruction of distinct vertebrate gastrulation modes via modulation of key cell behaviors in the chick embryo. Sci Adv 9(1):eabn5429
Cuissa JRC, Steiner O (2022) Innovative and automated method for vortex identification I. description of the SWIRL algorithm. Astron Astrophys 668:A118
Etournay R, Merkel M, Popović M, Brandl H, Dye NA, Aigouy B, Salbreux G, Eaton S, Jülicher F (2016) TissueMiner: a multiscale analysis toolkit to quantify how cellular processes create tissue dynamics. elife 5:e14334
Feroze R, Shawky JH, von Dassow M, Davidson LA (2015) Mechanics of blastopore closure during amphibian gastrulation. Dev Biol 398(1):57–67
Fung YC (1981) Biomechanics: mechanical properties of living tissues. Springer-Verlag, Berlin
Honda H (1983) Geometrical models for cells in tissues. Int Rev Cytol 81:191–248
Ishihara S, Sugimura K (2012) Bayesian inference of force dynamics during morphogenesis. J Theor Biol 313:201–211
Jackson TR, Kim HY, Balakrishnan UL, Stuckenholz C, Davidson LA (2017) Spatiotemporally controlled mechanical cues drive progenitor mesenchymal-to-epithelial transition enabling proper heart formation and function. Curr Biol 27(9):1326–1335
Kim Y, Hazar M, Vijayraghavan DS, Song J, Jackson TR, Joshi SD, Messner WC, Davidson LA, LeDuc PR (2014) Mechanochemical actuators of embryonic epithelial contractility. Proc Natl Acad Sci USA 111(40):14366–14371
Li X, Huebner RJ, Williams MLK, Sawyer J, Peifer M, Wallingford JB, Thirumalai D (2025) Emergence of cellular nematic order is a conserved feature of gastrulation in animal embryos. Nat Commun 16(1):5946
Masak G, Davidson LA (2025) Supracellular mechanics and counter-rotational bilateral flows orchestrate posterior morphogenesis. bioRxiv:689090
McCormick N, Lord J (2010) Digital image correlation. Mater Today 13(12):52–54
Nájera S, Guillermo P, Alex M, Steventon B, Weijer CJ, Serra M (2025) Control of tissue flows and embryo geometry in avian gastrulation. Nat Commun 16(1):5174
Olufsen SN, Andersen ME, Fagerholt E (2020) μDIC: an open-source toolkit for digital image correlation. SoftwareX 11:100391
Petitjean L, Reffay M, Grasland-Mongrain E, Poujade M, Ladoux B, Buguin A, Silberzan P (2010) Velocity fields in a collectively migrating epithelium. Biophys J 98(9):1790–1800
Saalfeld S, Cardona A, Hartenstein V, Tomančák P (2010) As-rigid-as-possible mosaicking and serial section registration of large ssTEM datasets. Bioinformatics 26(12):i57–i63
Schroeder AB, Dobson ETA, Rueden CT, Tomančák P, Jug F, Eliceiri KW (2021) The ImageJ ecosystem: open-source software for image visualization, processing, and analysis. Protein Sci 30(1):234–249
Serra M, Streichan S, Chuai M, Weijer CJ, Mahadevan L (2020) Dynamic morphoskeletons in development. Proc Natl Acad Sci 117(21):11444–11449
Song J, Shawky JH, Kim Y, Hazar M, LeDuc PR, Sitti M, Davidson LA (2015) Controlled surface topography regulates collective 3D migration by epithelial-mesenchymal composite embryonic tissues. Biomaterials 58:1–9
Stepien TL, Lynch HE, Yancey SX, Dempsey L, Davidson LA (2019) Using a continuum model to decipher the mechanics of embryonic tissue spreading from time-lapse image sequences: an approximate Bayesian computation approach. PLoS One 14(6):e0218021
Su J, Brau RR, Jiang X, Whitesides GM, Lange MJ, So PT (2007) Geometric confinement influences cellular mechanical properties II – intracellular variances in polarized cells. Mol Cell Biomech 4(2):105–118
Sutton MA, Orteu J-J, Schreier HW (2009) Image correlation for shape, motion and deformation measurements: basic concepts, theory and applications. Springer

Tinevez JY, Perry N, Schindelin J, Hoopes GM, Reynolds GD, Laplantine E, Bednarek SY, Shorte SL, Eliceiri KW (2017) TrackMate: an open and extensible platform for single-particle tracking. Methods 115:80–90

Unser M (2008) Advanced image processing and analysis using ImageJ. In: 8th European light microscopy initiative meeting

Vedula SR, Hirata H, Nai MH, Brugues A, Toyama Y, Trepat X, Lim CT, Ladoux B (2014) Epithelial bridges maintain tissue integrity during collective cell migration. Nat Mater 13(1):87–96

Vogel S (2003) Comparative biomechanics: life's physical world. Princeton University Press, Princeton

Wang CW, Ka SM, Chen A (2014) Robust image registration of biological microscopic images. Sci Rep 4(1):6050

Weng S, Devitt CC, Nyaoga BM, Alvarado J, Wallingford JB (2025) PCP-dependent polarized mechanics in the cortex of individual cells during convergent extension. Dev Biol 523:59–67

Yang R, Li Y, Zeng D, Guo P (2022) Deep DIC: deep learning-based digital image correlation for end-to-end displacement and strain measurement. J Mater Process Technol 302:117474

Zamir EA, Czirok A, Rongish BJ, Little CD (2005) A digital image-based method for computational tissue fate mapping during early avian morphogenesis. Ann Biomed Eng 33(6):854–865

Zhou J, Kim HY, Wang JH, Davidson LA (2010) Macroscopic stiffening of embryonic tissues via microtubules, RhoGEF and the assembly of contractile bundles of actomyosin. Development 137(16):2785–2794

Zhou J, Pal S, Maiti S, Davidson LA (2015) Force production and mechanical accommodation during convergent extension. Development 142(4):692–701

Part II
Mechanical Effects on Cellular, Organellar, and Molecular Systems

Chapter 7
Network-Associated Mechanotransduction in Health and Disease

Béla Suki, Joseph K. Hall, and Erzsébet Bartolák-Suki

Abstract Mechanical forces are known to influence many cellular and extracellular processes in the body via mechanical signal transduction pathways, collectively called mechanotransduction. The corresponding mechanical information transmission is a multiscale process: starting at the nanoscale level of motor proteins, mechanical forces percolate through the cytoskeleton of the cell to create contractile forces in the extracellular matrix, where they are further transmitted to reach the tissue and organ levels. This force transmission is supported by various networks of simpler elements, such as collagen and elastin molecules that form fibrils, fibers, and fiber networks. The efficiency of force transmission, including the magnitude and direction of the force, heavily depends on the organization of the network structure. In the limit, when the transmission pathway through a network is broken, the force signal does not reach its target, and no response can happen. Thus, mechanical force and stress transmission require a connected or percolating network. This chapter reviews how emergent network phenomena contribute to mechanotransduction at various length scales. Sufficiently realistic intra- and extracellular network models should eventually be able to help understand disease pathogenesis and progression, as well as drug effects, with potential for guiding clinical treatments.

7.1 Introduction

A defining mechanism governing biological processes is the production and release of a messenger by one biological entity that can travel to and influence the activity of another. Such entities can be organs, tissues, cells, intracellular organelles, or even individual molecules, and such messages can be vocalizations, pheromones, hormones, or myriad other means of sending a signal. The notion that such a distant action can occur in biology was first recognized by Claude Bernard, who observed

---

B. Suki (✉) · J. K. Hall · E. Bartolák-Suki
Department of Biomedical Engineering, Boston University, Boston, MA, USA
e-mail: bsuki@bu.edu

M. Kloc, J. Wosik (eds.), *Mechanotransduction and Mechanosensing*, Results and Problems in Cell Differentiation 78, https://doi.org/10.1007/978-3-032-19509-8_7

in the mid-nineteenth century that the liver glands released a substance into the bloodstream that affected cells in other parts of the body, which he later identified as glucose. This opened a new chapter in biology, laying out the foundations of biochemical signaling. The discovery in the 1950s of cyclic adenosine monophosphate (Sutherland and Rall 1958), the first intracellular second messenger to be isolated and identified, established a key principle of cellular communication via second messengers that has become central to biology and medicine (Newton et al. 2016). Typically, second messengers such as lipid messengers, ions (e.g., calcium), gases (e.g., nitric oxide), and other free radicals are released at the cell membrane from where they quickly diffuse through the cytosol or the extracellular matrix (ECM). Once they reach their target, these messengers trigger further action such as biochemical signaling, ultimately leading to whole-cell responses, including altered gene expression or protein activity.

Relaying signals, however, can be achieved in other ways. For example, mechanical forces generated by contracting muscle cells are transmitted along tendons and ligaments to the bone, allowing motion and locomotion. The phenomenon of mechanical force-induced signaling was proposed by Julius Wolff in 1870 who observed that trabecular bone anatomy matched the topography of internal mechanical stresses (force over cross-sectional area) which led him to conclude that "... bone trabeculae will disappear in places where they are no longer in use due to the bone having been bent, and that in those directions where, in bone regeneration as well in bent bones, material is to be utilized anew, new bone trabeculae will have to grow" (Wolff 2010). In other words, mechanical stresses, such as those due to body weight, somehow inform the bone to remodel its ECM. As summarized in (Richardson and Keuck 2022), Wilhelm His suggested in the late 1800s that cells and tissues are subject to the laws of mechanics. These two fundamental propositions, Wolff's homeostasis and His' mechanical action on cells, laid the conceptual foundation of mechanobiology, loosely defined as an interdisciplinary field that investigates how mechanical factors such as stress or stiffness (stress over strain) influence biological behavior. Mechanical information transmission is undoubtedly a multiscale process: starting at the nanoscale level of motor proteins, mechanical forces percolate through the cell to generate contractile forces in the ECM, which are then transmitted to the tissues and organs. It is also bidirectional because tissue-level stretches can directly affect many fundamental cellular processes. For example, stretching type II alveolar epithelial cells in the lung via macroscopic breathing (due to forces outside the lung) releases surfactant via calcium signaling (Wirtz and Dobbs 1990). Other experiments demonstrated that cell division can also be stimulated by mechanical stresses (Curtis and Seehar 1978). Intracellularly, mechanical stimulation applied at the cell surface was shown to rapidly and structurally alter both the cytoskeletal proteins and the nucleus (Maniotis et al. 1997). This phenomenon, called mechanotransduction, is defined as the conversion of mechanical stimuli to biochemical signals and has become the basis of cellular and nuclear mechanobiology.

A critical difference between second messenger- and mechanical force-induced signaling is that the former spreads more or less uniformly via diffusion over short distances, whereas the latter requires a physical solid structure to reach the target that can be at a long distance from the source. In macroscopic tissues, the solid structures include the tendons, ligaments, bone, skin, etc. At the microscale, force transmission occurs via collagen fibrils and fibers as well as elastic fibers, which are elastic structures with large aspect ratio, guaranteeing efficient load transfer (Suki 2021). In the cell, various cytoskeletal proteins, such as actin filaments and microtubules, carry and transmit forces usually from cell surface receptors such as integrins to various organelles (Wang et al. 2001), including the nucleus (Maniotis et al. 1997) and mitochondria (Bartolak-Suki et al. 2017). Importantly, at every length scale, the relevant structure can also be considered a network of smaller-scale constituents. How mechanical forces spread in a network strongly depends on what the network links are made of and how they are interconnected (Ruiz-Franco and van Der Gucht 2022), a concept that may be called the mechanical communication network. The strength of a mechanical signal at the distal end of the communication network is determined not only by the magnitude of the force signal at the inlet of the network but also by the architecture and, hence, the specific connectivity of the network. A strongly connected network transmits forces in every direction, similar to the diffusion of chemical messengers, whereas a loosely connected network allows mechanical stimuli to propagate in an asymmetric fashion (Ruiz-Franco and van Der Gucht 2022).

In the limit when the mechanical transmission pathway through a network is broken, the force signal does not reach the target, and no response can happen. Thus, mechanical stress and force transmission require a connected network; hence, it is useful to consider how the transmission pathway is organized. Statistical physics and the science of complexity deal with the characterization of complex network structures (Albert and Barabasi 2002). Furthermore, the theory of percolation (Stauffer and Aharony 1992) provides a useful framework to characterize the conditions under which a network allows information to be transmitted through it. Percolation in a network occurs when one side of the network is linked to the opposite side via an unbroken chain of existing links. When links are added randomly to a network with a given probability, connecting the opposite boundaries of a system occurs at a well-defined probability, the percolation threshold (Stauffer and Aharony 1992). In a physical network, the force-transmitting elements must therefore form an unbroken percolating path to trigger a biochemical cascade at the distal end via mechanotransduction. For example, triggering gene expression via mechanotransduction requires that the actin filaments in the cell be highly polymerized, running from the membrane to the nucleus and forming a percolation pathway for force transmission. Myosin molecular motors create an active, gel-like percolating force-carrying substance (Alvarado et al. 2017) which clearly contributes to mechanotransduction. However, when the actin network is depolymerized, there is no percolation, and mechanical forces do not reach the nucleus. While the biochemical and signaling aspects of mechanotransduction are now well characterized (Jaalouk and Lammerding 2009; Kang and Robling 2014; Morgan et al. 2013; Sun et al. 2016; Vogel 2006), network-associated aspects of mechanotransduction have received relatively little attention (Traverso 2005;

Ingber 2006). Accordingly, the purpose of this chapter is to review the concepts of mechanotransduction from the perspective of percolation to better understand how the underlying network properties enable and augment signal transduction in health and in specific disease examples.

7.2 Mechanotransduction from a Network Perspective

Over the last 20 years, cell and molecular biology have been studied from the perspective of network theory (Barabasi and Oltvai 2004; Alon 2007; Liu et al. 2020). To create a biochemical network, biological entities such as proteins, second messengers, transcription factors, and genes can be considered nodes, whereas interactions such as binding, phosphorylation, activation, and inhibition can be thought of as edges of a network. If a reaction is thermodynamically significantly more probable in one direction, then the edges also represent directions, forming a directed graph. The general topology and dynamics of biomolecular networks are similar to those of other systems, such as the Internet or society, suggesting that, as Barabási and Oltvai put it (Barabasi and Oltvai 2004), "similar laws govern the development and function of most complex networks in nature." In addition, transcriptional and signaling networks appear to be constructed from a few simple information-processing connectivity patterns, called motifs, which seem to have been selected for multiple times in different species due to their useful biological functions (Alon 2007). Common motifs include positive and negative feedback loops that are often linked with diseases. Interestingly, systems that combine positive and negative feedback loops exhibit a rich spectrum of complex behavior such as monostability, bistability, excitability, and oscillations, which can transition into one another depending on the network's topology and the relative strengths of the loops (Tian et al. 2009). Another important concept in network theory is robustness, defined as the ability of a network to withstand perturbations without compromising function (Kitano 2004). This can be evaluated by randomly removing links from a percolated network and checking how this affects percolation and, therefore, signal transmission. In real networks, link removal may be achieved by pharmacological inhibition of interactions, mutations, elimination of chemical reactants or products, or physical damage. The larger the number of links that are required to be removed to destroy percolation, the more robust the system is.

Mechanotransduction initiated by stretching, compression, shear, pressure, touch, or exposure to stiffness triggers a cascade of biochemical reactions within the cell. Usually, the process starts at focal adhesions, which are complex molecular hubs linking the ECM to the plasma membrane, and affects ion channels, integrins, components of the cytoskeleton, including actin, microtubules, intermediate filaments, and many other proteins (Winograd-Katz et al. 2014). At the receiver end, the mechanical signal is transduced to a chemical one by protein conformational change (Vogel 2006) followed usually by phosphorylation (Hashimoto et al. 2019), allowing the cascade to progress inward through various pathways such as MAPK

(Martineau and Gardiner 2001), Hippo (YAP/TAZ) (Morgan et al. 2013), Wnt/β-catenin (Kang and Robling 2014), and calcium signaling (Judice et al. 2009). Activation of integrins and other receptors or opening of ion channel eventually reaches the nucleus, promoting or silencing genes by modulating transcription factors and altering DNA activity, subsequently modifying cellular growth, differentiation, motility, division, shape, and repair. For example, experiments demonstrated that integrin-focal adhesion signaling upregulates RUNX2 and other osteogenic genes in response to matrix mechanics, thereby controlling mesenchymal cell differentiation (Salasznyk et al. 2007). Just like other signaling pathways, mechanotransduction also has its own motifs related to the cell cycle, cell adhesion, and cell-cell and cell-ECM interactions, many of which can contribute to disease pathogenesis (Lu et al. 2024). Although the robustness of mechanotransduction has been less studied, the YAP/TAZ motif appears to qualify as robust because activated YAP governs cell behavior independent of physical perturbations (Dupont et al. 2011).

Another way mechanotransduction can proceed is by affecting and/or stimulating organelles, as evidenced by the direct mechanosensitivity of membrane-bound organelles such as the nucleus (Srivastava and Ehrlicher 2024), endoplasmic reticulum (Townson and Progida 2025), mitochondria (Alevriadou et al. 2017), and even lysosomes (Li et al. 2024). While mechanotransduction through the plasma membrane is well studied (see below), much less is known about how these internal organelles sense and transduce physical forces. Emerging evidence suggests that their membrane also harbors mechanosensitive pores and ion channels that respond to extracellular forces transmitted to these organelles via the cytoskeleton (Srivastava and Ehrlicher 2024; Townson and Progida 2025; Alevriadou et al. 2017; Li et al. 2024). For example, mitochondria form an extensive network within the cell. The topological structure of mitochondria is governed by both myosin motor activity and a microtubule-dependent irregular stretch pattern called fluctuation-driven mechanotransduction which also influences ATP production rate (Bartolak-Suki et al. 2015). Interestingly, the magnitude of fluctuations, which mimic blood pressure variability, can be optimally tuned so that the mitochondrial network reaches percolation and maximal metabolic efficiency (Bartolak-Suki and Suki 2020). This is just one example of how organelle-level mechanotransduction can impact the metabolic activity and stability of the cell. These organelle-level processes can also become pathological if the forces are nonphysiological, leading to undesired calcium regulation, aberrant gene and protein expressions, and disease.

7.3 Mechanical Network Phenomena Inside the Cell

Early experiments and models treated the cell as an elastic bag containing a viscous or viscoelastic fluid (Chien et al. 1970; Crick and Hughes 1950). For example, experiments using the micropipette aspiration technique—in which one end of a cell is sucked into a narrow glass tube while tracking its shape inside the tube—

suggested that simple continuum models may be applicable to describe the data (Hochmuth 2000). If a cell is subjected to compressive stress, the fluid pressure inside the cell will increase uniformly, including the pressure around the nucleus. In a sense, the pressure as a mechanical signal percolates through the cell body to reach the nucleus without attenuation, thereby triggering a biological response. For example, in single articular chondrocytes, mechanical forces lead to upregulation of aggrecan, type II collagen, and tissue inhibitor of metalloproteinase-1 caused by compressive pressure-induced strains at the scale of the whole cell that were transmitted to the nucleus nearly unattenuated (Leipzig and Athanasiou 2008). Although such pressure itself can directly up- or downregulate gene activity, the timescale for the response in terms of protein expression is hours if not days. However, in response to changes in the extracellular environment, such as blood pressure (Zeng et al. 2018) and blood flow (Li et al. 2014) in the vasculature, or a compressive load in cartilage (Lv et al. 2018), there are fast-acting first responders: mechanosensitive ion channels, such as the Piezo channels. Opening an ion channel allows ions to flow into the cytoplasm. For example, a calcium ion channel releases calcium into the cell, which in turn quickly diffuses throughout the cytoplasm. The extra calcium can then be taken up by organelles such as the mitochondria, regulating energy production and cell death (Duchen 2000) or influencing a host of cell functions such as actomyosin motors and cell contraction (Kuo and Ehrlich 2015). These are rapid response mechanotransduction events, occurring on timescales of tens of seconds to minutes and relatively short distances.

Although continuum mechanics-based models accounted for the observed cell membrane deformation, they lacked a crucial component of mechanotransduction, the cytoskeleton. Even in the case of compression- or tension-induced mechanotransduction via ion channels, the open or closed states of ion channels are strongly influenced by the actin cytoskeleton (Cantiello et al. 1991; Guharay and Sachs 1984). The so-called stretch-activated channels usually receive mechanical input via membrane tension or the cytoskeleton, such as actin filaments. Since the polymerized form of actin, F-actin, forms a network, it is expected that network effects cannot be neglected. This can be illustrated by examining force transmission in a simple network model of the cytoskeleton interacting with a single ion channel (Fig. 7.1). Briefly, the cell, the channel, and the actin network were represented by a hexagonal lattice of linear springs. To mimic cell shape, the intracellular actin stress fibers and the subcortical actin were modeled by removing selected springs from the inside of the network. The outer boundary of the network represented the plasma membrane with a layer of springs just below it corresponding to the peripheral actin network. In addition, two nearly straight sets of springs connecting the top and bottom boundaries were included to model stress fibers. To stabilize the network, bond bending via torsional springs, which resist changes in angle between neighboring springs, was also included in the model. Further details can be found elsewhere (Ito et al. 2009); here, only the relevant results are reported. When the torsional spring constant was low, representing stress fibers with small persistence length, the channel at the top boundary of the network did not open when the cell was stretched at the base, indicating that strong bond bending with straighter fibers

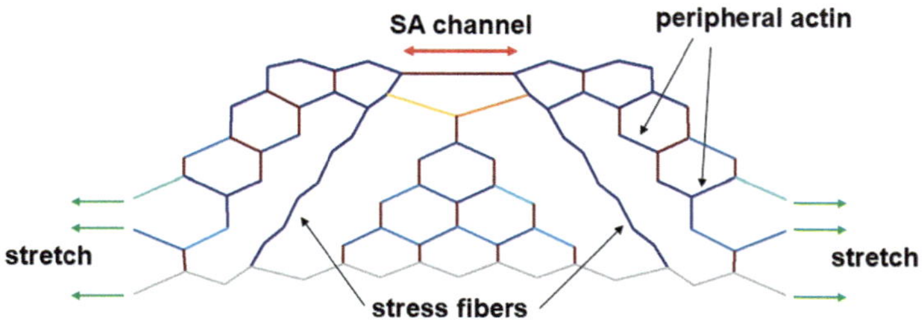

Fig. 7.1 A spring network model of the cell with a stretch-activated (SA) channel at the top, stress fibers, and peripheral actin network. The cell is stretched horizontally as shown by the green arrows. Colors show mechanical force increasing from blue to red. See text for further details. (Adapted with permission from Ito et al. (2009))

reinforces force transmission. More importantly, mimicking actin depolymerization eliminated the percolating force transmission pathway, thereby inhibiting channel opening through cell stretching. While these results may be specific to the network configuration, they are in agreement with experiments (Ito et al. 2009), clearly showing that a strong percolating network is needed to allow external stretch to open and activate mechanosensitive ion channels. Furthermore, percolation is also present between the plasma membrane and the nucleus. Mechanical forces from the ECM to the cell, transmitted through elastic fibers (Yanagisawa and Wagenseil 2020) and collagen fibrils and fibers (Suki et al. 2005), are sensed and transmitted to the cytoskeleton by integrins which are transmembrane proteins (Kolasangiani et al. 2022). Indeed, tugging on integrins demonstrated a direct force transmission pathway from focal adhesions at the cell membrane to the nucleus (Maniotis et al. 1997). This direct force transmission can also regulate gene expression.

Beyond percolation, does the strength of connectivity of the cytoskeleton play any role in force transmission? First, note that there are two additional cytoskeletal protein networks involved in mechanotransduction: the intermediate filaments that participate in force transmission to the nucleus and the microtubules that support and stabilize the nucleus during lateral compression (Maniotis et al. 1997). The three cytoskeletal networks—actin, intermediate filaments, and microtubules—also interact with each other and can undergo polymerization and depolymerization, effectively altering their connectivity. Interestingly, connectivity significantly influences how forces propagate through a network. For example, simulating a local perturbation in a diluted triangular network was used to explore how the induced force spreads through the network as a function of connectivity and bond bending (Ruiz-Franco and van Der Gucht 2022). Even though the analysis mimicked the contraction of a cell in the ECM, we can use the same results to demonstrate how force applied to an integrin receptor spreads through the cell. If the network is densely connected, the force or stress spreads equally in all directions, as in a continuum, decreasing with the square of the distance from the source. In this case, bond bending has little influence on force transmission. However, as the network is gradually diluted and its connectivity approaches the percolation threshold, the force propagates in discrete directions, and bond bending becomes necessary, in agreement with the results in Fig. 7.1. Furthermore, the stress now decreases inversely with distance, not with the square of the distance. This is important since larger forces can be transmitted over longer distances.

Instead of the widely used regular networks with pure tensile elements, another frequently referenced model of cell architecture is the tensegrity structure, which balances compression and tension resulting in a stable mechanical equilibrium (Ingber 2006; Ingber 1997). Tension on actin filaments is generated by the myosin motors, which tend to collapse the cell. Microtubules, however, act as compression-resistant elements that balance the inward forces on actin (Wang et al. 2001; Chen et al. 2010). The combined system is the tensegrity structure implicated in virtually all mechanobiological responses of the cell, including ion channels and mechanotransduction at focal adhesions (Ingber 2008). Actin fibers and microtubule filaments can span longer distances within the cell, ensuring that percolation for force transmission is achieved via long fibers and filaments. The presence of such long-distance connections fundamentally alters the concept of simple random percolation (Silveira et al. 2009). The polymerization of actin and tubulin grows the filaments in a given direction, leading to percolation much sooner than in regular random percolation. This saves material and accelerates the formation of percolating structures, with major implications for force transmission and hence mechanotransduction. Furthermore, these filamentous structures are not static; they can quickly undergo polymerization and depolymerization in response to external stimuli. For example, actin polymerization responds to stretching of the ECM and the cells within it (Albinsson et al. 2004). Similarly, microtubule polymerization/depolymerization is a key mechanism associated with cell reorientation in response to cyclic uniaxial stretching (Morioka et al. 2011). Since the cell can dynamically control the organization of the cytoskeleton, it effectively fine-tunes cytoskeletal connectivity to create or eliminate percolation, which, in turn, modulates force transmission and mechanotransduction with implications for gene activity, cell, ECM, and tissue health.

7.4 Network Effects in the ECM

The ways cells are mechanically linked to their microenvironment in the ECM is bidirectional: mechanical signals are not conveyed only one way—environment to cell—but cells are also responsible for generating and transmitting mechanical signals to and through their environment. Moreover, many cell types are actively involved in shaping and remodeling their environment. A limb will not move without the coordinated contraction of muscle cells; embryos will not develop correctly without carefully timed mechanical cues, wounds would not close without the collagen sutures that myofibroblasts weave and tug tightly, etc. Thus, cells mechanically and structurally affect and are mechanically and biochemically influenced by the environment they occupy. This bidirectional process is part of the homeostatic maintenance of the ECM.

Cellular mechanical communication via the environment must also be considered a network and has been studied using network theory (Suki 2021; Ruiz-Franco and van Der Gucht 2022; Ronceray et al. 2016). Many of the arguments discussed in the previous section regarding force transmission in cells can be applied to force transmission through the ECM, tissues, organs, and the body at large. In multicellular organisms, the main network that we consider is the ECM—the nonliving part of the tissue, largely composed of collagen, elastin, proteoglycans, and their side chains, the glycosaminoglycans (Suki 2021). Like spiders on a web, cells sense, react to, and transmit mechanical cues via the ECM that has profound long-term and multiscale impact on tissue behavior and health (Discher et al. 2005; Humphrey et al. 2014). In fact, the complex interplay of cells and their environment often leads to emergent phenomena, defined as properties or behaviors that cannot be explained by the smaller-scale simpler constituents of the system in question (Anderson 1972). Emergent behavior also has implications for disease pathogenesis. For example, the fractal-like patterns seen in fibrosis (Hall et al. 2024; Wellman et al. 2018) and the progression of emphysema at the apex of the lungs (Mondonedo and Suki 2017) are large-scale behaviors that can be attributed to the network amplification of cell-scale dynamic mechanical interactions between cells and the ECM.

The concept of tensegrity discussed above in relation to cells may also be applicable to the ECM, tissues, organs, and body. Tensegrity is a multiscale concept, capable of describing individual cells, in which the cytoskeleton and cell membrane act in opposition to maintain cell shape, through tissues and all the way to whole vertebrate bodies, where the skeleton and musculature act in compression and tension, respectively, to keep the body upright (Ingber 2008). The stressed nature of tensegrity structures lends itself to long-range force transmission that transcends scale: a large-scale deformation of the whole network sends a signal to the small members that make it up, and a change in a small member can have a long-range signal that affects the network as a whole. In this way, the tensegrity acts as both a structure and a conduit for bidirectional force transmission. Consequently, the ECM is an essential tensile element in a larger tensegrity structure and is therefore continuously under a preexisting stress, called the prestress. The types of stress the ECM encounters depend heavily on the tissue that it is a part of. Joint cartilage experiences intermittent high-impact compressive stresses that get transmitted from the bones; arteries receive cyclical pressure variations from the heart; lung tissue is continuously stretched and relaxed due to the contraction of the diaphragm and other respiratory muscles. While the standard tensegrity models tend to be symmetric and uniform, tissue is not. Instead, tissue is often modeled as a network of disordered fibers (Ruiz-Franco and van Der Gucht 2022; Ronceray et al. 2016; Hall et al. 2023). Because tissues are heterogeneous, their fibrous structure does not follow a rigid grid-like lattice such as squares or cubes. An interesting consequence of this is that while regular structures often show preferential direction of force transmission, random networks exhibit more isotropic (similar in all directions) behavior (Hall et al. 2023). In organs such as the heart and lungs, isotropic force transmission is essential for normal function. When directional force transmission is needed, a diluted network performs better (Ruiz-Franco and van Der Gucht 2022). An extreme form of a diluted network is a single fiber or a set of parallel fibrils, as in a tendon, capable of efficiently transferring force from one place directly to another.

When force is generated locally in a disordered network, such as the contractile force of a single cell or a group of cells, fibers can reorient within the network to be parallel to the direction of the strain, resulting in long-range fiber alignment and force transmission along emergent structures known as force chains (Ruiz-Franco and van Der Gucht 2022; Ronceray et al. 2016). The nature of these force chains changes depending on the density of the network and other conditions such as bending stiffness of the fibers. This gradual alignment is a recruitment process that eventually leads to force or stress percolation. Interestingly, stresses propagating within a nonlinearly elastic network can exceed those transmitted in a linearly elastic network due to an unexpected emergent network phenomenon of stress amplification (Ronceray et al. 2016). This long-range communication is a direct result of the network environment being capable of mechanical change. For this reason, studying cells under mechanically inaccurate conditions is detrimental to our understanding of cellular behavior. Petri dishes are static and do not have the same prestressed tensegrity behavior that living tissue has. Hence, without capturing the network environment of cells, we are not capturing their full behavior.

7.5 Network Effects in Disease Pathogenesis

The normal process of healing can be hijacked to create pathological processes that also depend on the mechanical cellular signals. For example, aberrant mechanotransduction can lead to developmental defects, cancer, atherosclerosis, muscular dystrophies, cardiac problems, fibrosis, and many more, all of which appear to have similarities in how force transmission from the ECM, through the cytoskeleton and to the nucleus, is disturbed (Jaalouk and Lammerding 2009). Consider, as an example, fibrosis, which is caused by excessive deposition (Kirk et al. 1986) and cross-linking (Jones et al. 2018) of collagen. Fibroblasts, for instance, are highly mechanically sensitive to the stiffness of the ECM they are on and will become more differentiated toward myofibroblasts as the microenvironment becomes stiffer; this can transition to a positive feedback loop that results in a self-sustaining fibrotic process that spreads from the initial fibrotic region across the tissue (Liu et al. 2010; Moore and Herzog 2013; Lin et al. 2017; Tschumperlin 2013). In this case, force transmission occurs along stiff collagen fibers that can affect cells in normal regions of the tissue at a distance from the fibrotic loci. In the lung, tissue remodeling leads to the formation of a fibrotic fractal structure that follows a percolation-like clustering caused by the migration of and collagen deposition by activated myofibroblasts (Hall et al. 2024; Wellman et al. 2018). This cellular feedback response on stiff matrix together with the enhanced force transmission is clearly a network phenomenon that results in tissue heterogeneity that is not captured in simplistic models of cell-ECM interactions.

In organs that undergo cyclic variations in stress and strain, stretching the ECM and the embedded cells also contributes to the fibrotic process. Many cell types are stretch sensitive: fibroblasts respond to stretch by releasing ECM-bound latent TGF-beta, which may trigger the transition from fibroblasts to myofibroblasts (Hinz 2009), and by triggering signaling along pathways such as YAP and MRTF-A (Cui et al. 2015), whereas endothelial cells activate ion channels upon stretching (Ito et al. 2009). Mechanosensing via stretching can also enhance cell activation, leading to increased collagen deposition (Tschumperlin et al. 2018). In the lungs, periodic stretching due to breathing provides a constant signal to fibroblasts, and it is reasonable to assume that there is an optimal range of tissue strains for normal homeostasis. Consequently, fibroblasts will deposit more collagen in response to supraphysiological stretch because they interpret the high strains as low stiffness that needs to be restored to normal levels. By combining stretch-related homeostatic repair with the positive feedback loop based on increased stiffness described above, a computational model that included migrating fibroblasts on an ECM network showed how physiologically accurate healthy homeostatic mechanotransduction can become aberrant due to network effects, leading to massive ECM remodeling. This model was also able to produce structures that mimicked those seen in human pulmonary fibrosis (Hall et al. 2024). Furthermore, the fibrotic clusters in this model appear to exhibit the same force-chain behaviors seen in other models (Ruiz-Franco and van Der Gucht 2022; Ronceray et al. 2016), demonstrating how the behavior of a small region can be amplified by mechanotransduction leading to either healing or pathogenic behavior (Hall et al. 2024). A follow-up study demonstrated that this model exhibits a network-associated bifurcation behavior: whether or not the entire network turns fibrotic can at times rest on the behavior of a microscopic fibrotic locus, represented by a single spring in the model (Israel et al. 2025). This structural bifurcation is illustrated in Fig. 7.2, raising the possibility of therapeutic intervention if clinical images of patients are available. While this is a highly simplified spring network model, the results highlight the need to consider network effects on mechanotransduction.

Another disease in which distant percolation-like force transmission contributes to pathogenesis and progression is pulmonary emphysema which is known to instigate at the apex of the lungs, where stresses on the alveolar walls are the greatest (West 1971). This is caused by gravity as the weight of the tissue within the lungs pulls more at the top than at the bottom, known as the slinky effect, when gravity precisely counteracts the tension in the slinky (Vanderbei 2017). Similarly, in the lung and possibly in other organs, gravity causes a stress gradient in the tissue. In this case, the mechanical environment at the top of the lungs is different than at the bottom due to network effects. As cells release various enzymes such as matrix metalloproteases, elastic and collagen fibers become digested, and, under the effects of static force due to gravity and dynamic forces created by breathing,

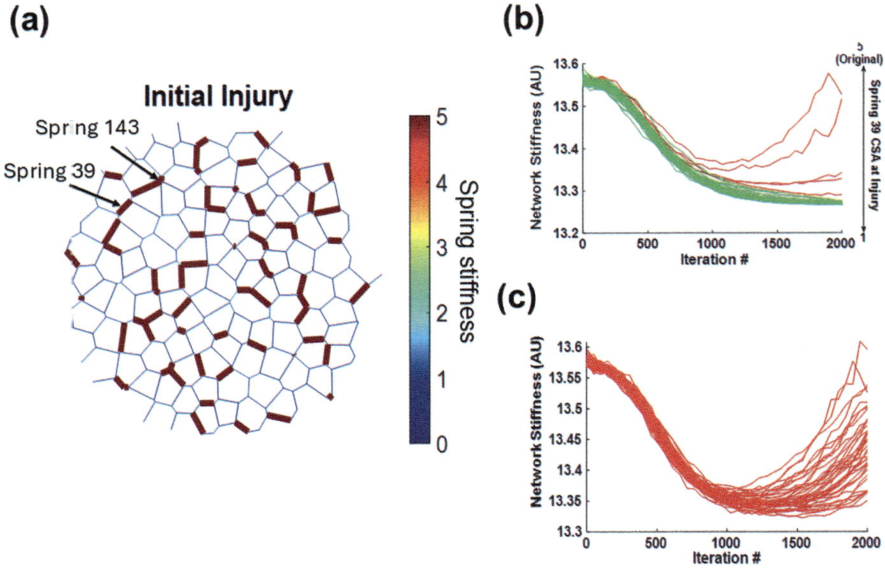

Fig. 7.2 Impact of a single spring's stiffness on the development of network fibrosis. (**a**) Schematic representation of the initial injured configuration of a network that turns fibrotic following approximately 500 iterations. Red springs represent local fibrosis with a spring stiffness five times higher than the blue springs. Springs 39 and 143 are part of the initial injury and are the two springs with the highest stiffness by iteration of 2000. (**b**) Trajectories of network stiffness over 2000 iterations with varied stiffness of spring 39 at time of initial injury, illustrating a bifurcation in the network's response as the stiffness of spring 39 decreases: above a critical value, the total network stiffness diverges, whereas below the critical value, the network heals. (**c**) Trajectories for spring 143 across all network simulations, depicting how altering its stiffness has no influence on the overall network fibrosis development. In panels (**b**) and (**c**), red curves indicate simulations where the network ultimately becomes fibrotic with diverging network stiffness, whereas green curves indicate simulations where the network heals. (Adapted with permission from Israel et al. (2025))

they eventually rupture (Kononov et al. 2001). Following rupture, the fiber ends expose chemotactic fragments that attract migrating inflammatory cells, which in turn release cytokines and additional enzymes (Houghton et al. 2006; Hunninghake et al. 1981). Furthermore, the mechanical force carried by the fiber before rupture is redistributed in the neighborhood, altering local mechanotransduction and potentiating further rupture in a positive feedback loop that defines emphysema progression (Suki et al. 2013). The lesson from this example is that in larger organs, network effects combined with the directional stress gradient of gravity, a phenomenon rarely considered in mechanobiology, create the conditions for preferential force transmission, which, in the case of emphysema, is directly responsible for the clinical phenotype.

7.6 Future Directions

In this chapter, we have highlighted how network phenomena both within the cell and in the ECM profoundly affect homeostasis and pathogenesis of disease. A better understanding of the intricate network effects at all length scales will help uncover mechanisms underlying the development of complex spatial patterns observed in a variety of diseases such as fibrosis. There is a need for improved technology that studies biological systems in their native dynamic mechanical environments. The standard of the static test has had its day; we have grown beyond the confines of the Petri dish and need to fully integrate the roles of cellular and network mechanics into future studies. Furthermore, new data will need to be analyzed and interpreted using advanced computational models potentially aided by machine learning that can capture the essence of the complexity of network phenomena in cell and ECM biology. Ultimately, sufficiently realistic intra- and extracellular network models should be able to not only study diseases but also explore drug effects on disease progression and drug side effects with the potential to guide clinical treatments.

References

Albert R, Barabási AL (2002) Statistical mechanics of complex networks. Rev Mod Phys 74:47–97

Albinsson S, Nordstrom I, Hellstrand P (2004) Stretch of the vascular wall induces smooth muscle differentiation by promoting actin polymerization. J Biol Chem 279(33):34849–34855

Alevriadou BR et al (2017) Mitochondrial ca(2+) transport in the endothelium: regulation by ions, redox signalling and mechanical forces. J R Soc Interface 14(137):20170672

Alon U (2007) Network motifs: theory and experimental approaches. Nat Rev Genet 8(6):450–461

Alvarado J et al (2017) Force percolation of contractile active gels. Soft Matter 13(34):5624–5644

Anderson PW (1972) More is different. Science 177(4047):393–396

Barabási AL, Oltvai ZN (2004) Network biology: understanding the cell's functional organization. Nat Rev Genet 5(2):101–113

Bartolak-Suki E, Suki B (2020) Tuning mitochondrial structure and function to criticality by fluctuation-driven mechanotransduction. Sci Rep 10(1):407

Bartolák-Suki E et al (2015) Fluctuation-driven mechanotransduction regulates mitochondrial-network structure and function. Nat Mater 14(10):1049–1057

Bartolák-Suki E et al (2017) Regulation of mitochondrial structure and dynamics by the cytoskeleton and mechanical factors. Int J Mol Sci 18(8):1812

Cantiello HF et al (1991) Actin filaments regulate epithelial Na+ channel activity. Am J Phys 261(5 Pt 1):C882–C888

Chen TJ et al (2010) Complexity of the tensegrity structure for dynamic energy and force distribution of cytoskeleton during cell spreading. PLoS One 5(12):e14392

Chien S et al (1970) Shear-dependent deformation of erythrocytes in rheology of human blood. Am J Physiol Legacy Content 219(1):136–142

Crick FHC, Hughes AFW (1950) The physical properties of cytoplasm: a study by means of the magnetic particle method part I. Experimental. Exp Cell Res 1(1):37–80

Cui Y et al (2015) Cyclic stretching of soft substrates induces spreading and growth. Nat Commun 6:6333

Curtis AS, Seehar GM (1978) The control of cell division by tension or diffusion. Nature 274(5666):52–53

Discher DE, Janmey P, Wang YL (2005) Tissue cells feel and respond to the stiffness of their substrate. Science 310(5751):1139–1143
Duchen MR (2000) Mitochondria and calcium: from cell signalling to cell death. J Physiol 529(Pt 1):57–68
Dupont S et al (2011) Role of YAP/TAZ in mechanotransduction. Nature 474(7350):179–183
Guharay F, Sachs F (1984) Stretch-activated single ion channel currents in tissue-cultured embryonic chick skeletal muscle. J Physiol 352:685–701
Hall JK et al (2023) Predicting alveolar ventilation heterogeneity in pulmonary fibrosis using a non-uniform polyhedral spring network model. Front Netw Physiol 3:1124223
Hall JK et al (2024) Elucidating the interaction between stretch and stiffness using an agent-based spring network model of progressive pulmonary fibrosis. Front Netw Physiol 4:1396383
Hashimoto Y et al (2019) Mechanical force induces phosphorylation-mediated Signaling that underlies tissue response and robustness in Xenopus embryos. Cell Syst 8(3):226–241 e7
Hinz B (2009) Tissue stiffness, latent TGF-beta1 activation, and mechanical signal transduction: implications for the pathogenesis and treatment of fibrosis. Curr Rheumatol Rep 11(2):120–126
Hochmuth RM (2000) Micropipette aspiration of living cells. J Biomech 33(1):15–22
Houghton AM et al (2006) Elastin fragments drive disease progression in a murine model of emphysema. J Clin Invest 116(3):753–759
Humphrey JD, Dufresne ER, Schwartz MA (2014) Mechanotransduction and extracellular matrix homeostasis. Nat Rev Mol Cell Biol 15(12):802–812
Hunninghake GW et al (1981) Elastin fragments attract macrophage precursors to diseased sites in pulmonary emphysema. Science 212(4497):925–927
Ingber DE (1997) Tensegrity: the architectural basis of cellular mechanotransduction. Annu Rev Physiol 59:575–599
Ingber DE (2006) Cellular mechanotransduction: putting all the pieces together again. FASEB J 20(7):811–827
Ingber DE (2008) Tensegrity-based mechanosensing from macro to micro. Prog Biophys Mol Biol 97(2–3):163–179
Israel E et al (2025) Bifurcation in the healing or fibrotic response in a network model of fibrosis: role of the initial injury structure. Front Netw Physiol 5:1589216
Ito S et al (2009) Actin cytoskeleton regulates stretch-activated Ca2+ influx in human pulmonary microvascular endothelial cells. Am J Respir Cell Mol Biol
Jaalouk DE, Lammerding J (2009) Mechanotransduction gone awry. Nat Rev Mol Cell Biol 10(1):63–73
Jones MG et al (2018) Nanoscale dysregulation of collagen structure-function disrupts mechano-homeostasis and mediates pulmonary fibrosis. elife:7
Judice CC, Marin TM, Franchini KG (2009) Calcium and the mechanotransduction in cardiac myocytes. Front Biosci (Elite Ed) 1(1):189–199
Kang KS, Robling AG (2014) New insights into Wnt-Lrp5/6-beta-catenin Signaling in Mechanotransduction. Front Endocrinol (Lausanne) 5:246
Kirk JM et al (1986) Biochemical evidence for an increased and progressive deposition of collagen in lungs of patients with pulmonary fibrosis. Clin Sci (Lond) 70(1):39–45
Kitano H (2004) Biological robustness. Nat Rev Genet 5(11):826–837
Kolasangiani R, Bidone TC, Schwartz MA (2022) Integrin conformational dynamics and mechanotransduction. Cells 11:22
Kononov S et al (2001) Roles of mechanical forces and collagen failure in the development of elastase-induced emphysema. Am J Respir Crit Care Med 164(10 Pt 1):1920–1926
Kuo IY, Ehrlich BE (2015) Signaling in muscle contraction. Cold Spring Harb Perspect Biol 7(2):a006023
Leipzig ND, Athanasiou KA (2008) Static compression of single chondrocytes catabolically modifies single-cell gene expression. Biophys J 94(6):2412–2422
Li J et al (2014) Piezo1 integration of vascular architecture with physiological force. Nature 515(7526):279–282

Li K et al (2024) Drosophila TMEM63 and mouse TMEM63A are lysosomal mechanosensory ion channels. Nat Cell Biol 26(3):393–403
Lin YC et al (2017) Simultaneously targeting myofibroblast contractility and extracellular matrix cross-linking as a therapeutic concept in airway fibrosis. Am J Transplant 17(5):1229–1241
Liu F et al (2010) Feedback amplification of fibrosis through matrix stiffening and COX-2 suppression. J Cell Biol 190(4):693–706
Liu X et al (2020) Robustness and lethality in multilayer biological molecular networks. Nat Commun 11(1):6043
Lu M et al (2024) Mechanical network motifs as targets for mechanomedicine. Drug Discov Today 29(10):104145
Lv M et al (2018) Calcium signaling of in situ chondrocytes in articular cartilage under compressive loading: roles of calcium sources and cell membrane ion channels. J Orthop Res 36(2):730–738
Maniotis AJ, Chen CS, Ingber DE (1997) Demonstration of mechanical connections between integrins, cytoskeletal filaments, and nucleoplasm that stabilize nuclear structure. Proc Natl Acad Sci USA 94(3):849–854
Martineau LC, Gardiner PF (2001) Insight into skeletal muscle mechanotransduction: MAPK activation is quantitatively related to tension. J Appl Physiol (1985) 91(2):693–702
Mondonedo JR, Suki B (2017) Predicting structure-function relations and survival following surgical and bronchoscopic lung volume reduction treatment of emphysema. PLoS Comput Biol 13(2):e1005282
Moore MW, Herzog EL (2013) Regulation and relevance of myofibroblast responses in idiopathic pulmonary fibrosis. Curr Pathobiol Rep 1(3):199–208
Morgan JT, Murphy CJ, Russell P (2013) What do mechanotransduction, hippo, Wnt, and TGFbeta have in common? YAP and TAZ as key orchestrating molecules in ocular health and disease. Exp Eye Res 115:1–12
Morioka M et al (2011) Microtubule dynamics regulate cyclic stretch-induced cell alignment in human airway smooth muscle cells. PLoS One 6(10):e26384
Newton AC, Bootman MD, Scott JD (2016) Second messengers. Cold Spring Harb Perspect Biol 8(8):a005926
Richardson MK, Keuck G (2022) The revolutionary developmental biology of Wilhelm His, *Sr.* Biol Rev 97(3):1131–1160
Ronceray P, Broedersz CP, Lenz M (2016) Fiber networks amplify active stress. Proc Natl Acad Sci USA 113(11):2827–2832
Ruiz-Franco J, van Der Gucht J (2022) Force transmission in disordered fibre networks. Front Cell Dev Biol 10:931776
Salasznyk RM et al (2007) Focal adhesion kinase signaling pathways regulate the osteogenic differentiation of human mesenchymal stem cells. Exp Cell Res 313(1):22–37
Silveira PSP et al (2009) Percolation in a network with long-range connections: implications for cytoskeletal structure and function. Phys A Stat Mech Appl 388(8):1521–1526
Srivastava LK, Ehrlicher AJ (2024) Sensing the squeeze: nuclear mechanotransduction in health and disease. Nucleus 15(1):2374854
Stauffer D, Aharony A (1992) Introduction to percolation theory, 2nd edn. Taylor & Francis/x, London/Washington, DC, p 181
Suki B (2021) Structure and function of the extracellular matrix: a multiscale quantitative approach. Academic Press
Suki B et al (2005) Biomechanics of the lung parenchyma: critical roles of collagen and mechanical forces. J Appl Physiol 98(5):1892–1899
Suki B et al (2013) Emphysema and mechanical stress-induced lung remodeling. Physiology (Bethesda) 28(6):404–413
Sun Z, Guo SS, Fassler R (2016) Integrin-mediated mechanotransduction. J Cell Biol 215(4):445–456
Sutherland EW, Rall TW (1958) Fractionation and characterization of a cyclic adenine ribonucleotide formed by tissue particles. J Biol Chem 232(2):1077–1091

Tian XJ et al (2009) Interlinking positive and negative feedback loops creates a tunable motif in gene regulatory networks. Phys Rev E Stat Nonlinear Soft Matter Phys 80(1 Pt 1):011926

Townson J, Progida C (2025) The emerging roles of the endoplasmic reticulum in mechanosensing and mechanotransduction. J Cell Sci 138(4):JCS263503

Traverso S (2005) Cytoskeleton as a fractal percolation cluster: some biological remarks. In: Fractals in biology and medicine. Birkhäuser, Basel

Tschumperlin DJ (2013) Fibroblasts and the ground they walk on. Physiology (Bethesda) 28(6):380–390

Tschumperlin DJ et al (2018) Mechanosensing and fibrosis. J Clin Invest 128(1):74–84

Vanderbei RJ (2017) The falling slinky. Am Math Mon 124(1):24–36

Vogel V (2006) Mechanotransduction involving multimodular proteins: converting force into biochemical signals. Annu Rev Biophys Biomol Struct 35:459–488

Wang N et al (2001) Mechanical behavior in living cells consistent with the tensegrity model. Proc Natl Acad Sci USA 98(14):7765–7770

Wellman TJ et al (2018) Topographic distribution of idiopathic pulmonary fibrosis: a hybrid physics- and agent-based model. Physiol Meas 39(6):064007

West JB (1971) Distribution of mechanical stress in the lung, a possible factor in localisation of pulmonary disease. Lancet 1(7704):839–841

Winograd-Katz SE et al (2014) The integrin adhesome: from genes and proteins to human disease. Nat Rev Mol Cell Biol 15(4):273–288

Wirtz HR, Dobbs LG (1990) Calcium mobilization and exocytosis after one mechanical stretch of lung epithelial cells. Science 250(4985):1266–1269

Wolff J (2010) The classic: on the inner architecture of bones and its importance for bone growth. 1870. Clin Orthop Relat Res 468(4):1056–1065

Yanagisawa H, Wagenseil J (2020) Elastic fibers and biomechanics of the aorta: insights from mouse studies. Matrix Biol 85-86:160–172

Zeng WZ et al (2018) PIEZOs mediate neuronal sensing of blood pressure and the baroreceptor reflex. Science 362(6413):464–467

Chapter 8
Influence of Magneto-Mechanical Stress on Cellular Membranes, Ion Channel Activity, and Gene Expression Dynamics

Vitalii Zablotskii and Tatyana Polyakova

Abstract This chapter elucidates the fundamental role of magneto-mechanical stresses as a primary mechanism through which static and low-frequency magnetic fields influence cellular processes. We demonstrate that magnetic forces, particularly when amplified by magnetic nanoparticles (MNPs), exert precise control over cellular mechanics. At the membrane level, these forces induce subtle deformations that directly modulate the gating of mechanosensitive ion channels, such as Piezo1, thereby regulating Ca^{2+} and K^{+} fluxes and initiating downstream signaling cascades that govern gene expression, metabolism, and cytoskeletal remodeling. At higher force thresholds, magneto-mechanical stress triggers significant structural alterations, including membrane blebbing due to cortical detachment, which can act as a precursor to apoptosis—an effect with pronounced implications for cancer therapeutics. Furthermore, these magnetic stresses are transmitted through the cytoskeletal network, altering the tension in F-actin, microtubules, and intermediate filaments to remodel cell shape, motility, and adhesion, thereby influencing critical processes like cell migration and differentiation. We demonstrate that the noninvasive and selective nature of this magneto-mechanical approach offers a highly versatile and potent strategy for biomedical applications.

8.1 Introduction

The search for mechanisms governing the interaction between magnetic fields (MFs) and living cells remains one of the most persistent and long-standing challenges in modern biology. While various intriguing hypotheses regarding these interactions have been proposed and explored in the literature (Zhang 2023), many fundamental aspects concerning the cellular response to MFs have yet to be fully elucidated.

V. Zablotskii (✉) · T. Polyakova
Institute of Physics of the Czech Academy of Sciences, Prague, Czech Republic
e-mail: zablot@fzu.cz

M. Kloc, J. Wosik (eds.), *Mechanotransduction and Mechanosensing*, Results and Problems in Cell Differentiation 78, https://doi.org/10.1007/978-3-032-19509-8_8

A living cell can be viewed fundamentally as an electrical entity, built upon a dense network of charges, ions, electric fields, and currents that operate across every level of cellular organization (Funk and Scholkmann 2023; Goldmann 2008). Because magnetic fields act on moving charges, and because spatially nonuniform magnetic fields exert forces on intracellular structures, it is fully expected that magnetic exposure produces a diverse spectrum of biological effects observed both in single cells and in whole organisms (Zhang 2023). From the smallest mobile ions such as potassium, sodium, calcium, and chloride to essential macromolecules like DNA, actin, microtubules, and proteins, electrical charge distributions and their interactions guide a large portion of biochemical and biophysical processes within the cell. The collective, highly synchronized operation of thousands of ion channels generates the membrane potential, an electrical voltage that arises from the difference in electric potential between charges on the inner and outer leaflets of the plasma membrane. Importantly, electrical potentials are not limited to the cell surface; membranes of intracellular organelles also maintain pronounced voltages, including the mitochondrial membrane (−140 mV), the endoplasmic reticulum (−45 mV), the nuclear envelope (−15 mV), lysosomes (+19 mV), and phagosomes (+27 mV) (Funk and Scholkmann 2023). At equilibrium, the value of a membrane potential is shaped not only by the balance of ion fluxes (primarily those of K^+, Na^+, and Cl^-) but also by the physiological condition of the cell, its developmental state, and its electrical coupling with neighboring cells through gap-junction channels.

Moreover, in the cell nucleus, chromatin positioning is directed by a weak radial electric field generated by the nuclear membrane potential. This radial electric field is believed to bias the movement of charged macromolecules within the gel-like nucleoplasm, similar to electrophoresis (Keydar n.d.). Surprisingly, recent numerical modeling has shown that the optimal distribution of receptors on the membrane of a cell of arbitrary shape, which maximizes the overall flux of molecules entering the cell, precisely corresponds to the distribution of charge density on an ideal conductor of the same shape as the cell (Wu et al. 2025). This fact reveals not only the evolutionary origin of receptor localization but also their possible electrostatic interaction.

Summarizing, a living cell functions as an electromagnetic unit, where the interaction of electrical charges drives intracellular molecular motors and various biochemical processes. From this perspective, even tiny magnetic or electric forces can alter the cell's machinery. For instance, the Coulomb force responsible for opening membrane channels is approximately 10 piconewtons (pN) (Wu et al. 2016).

This chapter aims to describe how magnetic forces influence cellular machinery, with particular emphasis on the physical mechanisms that underlie the biological effects of magnetic fields.

8.2 Magnetic Forces in Cells

When biological systems are exposed to a magnetic field, several distinct types of magnetic forces may act on tissues, cells, subcellular structures, and mobile charged molecules and ions. These include (i) the Lorentz force, $\boldsymbol{F_L} = q[\boldsymbol{vB}]$, which depends on the ion's charge q, its velocity $\boldsymbol{v}$, and the magnetic induction $\boldsymbol{B}$); (ii) the magnetic gradient force, $F_{\nabla B} \propto \nabla B^2$ (Hinds et al. 2001; Zablotskii et al. 2016a), which scales with the spatial gradient of the magnetic field; and (iii) the concentration gradient magnetic force, $F_{\nabla n} \propto B^2 \nabla n$, where$\nabla n$ denotes the gradient in the concentration of the diamagnetic and paramagnetic species and ∇ is the differential operator *nabla* (Leventis and Gao 2002; Dunne et al. 2011; Waskaas 1993). The potential involvement of these forces in cellular processes is examined in this review.

Unlike electric fields, magnetic fields are not attenuated by biological tissue and can penetrate deeply into the human body. As a result, the forces and effects induced by magnetic fields offer unique opportunities for controlling cellular movement, proliferation, and internal processes. These effects open up promising applications, ranging from micro- and nano-scale control, such as cell sorting and drug or gene delivery (Zborowski and Chalmers 2011), to influencing the behavior of animals (Huang et al. 2010) and even humans (Oliveira et al. 2010).

8.3 Cells in Soft Mode

In general, the pronounced effects of applications of magnetic fields to living cells are expected for cells in so-called "soft mode" state (Zablotskii et al. 2018). The concept of the "soft mode" is utilized to characterize cells in a transitional state where their mechanical elastic moduli are small, thereby enabling substantial deformation of the entire cell body by relatively minor mechanical forces. In such a state, magnetic forces are sufficient to induce modifications within the cellular machinery and elicit dramatic cellular responses. There are four examples of cells in soft mode.

8.3.1 Differentiating Stem Cells

During the differentiation of stem cells, the transition between cellular states occurs smoothly, without abrupt alterations in mechanical properties. This transition phase, which typically spans several days, maintains the cells in a mechanically soft state. Consequently, small external compressive or tensional forces can influence the differentiation process. The cells then transduce these mechanical stimuli to regulate the stem cell differentiation pathway and gene expression. For example, changes in shear stress (1–5 Pa) are associated with the widescale reprogramming of gene expression in endothelial cells (Dai et al. 2004). Generally, stem cells, when considered as a "material," are inherently softer than their mature, differentiated counterparts, making them highly responsive to the application of low mechanical stress (Holle and Engler 2010; MacQueen et al. 2013).

8.3.2 Mitosis

Mitosis is characterized by the forces generated by polymerization dynamics and motor-protein activity, which are essential for moving and equally separating chromosome copies between the dividing daughter cells (Mitchison and Salmon 2001). The dynamic instability of microtubules and the tiny mechanical forces accompanying mitosis are hallmarks of the soft mode. In this scenario, the presence of small intracellular stresses permits magnetic forces to alter cellular mechanics. For instance, nanoparticle-mediated magnetic forces, applied to HeLa cells via a gradient magnetic field, resulted in the biasing of the metaphase plate during mitosis under lower magnetic gradients and reduced intracellular mechanical forces (Tseng et al. 2012). Furthermore, exposure of cancerous and normal human cells to a 27 T static magnetic field for 4 h was shown to change the orientation of the mitotic spindle (Zhang et al. 2017).

8.3.3 Metastasizing Cancer Cells

Metastasizing cancer cells exhibit an adaptive capacity, adjusting their nuclear size, cell deformability, cell-to-substrate adhesion, and tissue invasion efficiency. Measurements using a magnetic tweezer system have shown that cancer cells with the highest migratory and invasive potential are several times less stiff than those with the lowest potential (Swaminathan et al. 2011). Theoretical modeling suggests that alterations in the mechanical properties of cancer cells (e.g., changes in the cell actin cortex elastic modulus and osmotic pressure) correlate with uncontrolled division and migration, with various biochemical factors driving cancer progression through this mechanism (Katira et al. 2012). Similar to stem cells, the adaptive mechanics of cancer cells involve their own "soft mode," which facilitates their dissemination through interstitial tissue. Crucially, immediately following mitosis, the cancer cell nucleus can deform up to 90% to navigate and penetrate small pores, a mechanism vital for the interstitial dissemination of metastatic cells (Panagiotakopoulou et al. 2016).

8.3.4 Early Embryo Development

It is known that the fertilized egg develops into an early mammalian embryo, called a blastocyst, after several rounds of cell division and two key lineage choices about cell types. First, the cells divide into an inner cell mass surrounded by a layer of polarized trophectoderm cells. In the second lineage, the inner cell mass splits into two groups: the epiblast (EPI), which becomes the embryo, and the primitive endoderm (PrE), which forms the extraembryonic yolk sac (Pauklin et al. 2011; Chazaud et al. 2006). Recent studies reveal that the second lineage segregation is driven by differential elasticity between PrE-primed cells and EPI-primed cells: pPrE cells exhibit a significantly higher Young's modulus ($E = 58.25 \pm 38.92$ Pa) than pEPI cells ($E = 8.54 \pm 6.78$ Pa) (Ritter et al. 2025). It is important to highlight that the Young's modulus of embryonic cells is notably low, indicating that these cells exhibit a fluid like state. Consequently, during this stage of embryonic development, even minor external electric or magnetic fields can significantly influence cell differentiation and the fate of the blastocyst.

Thus, embryonic cells are mechanically soft and have low or nearly zero membrane potential (Levin 2009). So it is expected that even small magnetic gradient forces can impact the development of organisms. Indeed, the early embryonic development of the leopard frog (*Rana pipiens*) was significantly hindered by a 1 T magnetic field with a gradient of 84 T/m (Neurath 1968). This phenomenon is explained by the presence of ferritin aggregates in the tissue, which have a positive magnetic susceptibility ($\Delta\chi = 1.2 \times 10^{-5}$ in SI units). These aggregates can be moved within the cell by the magnetic gradient force at a speed of approximately 300 μm/s (Neurath 1968). Since ferritin is essential for hemoglobin formation in the developing embryo, its movement could disrupt its proper utilization.

The mechanical properties of lipid cell membranes represent a special case of soft-mode behavior. In the general case, the shear moduli of the cell membrane and cytoskeleton depend on the rate at which force is applied and the duration of its action. For example, NIH 3 T3 fibroblasts exhibited a viscoelastic timescale of $\approx$1 s and a shear modulus of $\approx$1000 Pa (Karcher et al. 2003). Micropipette measurements in endothelial cells yield Young's moduli in the range of 10^2–10^3 Pa (Theret et al. 1988).

Harnessing mechanical regulation of cellular function raises the question of how to apply physically meaningful and controllable mechanical stress to a cell. Several approaches exist, including (i) applying of hydrostatic (Cheng et al. 2009) or osmotic pressure (Montel et al. 2011), (ii) generating shear stress, for example, through fluid flow (Berrout et al. 2012; Davies 1995), or (iii) transmitting mechanical forces directly to the cell membrane using a micro-disk oscillating in an AC magnetic field (Kim et al. 2010), and (iv) inducing mechanical tension via nanoparticles (Tseng et al. 2012).

8.4 Effects of the Lorentz Force

The Lorentz force. Because all cells and their building blocks are charged systems in motion, physical explanations of biomagnetic effects frequently rely on a role of the Lorentz force:

$$\vec{F} = q\left[\vec{v}\ \vec{B}\right] \tag{8.1}$$

or its volume density given as

$$\vec{f} = \left[\vec{j}\ \vec{B}\right], \tag{8.2}$$

where q and $\boldsymbol{v}$ are the particle charge and velocity and $\boldsymbol{j}$ is the current density, for example, in a membrane ion channel. We show that this force is operative when the external magnetic field is strong enough, in cells or tissue. Indeed, the cell machinery is mainly driven by the electrical forces exerted on moving ions and molecules, $\boldsymbol{F}_\mathrm{E} = q\boldsymbol{E}$, where $\boldsymbol{E}$ is the electric field strength. In the presence of a uniform magnetic field, both Lorentz and electric forces may affect the motion of biological particles. However, we demonstrate below that the Lorentz force is much smaller than the electric force, $F << F_\mathrm{E}$. By equalizing the electric and Lorentz force magnitude, $qE = qvB$, one can find that the magnetic field induction magnitude should be $B = E/\mathrm{v}$. In the outer cell membrane $E \approx (10^7–10^8)$ V/m, while in the cell nucleus the electric field strength varies from zero to $E \approx 3\bullet10^3$ V/m at the periphery (Keydar n.d.). Typical ion diffusion velocities are around 1–10 μm/s under physiological conditions. Inserting these velocity magnitudes and the electric field strengths into $B = E/\mathrm{v}$, one can conclude that the Lorentz force can be the same order of magnitude as the intracellular electric forces at extremely high magnetic fields. Thus, the direct influence of the Lorentz force on intracellular moving ions is unlikely. Nevertheless, an influence of the Lorentz force on channel gating seems to be possible. Indeed, a working membrane ion channel carries an electrical current, I. If a static magnetic field is applied parallel to the cell membrane, the ion current creates lateral shear stress, $s = IB/2r_{ch}$ (Zablotskii et al. 2013), where r_{ch} is the channel radius. For a high conductance ion channel $I = 30$ pA with $r_{ch} = 0.2$ nm in a static magnetic field $B = 1$ T, the estimation gives $s = 10^{-11}/(2\bullet0.2\cdot10^{-9}) = 0.075$ Pa. For a high magnetic field $B > 10$ T, the membrane stress is of the order $s = 1$ Pa which is comparable with physiological shear stress driving mechanosensitive ion channels. For example, the wall shear stress (WSS) induced by hydrodynamic blood flow in endothelial cells typically ranges from $\tau_h = 1$ Pa for arteries to $\tau_h = 0.1$ Pa for capillaries (Roux et al. 2020).

Shear stress during the determination of stem cell fate in utero was estimated to be less than 0.5 Pa (see Ref. (Song et al. 2012) and references therein). Therefore, mechanical stress of tens of Pa or even the smallest shear stress (0.02–0.2 Pa) may be sufficient to affect cell functionality, either by disrupting the balance between internal and external pressures and thereby controlling cell volume (Jiang and

Sean 2013) or by activating stretch-activated gating of ion channels (Davies 1995; Bouffanais et al. 2013; Olesen et al. 1988). The ion channels P2X4, TRP, and Piezo1 exposure to the stress 1.5 Pa lead to mediated the influx of extracellular calcium (Lampugnani 2012).

High static magnetic fields can enforce their effects on ion channels by synchronizing channel activity. This synchronization may occur because a magneto-mechanical shear stress ($s = IB/2r_{ch}$) is raised when the magnetic field lies within the cell membrane plane. Indeed, a cell may contain thousands of ion channels, and the likelihood of any one of them being open at any given moment typically ranges from a few to several tens of percent (Hughes et al. 2008; Christensen and Corey 2007). Magneto-mechanical stress ($s = IB/2r_{ch}$) spreads throughout the cell membrane, triggering the opening or closing of mechanosensitive ion channels. This action, in turn, generates membrane hyperpolarization or depolarization, which can have numerous consequences for the whole-cell machinery. This mechanism may explain the antidepressive effect of ultrahigh static magnetic fields (7–33 T) observed in mice (Lv et al. 2022). It was observed that oxytocin levels in the brains of mice were significantly elevated in two experimental groups. In the group exposed to a static magnetic field of 11.1–33 T for 1 h, the oxytocin level was measured at 25.16 ± 2.81 pg/mL, which represents a 165% increase. Similarly, the 7 T group exhibited an oxytocin level of 30.92 ± 2.27 pg/mL, reflecting a 36% increase compared to their respective sham groups. The mechanism behind this effect may work as follows: Neurons in the hypothalamic regions can be activated by changes in osmotic pressure (Leng et al. 1999). A high static magnetic field can exert stress on the cell membrane due to the magnetic forces acting on active ion channels, which we refer to as magnetic pressure. If the pressure is sufficiently strong, it can alter and synchronize mechanosensitive ion channel activity and ion transport across the cell membrane. Consequently, it is possible that these MF-induced alterations in cell pressure balance, mediated through mechanosensitive ion channels, may influence oxytocin levels.

8.4.1 Effect of the Lorentz Force on DNA Synthesis

DNA is a highly charged macromolecule, with each phosphate group in its sugar-phosphate backbone carrying a unit negative charge (Hedley et al. 2024). As DNA rotates, the Lorentz force acts on its negatively charged phosphate groups. This force can either enhance or impede the DNA's natural rotation within the nucleus, depending on the field's direction relative to its symmetry axis. The variation in angular velocities is proportional to both the induction of the magnetic field and the charge density of the DNA. The change in angular velocities is proportional to both the induction of the magnetic field and the charge density of the DNA, σ, and inversely proportional to the mass density (ρ) and radius of DNA (a): $\Delta\omega = \pm(B\sigma/\rho a)$ (Yang et al. 2020). Thus, depending on the mutual orientation, an applied SMF, and DNA helix axis, the SMF can either accelerate or slow down DNA rotation during its replication. A decrease in DNA rotation results in a delay in DNA replication. On the other hand, an increase in DNA rotation does not automatically result in faster DNA replication and transcription, as there are various other speed-limiting factors that play a role in these processes (Mirkin and Mirkin 2007). Furthermore, speeding up DNA rotation can generate extra negative or positive supercoils, which may also reduce the replication and synthesis rates (Koster et al. 2005). Since the Lorentz force changes direction when a static magnetic field (SMF) switches from upward to downward, DNA synthesis rates differ between these two orientations, leading to unequal rates of cell proliferation depending on the direction of the magnetic field (Yang et al. 2020; Yang et al. 2021).

The study examined the effects of two opposing directions of SMFs on mice with lung cancer that were treated with 9.4 T static MFs (SMFs) for a total of 88 h (Yang et al. 2021). The results indicated that the upward 9.4 T SMF significantly inhibited the growth of A549 tumors, while the downward 9.4 T SMF did not have a similar effect. Additionally, in vitro analysis demonstrated that a 24-h treatment with the upward 9.4 T SMF not only inhibited A549 DNA synthesis but also significantly increased levels of reactive oxygen species (ROS) and P53 and caused an arrest in the G2 phase of the cell cycle (Yang et al. 2021).

8.5 Effects of Magnetic Concentration Gradient Forces

Let us examine the situation in which a uniform static magnetic field (with zero gradient) is applied to a cell or tissue. Under these conditions, the concentration gradient magnetic force can act on diamagnetic and paramagnetic ions, either facilitating or opposing their movement within the cell. The volume density of this concentration gradient magnetic force is given by (Hinds et al. 2001).

$$\vec{f} = \frac{\chi B^2}{2\mu_0} \vec{\nabla} n, \tag{8.3}$$

where n is the molar concentration of ions with the molar magnetic susceptibility χ and μ_0 is the vacuum permeability. In the specific case where the concentration varies only along a single coordinate, x, the force is given by

$$f(x) = \frac{\chi B^2}{2\mu_0} \frac{dn(x)}{dx} \tag{8.4}$$

It is important to emphasize that living cells contain hundreds of diamagnetic and paramagnetic species, many of which can exhibit significant concentration gradients (Sear 2019). In this case, a concentration gradient magnetic force (Eq. (8.3)) comes into play and interferes with intracellular forces in the presence of a very high static magnetic field. For example, a high static magnetic field (larger than 10 T) can accelerate the diffusion of diamagnetic species while slowing the diffusion of paramagnetic biologically active molecules, as well as affecting protein transport in cells (Zablotskii et al. 2021, 2022).

8.6 Effects of Magnetic Gradient Forces

An external magnetic field acts on a magnetic dipole (which is also treated as an elementary current loop) in two ways. First, the magnetic field exerts a torque on the dipole, attempting to align it parallel to the vector of the magnetic field induction. Second, an inhomogeneous magnetic field applies a force to the magnetic dipole, proportional to the magnitude of its magnetic moment (p) and the gradient of the magnetic field induction: $\boldsymbol{F} = (\boldsymbol{p}\nabla)\boldsymbol{B}$. It is important to emphasize that the direction of the magnetic force acting on a magnetic dipole in a nonuniform field does not coincide with either the magnetic induction vector or the magnetic moment vector. The direction of the magnetic force vector aligns only with the direction of the infinitesimal increment of the magnetic induction vector $\boldsymbol{B}$, taken in the direction of the magnetic moment vector $\boldsymbol{p}$ at the location of the dipole. When the magnetic moment is aligned with the magnetic field, the magnetic force is given by $\boldsymbol{F} = p\nabla B$, and the force acts in the direction of the magnetic gradient vector.

To illustrate the magnetic forces exerted on cells, let us consider a cell with volume V and magnetic susceptibility χ in the presence of a gradient magnetic field. If the cell is placed in a diamagnetic medium with susceptibility χ_m, the magnetic gradient force will also be exerted on the medium. Thus, the force exerted on the cell in the medium is

$$\vec{F} = \frac{\Delta\chi V}{2\mu_0}\nabla B^2, \tag{8.5}$$

where $\Delta\chi=\chi-\chi_m$ is the difference in the magnetic susceptibilities, μ_0 is the vacuum permeability, B is the magnetic induction modulus, and ∇ is the differential operator nabla. Cells, and typical cell media, are generally considered to be diamagnetic. Depending on the concentrations of intracellular and extracellular diamagnetic and paramagnetic molecules, a cell may be either less or more diamagnetic than the surrounding medium ($\Delta\chi>0$ or $\Delta\chi<0$). Thus, for $\Delta\chi > 0$, the cells are attracted to areas with the highest magnetic field gradient, while for $\Delta\chi < 0$, the cells are repelled from them. When the magnetic gradient direction is parallel or antiparallel to the substrate normal, the magnetic force (Eq. (8.1)) acts as either gravity or buoyancy. If the magnetic gradient force pushes a cell toward the substrate, additional hydrostatic pressure will squeeze the cell body. A sufficiently large magnetic gradient force exerted in the direction opposite to gravity can cause diamagnetic levitation of cells and even small animals, such as frogs (Berry and Geim 1997; Valles et al. 1997; Kauffmann et al. 2011).

The magnetic susceptibilities of various cells, cellular components, and tissues are provided in (Schenck 1996; Luo et al. 2010; He and Yablonskiy 2009; Senftle and Hambright 1995). By considering a cell as a magnetic material composed of micro- and nanosized structural elements with different magnetic susceptibilities (both negative and positive relative to the medium), one can calculate the forces acting on the cell and its components using Eq. (8.5).

To understand the potential effects of magnetic forces on cells, it is essential to compare the magnitude of intracellular and intercellular forces with that of magnetic gradient forces, which typically range from 0.1 pN to several hundred pN (Zablotskii et al. 2014a, 2016a, b, 2018). Intracellular and intercellular mechanical forces vary widely, from 1 pN to several hundred nN (Meenakshi et al. 2017). Therefore, magnetic forces are comparable in magnitude to the forces driving cellular machinery, making magnetic fields capable of influencing numerous intracellular processes and cell functions. A comprehensive overview of the application of magnetic forces, spanning from 0.01 fN (for controlling cell receptors) to dozens of nN (for magneto-mechanical stimulation and regulation of intracellular processes), is available in (Monzel et al. 2017; Torbati et al. 2022).

To illustrate the potential role of magnetic gradient forces in biological systems, let us compare their volume density with that of gravity. From Eq. (2) the volume density of the magnetic gradient force is $f_m = \chi(gradB^2)/2\mu_0$. It should be compared with the volume density of the gravitational force, $f_g = \rho g \approx 10^4$ N m^{-3}. The ratio between these two densities is $f_m/f_g = \chi$ grad(B^2) $(2\rho g\mu_0)^{-1}$. Assuming χ is on the order of the diamagnetic susceptibility of water (-9×10^{-6} in SI), $B = 1$ T and $\rho = 10^3$ kg m^{-3}, the ratio of magnetic to gravitational force is approximately unity, $f_m/f_g \approx 1$ at $B \approx 1.4 \times 10^3$ T m^{-1}. Thus, in moderate magnetic fields with sufficiently high gradients, the resulting magnetic forces can reach magnitudes comparable to gravity and may therefore be strong enough to influence cellular processes.

When a cell is subjected to a gradient magnetic field, a complex internal stress pattern arises due to differences in the magnetic susceptibilities of its organelles. In this scenario, the total magnetic gradient force exerted on the cell is the sum of the local intrinsic magnetic forces (Zablotskii et al. 2016a):

$$\vec{F}_{tot} = \frac{1}{2\mu_0}\sum_i \Delta\chi_i \Delta V_i \nabla B^2(x_i, y_i, z_i), \tag{8.6}$$

where $\Delta\chi = \chi_i - \chi_{cyt}$, χ_i represents the magnetic susceptibility of an intracellular component, χ_{cyt} is the magnetic susceptibility of the cytosol, and ΔV_i denotes the volume of an intracellular component with the coordinates (x_i,y_i,z_i). It is important to note that depending on whether $\Delta\chi$ is positive or negative, the local magnetic gradient forces within a cell can have opposite directions. As a result, these forces produce a complex pattern of magnetic stress within the cell. Since the differences in magnetic susceptibilities among cellular components are relatively small, the resulting local magnetic stresses are also small, say less than

1 Pa (Zablotskii et al. 2016a). In low-gradient static magnetic fields, the magnetic forces (Eq. (8.6)) are weaker than the intrinsic intracellular forces that govern cellular functions. However, prolonged application of a magnetic gradient field to intracellular components or the cell membrane can subtly alter cell functions and effectively modify cell fate, similar to the effects observed in high-gradient static magnetic fields (Zablotskii et al. 2018).

It is known that living cells can sense mechanical cues from their surroundings, such as mechanical forces, pressure, shear stress, and stretch. This ability to detect and respond to mechanical signals is an essential aspect of cell behavior and function, influencing processes like cell growth, differentiation, and migration. Mechanotransduction is the term used to describe how cells convert these mechanical cues into biochemical signaling pathways. From the perspective of mechanotransduction, the spatial arrangement of local internal magnetic gradient forces (Eq. (8.6)) creates a complex network of mechanical cues. These cues can influence various cellular functions across different pathways (Suriyagandhi et al. 2025).

It should be noted that, as a rule, the magnetic fields generated by electromagnets or permanent magnets are inhomogeneous, i.e., gradient. For example, the magnetic field inside a long solenoid is nearly uniform; however, at the edge of the solenoid, the field decreases by half and exhibits both axial and radial gradients. The magnetic fields generated by permanent magnets are highly inhomogeneous near the edges of the magnets, and the distribution of the field dramatically changes with varying distances from the surface of the magnet. At the edge of a centimeter-sized permanent magnet, the magnetic gradient value is on the order of 10^3 T/m (Zablotskii and Polyakova 2025). It is important to emphasize that the spatial distribution of the magnetic field, the direction of the magnetic gradient, and its magnitude depend not only on the geometry of the permanent magnet but also on its size. It is known that when all linear dimensions of the magnet are reduced by a factor of k, while maintaining all its magnetic characteristics, the magnitude of the magnetic gradient increases by a factor of k (Cugat et al. 2003). Therefore, the use of micromagnets allows for the generation of magnetic fields with local gradient magnitudes of up to 10^6 T/m (Osman et al. 2013; Dumas-Bouchiat et al. 2010). For 8-nm-diameter ferritin particle, the maximum magnetic field gradient on the surface of the particle is $4.9 \cdot 10^8$ T/m (Barbic 2019).

The investigation reported in (Zablotskii et al. 2014b) examined the impact of spatially modulated high-gradient magnetic fields (HGMFs) on the cellular functions of human THP-1 leukemia cells. Exposure to arrays of micrometer-sized magnets, which generate the HGMFs, was found to elicit several key responses: (i) cell swelling, (ii) sustained elevation of reactive oxygen species (ROS) production, (iii) inhibition of cell proliferation, and (iv) induction of apoptosis. These effects were observed in THP-1 monocytic leukemia cells without the introduction of chemical or pharmacological agents.

Since magnetic fields are not attenuated by living tissues, magnetic gradient forces can affect the entire organism, both in a healthy state and in disease. For example, study (Yu et al. 2023) investigated the effects of high static magnetic fields (SMFs) on type 1 diabetic (T1D) and type 2 diabetic (T2D) mice exposed to SMFs ranging from 1.0 to 9.4 T, including both gradient and uniform fields. Results indicated that a 14-h exposure to gradient SMFs at 55.5 T/m (1.0 to 8.6 T) had detrimental effects on both T1D and T2D mice. These effects included damage to the spleen, liver, and kidneys, as well as increased levels of glycosylated serum protein, blood glucose, inflammation, and anxiety. Conversely, exposure to 9.4 T low-gradient SMFs (<10 T/m) did not cause similar negative effects. In normal T1D mice, exposure to gradient SMFs (>10 T/m) led to increased malondialdehyde levels and decreased superoxide dismutase activity. In severe T1D mice, this exposure significantly heightened tissue damage and lowered survival rates. In vitro studies indicated that gradient high SMFs resulted in increased production of reactive oxygen species and apoptosis while decreasing the number and proliferation of MS-1 cells.

Study (Zablotskii et al. 2014c) has demonstrated that the magnetic gradient forces applied via low-frequency oscillating gradient magnetic fields significantly influence the adipogenic differentiation of mesenchymal stem cells. A similar effect was found under the application of mechanical vibration (mechanical forces of inertia) with the same frequency, 4 Hz. The underlying mechanism involves the transmission of mechanical stress to the cell's cytoskeleton. This stress leads to F-actin remodeling, a critical structural change, which subsequently causes the downregulation of key adipogenic genes. Specifically, this effect has been linked to the suppression of genes such as adiponectin, PPARc, and AP2. There are mechanistic parallels between low-intensity mechanical and magnetic vibrations. Low-intensity mechanical vibrations, defined as having accelerations less than that of free fall ($g = 9.8\ \mathrm{m/s^2}$), share common mechanisms of action with magnetic vibrations. A key piece of evidence for this shared action is the observed effect on cellular differentiation. For example, low-magnitude mechanical signals have been demonstrated to inhibit the differentiation of GFP marrow cells into adipocytes (Rubin et al. 2007), just as low-frequency magnetic vibrations suppress the differentiation of mesenchymal stem cells into adipocytes (Zablotskii et al. 2014c). This correlation is mechanistically plausible, as the magnetic forces exerted on cells and their intracellular organelles fundamentally function as mechanical forces. These forces induce deformation of the cell membrane, which subsequently influences the expression and activity of mechanosensitive ion channels. Alterations in these channels, in turn, modify the membrane potential (Gorobets et al. 2024) and ultimately impact the cellular differentiation pathway and gene expression (Zablotskii et al. 2018, 2025a).

Exposure of macrophages to a gradient magnetic field of 10^4 T/m results in significant elongation of the cells and substantially impacts their molecular structures and organelles. Immunostaining and Western blot analysis revealed that the magnetic gradient force induces rearrangement of the actin cytoskeleton, the Golgi apparatus, and the TRPM2 cation channel receptor in macrophages, while also altering the expression of various molecular markers associated with these immune cells (Wosik et al. 2018).

The ability to alter cell membrane potential through the application of high-gradient or spatiotemporal magnetic fields has garnered considerable interest (Zablotskii et al. 2021, 2016b, 2023; Rubio Ayala et al. 2018; Ji et al. 2023). This is particularly relevant given that cellular function is fundamentally influenced by electric membrane potential, which regulates intracellular physiology and facilitates signal propagation from motor neurons to muscle fibers, ultimately leading to muscle contraction. These observations suggest the potential for utilizing magnetic fields with both spatial and temporal gradients to control muscle contraction. In a recent study (Rubio Ayala et al. 2018), researchers employed complex spatiotemporal magnetic fields (17–70 mT) to modulate intracellular signaling in skeletal muscle cells. By varying the amplitude and frequency of the magnetic field (1–10 Hz), they achieved transient depolarization of cellular membranes. This process was characterized by (i) sodium ion (Na^+) influx via voltage-gated sodium channels (VGSC), (ii) an increase in cytosolic calcium levels, and (iii) VGSC- and ryanodine receptor-dependent enhancement of actin polymerization. Notably, these ion fluxes occurred exclusively during the application of the magnetic field, returning to baseline levels once the field was deactivated. Importantly, the 30-s activation cycle could be repeated multiple times without diminishing signal intensity. In contrast, static magnetic fields of equivalent strength did not produce any significant effect on myotube calcium (Ca^{2+}) levels. These findings may pave the way for innovative strategies employing complex magnetic fields to enhance skeletal muscle function in conditions such as myopathies.

F-actin (filamentous actin) is known to be sensitive to mechanical stimuli and plays a critical role in processes such as cell adhesion, migration, and cancer metastasis. However, there are very few methods available to modify F-actin dynamics with minimal cellular toxicity. In contrast, magnetic fields serve as a noninvasive and reversible physical tool that can easily penetrate cells and influence F-actin behavior.

A study (Ji et al. 2023) demonstrated that exposure to a nonuniform rotating magnetic field (0.1–0.4 T at 4.2 Hz) can decrease F-actin formation both in vitro and in vivo, leading to reduced migration, invasion, and attachment of breast cancer cells. The effects of these rotating magnetic fields were significantly different in breast cancer cells compared to noncancerous cells, particularly regarding the number of F-actin filaments and their recovery once the magnetic field was removed. Furthermore, utilizing an intermittent treatment approach, it was found that low-frequency rotating magnetic fields substantially reduced breast cancer metastasis in mice, extending their survival by 31.5–46.0% while also improving overall physical condition (Ji et al. 2023). These findings indicate that low-frequency rotating magnetic fields can serve not only as a research tool to perturb F-actin dynamics but also as a potential therapeutic strategy to inhibit breast cancer metastasis through modulation of F-actin, all while minimally affecting normal cells. This suggests the promising potential for developing these fields into temporally controlled, noninvasive, and highly penetrating physical treatments for metastatic cancer.

Gradient magnetic fields significantly impact plant growth and development (Maffei 2014). In a study (Shabrangy and Luschnig 2025), the plant model

Arabidopsis thaliana was utilized to investigate root growth responses to both static uniform magnetic fields and gradient magnetic fields. The exposure to gradient magnetic fields altered root curvature and auxin signaling, even under conditions with reduced gravitational effects. Through comparative transcriptomic analysis, it was found that in *Arabidopsis* genes involved in auxin synthesis and signaling were upregulated in response to gradient magnetic field exposure, while no such effect was observed with uniform magnetic fields (Zhou et al. 2024). The study also revealed that a gradient magnetic field generated by permanent magnets helped plants maintain iron homeostasis.

8.7 Cell Membrane Pore Formation and Ion Channel Activity in Gradient Magnetic Fields

The structural integrity of lipid membranes is essential for cellular survival under the influence of external factors such as electromagnetic fields, chemical agents, and temperature variations. Electroporation and magnetoporation—processes in which cells are exposed to pulses of electric or magnetic fields—have become widely applied techniques in biology, biotechnology, and medicine (Rems and Miklavčič 2016; Novickij et al. 2016).

The nucleation of a pore within a lipid membrane results from the interplay of three competing forces. The first is the interfacial force, which arises from the energetic cost associated with maintaining a boundary between the membrane and its surrounding medium. The second is the electrostatic contribution, stemming from changes in the energy of the charged lipid membrane upon pore formation. Because the pore lacks both interfacial and surface charge densities, the creation of a pore reduces the total interfacial and electrostatic energy of the system. Opposing these effects is the line tension—the energetic cost of generating an edge along the pore's perimeter.

The balance among these forces determines the energetic barrier to pore formation and defines a critical pore radius, r_{cr}. Pores with radii $r < r_{cr}$ are energetically unfavorable and tend to close, whereas those with $r > r_{cr}$ expand spontaneously (Polyakova et al. 2017).

In the absence of a magnetic field, the critical pore radius is influenced by the interfacial tension between the lipid membrane and the surrounding medium, the cell membrane potential, and the membrane's electric capacitance per unit area. However, when a gradient magnetic field is applied to the cell membrane, it induces shear stress, which changes the membrane tension and alters the membrane potential. Consequently, the critical pore radius reduces to that of the ion channel radius, resulting in either the closing or collapsing of a number of working ion channels in the cell membrane (Polyakova et al. 2017). Thus, in a high-enough gradient magnetic field with a gradient value greater than 1 kT/m, the magneto-mechanical stress in the cell membrane can change channel selectivity and decrease ion channel activity.

8.8 Interaction of Magnetic Fields with Biogenic and Artificial Magnetic Nanoparticles on Cell Membranes

As it was mentioned above, the magnetic gradient forces acting in the cell membrane may affect cell functions in two main ways: changing the resting membrane potential and locally disturbing the channel gating mechanism. The second mechanism is more effective when a magnetic nanoparticle is directly bound to a membrane channel, especially to mechanosensitive or ligand-gated channels. There are two possible cases: one magnetic nanoparticle bound to a channel in combination with a source of high-gradient magnetic field (Fig. 8.1a) and two interacting magnetic nanoparticles bound to a channel (Fig. 8.1b and c).

To start, we examine a magnetic nanoparticle that is coupled to a channel domain within a magnetic field (MF) that has a gradient aligned with the membrane (see

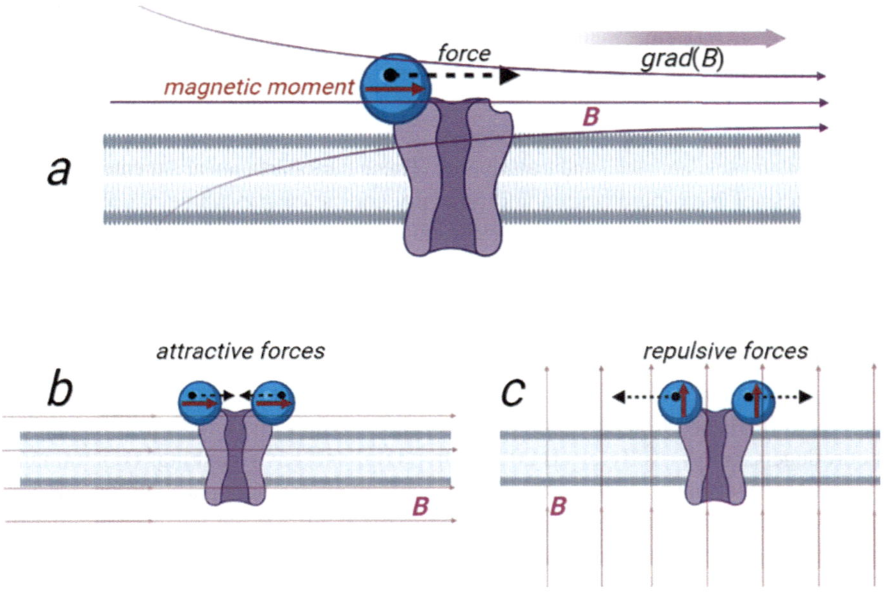

Fig. 8.1 Illustration of an ion channel in the cell membrane subjected to magnetic forces from a magnetic nanoparticle and an externally applied gradient magnetic field. (**a**) Upon activation of the gradient magnetic field, the ion channel closes. (**b** and **c**) The interactions between two magnetic nanoparticles bound to the channel body facilitate its opening and closing in the presence of a uniform magnetic field. Reproduced with permission from (Zablotskii et al. 2023)

Fig. 8.1a). In this scenario, the force exerted on the particle is also aligned with the membrane, as illustrated in Fig. 1a. If this force is comparable to the forces that regulate the opening and closing of the channel, it can disrupt the channel's functioning. For instance, a magnetite (Fe_3O_4) nanoparticle with a radius of 5 nm in a magnetic field with a gradient of approximately 10^7 T/m experiences a magnetic force on the order of 10 pN, which is sufficient to cause the opening or closing of an ion channel (Zablotskii et al. 2023).

When two magnetite or ferritin nanoparticles are attached to a mechanosensitive ion channel (see Fig. 8.1b and c), one nanoparticle experiences a high-gradient magnetic field generated by the other. Depending on the magnetic field's orientation, these nanoparticles can either attract (Fig. 8.1b) or repel each other (Fig. 8.1c) with forces of up to 100 pN (Barbic 2019; Zablotskii et al. 2023). This mechanism provides a means to control the expression of mechanosensitive channels through external magnetic fields and nanoparticles connected to the channel.

For example, magnetically controlled channel gating can be accomplished using a uniform rotating moderate magnetic field, which allows for the periodic switching of ion channels between open and closed states. In fact, when the magnetic field direction changes, the nanoparticles' magnetic moments align either parallel (Fig. 8.1b) or perpendicular to the cell membrane (Fig. 8.1c). In the parallel alignment, the attractive force between the nanoparticles causes the channel to close, while in the perpendicular alignment, the repulsive force opens the channel.

Living organisms serve as carriers of naturally occurring magnetic nanoparticles, referred to as biogenic magnetic nanoparticles (BMNs) (Kobayashi et al. 1997). Since their initial discovery in magnetotactic bacteria in 1975 (Blakemore 1975), BMNs have been the subject of extensive research. Subsequent studies have identified BMNs across a diverse range of organisms spanning all three domains of life: prokaryotes, archaea, and eukaryotes. These nanoparticles are formed through the process of biomineralization and typically consist of crystalline nanocrystalline forms of antiferromagnets or ferrites, such as magnetite, maghemite, or greigite (Gorobets et al. 2017). Within an organism, BMNs tend to self-assemble into chains that are primarily situated on the lipid membranes of various cell types, including endothelial (Gorobets et al. 2022), cardiac (Darmenko et al. 2018), and other cell types (Gorobets et al. 2024; Brem et al. 2005; Van de Walle et al. 2019). The sources of biogenic magnetic nanoparticles include aging (where mitochondrial dysfunction results in increased iron accumulation in various tissues, leading to iron-dependent neurodegeneration (Kelly et al. 2023)), hypoxia, and environmental contaminants (Gieré 2016; Maher 2019). The injection of magnetic nanoparticles as contrast agents during magnetic resonance imaging (MRI) also leads to an accumulation of magnetic nanoparticles in human organs (Winkler et al. 2018; Avasthi et al. 2020; Kloc et al. 2025). In pathology, biogenic magnetic nanoparticles are found in abundance within tumor tissue, particularly located at the cell membrane (Gorobets et al. 2024; Kobayashi et al. 1997; Brem et al. 2006). Moreover, in living organisms, BMN chains can be located within capillary walls (Gorobets et al. 2023), as schematically shown in Fig. 8.2.

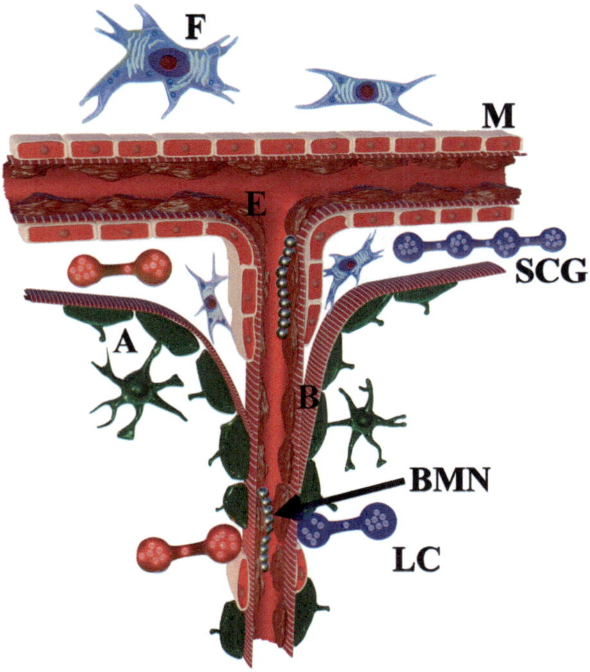

Fig. 8.2 Chains of biogenic magnetic nanoparticles on the walls of the capillaries of animals and dual innervation of the cerebral blood vessels: *A* astrocyte, *B* basement membrane, *E* endothelium, *F* fibrocyte, *LC* locus coeruleus, *M* muscle cell, *SCG* superior cervical ganglion (Itakura et al. 1977), *BMN* a chain of biogenic magnetic nanoparticles. Reprinted from (Gorobets et al. 2023), used under CC BY 4.0

Given the dual autonomic innervation of cerebral blood vessels (Fig. 8.2), chains of biogenic magnetic nanoparticles located along capillary walls can act as mediators of magneto-mechanical stress on specific brain regions and the central nervous system. The potential biological and medical implications of magnetic field effects on endothelial cells, arising from mechanical forces exerted on chains of biogenic and/or artificial magnetic nanoparticles attached to capillary wall membranes, are analyzed below.

8.9 Modulation of Mechanosensitive Ion Channel Activity by Magnetic Forces

The activity of ion channels in endothelial cells is profoundly influenced by mechanical forces. A primary regulator is shear stress from blood flow, to which these channels are exceptionally sensitive. They respond to stresses below 1 Pa (0.00750 mmHg)—a value 16,000 times smaller than arterial blood pressure.

This extreme sensitivity enables a tight control mechanism of membrane ion channel activity in endothelial cells. An applied magnetic field can induce an additional shear stress within the cell membrane. Depending on its direction, this magnetic stress can work with or against the natural stress from blood flow. Consequently, by generating a targeted magneto-mechanical shear stress of several Pascals, it is possible to externally regulate the operation of these ion channels (Gorobets et al. 2024). This approach exploits the fundamental electrical-mechanical coupling within the cell to achieve magnetic control (Torbati et al. 2022). This approach represents a robust platform for developing new electroceuticals and magnetic-based therapies that target ion channels (Kofman and Levin 2024; Zablotskii et al. 2025b).

In arteries and capillaries, a magnetic force acting on chains of magnetic nanoparticles creates additional shear stress in the membrane, which, together with the blood flow, regulates mechanosensitive calcium channels. To take into account the magnetic force acting on MNPs, we add the magnetic-field-induced shear stress to that induced by blood flow as

$$\tau = \tau_h \pm \tau_{mag}, \tag{8.7}$$

where τ_h represents the wall shear stress (WSS) induced by hydrodynamic blood flow on endothelial cells, with typical values of $\tau_h = 1$ Pa for artery and $\tau_h = 0.1$ Pa for capillary (Roux et al. 2020), and τ_{magn} is the WSS induced by the magnetic forces acting on the chain of magnetic nanoparticles embedded in the cell membrane. Thus, the chains of BMNs located on endothelial cell membranes act as magnetic-field-responsive effectors, regulating mechanosensitive ion channels. Figure 8.3 schematically illustrates this mechanism, showing a chain of magnetic nanoparticles positioned on a cell membrane, near ion channels, and subjected to both gradient and uniform magnetic fields.

In an alternating gradient MF with the frequency ω, the magnetic shear stress acting on the cell membrane in the vicinity of the MNP chain is (Gorobets et al. 2024).

$$\tau_{mag} = \frac{4}{3} rMGsin(\theta)\cos(\omega t), \tag{8.8}$$

where r and M are the MNP radius and magnetization, respectively, with M as a function of the applied field, G is the magnetic field gradient, Θ is the angle between the magnetic gradient and the normal to the membrane, and t is time. In Eq. (8.8),

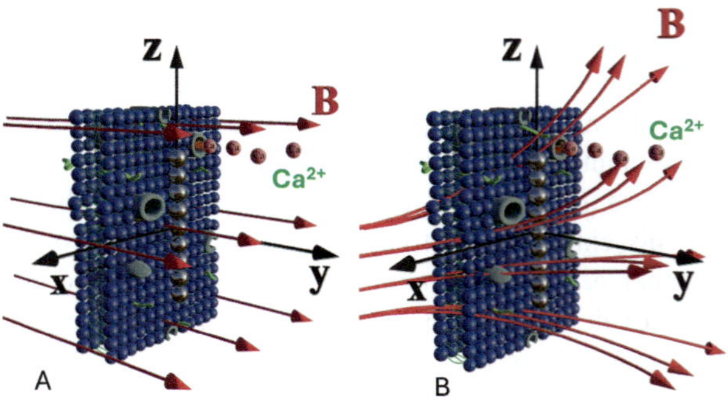

Fig. 8.3 Schematic of a chain of magnetic nanoparticles on a cell membrane under the influence of (**a**) gradient and (**b**) uniform magnetic field. Two calcium ion channels are depicted in the vicinity of the chain. The red arrows are magnetic lines. Reprinted from (Gorobets et al. 2024), used under CC BY 4.0

the applied magnetic field is an oscillating harmonic function of time, while for a static MF $\omega = 0$.

The investigation of biological effects resulting from rotating, spatially uniform magnetic fields is a highly active research area, driven by their potential for medical applications like cancer therapy through magnetic manipulation of cell membranes and subsequent apoptotic induction (Mazuel et al. 2017; Wei and Wang 2023).

In spatially homogeneous rotating magnetic fields, magnetic nanoparticle chains positioned on the cell membrane experience a magnetic torque. When a chain of magnetic nanoparticles on the cell membrane is under the influence of a uniform rotating magnetic field, the magnetic shear stress is

$$\tau_{mag} = \frac{\pi}{6N} MBsin\left(\theta_0\right) \sin\left(\gamma\right) \cos\left(\omega t\right), \tag{8.9}$$

where θ_0 is the angle between the normal to the cell membrane and the MF direction, γ is the angle between the MF in-plane component and the magnetization of the chain of magnetic nanoparticles, and N is the nanoparticle number in the chain.

Magnetic nanoparticles targeted to ion channels on the cell membrane enable precise control of the channel gating mechanism. Specifically, the torque exerted by a 200-nm nanoparticle in a rotating magnetic field can directly open Piezo1 channels, modulating an influx of Ca^{2+} ions (Choi et al. 2024).

The magnetic stress given by Eq. (8.8) or Eq. (8.9) contributes to the energy that drives ion channel gating and changes the channel's probability of being open or closed. Assuming two-state Boltzmann statistics, the probability that the channel is in the open state is given by (Reeves et al. 2008).

$$W_{open} = \left(1 + Exp\left(\frac{\Delta G}{k_B T}\right)\right)^{-1}, \tag{8.10}$$

where ΔG is the total free energy difference between closed and open states, k_B is the Boltzmann constant, and T is temperature. Thus, the interaction energy between a MF and an MNP chain posed on the cell membrane contributes to the total energy ΔG, thereby changing the activity of membrane ion channels (Gorobets et al. 2023).

Theoretical modeling has shown that, depending on the orientation and frequency of the applied field, this torque can either modulate the activity of mechanosensitive ion channels or induce membrane blebbing (Gorobets et al. 2023). Membrane blebbing arises from localized negative pressure generated in the region of the membrane adjacent to the MNP chains. Notably, blebbing represents a critical pre-apoptotic state in cancer cells. Thus, magnetically induced apoptosis in cancer cells may proceed through a blebbing-mediated pathway.

Model estimates indicate that rotating MFs with amplitudes on the order of 0.1–1 T can initiate membrane blebbing (Gorobets et al. 2023). Consequently, the application of such fields may be used to trigger apoptosis in cancer cells. It has further been demonstrated that the most efficient induction of blebbing-driven apoptosis is achieved using rotating MFs with frequencies in the range of 0.02–

0.5 Hz (Gorobets et al. 2023), which correspond to the inverse of the typical time scale of bleb formation (Charras et al. 2008).

In a general case, the static magnetic gradient force acting on a MNP chain is

$$\vec{F} = N\left(\vec{p}\,\vec{\nabla}\right)\vec{B}. \tag{8.11}$$

For a chain with a nanoparticle magnetic moment $\vec{p} = (0, 0, p)$ and a magnetic field with the spatial distribution $\vec{B}\ (x, y, z) = \vec{i}\ const - \vec{j}\,Gy + \vec{k}\,Gz$, which is a typical of the magnetic field between the z-coils of MRI scanners, the magnetic gradient force has only a z-component that is parallel to the cell membrane and equal to $F_z = pG$ (Gorobets et al. 2023). This force produces shear stress in the cell membrane, assumed to be aligned with the z-axis. However, when the nanoparticle magnetic moments are oriented perpendicular to the cell membrane, the same gradient static magnetic field exerts a force $F_y = -pG$, which creates negative pressure on the cell membrane. When the direction of the magnetic field gradient is reversed, the force also reverses and generates positive pressure on the cell membrane. Thus, in such a case, the local magnetic pressure exerted by an MNP chain on the cell membrane is given by (Gorobets et al. 2023)

$$P_{mag} = \pm\frac{4}{3}\ rMG. \tag{8.12}$$

This pressure (Eq. (8.12)) can disrupt the cell's pressure balance and therefore lead to cell swelling, compression, or even membrane rupture. Let us estimate the critical magnetic gradient value (G_{cr}) that may cause rupture of the endothelia cell membrane.

Membrane rupture in endothelial cells occurs at a critical compressive stress of $P_{cr} = (12.4 \pm 0.6)$ kPa (Theret et al. 1988). Thus, the membrane loses its integrity when $P_{mag} > P_{cr}$. As it follows from Eq. (8.12), the critical magnetic gradient can be estimated $G_{cr} = 3P_{mag}/(4rM)$. For magnetite nanoparticles with a saturation magnetization of $M = 510$ kA/m and radii of $r = 10$ nm, 50 nm, and 250 nm, the estimates give $G_{cr} = 1.8\bullet10^6$ T/m, $3.6\bullet10^5$ T/m, and $7.3\bullet10^4$ T/m, respectively. These critical gradients are overestimated as the calculations did not include the effects of the electrostatic pressure of the cell membrane (Eq. (8.1)) and the tension induced by the cortex on the cell membrane. As mentioned above, these magnetic gradients are achievable in laboratory settings.

It is worth noting that disturbances in the cell's pressure balance and membrane area can also modify ion fluxes across the membrane and, in turn, influence ion channel activity. It is interesting to compare the magnetic pressure exerted on a cell by an external magnetic field with the electrostatic pressure on the cell membrane.

Considering a cell as a spherical capacitor with energy $W = cU^2/2$, where c is the capacitance and U is the voltage (which represents the membrane potential), the electrostatic pressure on the cell membrane can be easily derived as follows:

$$P_{el} = \frac{\varepsilon\varepsilon_0}{Rd}U^2, \tag{8.13}$$

where $\varepsilon = (2\text{–}10)$ is the membrane permittivity, $\varepsilon_0 = 8.854\ 10^{-12}$ F/m is the vacuum permittivity, R is the cell radius, and d is the membrane thickness. From Eq. (13) we obtain radial stress as $s_r = P/2$ (Landau 1959):

$$s_r = \frac{\varepsilon\varepsilon_0}{2Rd}U^2 \tag{8.14}$$

and tangential stress $s_t = PR/2d$ (Landau 1959):

$$s_t = \frac{\varepsilon\varepsilon_0}{4d^2}U^2. \tag{8.15}$$

Electrostatic surface tension cell membrane of a spherical cell is

$$\sigma_{el} = \frac{\varepsilon\varepsilon_0}{2d}U^2. \tag{8.16}$$

Estimations from Eqs. (8.13, 8.14, 8.15, and 8.16) for the following set of the parameters, $\varepsilon = 5$; $U = 70$ mV; $d = 5$ nm; $R = 5$ μm, give $P_{el} = 8.68$ Pa, $s_r = 4.34$ Pa, $s_t = 4.34$ kPa, and $\sigma_{el} = 2.17 \bullet 10^{-5}$ N/m. The latter is comparable to the membrane tension of THP-1 human leukemia cells, which ranges within $\sigma = (1.4\text{–}9.1) \times 10^{-5}$ N/m (Richelme et al. 1997). This suggests that, for cells in a soft state, the electrostatic contribution to membrane tension can play a significant role in regulating cell stiffness and rigidity. For example, the stiffness of metastatic cancer cells, which have a low membrane potential (Zablotskii et al. 2025b), is 73% lower than that of benign mesothelial cells in the same sample (Cross et al. 2007).

We examine the activity of membrane channels in endothelial cells exposed to static gradient magnetic fields. Because a magnetic gradient force induces stress and deformation in the cell membrane, mechanosensitive (stretch-activated) and voltage-gated ion channels may be modulated by such forces. The underlying mechanism can be described qualitatively as follows: an applied MF, acting through MNP chains, deforms the cell membrane. If the deformation is sufficiently large, it alters the activity of stretch-activated ion channels. Changes in ion channel activity can then trigger intracellular responses that may disrupt, or even destabilize, cellular homeostasis.

To quantitatively assess the feasibility of this mechanism, we calculate the magnetic pressure using Eq. (8.8) and compare it with the characteristic pressures required to modulate mechanosensitive channels in endothelial cell membranes. Modulation can occur when the magnetic pressure given by Eq. (8.12) matches the activation or saturation pressure (P_a) of these channels. Reported activation/saturation pressures for various mechanosensitive ion channels in endothelial cell membranes range from 0.01 Pa to 1.5 Pa (Gorobets et al. 2023; Barakat et al. 2006; Li et al. 2022). From the equations $P_{mag} = P_a$ and Eq. (8.12), we calculate

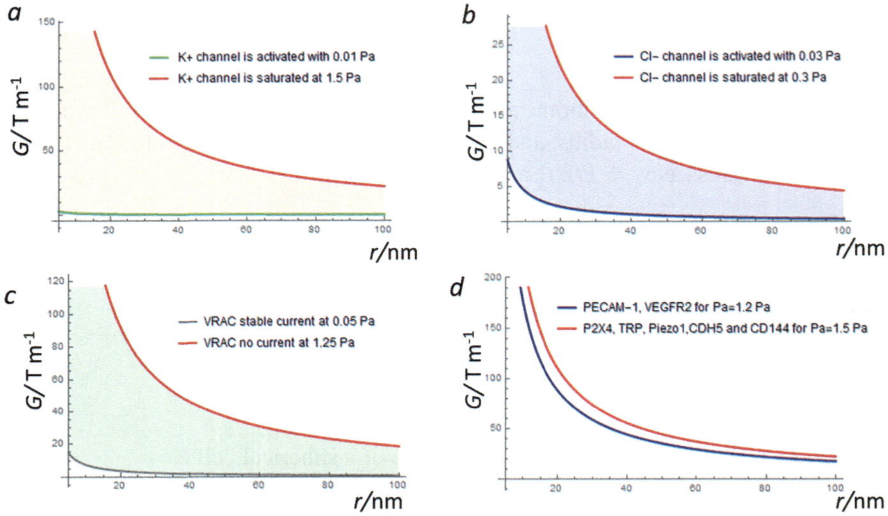

Fig. 8.4 Relationship between the magnetic field gradient (in T/m) and the nanoparticle radius (in nm) at which the magnetic pressure (Eq. (8.8)) equals the activation/saturation pressure P_a of different mechanosensitive ion channel types in endothelial cell membranes. The legend provides the P_a values and corresponding channel types. Panels (**a–c**): The filled areas indicate the ranges of magnetic field gradient and nanoparticle size where the magnetic pressure is strong enough to modulate the ion channel activity. Panel (**d**): This panel illustrates distinct intracellular signaling pathways activated upon reaching specific shear stress thresholds. At the lower shear stress threshold (1.2 Pa, blue curve): The mechanical stress triggers tyrosine phosphorylation of PECAM-1. This event is a critical upstream signal that modulates the activation of the serine/threonine kinase Akt and its downstream target, endothelial nitric oxide synthase (eNOS) (Lampugnani 2012). At the higher shear stress threshold (1.5 Pa, red curve): Two key biomechanical responses are observed: A rapid reduction in tension across VE-cadherin molecules, which occurs in parallel with a broader decrease in total tension at the cell-cell junctions over a 24-h period (Lampugnani 2012). The activation of ion channels, including P2X4, TRP, and Piezo1, which mediate the influx of extracellular calcium in response to the applied stress (Lampugnani 2012). Reprinted from (Gorobets et al. 2023), used under CC BY 4.0

the critical values of the magnetic field gradient as a function of the nanoparticle radius. The results of the calculations are shown in Fig. 8.4. Figure 8.4 illustrates that a magnetic field with the gradient of only a few tens of T/m is sufficient to significantly alter the gating of the ion channel across a wide range of magnetic nanoparticle sizes.

8.10 Effects of Magnetic Fields on Gene Expression

The process of gene expression is dynamically shaped by the cellular environment, including intracellular chemical, mechanical, thermal, and bioelectrical cues. Magnetic forces do not act directly on genes, but when they generate mechanical stresses in the cell membrane and interior, these stresses can be transmitted through the cytoskeleton to the nucleus, where they may influence nuclear mechanics and thereby modulate gene expression. In this context, magneto-mechanical stimuli function as important regulatory signals that affect the operation of the broader cellular machinery (Torbati et al. 2022; Yusko and Asbury 2014). In contrast to biochemical and electrical cues, mechanical signals propagate through the physical transmission of forces via mechanically stiff cellular and extracellular structures. The mechanotransduction process subsequently converts mechanical inputs into biochemical and bioelectric signals, thereby providing an additional route through which magneto-mechanical forces can influence gene expression.

As outlined above, magnetic forces in cells can generate mechanical stresses such as shear, positive or negative pressure, and tension on the cell membrane. These magneto-mechanical cues are primarily sensed by protein structures including integrins in focal adhesions, cadherins in cell-cell junctions, and mechanosensitive ion channels. The resulting forces deform the cytoskeleton, composed of actin filaments and microtubules, which in turn transmit these mechanical signals to intracellular components and to the nucleus. Ultimately, such forces can alter nuclear shape, thereby affecting chromatin organization and modulating gene expression (Martino et al. 2018).

In addition to the process of mechanotransduction, magnetically controlled mechanical signals are potent activators of internal cellular communication. The application of magneto-mechanical stress rapidly induces alterations in ion flux and changes in the cell's membrane potential. These shifts serve as the initial triggers for complex signaling cascades within the cell (Zablotskii et al. 2025b). The mechanical stress generated by magnetic fields elevates the concentration of several crucial intracellular messengers, including Ca^{2+} ions (Gorobets et al. 2024), ROS (Wang and Zhang 2017), and cyclic AMP. These elevated messenger levels then initiate the next step in the signaling chain. These messengers activate mechanosensitive transcription factors such as YAP/TAZ, NF-κB, or SRF, which then move to the nucleus to regulate gene expression (Finch-Edmondson and Sudol 2016). Thus, transcription factors play a central role in mechanotransduction and can mediate magnetically induced changes in gene expression. Furthermore, studies have shown that magnetic fields can directly enhance the DNA-binding activity of transcription factors like HSF, AP-1, AP-2, and SP-1 across various human cell lines (Lin et al. 1998).

Another pathway through which magnetic fields can influence gene expression is the modulation of calcium signaling by low-frequency magnetic fields. Intracellular calcium (Ca^{2+}) oscillations are tightly linked to the activation of specific transcription factors, with their frequency, amplitude, and duty cycle determining downstream gene regulatory outcomes. These oscillation frequencies vary widely among cell types from approximately 1 mHz in endothelial cells to nearly 1 Hz in cardiomyocytes and neurons (Smedler and Uhlen 1840; Colella et al. 2008). Magnetic fields operating within the millihertz range have been shown to interact with calcium ion channels of the endothelial cells, altering intracellular calcium concentrations and the signaling cascades they govern (Gorobets et al.

2024). Furthermore, applying a time-varying magnetic field allows modulation of oscillatory calcium dynamics and can shift the activation of major calcium-responsive pathways, including NF-κB, MAPK, NFAT, and glycogen phosphorylase kinase, thereby influencing metabolic activity in endothelial cells (Gorobets et al. 2024).

Extensive experimental evidence and the underlying intracellular mechanisms by which low- to high-intensity magnetic fields modulate gene expression have been summarized in (Zablotskii et al. 2025a). The effects of high magnetic fields on gene activity have been reported in both uniform and gradient magnetic field configurations. However, as noted in (Zablotskii et al. 2025a), while numerous individual studies document MF-mediated regulation of gene expression, only a limited number of systematic investigations directly compare cellular responses under homogeneous versus gradient fields, making it difficult to draw broad generalizations. For instance, comparative transcriptomic analysis in plants demonstrated that signal transduction genes were upregulated following exposure to a gradient static magnetic field, whereas no significant changes were observed under a uniform field (Zhou et al. 2024).

Overall, the impact of magnetic fields on gene expression is influenced by multiple factors, including field strength and orientation, the magnitude and direction of magnetic gradients, exposure duration, cell type, and specific experimental or physiological conditions. For example, identical static magnetic fields applied to similar cell types may induce either upregulation or downregulation of gene expression depending on the cells' physiological state (Zablotskii et al. 2025a). Notably, the most pronounced effects are often observed under strong, high-gradient magnetic fields, particularly in conditions that facilitate magnetic levitation (Wang et al. 2015).

The mechanisms and processes described above are illustrated in Fig. 8.5. The effectiveness of any of these mechanisms is significantly enhanced by the presence of biogenic or artificial magnetic nanoparticles on cell membranes.

8.11 Conclusions

Static and low-frequency magnetic fields can affect many molecular biological processes by exerting magnetic forces that alter the function of the cell's internal force network. By acting on the cell membrane, magnetic gradient forces can change the membrane surface tension, create pores, and affect the expression of ion channels. Magneto-mechanical stresses play a pivotal role in mediating the biological effects of magnetic fields, particularly when magnetic nanoparticles or magnetically responsive cellular components are present on the cell membrane or in the cell interior. Magneto-mechanical forces can modulate the gating of mechanosensitive ion channels embedded within the plasma membrane. Subtle membrane deformations, induced by the torque or translational force exerted on magnetic nanoparticles, are sufficient to alter the channel's conformational state,

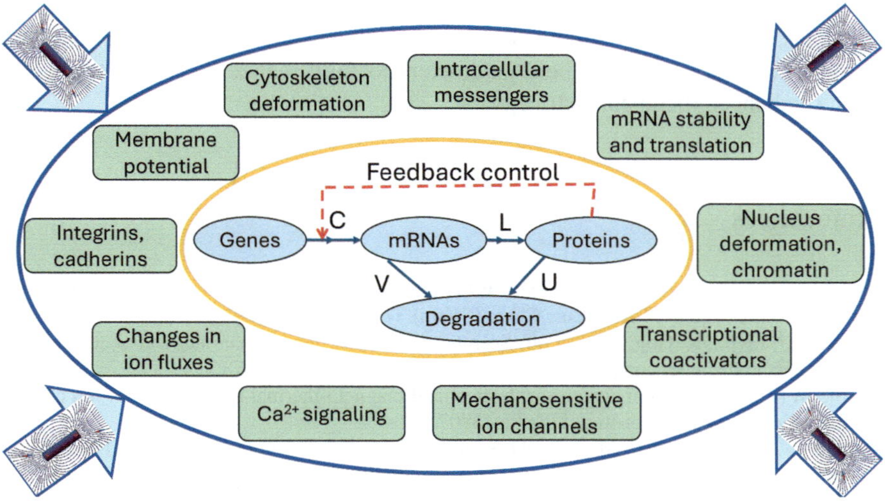

Fig. 8.5 Mechanisms of the magnetic field's influence on gene expression are shown within the framework of the dynamic transcription model (Chen et al. 1999). The thick arrows represent the magnetic forces acting on the cell membrane and its interior. Adapted from (Zablotskii et al. 2025a), used under CC BY 4.0

thereby regulating ion fluxes such as Ca2+ or K+. This regulation of ion homeostasis consequently initiates downstream signaling cascades that control crucial cellular processes, including gene expression, cytoskeletal remodeling, metabolism, and programmed cell death.

At higher levels of magnetic torque or force, magneto-mechanical stresses can induce noticeable structural changes in the plasma membrane, including bleb formation. Blebbing results from localized membrane detachment from the actin cortex and the development of negative pressure zones near MNP chains. This process can act as a precursor to apoptosis, particularly in cancer cells, where membrane integrity and cytoskeletal dynamics are already compromised. Hence, magneto-mechanical stresses can serve as a trigger for controlled cell death.

Magneto-mechanical stresses transmitted through the cytoskeleton alter F-actin, microtubule, and intermediate filament tension, leading to the remodeling of cell shape, motility, adhesion, and intracellular transport. Consequently, these perturbations modulate critical processes including cell migration, mitotic progression, and differentiation.

The noninvasive nature and high selectivity of magnetic fields, particularly when combined with magnetic nanoparticles, position biomagnetic technologies as a highly promising tool for advanced biomedical research and therapeutic applications.

In summary, the utilization of magnetic fields to remotely induce mechanical stresses offers a highly potent and versatile strategy for advanced biomedical applications. This magneto-mechanical approach leverages the noninvasive selective control of cellular mechanics to overcome limitations inherent in traditional therapies. Future efforts should focus on optimizing the integration of magnetic nanoparticles and magnetic field parameters to fully realize the clinical potential of this enhanced therapeutic strategy.

References

Avasthi A, Caro C, Pozo-Torres E, Leal MP, García-Martín ML (2020) Magnetic nanoparticles as MRI contrast agents. Topics Curr Chem (Cham) 378:40

Barakat AI, Lieu DK, Gojova A (2006) Secrets of the code: do vascular endothelial cells use ion channels to decipher complex flow signals? Biomaterials 27:671–678

Barbic M (2019) Possible magneto-mechanical and magneto-thermal mechanisms of ion channel activation in magnetogenetics. elife 8:e45807

Berrout J, Jin M, Mamenko M, Zaika O, Pochynyuk O, O'Neil RG (2012) Function of transient receptor potential cation channel subfamily V member 4 (TRPV4) as a mechanical transducer in flow-sensitive segments of renal collecting duct system, (vol 287, pg 8782). J Biol Chem 287(2012):42454–42454

Berry MV, Geim AK (1997) Of flying frogs and levitrons. Eur J Phys 18:307

Blakemore R (1975) Magnetotactic bacteria. Science 190:377–379

Bouffanais R, Sun J, Yue DKP (2013) Physical limits on cellular directional mechanosensing. Phys Rev E 87:052716

Brem F, Hirt AM, Simon C, Wieser H-G, Dobson J (2005) Characterization of iron compounds in tumour tissue from temporal lobe epilepsy patients using low temperature magnetic methods. Biometals 18:191–197

Brem F, Hirt AM, Winklhofer M, Frei K, Yonekawa Y, Wieser HG, Dobson J (2006) Magnetic iron compounds in the human brain: a comparison of tumour and hippocampal tissue. J R Soc Interface 3:833–841

Charras GT, Coughlin M, Mitchison TJ, Mahadevan L (2008) Life and times of a cellular bleb. Biophys J 94:1836–1853

Chazaud C, Yamanaka Y, Pawson T, Rossant J (2006) Early lineage segregation between epiblast and primitive endoderm in mouse blastocysts through the Grb2-MAPK pathway. Dev Cell 10:615–624

Chen T, He HL, Church GM (1999) Modeling gene expression with differential equations, pacific symposium on Biocomputing. Pac Symp Biocomput:29–40

Cheng G, Tse J, Jain RK, Munn LL (2009) Micro-environmental mechanical stress controls tumor spheroid size and morphology by suppressing proliferation and inducing apoptosis in cancer cells. PLoS One 4:e4632

Choi S-H, Shin J, Park C, Lee J-u, Lee J, Ambo Y, Shin W, Yu R, Kim J-Y, Lah JD, Shin D, Kim G, Noh K, Koh W, Lee CJ, Lee J-H, Kwak M, Cheon J (2024) In vivo magnetogenetics for cell-type-specific targeting and modulation of brain circuits. Nat Nanotechnol 19:1333–1343

Christensen AP, Corey DP (2007) TRP channels in mechanosensation: direct or indirect activation? Nat Rev Neurosci 8:510–521

Colella M, Grisan F, Robert V, Turner JD, Thomas AP, Pozzan T (2008) Ca oscillation frequency decoding in cardiac cell hypertrophy:: Role of calcineurin/NFAT as Ca signal integrators. P Natl Acad Sci USA 105:2859–2864

Cross SE, Jin YS, Rao J, Gimzewski JK (2007) Nanomechanical analysis of cells from cancer patients. Nat Nanotechnol 2:780–783

Cugat O, Delamare J, Reyne G (2003) Magnetic micro-actuators and systems (MAGMAS). IEEE T Magn 39:3607–3612

Dai G, Kaazempur-Mofrad MR, Natarajan S, Zhang Y, Vaughn S, Blackman BR, Kamm RD, García-Cardeña G, Gimbrone MA Jr (2004) Distinct endothelial phenotypes evoked by arterial waveforms derived from atherosclerosis-susceptible and -resistant regions of human vasculature. Proc Natl Acad Sci USA 101:14871–14876

Darmenko YA, Gorobets O, Gorobets S, Sharai IV, Lazarenko OM (2018) Detection of biogenic magnetic nanoparticles in human aortic aneurysms. Acta Phys Pol A 133:738–741

Davies PF (1995) Flow-mediated endothelial mechanotransduction. Physiol Rev 75:519–560

Dumas-Bouchiat F, Zanini LF, Kustov M, Dempsey NM, Grechishkin R, Hasselbach K, Orlianges JC, Champeaux C, Catherinot A, Givord D (2010) Thermomagnetically patterned micromagnets. Appl Phys Lett 96

Dunne P, Mazza L, Coey JMD (2011) Magnetic structuring of electrodeposits. Phys Rev Lett 107:024501

Finch-Edmondson M, Sudol M (2016) Framework to function: mechanosensitive regulators of gene transcription. Cell Mol Biol Lett 21:28

Funk RHW, Scholkmann F (2023) The significance of bioelectricity on all levels of organization of an organism. Part 1: from the subcellular level to cells. Prog Biophys Mol Biol 177:185–201

Gieré R (2016) Magnetite in the human body: biogenic vs. anthropogenic. Proc Natl Acad Sci 113:11986–11987

Goldmann WH (2008) Actin: a molecular wire, an electrical cable? Cell Biol Int 32:869–870

Gorobets O, Gorobets S, Koralewski M (2017) Physiological origin of biogenic magnetic nanoparticles in health and disease: from bacteria to humans. Int J Nanomedicine 12:4371–4395

Gorobets S, Gorobets O, Gorobets Y, Bulaievska M (2022) Chain-like structures of biogenic and nonbiogenic magnetic nanoparticles in vascular tissues. Bioelectromagnetics 43:119–143

Gorobets O, Gorobets S, Sharai I, Polyakova T, Zablotskii V (2023) Interaction of magnetic fields with biogenic magnetic nanoparticles on cell membranes: physiological consequences for organisms in health and disease. Bioelectrochemistry 151

Gorobets O, Gorobets S, Polyakova T, Zablotskii V (2024) Modulation of calcium signaling and metabolic pathways in endothelial cells with magnetic fields, nanoscale. Advances 6:1163–1182

He X, Yablonskiy DA (2009) Biophysical mechanisms of phase contrast in gradient echo MRI. Proc Natl Acad Sci USA 106:13558–13563

Hedley JG, Coshic K, Aksimentiev A, Kornyshev AA (2024) Electric field of DNA in solution: who is in charge? Phy Rev X 14

Hinds G, Coey JMD, Lyons MEG (2001) Influence of magnetic forces on electrochemical mass transport. Electrochem Commun 3:215–218

Holle AW, Engler AJ (2010) Cell rheology: stressed-out stem cells. Nat Mater 9:4–6

Huang H, Delikanli S, Zeng H, Ferkey DM, Pralle A (2010) Remote control of ion channels and neurons through magnetic-field heating of nanoparticles. Nat Nanotechnol 5:602

Hughes S, McBain S, Dobson J, El Haj AJ (2008) Selective activation of mechanosensitive ion channels using magnetic particles. J R Soc Interface 5:855–863

Itakura T, Yamamoto K, Tohyama M, Shimizu N (1977) Central dual innervation of arterioles and capillaries in the brain. Stroke 8:360–365

Ji X, Tian X, Feng S, Zhang L, Wang J, Guo R, Zhu Y, Yu X, Zhang Y, Du H, Zablotskii V, Zhang X (2023) Intermittent F-actin perturbations by magnetic fields inhibit breast cancer metastasis. Research 6:0080

Jiang H, Sean X (2013) Sun, cellular pressure and volume regulation and implications for cell mechanics. Biophys J 105:609–619

Karcher H, Lammerding J, Huang H, Lee RT, Kamm RD, Kaazempur-Mofrad MR (2003) A three-dimensional viscoelastic model for cell deformation with experimental verification. Biophys J 85:3336–3349

Katira P, Zaman MH, Bonnecaze RT (2012) How changes in cell mechanical properties induce cancerous behavior. Phys Rev Lett 108:028103

Kauffmann P, Ith A, O'Brien D, Gaude V, Boue F, Combe S, Bruckert F, Schaack B, Dempsey NM, Haguet V, Reyne G (2011) Diamagnetically trapped arrays of living cells above micromagnets. Lab Chip 11:3153–3161

Kelly CJ, Couch RK, Ha VT, Bodart CM, Wu J, Huff S, Herrel NT, Kim HD, Zimmermann AO, Shattuck J, Pan YC, Kaeberlein M, Grillo AS (2023) Iron status influences mitochondrial disease progression in complex I-deficient mice. elife 12

Keydar I (n.d.) The electric genome: a mechanistic hypothesis for nuclear architecture and gene expression bias: chromatin charge, modulated by genetic and epigenetic features, establishes a reproducible radial nuclear architecture that directs transcription. BioEssays:e70076. https://doi.org/10.1002/bies.70076(2025)

Kim D-H, Rozhkova EA, Ulasov IV, Bader SD, Rajh T, Lesniak MS, Novosad V (2010) Biofunctionalized magnetic-vortex microdiscs for targeted cancer-cell destruction. Nat Mater 9:165–171

Kloc M, Halasa M, Wosik J, Ghobrial RM (2025) Gadolinium-based MRI contrast agent effects on calcium signaling and actin-dependent cell functions. Magn Med 1:100004

Kobayashi A, Yamamoto N, Kirschvink J (1997) Studies of inorganic crystals in biological tissue: magnetite in human tumor, 粉体および粉末冶金, vol 44, pp 294–300

Kofman K, Levin M (2024) Bioelectric pharmacology of cancer: a systematic review of ion channel drugs affecting the cancer phenotype. Prog Biophys Mol Biol 191:25–39

Koster DA, Croquette V, Dekker C, Shuman S, Dekker NH (2005) Friction and torque govern the relaxation of DNA supercoils by eukaryotic topoisomerase IB. Nature 434:671–674

Lampugnani MG (2012) Endothelial cell-to-cell junctions: adhesion and signaling in physiology and pathology. Cold Spring Harb Perspect Med 2

Landau LD, Lifshitz EM (1959) Theory of elasticity. Pergamon Press

Leng G, Brown CH, Russell JA (1999) Physiological pathways regulating the activity of magnocellular neurosecretory cells. Prog Neurobiol 57:625–655

Leventis N, Gao X (2002) Nd−Fe−B permanent magnet electrodes. Theoretical evaluation and experimental demonstration of the paramagnetic body forces. J Am Chem Soc 124:1079–1088

Levin M (2009) Bioelectric mechanisms in regeneration: unique aspects and future perspectives. Semin Cell Dev Biol 20:543–556

Li H, Zhou WY, Xia YY, Zhang JX (2022) Endothelial Mechanosensors for Atheroprone and Atheroprotective shear stress signals. J Inflamm Res 15:1771–1783

Lin H, Han L, Blank M, Head M, Goodman R (1998) Magnetic field activation of protein-DNA binding. J Cell Biochem 70:297–303

Luo J, He X, d'Avignon DA, Ackerman JJH, Yablonskiy DA (2010) Protein-induced water H-1 MR frequency shifts: contributions from magnetic susceptibility and exchange effects. J Magn Reson 202:102–108

Lv Y, Fan Y, Tian X, Yu B, Song C, Feng C, Zhang L, Ji X, Zablotskii V, Zhang X (2022) The anti-depressive effects of ultra-high static magnetic field. J Magn Reson Imaging 56:354–365

MacQueen L, Sun Y, Simmons CA (2013) Mesenchymal stem cell mechanobiology and emerging experimental platforms. J R Soc Interface 10:20130179

Maffei ME (2014) Magnetic field effects on plant growth, development, and evolution. Front Plant Sci 5:445

Maher BA (2019) Airborne magnetite- and iron-rich pollution nanoparticles: potential Neurotoxicants and environmental risk factors for neurodegenerative disease, including Alzheimer's disease. J Alzheimer's Disease 71:361–375

Martino F, Perestrelo AR, Vinarský V, Pagliari S, Forte G (2018) Cellular Mechanotransduction: from tension to function. Front Physiol 9:824

Mazuel F, Mathieu S, Di Corato R, Bacri JC, Meylheuc T, Pellegrino T, Reffay M, Wilhelm C (2017) Forced- and self-rotation of magnetic Nanorods assembly at the cell membrane: a biomagnetic torsion pendulum. Small 13

Meenakshi P, Florian R, Christoph FS (2017) Molecular force sensors to measure stress in cells. J Phys D Appl Phys 50:233001

Mirkin EV, Mirkin SM (2007) Replication fork stalling at natural impediments. Microbiol Mol Biol Rev 71:13–35

Mitchison TJ, Salmon ED (2001) Mitosis: a history of division. Nat Cell Biol 3:E17–E21

Montel F, Delarue M, Elgeti J, Malaquin L, Basan M, Risler T, Cabane B, Vignjevic D, Prost J, Cappello G, Joanny J-F (2011) Stress clamp experiments on multicellular tumor spheroids. Phys Rev Lett 107:188102

Monzel C, Vicario C, Piehler J, Coppey M, Dahan M (2017) Magnetic control of cellular processes using biofunctional nanoparticles. Chem Sci 8:7330–7338

Neurath PW (1968) High gradient magnetic field inhibits embryonic development of frogs. Nature 219:1358–1359

Novickij V, Grainys A, Lastauskienė E, Kananavičiūtė R, Pamedytytė D, Kalėdienė L, Novickij J, Miklavčič D (2016) Pulsed electromagnetic field assisted in vitro electroporation: a pilot study. Sci Rep-Uk 6:33537

Olesen S-P, Claphamt D, Davies P (1988) Haemodynamic shear stress activates a K+ current in vascular endothelial cells. Nature 331:168–170

Oliveira FTP, Diedrichsen J, Verstynen T, Duque J, Ivry RB (2010) Transcranial magnetic stimulation of posterior parietal cortex affects decisions of hand choice. Proc Natl Acad Sci 107:17751–17756

Osman O, Toru S, Dumas-Bouchiat F, Dempsey NM, Haddour N, Zanini L-F, Buret F, Reyne G, Frénéa-Robin M (2013) Microfluidic immunomagnetic cell separation using integrated permanent micromagnets. Biomicrofluidics 7

Panagiotakopoulou M, Bergert M, Taubenberger A, Guck J, Poulikakos D, Ferrari A (2016) A Nanoprinted model of interstitial cancer migration reveals a link between cell deformability and proliferation. ACS Nano 10:6437–6448

Pauklin S, Pedersen RA, Vallier L (2011) Mouse pluripotent stem cells at a glance. J Cell Sci 124:3727–3732

Polyakova T, Zablotskii V, Dejneka A (2017) Cell membrane pore formation and change in Ion Channel activity in high-gradient magnetic fields. IEEE Magn Lett 8:1–5

Reeves D, Ursell T, Sens P, Kondev J, Phillips R (2008) Membrane mechanics as a probe of ion-channel gating mechanisms. Phys Rev E 78:041901

Rems L, Miklavčič D (2016) Tutorial: electroporation of cells in complex materials and tissue. J Appl Phys 119

Richelme F, Benoliel AM, Bongrand P (1997) Aspiration of THP1 into a micropipette. Mechanical deformation of monocytic THP-1 cells: occurrence of two sequential phases with differential sensitivity to metabolic inhibitors. Exp Biol Online 2:1–14

Ritter CM, Ma T, Leijnse N, Farhangi Barooji Y, Hamilton W, Brickman JM, Doostmohammadi A, Oddershede LB (2025) Differential elasticity affects lineage segregation of embryonic stem cells. Phys Rev Lett 134:168401

Roux E, Bougaran P, Dufourcq P, Couffinhal T (2020) Fluid shear stress sensing by the endothelial layer. Front Physiol 11

Rubin CT, Capilla E, Luu YK, Busa B, Crawford H, Nolan DJ, Mittal V, Rosen CJ, Pessin JE, Judex S (2007) Adipogenesis is inhibited by brief, daily exposure to high-frequency, extremely low-magnitude mechanical signals. P Natl Acad Sci USA 104:17879–17884

Rubio Ayala M, Syrovets T, Hafner S, Zablotskii V, Dejneka A, Simmet T (2018) Spatiotemporal magnetic fields enhance cytosolic Ca2+ levels and induce actin polymerization via activation of voltage-gated sodium channels in skeletal muscle cells. Biomaterials 163:174–184

Schenck JF (1996) The role of magnetic susceptibility in magnetic resonance imaging: MRI magnetic compatibility of the first and second kinds. Med Phys 23:815–850

Sear RP (2019) Diffusiophoresis in cells: a general nonequilibrium, nonmotor mechanism for the metabolism-dependent transport of particles in cells. Phys Rev Lett 122:128101

F.E. Senftle, W.P. Hambright, Magnetic susceptibility of biological materials, in: M.F. Barnothy (Ed.) Biological effects of magnetic fields: Volume 2, Springer Boston, 1995, pp. 261–306

Shabrangy A, Luschnig C (2025) Magnetic fields impact on PIN-FORMED protein polarity in Arabidopsis thaliana. Physiol Plant 177:e70274

Smedler E, Uhlen P (1840) Frequency decoding of calcium oscillations. BBA-Gen Subjects 2014:964–969
Song MJ, Brady-Kalnay SM, McBride SH, Phillips-Mason P, Dean D, Knothe Tate ML (2012) Mapping the Mechanome of live stem cells using a novel method to measure local strain fields in situ at the fluid-cell Interface. PLoS One 7:e43601
Suriyagandhi V, Ma Y, Paparozzi V, Guarnieri T, Di Pietro B, Dimitri GM, Tieri P, Sala C, Lai D, Nardini C (2025) Mechanotransduction and inflammation: an updated comprehensive representation. Mechanobiol Med 3:100112
Swaminathan V, Mythreye K, O'Brien ET, Berchuck A, Blobe GC, Superfine R (2011) Mechanical stiffness grades metastatic potential in patient tumor cells and in cancer cell lines. Cancer Res 71:5075–5080
Theret DP, Levesque MJ, Sato M, Nerem RM, Wheeler LT (1988) The application of a homogeneous half-space model in the analysis of endothelial cell micropipette measurements. J Biomech Eng 110:190–199
Torbati M, Mozaffari K, Liu L, Sharma P (2022) Coupling of mechanical deformation and electromagnetic fields in biological cells. Rev Mod Phys 94:025003
Tseng P, Judy JW, Di Carlo D (2012) Magnetic nanoparticle-mediated massively parallel mechanical modulation of single-cell behavior. Nat Methods 9:1113-+
Valles JM, Lin K, Denegre JM, Mowry KL (1997) Stable magnetic field gradient levitation of Xenopus laevis: toward low-gravity simulation. Biophys J 73:1130–1133
Van de Walle A, Plan Sangnier A, Abou-Hassan A, Curcio A, Hémadi M, Menguy N, Lalatonne Y, Luciani N, Wilhelm C (2019) Biosynthesis of magnetic nanoparticles from nano-degradation products revealed in human stem cells. Proc Natl Acad Sci 116:4044–4053
Wang H, Zhang X (2017) Magnetic fields and reactive oxygen species. Int J Mol Sci 18
Wang Y, Chen ZH, Yin C, Ma JH, Li DJ, Zhao F, Sun YL, Hu LF, Shang P, Qian AR (2015) GeneChip expression profiling reveals the alterations of energy metabolism related genes in osteocytes under large gradient high magnetic fields. PLoS One 10:e0116359
Waskaas M (1993) Short-term effects of magnetic fields on diffusion in stirred and unstirred paramagnetic solutions. J Phys Chem 97:6470–6476
Wei Y, Wang X (2023) Biological effects of rotating magnetic field: a review from 1969 to 2021. Prog Biophys Mol Biol 178:103–115
Winkler SA, Schmitt F, Landes H, de Bever J, Wade T, Alejski A, Rutt BK (2018) Gradient and shim technologies for ultra high field MRI. NeuroImage 168:59–70
Wosik J, Chen W, Qin K, Ghobrial RM, Kubiak JZ, Kloc M (2018) Magnetic field changes macrophage phenotype. Biophys J 114:2001–2013
Wu J, Goyal R, Grandl J (2016) Localized force application reveals mechanically sensitive domains of Piezo1. Nat Commun 7:12939
Wu D, Mao S, Lin J (2025) Optimal distributions of receptors on arbitrarily shaped cell surfaces. Phys Rev Lett 135:208401
Yang X, Li Z, Polyakova T, Dejneka A, Zablotskii V, Zhang X (2020) Effect of static magnetic field on DNA synthesis: the interplay between DNA chirality and magnetic field left-right asymmetry. FASEB BioAdv 2:254–263
Yang X, Song C, Zhang L, Wang J, Yu X, Yu B, Zablotskii V, Zhang X (2021) An upward 9.4 T static magnetic field inhibits DNA synthesis and increases ROS-P53 to suppress lung cancer growth. Transl Oncol 14:101103
Yu B, Song C, Feng C-L, Zhang J, Wang Y, Zhu Y-M, Zhang L, Ji X-M, Tian X-F, Cheng G-F, Chen W-L, Zablotskii V, Wang H, Zhang X (2023) Effects of gradient high-field static magnetic fields on diabetic mice. Zool Res 44:249
Yusko EC, Asbury CL (2014) Force is a signal that cells cannot ignore. Mol Biol Cell 25:3717–3725
Zablotskii V, Polyakova T (2025) Permanent magnets in magnetic medicine: applications and advances. Magnetic Med 1:100024
Zablotskii V, Dejneka A, Kubinová S, Le-Roy D, Dumas-Bouchiat F, Givord D, Dempsey NM, Syková E (2013) Life on magnets: stem cell networking on micro-magnet arrays. PLoS One 8

Zablotskii V, Lunov O, Novotna B, Churpita O, Trosan P, Holan V, Sykova E, Dejneka A, Kubinova S (2014a) Down-regulation of adipogenesis of mesenchymal stem cells by oscillating high-gradient magnetic fields and mechanical vibration. Appl Phys Lett 105:103702

Zablotskii V, Syrovets T, Schmidt ZW, Dejneka A, Simmet T (2014b) Modulation of monocytic leukemia cell function and survival by high gradient magnetic fields and mathematical modeling studies. Biomaterials 35:3164–3171

Zablotskii V, Lunov O, Novotná B, Churpita O, Trosan P, Holán V, Syková E, Dejneka A, Kubinová S (2014c) Down-regulation of adipogenesis of mesenchymal stem cells by oscillating high-gradient magnetic fields and mechanical vibration. Appl Phys Lett 105

Zablotskii V, Lunov O, Kubinova S, Polyakova T, Sykova E, Dejneka A (2016a) Effects of high-gradient magnetic fields on living cell machinery. J Phys D Appl Phys 49:493003

Zablotskii V, Polyakova T, Lunov O, Dejneka A (2016b) How a high-gradient magnetic field could affect cell life. Sci Rep-Uk 6:37407

Zablotskii V, Polyakova T, Dejneka A (2018) Cells in the non-uniform magnetic world: how cells respond to high-gradient magnetic fields. BioEssays 40:1800017

Zablotskii V, Polyakova T, Dejneka A (2021) Modulation of the cell membrane potential and intracellular protein transport by high magnetic fields. Bioelectromagnetics 42:27–36

Zablotskii V, Polyakova T, Dejneka A (2022) Effects of high magnetic fields on the diffusion of biologically active molecules. Cells-Basel 11:81

Zablotskii V, Polyakova T, Dejneka A (2023) Controlling cell membrane potential with static nonuniform magnetic fields. In: Zhang X (ed) Biological effects of static magnetic fields. Springer, Singapore, pp 113–131

Zablotskii V, Gorobets O, Gorobets S, Polyakova T (2025a) Effects of static and low-frequency magnetic fields on gene expression. J Magn Reson Imaging 62:317–334

Zablotskii V, Polyakova T, Dejneka A (2025b) Exploring Ion Channel magnetic pharmacology: are magnetic cues a viable alternative to Ion Channel drugs? BioEssays 47:e202400200. https://doi.org/10.1002/bies.202400200(2024)

Zborowski M, Chalmers JJ (2011) Rare cell separation and analysis by magnetic sorting. Anal Chem 83:8050–8056

Zhang X (2023) Biological effects of static magnetic fields. Springer, Singapore

Zhang L, Hou YB, Li ZY, Ji XM, Wang Z, Wang HZ, Tian XF, Yu FZ, Yang ZY, Pi L, Mitchison TJ, Lu QY, Zhang X (2017) 27 T ultra-high static magnetic field changes orientation and morphology of mitotic spindles in human cells. elife 6

Zhou X, Zhang L, Zhang P, Xu H, Song J, Chang Y, Cai T, Xie C (2024) Comparative transcriptomic analysis revealed important processes underlying the static magnetic field effects on Arabidopsis. Front Plant Sci 15:1390031

Chapter 9
Setting Up to Divide: Nuclear Mechanotransduction in the Regulation of Mitosis

Jorge G. Ferreira

Abstract As the largest and stiffest cellular organelle, the nucleus is continuously under the influence of mechanical forces from external and internal sources. These forces trigger an adaptive response at the nuclear level that include changes in nucleocytoplasmic transport, chromatin architecture, or transcriptional activity, ultimately driving cell proliferation or differentiation. In preparation for mitosis, nuclear structure and organization is modified to ensure that chromosomes are efficiently captured and correctly separated. How these modifications impact the mechanical response of the nucleus remains a largely unexplored area. One emerging concept is that changes in the mechanical properties of the nucleus during early mitosis are required to ensure the timing and accuracy of chromosome segregation. This review will discuss the contribution of key nuclear components to nuclear mechanics and their relevance for chromosome segregation fidelity.

9.1 Introduction

Cells and tissues are continuously under the influence of physical forces that shape their architecture and function. When these forces act at the cellular level, they are translated into a biochemical signal, in a process known as mechanotransduction. This process starts at the cell membrane, where adhesion complexes sense the cues generated by the extracellular matrix or by surrounding cells. These signals are then transmitted by the cytoskeleton until they reach the nucleus, modifying nuclear structure and organization (Lammerding 2011; Maurer and Lammerding 2019). Over the recent years, it has become clear that force sensing is not exclusive to cytoplasmic structures and that the nucleus is, in itself, a mechanosensitive organelle (Niethammer 2021). This mechanical response is defined in combination by the

J. G. Ferreira (✉)
Departamento de Biomedicina, Faculdade de Medicina da Universidade do Porto, Porto, Portugal

Instituto de Investigação e Inovação em Saúde (i3S), Universidade do Porto, Porto, Portugal
e-mail: jferreir@i3s.up.pt

M. Kloc, J. Wosik (eds.), *Mechanotransduction and Mechanosensing*, Results and Problems in Cell Differentiation 78, https://doi.org/10.1007/978-3-032-19509-8_9

nuclear membranes (Agrawal and Lele 2019), the nuclear lamina (Stephens et al. 2017), and chromatin (Schreiner et al. 2015; Stephens et al. 2017) and can lead to structural changes ranging from nuclear membrane and/or nuclear pore stretching to conformational changes in protein structure or chromatin modifications (Kirby and Lammerding 2018; Niethammer 2021). Ultimately, these can modify cell signaling pathways and control cell cycle progression (Aureille et al. 2019; Dantas et al. 2022; Uroz et al. 2018). However, the mechanistic relationship between these nuclear mechanosensitive pathways and cell division is still unclear.

Mitosis is the process by which cells segregate their genome, giving rise to two identical daughter cells. To achieve this, cells undergo an extensive cytoplasmic and nuclear reorganization (Champion et al. 2017; Lima and Ferreira 2024), required to assemble a microtubule-based mitotic spindle that will capture and eventually segregate chromosomes. Most of the cellular changes involved in this reorganization are regulated by kinases such as CDK1 and Plk1 (Gavet and Pines 2010a, b; Gheghiani et al. 2017) and occur during a short period of the cell cycle as cells transition from G2 phase to mitosis (G2-M). During this transition, chromosomes condense (Antonin and Neumann 2016), nuclear envelope (NE) tension increases (Dantas et al. 2022), and the nuclear lamina and nuclear pore complexes (NPCs) disassemble (Kutay et al. 2021). These widespread changes in nuclear organization and structure are required for NE permeabilization (NEP) and breakdown (NEB) so that chromosomes can eventually be captured by microtubules. Whether these changes impact the mechanical response of the nucleus to order early mitotic events remains to be determined.

Here, we describe and discuss the molecular mechanisms that impact nuclear mechanics, structure, and function during the preparation for mitosis. We identify the main molecular players that determine nuclear mechanotransduction, focusing on nuclear membranes, the nuclear lamina, and the interaction of chromatin with the nuclear envelope, and discuss their critical role in determining the timing and robustness of cell division. Lastly, we discuss how alterations in these molecular players lead to significant changes in nuclear mechanics, with deleterious consequences for mitosis and chromosomal stability.

9.2 Nuclear Envelope Structure

The nuclear envelope (NE) is a highly specialized membrane system that compartmentalizes the genetic material and regulates transport between the nucleus and the cytoplasm. It is composed of a double membrane system, termed the outer nuclear membrane (ONM) and the inner nuclear membrane (INM), a nuclear lamina underlying the INM, and a significant number of membrane-associated proteins. The ONM is continuous with the endoplasmic reticulum (ER), which results in a similar protein composition between the two structures. On the other hand, the INM faces the nucleoplasm and contains a unique set of membrane proteins (De Magistris and Antonin 2018). The two membranes fuse at discrete sites where nuclear pore complexes (NPCs) assemble, allowing for bidirectional transport between the nucleus and the cytoplasm. NPCs are composed of different nucleoporins (Nups), which organize into cytoplasmic, nucleoplasmic, and inner rings around a central channel (Hampoelz et al. 2019; Schuller et al. 2021; Zimmerli et al. 2021). Additional contact points across the two membranes are established by the Linker of Nucleoskeleton and Cytoskeleton (LINC) complex, which links the cytoskeleton to the nuclear interior (Lombardi et al. 2011; Starr and Fridolfsson 2010). LINC complexes are composed of Nuclear Envelope SPectRIN repeat proteins (nesprins) that are embedded on the ONM due to the presence of a Klarsicht/ANC-1/Syne Homology (KASH) domain and Sad1/UNC84 (SUN) domain-containing proteins at the INM. Nesprins interact with different cytoskeletal components via their N-terminal domains and with SUN proteins in the perinuclear space, via their C-terminal domains (Mellad et al. 2011). Localization of SUN proteins on the INM enables coupling of the complex to internal nuclear structures such as chromatin or NPCs (McGillivary et al. 2023), establishing a functional link between the cytoplasm and the nucleoplasm. In metazoans, the NE is further reinforced by the nuclear lamina, a dense, proteinaceous meshwork underlying the INM. This structure is composed of intermediate filament proteins known as lamins. All lamins share a conserved tripartite structure, consisting of a long central α-helical rod domain flanked by globular N-terminal (head) and C-terminal (tail) domains (Dechat et al. 2008). A-type lamins comprise lamin A and lamin C and are expressed through the *LMNA* gene. B-type lamins are encoded by the *LMNB1* gene, which expresses lamin B1, and the *LMNB2* gene giving rise to lamin B2 (Xie and Burke 2016). Comprehensive reviews of NE structure and dynamics, which will not be directly addressed here, can be found elsewhere (Hetzer 2010; Turkmen et al. 2023; Ungricht and Kutay 2017).

9.2.1 Nuclear Envelope Mechanics

Over the past years, it has become evident that the nucleus does not merely function as a passive repository of genetic information but has the ability to sense and transmit physical forces, acting as a mechanosensitive organelle (Niethammer 2021). Indeed, applying force on the nucleus is sufficient to trigger a downstream response that involves NE unfolding and nuclear stiffening (Guilluy et al. 2014; Tang et al. 2023). This mechanical response of the nucleus is largely dependent on chromatin (Stephens et al. 2017, 2019a), the nuclear lamina (Guilluy et al. 2014; Stephens et al. 2017), and NE membranes (Agrawal and Lele 2019). However, it is also influenced by the way these components interact with the cytoskeleton. Therefore, how the different components of the NE contribute to the overall mechanical response of the nucleus remains an open question, requiring an integrated approach that combines experimental (Hobson et al. 2021; Maurer and Lammerding 2019) and computational tools (Hobson and Stephens 2020).

Due to their inherent structure and continuity with the ER (Bragulat-Teixidor et al. 2024; English and Voeltz 2013; Fricker et al. 1997), NE membranes are often folded, creating a membrane reservoir that buffers nuclear shape changes. Through nuclear swelling or compression, these membranes can unfold, creating tension (Enyedi and Niethammer 2017), which decreases lipid packing (Zhang et al. 2006) and results in membrane stretching (Dahl et al. 2004; Enyedi et al. 2016). Whether this tension is equally distributed between the INM and ONM remains to be determined. However, the loosely packed lipid content of the INM (Lee et al. 2024) favors the stretch-induced recruitment and insertion of enzymes such as the cytosolic phospholipase A2 (cPLA2) specifically to this membrane (Enyedi et al. 2013, 2016; Lomakin et al. 2020; Venturini et al. 2020), suggesting that the force-sensing mechanisms might be preferentially located at the INM. Nevertheless, these increases in membrane tension can act as a sensor for nuclear deformation, leading to a downstream adaptive cellular response (Lomakin et al. 2020; Venturini et al. 2020).

The nuclear lamina is a well-established regulator of nuclear shape and mechanics (Davidson and Lammerding 2014; Lammerding et al. 2004, 2006). Due to their spatial organization (Nmezi et al. 2019), lamins are suggested to impart different mechanical properties to the nucleus. Accordingly, lamin B1 was proposed to stabilize nuclear shape, localizing mainly to regions of low curvature (Nmezi et al. 2019). Moreover, B-type lamins mainly contribute to the elastic response of the nucleus (Panorchan et al. 2004), mediating the instantaneous recovery from deformation. A-type lamins, on the other hand, are responsible for the viscoelastic behavior of the nucleus (Banerjee et al. 2013), impacting overall nuclear stiffness and mechanics (Lammerding et al. 2004, 2006; Stephens et al. 2017; Swift et al. 2013). Accordingly, cells expressing high levels of lamin A exhibit increased nuclear stiffness, thereby impairing nuclear deformability (Rowat et al. 2013; Shin et al. 2013). Taken together, these observations suggest that lamin B1 acts as an elastic shell that could respond rapidly to nuclear strains, whereas lamin A/C would behave as a viscous element, providing long-term resistance to mechanical stress. In combination, these different networks allow the nucleus to manage mechanical deformations of different magnitudes.

Apart from lamins, various other proteins that are either membrane embedded or associated with the INM were shown to mediate the mechanical response of the nucleus. Notably, the LINC complex composed of SUN and KASH-domain-containing proteins ensures the transmission of forces between the cytoskeleton and the nuclear interior (Chang et al. 2015; Lombardi et al. 2011; Stewart-Hutchinson et al. 2008). Due to its localization across the NE membranes, the LINC complex is ideally placed to directly mediate outside-in and inside-out nuclear mechanical signals. In addition, this complex might also play an indirect role in the regulation of NE mechanics by retaining additional INM or transmembrane NE proteins. Accordingly, emerin localization and oligomerization on the INM depend on interactions with SUN1 and lamin A/C (Berk et al. 2013; Fernandez et al. 2022), which is crucial for nuclear adaptation against mechanical forces (Fernandez et al. 2022) and maintenance of NE elasticity (Rowat et al. 2006). Another relevant example involves lamin B receptor (LBR), an integral membrane protein that is anchored to the INM and associates with the nuclear lamina (Buxboim et al. 2017; Olins et al. 2010). Expression of LBR was shown to anticorrelate with lamin A levels (Buxboim et al. 2017; Solovei et al. 2013), suggesting that higher expression of LBR is associated with softer tissues. Moreover, overexpression of LBR leads to an excess of nuclear membranes (Dantas et al. 2022; Lomakin et al. 2020; Lu et al. 2010), which could, in principle, affect NE tension and disrupt membrane-mediated mechanotransduction.

9.3 Coupling of Chromatin to the Nuclear Envelope

Within the interphase nucleus, chromosomes are spatially organized in distinct domains (Cremer and Cremer 2010; Rowley and Corces 2018; van Steensel and Belmont 2017) that dynamically associate with the NE and regulate nuclear architecture, transcriptional activity, and DNA replication and repair. Over the years, there has been considerable interest in understanding how these chromatin-NE attachments are established. Chromatin association with the NE can occur through multiple mechanisms involving the nuclear lamina (Dauban et al. 2025; Guelen et al. 2008; Schreiner et al. 2015), NPCs (Liang and Hetzer 2011; Van de Vosse et al. 2013), or INM-resident proteins such as LBR (Polioudaki et al. 2001; Solovei et al. 2013), LAP2 (Brachner and Foisner 2011; Foisner and Gerace 1993), and SUN1 and 2 (Ding et al. 2007; Schmitt et al. 2007). In this review, we will briefly describe the main molecular players involved in this process and explain their relevance in the context of NE mechanics.

9.3.1 Lamina-Associated Domains

Large parts of the genome associate with the nuclear lamina, forming Lamina-Associated Domains (LADs) (van Steensel and Belmont 2017). These LADs associate with the lamina through multiple regions (Dauban et al. 2025) in a cell cycle-dependent manner (Kind et al. 2013; van Schaik et al. 2020) and possess characteristics of heterochromatin, with low gene density and enrichment in histone posttranslational modifications (PTMs) such as histone H3K9 dimethylation (H3K9me2) and histone H3K9 trimethylation (H3K9me3) (Guelen et al. 2008; Wen et al. 2009). Since chromatin compaction and histone PTMs are key determinants for nuclear mechanics (Stephens et al. 2017, 2018, 2019a, b), the presence of this heterochromatin shell near the NE likely contributes to nuclear mechanics, shape, and integrity (Manning et al. 2025; Stephens et al. 2018).

9.3.2 Chromatin Binding to Proteins of the INM

Additional chromatin links to the NE also contribute to nuclear mechanics, independently of LADs. Accordingly, histone PTMs such as H3K9me3 can directly contribute to nuclear mechanics (Manning et al. 2025) independently of lamins. Moreover, LBR was shown to directly bind heterochromatin (Makatsori et al. 2004; Solovei et al. 2013) by recognizing specific modifications on histone H4 (Hirano et al. 2012; Polioudaki et al. 2001), thereby promoting transcriptional silencing through chromatin compaction and the localized recruitment of transcriptional repressors (Hirano et al. 2012; Solovei et al. 2013). Therefore, it is possible that LBR could potentially regulate nuclear mechanics by tethering heterochromatin to the nuclear periphery (Hirano et al. 2012; Makatsori et al. 2004). In agreement with this possibility, overexpression of LBR, which leads to excess NE membrane production (Dantas et al. 2022; Lomakin et al. 2020; Lu et al. 2010), decreases NE tension (Dantas et al. 2022; Lomakin et al. 2020) and disrupts nuclear mechanotransduction. Additionally, heterochromatin protein 1α (HP1α), a well-known heterochromatin interactor (Singh et al. 1991; Wreggett et al. 1994) that associates with the NE (Kourmouli et al. 2000), is involved in the crosslinking and compaction of chromatin (Keenen et al. 2021; Strom et al. 2021), contributing to overall nuclear mechanics (Strom et al. 2021; Williams et al. 2024). Notably, HP1α was also shown to interact with LBR (Ye et al. 1997; Ye and Worman 1996), suggesting that multiple molecular players and pathways act in parallel to regulate NE mechanics.

9.3.3 Chromatin Association with Nuclear Pore Complexes

Chromatin can also interact with NPCs by binding to specific Nups. Accordingly, yeast Nup170 interacts with chromatin at telomeric regions and is required for its association with the NE (Van de Vosse et al. 2013). Furthermore, in *Drosophila*, NPC-linked Elys was shown to anchor peripheral chromatin to the NE (Doronin et al. 2024). These chromatin-NPC interactions correlate with specific patterns of histone PTMs (Fiserova et al. 2017) and are required for gene expression and regulation (Buchwalter et al. 2019). It is tempting to speculate that these interactions could modulate NE mechanics, providing a simple way to couple gene expression with the mechanical state of the nucleus. Irrespective of the mode of coupling, it is clear that chromatin tethering to the NE is an essential component contributing to the global mechanical response of the nucleus (Attar et al. 2025; Schreiner et al. 2015). Still, how these chromatin-NE interactions are cell cycle regulated remains largely unexplored, although there are indications that specific chromosomal regions associate with the NE in a cell cycle-dependent manner in some model systems. Accordingly, in fission yeast, centromeres and telomeres were shown to associate with the spindle pole body (SPB) near the NE (Funabiki et al. 1993) through an interaction with the SUN domain protein Sad1 (Fernandez-Alvarez et al. 2016; Hou et al. 2012). This centromeric association is required to establish centromere identity (Wu et al. 2022) and ensure successful cell division (Fernandez-Alvarez et al. 2016). Whether similar control mechanisms exist in higher eukaryotes remains largely unexplored although some reports indicate differential cell cycle distribution of centromeres (Bartholdi 1991) and telomeres (Crabbe et al. 2012; Keller et al. 2024). Understanding whether and how these interactions are regulated is particularly relevant during the initial stage of mitosis, when chromosome condensation triggers a massive reorganization of chromatin within the nucleus.

9.4 Mitotic Chromosome Condensation

As cells prepare to divide, chromatin is reorganized into condensed, discrete chromosomes. Condensation of mitotic chromosomes is a complex process that involves both chromatin compaction and chromosome shaping (Hibino et al. 2024; Naumova et al. 2013; Samejima et al. 2012; Schneider et al. 2022; Zhiteneva et al. 2017). The initial compaction of chromatin takes place during early prophase and is thought to depend mainly on condensin activity (Hirota et al. 2004; Samejima et al. 2012). Condensins are composed of Kleisin and SMC proteins, which are enriched in the axial region of sister chromatids (Hirano and Mitchison 1994; Maeshima and Laemmli 2003). Here, they are proposed to organize chromatin in nested loop arrays emanating from a central scaffold (Gibcus et al. 2018; Walther and Ellenberg 2018). Therefore, it is not surprising that depletion of the condensin complex SMC4-MIX1 in *C. elegans* causes an aberrant mitotic chromosome structure, which impacts chromosome segregation (Hagstrom et al. 2002). Moreover, depletion of the condensin subunit ScII/SMC2 in chicken DT40 cells revealed that condensins promote the association of nonhistone proteins, such as the DNA decatenating enzyme topoisomerase II and the chromosomal passenger complex component INCENP, with mitotic chromosomes, to ensure their structural integrity (Hudson et al. 2003). Interestingly, topoisomerase II had already been described as a major component of the mitotic chromosome scaffold (Adolph et al. 1977; Earnshaw et al. 1985; Lewis and Laemmli 1982) and proposed to regulate chromosome condensation. The use of temperature-sensitive mutants of topoisomerase II in *S. pombe* revealed that its activity is necessary for the process of chromosome condensation and chromatid disjunction (Uemura et al. 1987). More recent work in chicken DT40 cells indicated that topoisomerase II is required to induce an axial compaction of chromosomes that, together with condensins and the chromokinesin Kif4A, define the characteristic chromosome shape (Meijering et al. 2022; Samejima et al. 2012).

9.4.1 Impact of Chromosome Condensation on Nuclear Mechanics

In addition to the well-established pathways of mitotic chromosome formation, chromatin compaction was also proposed to depend on specific PTMs of histone tails such as deacetylation, phosphorylation, and methylation (Biggs et al. 2019; Hibino et al. 2024; Wilkins et al. 2014; Zhiteneva et al. 2017). Phosphorylation of histone H3 on Ser10 (H3S10ph) mediated by Aurora B kinase is one of the first modifications to be detected on chromosomes during the early stages of condensation (Garcia et al. 2005; Hendzel et al. 1997). Moreover, in *Saccharomyces cerevisiae*, it was proposed that following H3S10 phosphorylation, a histone deacetylase (HDAC) is recruited to chromatin that will result in the deacetylation of histone H4 on Lys16 (H4K16ac). This would enable the interaction between histone H4 and the dimer interface of H2A-H2B in the adjacent nucleosome. In combination with CDK1-mediated phosphorylation of condensins I and II (Abe et al. 2011; Shintomi et al. 2015), these events would promote global chromatin compaction (Shogren-Knaak et al. 2006; Wilkins et al. 2014; Zhiteneva et al. 2017). Notably, many of these histone PTMs are enriched near the NE (Chi et al. 2007) and were associated with nuclear shape and mechanics (Chi et al. 2022; Stephens et al. 2019b). Together, these observations suggest that mitotic chromosome condensation could lead to changes in the interactions of chromatin with the NE and influence the mechanical properties of the nucleus. This hypothesis is supported by observations showing condensation-dependent stiffening of chromosomes (Man et al. 2021). This increased stiffness depends on condensins (Sun et al. 2018) that organize DNA around a rigid core (Houchmandzadeh and Dimitrov 1999) and requires chromatin crosslinking mediated by HP1α (Strom et al. 2021). Accordingly, stiffness measurements using micropipette aspiration on chromosomes isolated from HeLa cells reported a Young's modulus for mitotic chromosomes in the range of 440 Pa, in contrast with a Young's modulus in the range of 50 Pa for cells depleted of condensins (Sun et al. 2018). On the other hand, interphase chromosomes were predicted to be tenfold softer than mitotic chromosomes (Ruben et al. 2023). Notably, chromosome stiffness and stability were also shown to depend on histone modifications such as hypermethylation or deacetylation (Biggs et al. 2019; Schneider et al. 2022). This has led to the hypothesis that the material properties of mitotic chromosomes (and consequently their degree of condensation) could be sensitive to the physicochemical properties of the environment (Witt et al. 2024). Overall, these observations suggest that chromosome stiffness is regulated by multiple components acting in parallel. Considering the extreme nuclear changes that take place during the transition to mitosis (Champion et al. 2017), it would be relevant to determine whether mitotic chromosome condensation somehow modulates nuclear mechanics or is sensitive to the physicochemical environment of the nucleus at this stage.

9.5 Regulation of Mitotic Entry

In higher eukaryotes, irreversible commitment to mitosis requires extensive nuclear remodeling to allow microtubules to interact with chromosomes (Kutay and Hetzer 2008). To achieve this, cells rely on the activity of cyclin-CDK complexes (Malumbres and Barbacid 2009). These complexes are part of an intricate biochemical network that ensures timely and accurate progression through the cell cycle (Malumbres and Barbacid 2009). While the biochemical mechanisms remain as the most studied aspect of mitotic regulation, there is now compelling evidence supporting a direct link between the cell cycle machinery and physical forces (Aureille et al. 2019; Chen et al. 1997; Dantas et al. 2022; Huang et al. 1998; Nunes et al. 2019; Uroz et al. 2018; Vitiello et al. 2019). One key moment occurs during the transition from G2 phase to mitosis (G2-M), when cells need to reorganize their cytoplasmic and nuclear structure (Champion et al. 2017), under the coordination of cyclin B1-CDK1 (Akopyan et al. 2025; Gavet and Pines 2010a, b). At this point, several key nuclear proteins are phosphorylated by mitotic kinases such as CDK1 or Plk1, which trigger chromosome condensation (Abe et al. 2011), NPCs dismantling (Laurell et al. 2011; Linder et al. 2017), and disassembly of the nuclear lamina (Heald and McKeon 1990; Peter et al. 1990) to ensure an efficient mitosis. As we mentioned in the previous sections, many of these components are required to ensure the mechanical stability of the nucleus during interphase. Therefore, it is possible that these cyclin B1-CDK1-induced changes modify the mechanical properties of the nucleus and have downstream consequences for mitosis.

9.5.1 *Nuclear Mechanotransduction in the Prophase Nucleus*

Recent work has demonstrated that nuclear tension during prophase regulates the nuclear translocation of cyclin B1 (Dantas et al. 2022) to ensure timely nuclear reorganization and mitotic entry. In this context, it is tempting to speculate that the NE could act as a mechanosensor, detecting the optimal tension required to trigger mitotic entry. Indeed, decreasing nuclear tension is sufficient to delay cyclin B1 translocation (Dantas et al. 2022), whereas increasing nuclear tension induces premature translocation and mitotic entry, with consequences for chromosome segregation fidelity (Dantas et al. 2022; Furuno et al. 1999). The identity of the molecular player(s) that define(s) nuclear behavior during prophase is(are) currently unknown. However, one possibility is that this mechanical signal comes from the chromosomes, which could act as an early source of mechanical instability in the nucleus, given that chromosome condensation is one of the earliest signs of mitotic commitment (McIntosh 2016). Furthermore, it is well known that the mechanical behavior of chromosomes changes during the condensation process (Man et al. 2021), and this increase in stiffness is crucial for correct genome segregation (Batty and Gerlich 2019; Ribeiro et al. 2009; Sun et al. 2018). Therefore, this nuclear mechanotransduction pathway would enable cells to sense the degree of chromosome condensation and define the exact timing of mitotic entry accordingly.

9.5.2 *Force Imposed on the Nuclear Envelope during Mitotic Entry*

Irreversible mitotic commitment involves an initial permeabilization of the NE (NEP) induced by NPC dismantling (Laurell and Kutay 2011), followed by a disassembly of the nuclear lamina and removal of nuclear membranes from chromosomes. To achieve this, cells employ microtubule-generated forces that pull on nuclear membranes (Beaudouin et al. 2002; Salina et al. 2002), ultimately resulting in nuclear envelope breakdown (NEB). One important player in this process is the molecular motor dynein, which associates with the specific Nups on NPCs during the early stages of mitosis (Bolhy et al. 2011; Hu et al. 2013; Nunes et al. 2020; Splinter et al. 2010). Following activation by CDK1 (Baffet et al. 2015), specific dynein adaptors such as BicD2 or CENP-F associate with Nup358/RanBP2 (Splinter et al. 2010) and Nup133 (Bolhy et al. 2011), respectively, enabling dynein binding. Once at NPCs, dynein then generates forces on microtubules that tether and position centrosomes close to the NE (Nunes et al. 2020), allowing efficient spindle assembly and timely NEB. Notably, the temporal regulation of dynein loading was recently shown to depend on the LINC complex (Lima et al. 2024), suggesting that this process may depend on forces generated at the NE. However, whether dynein loading is sensitive to mechanical signals from the nuclear interior or the cytoskeleton remains unknown.

One of the last steps in mitotic commitment involves the disassembly of the nuclear lamina and the removal of NE membranes from chromosomes. Nuclear lamina disassembly involves CDK1-mediated phosphorylation in the N-terminal head region (Heald and McKeon 1990; Peter et al. 1990; Ward and Kirschner 1990), possibly disrupting interactions between lamin filaments (Jeong et al. 2022). Taking into consideration the extensive contribution of lamin A to the mechanical properties of the interphase nucleus (Buxboim et al. 2017; Davidson and Lammerding 2014; Lammerding et al. 2004, 2006; Stephens et al. 2017), it is possible that the dissolution of the nuclear lamina impacts nuclear mechanics during the initial stages of mitosis. Notably, nuclear lamina removal occurs near simultaneously with mitotic chromosome condensation. These observations suggest that a global modification in the mechanical properties of the nucleus is required for an efficient NEB (Beaudouin et al. 2002). However, little is known about the spatiotemporal regulation of nuclear mechanics during the G2-M transition and therefore, this hypothesis remains largely untested.

9.6 Impact of Disrupting Nuclear Mechanics for Mitotic Fidelity

While the nucleus needs to be disassembled to allow the interaction between chromosomes and microtubules, there is ample evidence correlating errors in chromosome segregation with the disruption of specific nuclear components. However, a direct causal relationship between impairment in nuclear mechanics and mitotic fidelity is still lacking. Nevertheless, changes in the levels of nuclear lamina components lamin A (Pratt et al. 2011) and lamin B2 (Kuga et al. 2014; Ranade et al. 2017) have been previously associated with mitotic defects. Furthermore, expression of a mutant form of lamin A, normally found in Hutchinson-Gilford progeria syndrome, is sufficient to generate lagging chromosomes and increase the number of binucleated cells (Cao et al. 2007). Whether these defects in chromosome segregation are strictly dependent on lamin levels or on the mechanical role of the nuclear lamina remains unclear.

Mitotic robustness and fidelity are also dependent on NE-associated proteins. For instance, a complex composed of lamin A-Lap2β-BAF was shown to mediate the interaction of the mitotic spindle with the cell cortex regulating spindle orientation (Qi et al. 2015). On the other hand, depletion of INM proteins emerin and man1 in *C. elegans* leads to severe chromosome segregation defects (Liu et al. 2003). In addition, SUN proteins regulate the efficiency of mitotic spindle assembly by facilitating the removal of NE membranes from chromatin (Turgay et al. 2014) and ensuring the correct positioning of centrosomes during prophase (Lima et al. 2024; Stiff et al. 2020), whereas the LINC complex was proposed to regulate mitotic progression and chromosome segregation fidelity (Belaadi et al. 2022; Lima et al. 2024; Stiff et al. 2020). The multiple mitotic roles attributed to INM proteins might arise from indirect effects exerted on other subcellular structures and not due to direct interactions between these proteins and the chromosome segregation machinery. Accordingly, the LINC complex facilitates the removal of NE membranes from chromosomes (Turgay et al. 2014) to assist in chromosome capture (Champion et al. 2019), without interacting directly with mitotic chromosomes or the spindle (Lima et al. 2024).

Many kinetochore components localize to NPCs during the cell cycle until the kinetochore is assembled (Bangs et al. 1998; Berto et al. 2018; Rodriguez-Bravo et al. 2014). This dynamic relocalization pattern is shared by some Nups that transfer from NPCs to kinetochores after NEB (Belgareh et al. 2001; Rasala et al. 2006; Zuccolo et al. 2007). Together, these data suggest a functional relationship between NPCs and the mitotic machinery. Therefore, it is not surprising that the Nup107 complex was shown to regulate the localization of Aurora B at centromeres (Platani et al. 2009) or that Tpr localization at NPCs is necessary to maintain a robust Spindle Assembly Checkpoint response (Schweizer et al. 2013). Moreover, Nup155 depletion in *C. elegans* (Franz et al. 2005) or disruption of Nup358 in HeLa cells (Salina et al. 2003) results in chromosome segregation errors. Finally, force generation at the NE during prophase depends on the association of dynein with either Nup358/RanBP2 (Splinter et al. 2010) or Nup133 (Bolhy et al. 2011). Disruption of this association leads to centrosome detachment from the NE and delayed spindle assembly (Bolhy et al. 2011; Nunes et al. 2020), likely due to a delay in NEB (Beaudouin et al. 2002; Salina et al. 2002).

Elucidating whether the changes in mitotic fidelity are directly due to an impact on nuclear mechanics is of crucial importance, considering that chromosome missegregation was shown to trigger a mechanosensitive NE checkpoint (Herve et al. 2025) and that defects in nuclear structure and mechanotransduction have been previously associated with aneuploidy and cancer (Capo-chichi et al. 2011, 2016; Isermann and Lammerding 2013).

References

Abe S, Nagasaka K, Hirayama Y, Kozuka-Hata H, Oyama M, Aoyagi Y, Obuse C, Hirota T (2011) The initial phase of chromosome condensation requires Cdk1-mediated phosphorylation of the CAP-D3 subunit of condensin II. Genes Dev 25(8):863–874. https://doi.org/10.1101/gad.2016411

Adolph KW, Cheng SM, Paulson JR, Laemmli UK (1977) Isolation of a protein scaffold from mitotic HeLa cell chromosomes. Proc Natl Acad Sci USA 74(11):4937–4941. https://doi.org/10.1073/pnas.74.11.4937

Agrawal A, Lele TP (2019) Mechanics of nuclear membranes. J Cell Sci 132(14). https://doi.org/10.1242/jcs.229245

Akopyan K, Hao Z, Lindqvist A (2025) Preparation for mitosis requires gradual CDK1 activation. iScience 28(5):112292. https://doi.org/10.1016/j.isci.2025.112292

Antonin W, Neumann H (2016) Chromosome condensation and decondensation during mitosis. Curr Opin Cell Biol 40:15–22. https://doi.org/10.1016/j.ceb.2016.01.013

Attar AG, Paturej J, Sariyer OS, Banigan EJ, Erbas A (2025) Peripheral heterochromatin tethering is required for chromatin-based nuclear mechanical response. Nucleic Acids Res 53(15). https://doi.org/10.1093/nar/gkaf763

Aureille J, Buffiere-Ribot V, Harvey BE, Boyault C, Pernet L, Andersen T, Bacola G, Balland M, Fraboulet S, Van Landeghem L, Guilluy C (2019) Nuclear envelope deformation controls cell cycle progression in response to mechanical force. EMBO Rep 20(9):e48084. https://doi.org/10.15252/embr.201948084

Baffet AD, Hu DJ, Vallee RB (2015) Cdk1 activates pre-mitotic nuclear envelope dynein recruitment and apical nuclear migration in neural stem cells. Dev Cell 33(6):703–716. https://doi.org/10.1016/j.devcel.2015.04.022

Banerjee A, Rathee V, Krishnaswamy R, Bhattacharjee P, Ray P, Sood AK, Sengupta K (2013) Viscoelastic behavior of human lamin A proteins in the context of dilated cardiomyopathy. PLoS One 8(12):e83410. https://doi.org/10.1371/journal.pone.0083410

Bangs P, Burke B, Powers C, Craig R, Purohit A, Doxsey S (1998) Functional analysis of Tpr: identification of nuclear pore complex association and nuclear localization domains and a role in mRNA export. J Cell Biol 143(7):1801–1812. https://doi.org/10.1083/jcb.143.7.1801

Bartholdi MF (1991) Nuclear distribution of centromeres during the cell cycle of human diploid fibroblasts. J Cell Sci 99(Pt 2):255–263. https://doi.org/10.1242/jcs.99.2.255

Batty P, Gerlich DW (2019) Mitotic chromosome mechanics: how cells segregate their genome. Trends Cell Biol. https://doi.org/10.1016/j.tcb.2019.05.007

Beaudouin J, Gerlich D, Daigle N, Eils R, Ellenberg J (2002) Nuclear envelope breakdown proceeds by microtubule-induced tearing of the lamina. Cell 108(1):83–96

Belaadi N, Pernet L, Aureille J, Chadeuf G, Rio M, Vaillant N, Vitiello E, Lafanechere L, Loirand G, Guilluy C (2022) SUN2 regulates mitotic duration in response to extracellular matrix rigidity. Proc Natl Acad Sci USA 119(45):e2116167119. https://doi.org/10.1073/pnas.2116167119
Belgareh N, Rabut G, Bai SW, van Overbeek M, Beaudouin J, Daigle N, Zatsepina OV, Pasteau F, Labas V, Fromont-Racine M, Ellenberg J, Doye V (2001) An evolutionarily conserved NPC subcomplex, which redistributes in part to kinetochores in mammalian cells. J Cell Biol 154(6):1147–1160. https://doi.org/10.1083/jcb.200101081
Berk JM, Tifft KE, Wilson KL (2013) The nuclear envelope LEM-domain protein emerin. Nucleus 4(4):298–314. https://doi.org/10.4161/nucl.25751
Berto A, Yu J, Morchoisne-Bolhy S, Bertipaglia C, Vallee R, Dumont J, Ochsenbein F, Guerois R, Doye V (2018) Disentangling the molecular determinants for Cenp-F localization to nuclear pores and kinetochores. EMBO Rep 19(5). https://doi.org/10.15252/embr.201744742
Biggs R, Liu PZ, Stephens AD, Marko JF (2019) Effects of altering histone posttranslational modifications on mitotic chromosome structure and mechanics. Mol Biol Cell 30(7):820–827. https://doi.org/10.1091/mbc.E18-09-0592
Bolhy S, Bouhlel I, Dultz E, Nayak T, Zuccolo M, Gatti X, Vallee R, Ellenberg J, Doye V (2011) A Nup133-dependent NPC-anchored network tethers centrosomes to the nuclear envelope in prophase. J Cell Biol 192(5):855–871. https://doi.org/10.1083/jcb.201007118
Brachner A, Foisner R (2011) Evolvement of LEM proteins as chromatin tethers at the nuclear periphery. Biochem Soc Trans 39(6):1735–1741. https://doi.org/10.1042/BST20110724
Bragulat-Teixidor H, Ishihara K, Szucs GM, Otsuka S (2024) The endoplasmic reticulum connects to the nucleus by constricted junctions that mature after mitosis. EMBO Rep 25(7):3137–3159. https://doi.org/10.1038/s44319-024-00175-w
Buchwalter A, Kaneshiro JM, Hetzer MW (2019) Coaching from the sidelines: the nuclear periphery in genome regulation. Nat Rev Genet 20(1):39–50. https://doi.org/10.1038/s41576-018-0063-5
Buxboim A, Irianto J, Swift J, Athirasala A, Shin JW, Rehfeldt F, Discher DE (2017) Coordinated increase of nuclear tension and lamin-A with matrix stiffness outcompetes lamin-B receptor that favors soft tissue phenotypes. Mol Biol Cell 28(23):3333–3348. https://doi.org/10.1091/mbc.E17-06-0393
Cao K, Capell BC, Erdos MR, Djabali K, Collins FS (2007) A lamin A protein isoform overexpressed in Hutchinson-Gilford progeria syndrome interferes with mitosis in progeria and normal cells. Proc Natl Acad Sci USA 104(12):4949–4954. https://doi.org/10.1073/pnas.0611640104
Capo-chichi CD, Cai KQ, Simpkins F, Ganjei-Azar P, Godwin AK, Xu XX (2011) Nuclear envelope structural defects cause chromosomal numerical instability and aneuploidy in ovarian cancer. BMC Med 9:28. https://doi.org/10.1186/1741-7015-9-28
Capo-Chichi CD, Yeasky TM, Smith ER, Xu XX (2016) Nuclear envelope structural defect underlies the main cause of aneuploidy in ovarian carcinogenesis. BMC Cell Biol 17(1):37. https://doi.org/10.1186/s12860-016-0114-8
Champion L, Linder MI, Kutay U (2017) Cellular reorganization during mitotic entry. Trends Cell Biol 27(1):26–41. https://doi.org/10.1016/j.tcb.2016.07.004
Champion L, Pawar S, Luithle N, Ungricht R, Kutay U (2019) Dissociation of membrane-chromatin contacts is required for proper chromosome segregation in mitosis. Mol Biol Cell 30(4):427–440. https://doi.org/10.1091/mbc.E18-10-0609
Chang W, Worman HJ, Gundersen GG (2015) Accessorizing and anchoring the LINC complex for multifunctionality. J Cell Biol 208(1):11–22. https://doi.org/10.1083/jcb.201409047
Chen CS, Mrksich M, Huang S, Whitesides GM, Ingber DE (1997) Geometric control of cell life and death. Science 276(5317):1425–1428
Chi YH, Haller K, Peloponese JM Jr, Jeang KT (2007) Histone acetyltransferase hALP and nuclear membrane protein hsSUN1 function in de-condensation of mitotic chromosomes. J Biol Chem 282(37):27447–27458. https://doi.org/10.1074/jbc.M703098200

Chi YH, Wang WP, Hung MC, Liou GG, Wang JY, Chao PG (2022) Deformation of the nucleus by TGFbeta1 via the remodeling of nuclear envelope and histone isoforms. Epigenetics Chromatin 15(1):1. https://doi.org/10.1186/s13072-021-00434-3

Crabbe L, Cesare AJ, Kasuboski JM, Fitzpatrick JA, Karlseder J (2012) Human telomeres are tethered to the nuclear envelope during postmitotic nuclear assembly. Cell Rep 2(6):1521–1529. https://doi.org/10.1016/j.celrep.2012.11.019

Cremer T, Cremer M (2010) Chromosome territories. Cold Spring Harb Perspect Biol 2(3):a003889. https://doi.org/10.1101/cshperspect.a003889

Dahl KN, Kahn SM, Wilson KL, Discher DE (2004) The nuclear envelope lamina network has elasticity and a compressibility limit suggestive of a molecular shock absorber. J Cell Sci 117(Pt 20):4779–4786. https://doi.org/10.1242/jcs.01357

Dantas M, Oliveira A, Aguiar P, Maiato H, Ferreira JG (2022) Nuclear tension controls mitotic entry by regulating cyclin B1 nuclear translocation. J Cell Biol 221(12). https://doi.org/10.1083/jcb.202205051

Dauban L, Eder M, de Haas M, Franceschini-Santos VH, Yanez-Cuna JO, Martinovic M, van Schaik T, Leemans C, Teunissen H, Rademaker K, Martinez Ara M, Verkuilen M, de Wit E, van Steensel B (2025) Interactions between the genome and the nuclear lamina are multivalent and cooperative. Nat Struct Mol Biol. https://doi.org/10.1038/s41594-025-01655-w

Davidson PM, Lammerding J (2014) Broken nuclei—lamins, nuclear mechanics, and disease. Trends Cell Biol 24(4):247–256. https://doi.org/10.1016/j.tcb.2013.11.004

De Magistris P, Antonin W (2018) The dynamic nature of the nuclear envelope. Curr Biol 28(8):R487–R497. https://doi.org/10.1016/j.cub.2018.01.073

Dechat T, Pfleghaar K, Sengupta K, Shimi T, Shumaker DK, Solimando L, Goldman RD (2008) Nuclear lamins: major factors in the structural organization and function of the nucleus and chromatin. Genes Dev 22(7):832–853. https://doi.org/10.1101/gad.1652708

Ding X, Xu R, Yu J, Xu T, Zhuang Y, Han M (2007) SUN1 is required for telomere attachment to nuclear envelope and gametogenesis in mice. Dev Cell 12(6):863–872. https://doi.org/10.1016/j.devcel.2007.03.018

Doronin SA, Ilyin AA, Kononkova AD, Solovyev MA, Olenkina OM, Nenasheva VV, Mikhaleva EA, Lavrov SA, Ivannikova AY, Simonov RA, Fedotova AA, Khrameeva EE, Ulianov SV, Razin SV, Shevelyov YY (2024) Nucleoporin Elys attaches peripheral chromatin to the nuclear pores in interphase nuclei. Commun Biol 7(1):783. https://doi.org/10.1038/s42003-024-06495-w

Earnshaw WC, Halligan B, Cooke CA, Heck MM, Liu LF (1985) Topoisomerase II is a structural component of mitotic chromosome scaffolds. J Cell Biol 100(5):1706–1715. https://doi.org/10.1083/jcb.100.5.1706

English AR, Voeltz GK (2013) Endoplasmic reticulum structure and interconnections with other organelles. Cold Spring Harb Perspect Biol 5(4):a013227. https://doi.org/10.1101/cshperspect.a013227

Enyedi B, Niethammer P (2017) Nuclear membrane stretch and its role in mechanotransduction. Nucleus 8(2):156–161. https://doi.org/10.1080/19491034.2016.1263411

Enyedi B, Kala S, Nikolich-Zugich T, Niethammer P (2013) Tissue damage detection by osmotic surveillance. Nat Cell Biol 15(9):1123–1130. https://doi.org/10.1038/ncb2818

Enyedi B, Jelcic M, Niethammer P (2016) The cell nucleus serves as a mechanotransducer of tissue damage-induced inflammation. Cell 165(5):1160–1170. https://doi.org/10.1016/j.cell.2016.04.016

Fernandez A, Bautista M, Wu L, Pinaud F (2022) Emerin self-assembly and nucleoskeletal coupling regulate nuclear envelope mechanics against stress. J Cell Sci 135(6). https://doi.org/10.1242/jcs.258969

Fernandez-Alvarez A, Bez C, O'Toole ET, Morphew M, Cooper JP (2016) Mitotic nuclear envelope breakdown and spindle nucleation are controlled by interphase contacts between centromeres and the nuclear envelope. Dev Cell 39(5):544–559. https://doi.org/10.1016/j.devcel.2016.10.021

Fiserova J, Efenberkova M, Sieger T, Maninova M, Uhlirova J, Hozak P (2017) Chromatin organization at the nuclear periphery as revealed by image analysis of structured illumination microscopy data. J Cell Sci 130(12):2066–2077. https://doi.org/10.1242/jcs.198424

Foisner R, Gerace L (1993) Integral membrane proteins of the nuclear envelope interact with lamins and chromosomes, and binding is modulated by mitotic phosphorylation. Cell 73(7):1267–1279. https://doi.org/10.1016/0092-8674(93)90355-t

Franz C, Askjaer P, Antonin W, Iglesias CL, Haselmann U, Schelder M, de Marco A, Wilm M, Antony C, Mattaj IW (2005) Nup155 regulates nuclear envelope and nuclear pore complex formation in nematodes and vertebrates. EMBO J 24(20):3519–3531. https://doi.org/10.1038/sj.emboj.7600825

Fricker M, Hollinshead M, White N, Vaux D (1997) Interphase nuclei of many mammalian cell types contain deep, dynamic, tubular membrane-bound invaginations of the nuclear envelope. J Cell Biol 136(3):531–544. https://doi.org/10.1083/jcb.136.3.531

Funabiki H, Hagan I, Uzawa S, Yanagida M (1993) Cell cycle-dependent specific positioning and clustering of centromeres and telomeres in fission yeast. J Cell Biol 121(5):961–976. https://doi.org/10.1083/jcb.121.5.961

Furuno N, den Elzen N, Pines J (1999) Human cyclin A is required for mitosis until mid prophase. J Cell Biol 147(2):295–306. https://doi.org/10.1083/jcb.147.2.295

Garcia BA, Barber CM, Hake SB, Ptak C, Turner FB, Busby SA, Shabanowitz J, Moran RG, Allis CD, Hunt DF (2005) Modifications of human histone H3 variants during mitosis. Biochemistry 44(39):13202–13213. https://doi.org/10.1021/bi050906n

Gavet O, Pines J (2010a) Activation of cyclin B1–Cdk1 synchronizes events in the nucleus and the cytoplasm at mitosis. J Cell Biol 189(2):247–259. https://doi.org/10.1083/jcb.200909144

Gavet O, Pines J (2010b) Progressive activation of Cyclin B1-Cdk1 coordinates entry to mitosis. Dev Cell 18(4):533–543. https://doi.org/10.1016/j.devcel.2010.02.013

Gheghiani L, Loew D, Lombard B, Mansfeld J, Gavet O (2017) PLK1 activation in late G2 sets up commitment to mitosis. Cell Rep 19(10):2060–2073. https://doi.org/10.1016/j.celrep.2017.05.031

Gibcus JH, Samejima K, Goloborodko A, Samejima I, Naumova N, Nuebler J, Kanemaki MT, Xie L, Paulson JR, Earnshaw WC, Mirny LA, Dekker J (2018) A pathway for mitotic chromosome formation. Science 359(6376). https://doi.org/10.1126/science.aao6135

Guelen L, Pagie L, Brasset E, Meuleman W, Faza MB, Talhout W, Eussen BH, de Klein A, Wessels L, de Laat W, van Steensel B (2008) Domain organization of human chromosomes revealed by mapping of nuclear lamina interactions. Nature 453(7197):948–951. https://doi.org/10.1038/nature06947

Guilluy C, Osborne LD, Van Landeghem L, Sharek L, Superfine R, Garcia-Mata R, Burridge K (2014) Isolated nuclei adapt to force and reveal a mechanotransduction pathway in the nucleus. Nat Cell Biol 16(4):376–381. https://doi.org/10.1038/ncb2927

Hagstrom KA, Holmes VF, Cozzarelli NR, Meyer BJ (2002) C. elegans condensin promotes mitotic chromosome architecture, centromere organization, and sister chromatid segregation during mitosis and meiosis. Genes Dev 16(6):729–742. https://doi.org/10.1101/gad.968302

Hampoelz B, Andres-Pons A, Kastritis P, Beck M (2019) Structure and assembly of the nuclear pore complex. Annu Rev Biophys 48:515–536. https://doi.org/10.1146/annurev-biophys-052118-115308

Heald R, McKeon F (1990) Mutations of phosphorylation sites in lamin A that prevent nuclear lamina disassembly in mitosis. Cell 61(4):579–589

Hendzel MJ, Wei Y, Mancini MA, Van Hooser A, Ranalli T, Brinkley BR, Bazett-Jones DP, Allis CD (1997) Mitosis-specific phosphorylation of histone H3 initiates primarily within pericentromeric heterochromatin during G2 and spreads in an ordered fashion coincident with mitotic chromosome condensation. Chromosoma 106(6):348–360. https://doi.org/10.1007/s004120050256

Herve S, Scelfo A, Bersano Marchisio G, Grison M, Vaidziulyte K, Dumont M, Angrisani A, Keikhosravi A, Pegoraro G, Deygas M, G PFN, Mace AS, Gentili M, Williart A, Manel N, Piel M, Miroshnikova YA, Fachinetti D (2025) Chromosome mis-segregation triggers

cell cycle arrest through a mechanosensitive nuclear envelope checkpoint. Nat Cell Biol. https://doi.org/10.1038/s41556-024-01565-x
Hetzer MW (2010) The nuclear envelope. Cold Spring Harb Perspect Biol 2(3):a000539. https://doi.org/10.1101/cshperspect.a000539
Hibino K, Sakai Y, Tamura S, Takagi M, Minami K, Natsume T, Shimazoe MA, Kanemaki MT, Imamoto N, Maeshima K (2024) Single-nucleosome imaging unveils that condensins and nucleosome-nucleosome interactions differentially constrain chromatin to organize mitotic chromosomes. Nat Commun 15(1):7152. https://doi.org/10.1038/s41467-024-51454-y
Hirano T, Mitchison TJ (1994) A heterodimeric coiled-coil protein required for mitotic chromosome condensation in vitro. Cell 79(3):449–458. https://doi.org/10.1016/0092-8674(94)90254-2
Hirano Y, Hizume K, Kimura H, Takeyasu K, Haraguchi T, Hiraoka Y (2012) Lamin B receptor recognizes specific modifications of histone H4 in heterochromatin formation. J Biol Chem 287(51):42654–42663. https://doi.org/10.1074/jbc.M112.397950
Hirota T, Gerlich D, Koch B, Ellenberg J, Peters J-M (2004) Distinct functions of condensin I and II in mitotic chromosome assembly. J Cell Sci 117(Pt 26):6435–6445. https://doi.org/10.1242/jcs.01604
Hobson CM, Stephens AD (2020) Modeling of cell nuclear mechanics: classes, components, and applications. Cells 9(7). https://doi.org/10.3390/cells9071623
Hobson CM, Falvo MR, Superfine R (2021) A survey of physical methods for studying nuclear mechanics and mechanobiology. APL Bioeng 5(4):041508. https://doi.org/10.1063/5.0068126
Hou H, Zhou Z, Wang Y, Wang J, Kallgren SP, Kurchuk T, Miller EA, Chang F, Jia S (2012) Csi1 links centromeres to the nuclear envelope for centromere clustering. J Cell Biol 199(5):735–744. https://doi.org/10.1083/jcb.201208001
Houchmandzadeh B, Dimitrov S (1999) Elasticity measurements show the existence of thin rigid cores inside mitotic chromosomes. J Cell Biol 145(2):215–223. https://doi.org/10.1083/jcb.145.2.215
Hu DJ, Baffet AD, Nayak T, Akhmanova A, Doye V, Vallee RB (2013) Dynein recruitment to nuclear pores activates apical nuclear migration and mitotic entry in brain progenitor cells. Cell 154(6):1300–1313. https://doi.org/10.1016/j.cell.2013.08.024
Huang S, Chen CS, Ingber DE (1998) Control of cyclin D1, p27(Kip1), and cell cycle progression in human capillary endothelial cells by cell shape and cytoskeletal tension. Mol Biol Cell 9(11):3179–3193. https://doi.org/10.1091/mbc.9.11.3179
Hudson DF, Vagnarelli P, Gassmann R, Earnshaw WC (2003) Condensin is required for nonhistone protein assembly and structural integrity of vertebrate mitotic chromosomes. Dev Cell 5(2):323–336. https://doi.org/10.1016/s1534-5807(03)00199-0
Isermann P, Lammerding J (2013) Nuclear mechanics and mechanotransduction in health and disease. Curr Biol 23(24):R1113–R1121. https://doi.org/10.1016/j.cub.2013.11.009
Jeong S, Ahn J, Jo I, Kang SM, Park BJ, Cho HS, Kim YH, Ha NC (2022) Cyclin-dependent kinase 1 depolymerizes nuclear lamin filaments by disrupting the head-to-tail interaction of the lamin central rod domain. J Biol Chem 298(9):102256. https://doi.org/10.1016/j.jbc.2022.102256
Keenen MM, Brown D, Brennan LD, Renger R, Khoo H, Carlson CR, Huang B, Grill SW, Narlikar GJ, Redding S (2021) HP1 proteins compact DNA into mechanically and positionally stable phase separated domains. elife 10. https://doi.org/10.7554/eLife.64563
Keller D, Stinus S, Umlauf D, Gourbeyre E, Biot E, Olivier N, Mahou P, Beaurepaire E, Andrey P, Crabbe L (2024) Non-random spatial organization of telomeres varies during the cell cycle and requires LAP2 and BAF. iScience 27(4):109343. https://doi.org/10.1016/j.isci.2024.109343
Kind J, Pagie L, Ortabozkoyun H, Boyle S, de Vries SS, Janssen H, Amendola M, Nolen LD, Bickmore WA, van Steensel B (2013) Single-cell dynamics of genome-nuclear lamina interactions. Cell 153(1):178–192. https://doi.org/10.1016/j.cell.2013.02.028
Kirby TJ, Lammerding J (2018) Emerging views of the nucleus as a cellular mechanosensor. Nat Cell Biol 20(4):373–381. https://doi.org/10.1038/s41556-018-0038-y

Kourmouli N, Theodoropoulos PA, Dialynas G, Bakou A, Politou AS, Cowell IG, Singh PB, Georgatos SD (2000) Dynamic associations of heterochromatin protein 1 with the nuclear envelope. EMBO J 19(23):6558–6568. https://doi.org/10.1093/emboj/19.23.6558

Kuga T, Nie H, Kazami T, Satoh M, Matsushita K, Nomura F, Maeshima K, Nakayama Y, Tomonaga T (2014) Lamin B2 prevents chromosome instability by ensuring proper mitotic chromosome segregation. Oncogene 3:e94. https://doi.org/10.1038/oncsis.2014.6

Kutay U, Hetzer MW (2008) Reorganization of the nuclear envelope during open mitosis. Curr Opin Cell Biol 20(6):669–677. https://doi.org/10.1016/j.ceb.2008.09.010

Kutay U, Juhlen R, Antonin W (2021) Mitotic disassembly and reassembly of nuclear pore complexes. Trends Cell Biol 31(12):1019–1033. https://doi.org/10.1016/j.tcb.2021.06.011

Lammerding J (2011) Mechanics of the nucleus. Compr Physiol 1(2):783–807. https://doi.org/10.1002/cphy.c100038

Lammerding J, Schulze PC, Takahashi T, Kozlov S, Sullivan T, Kamm RD, Stewart CL, Lee RT (2004) Lamin A/C deficiency causes defective nuclear mechanics and mechanotransduction. J Clin Invest 113(3):370–378. https://doi.org/10.1172/JCI19670

Lammerding J, Fong LG, Ji JY, Reue K, Stewart CL, Young SG, Lee RT (2006) Lamins A and C but not lamin B1 regulate nuclear mechanics. J Biol Chem 281(35):25768–25780. https://doi.org/10.1074/jbc.M513511200

Laurell E, Kutay U (2011) Dismantling the NPC permeability barrier at the onset of mitosis. Cell Cycle 10(14):2243–2245. https://doi.org/10.4161/cc.10.14.16195

Laurell E, Beck K, Krupina K, Theerthagiri G, Bodenmiller B, Horvath P, Aebersold R, Antonin W, Kutay U (2011) Phosphorylation of Nup98 by multiple kinases is crucial for NPC disassembly during mitotic entry. Cell 144(4):539–550. https://doi.org/10.1016/j.cell.2011.01.012

Lee S, Le Roux AL, Mors M, Vanni S, Roca-Cusachs P, Bahmanyar S (2024) Amphipathic helices sense the inner nuclear membrane environment through lipid packing defects. bioRxiv. https://doi.org/10.1101/2024.11.14.623600

Lewis CD, Laemmli UK (1982) Higher order metaphase chromosome structure: evidence for metalloprotein interactions. Cell 29(1):171–181. https://doi.org/10.1016/0092-8674(82)90101-5

Liang Y, Hetzer MW (2011) Functional interactions between nucleoporins and chromatin. Curr Opin Cell Biol 23(1):65–70. https://doi.org/10.1016/j.ceb.2010.09.008

Lima JT, Ferreira JG (2024) Mechanobiology of the nucleus during the G2-M transition. Nucleus 15(1):2330947. https://doi.org/10.1080/19491034.2024.2330947

Lima JT, Pereira AJ, Ferreira JG (2024) The LINC complex ensures accurate centrosome positioning during prophase. Life Sci Alliance 7(4). https://doi.org/10.26508/lsa.202302404

Linder MI, Kohler M, Boersema P, Weberruss M, Wandke C, Marino J, Ashiono C, Picotti P, Antonin W, Kutay U (2017) Mitotic disassembly of nuclear pore complexes involves CDK1- and PLK1-mediated phosphorylation of key interconnecting nucleoporins. Dev Cell 43(2):141–156 e147. https://doi.org/10.1016/j.devcel.2017.08.020

Liu J, Lee KK, Segura-Totten M, Neufeld E, Wilson KL, Gruenbaum Y (2003) MAN1 and emerin have overlapping function(s) essential for chromosome segregation and cell division in Caenorhabditis elegans. Proc Natl Acad Sci USA 100(8):4598–4603. https://doi.org/10.1073/pnas.0730821100

Lomakin AJ, Cattin CJ, Cuvelier D, Alraies Z, Molina M, Nader GPF, Srivastava N, Sáez PJ, Garcia-Arcos JM, Zhitnyak IY, Bhargava A, Driscoll MK, Welf ES, Fiolka R, Petrie RJ, Silva NSD, González-Granado JM, Manel N, Lennon-Duménil AM et al (2020) The nucleus acts as a ruler tailoring cell responses to spatial constraints. Science 370(6514). https://doi.org/10.1126/science.aba2894

Lombardi ML, Jaalouk DE, Shanahan CM, Burke B, Roux KJ, Lammerding J (2011) The interaction between nesprins and sun proteins at the nuclear envelope is critical for force transmission between the nucleus and cytoskeleton. J Biol Chem 286(30):26743–26753. https://doi.org/10.1074/jbc.M111.233700

Lu X, Shi Y, Lu Q, Ma Y, Luo J, Wang Q, Ji J, Jiang Q, Zhang C (2010) Requirement for lamin B receptor and its regulation by importin beta and phosphorylation

in nuclear envelope assembly during mitotic exit. J Biol Chem 285(43):33281–33293. https://doi.org/10.1074/jbc.M110.102368
Maeshima K, Laemmli UK (2003) A two-step scaffolding model for mitotic chromosome assembly. Dev Cell 4(4):467–480. https://doi.org/10.1016/s1534-5807(03)00092-3
Makatsori D, Kourmouli N, Polioudaki H, Shultz LD, McLean K, Theodoropoulos PA, Singh PB, Georgatos SD (2004) The inner nuclear membrane protein lamin B receptor forms distinct microdomains and links epigenetically marked chromatin to the nuclear envelope. J Biol Chem 279(24):25567–25573. https://doi.org/10.1074/jbc.M313606200
Malumbres M, Barbacid M (2009) Cell cycle, CDKs and cancer: a changing paradigm. Nat Rev Cancer 9(3):153–166. https://doi.org/10.1038/nrc2602
Man T, Witt H, Peterman EJG, Wuite GJL (2021) The mechanics of mitotic chromosomes. Q Rev Biophys 54:e10. https://doi.org/10.1017/S0033583521000081
Manning G, Li A, Eskndir N, Currey M, Stephens AD (2025) Constitutive heterochromatin controls nuclear mechanics, morphology, and integrity through H3K9me3 mediated chromocenter compaction. Nucleus 16(1):2486816. https://doi.org/10.1080/19491034.2025.2486816
Maurer M, Lammerding J (2019) The driving force: nuclear mechanotransduction in cellular function, fate, and disease. Annu Rev Biomed Eng 21:443–468. https://doi.org/10.1146/annurev-bioeng-060418-052139
McGillivary RM, Starr DA, Luxton GWG (2023) Building and breaking mechanical bridges between the nucleus and cytoskeleton: regulation of LINC complex assembly and disassembly. Curr Opin Cell Biol 85:102260. https://doi.org/10.1016/j.ceb.2023.102260
McIntosh JR (2016) Mitosis. Cold Spring Harb Perspect Biol 8(9). https://doi.org/10.1101/cshperspect.a023218
Meijering AEC, Sarlos K, Nielsen CF, Witt H, Harju J, Kerklingh E, Haasnoot GH, Bizard AH, Heller I, Broedersz CP, Liu Y, Peterman EJG, Hickson ID, Wuite GJL (2022) Nonlinear mechanics of human mitotic chromosomes. Nature 605(7910):545–550. https://doi.org/10.1038/s41586-022-04666-5
Mellad JA, Warren DT, Shanahan CM (2011) Nesprins LINC the nucleus and cytoskeleton. Curr Opin Cell Biol 23(1):47–54. https://doi.org/10.1016/j.ceb.2010.11.006
Naumova N, Imakaev M, Fudenberg G, Zhan Y, Lajoie BR, Mirny LA, Dekker J (2013) Organization of the mitotic chromosome. Science 342(6161):948–953. https://doi.org/10.1126/science.1236083
Niethammer P (2021) Components and mechanisms of nuclear mechanotransduction. Annu Rev Cell Dev Biol 37:233–256. https://doi.org/10.1146/annurev-cellbio-120319-030049
Nmezi B, Xu J, Fu R, Armiger TJ, Rodriguez-Bey G, Powell JS, Ma H, Sullivan M, Tu Y, Chen NY, Young SG, Stolz DB, Dahl KN, Liu Y, Padiath QS (2019) Concentric organization of A- and B-type lamins predicts their distinct roles in the spatial organization and stability of the nuclear lamina. Proc Natl Acad Sci USA 116(10):4307–4315. https://doi.org/10.1073/pnas.1810070116
Nunes V, Dantas M, Vitiello E, Wang E, Carpi N, Balland M, Piel M, Aguiar P, Maiato H, Ferreira J (2019) Mechanosensitive nuclear asymmetries define a bipolar spindle scaffold to ensure mitotic fidelity [Preprint]. bioRxiv. https://doi.org/10.1101/526939
Nunes V, Dantas M, Castro D, Vitiello E, Wang I, Carpi N, Balland M, Piel M, Aguiar P, Maiato H, Ferreira JG (2020) Centrosome-nuclear axis repositioning drives the assembly of a bipolar spindle scaffold to ensure mitotic fidelity. Mol Biol Cell 31(16):1675–1690. https://doi.org/10.1091/mbc.E20-01-0047
Olins AL, Rhodes G, Welch DB, Zwerger M, Olins DE (2010) Lamin B receptor: multi-tasking at the nuclear envelope. Nucleus 1(1):53–70. https://doi.org/10.4161/nucl.1.1.10515
Panorchan P, Wirtz D, Tseng Y (2004) Structure-function relationship of biological gels revealed by multiple-particle tracking and differential interference contrast microscopy: the case of human lamin networks. Phys Rev E Stat Nonlinear Soft Matter Phys 70(4 Pt 1):041906. https://doi.org/10.1103/PhysRevE.70.041906

Peter M, Nakagawa J, Doree M, Labbe JC, Nigg EA (1990) In vitro disassembly of the nuclear lamina and M phase-specific phosphorylation of lamins by cdc2 kinase. Cell 61(4):591–602. https://doi.org/10.1016/0092-8674(90)90471-p

Platani M, Santarella-Mellwig R, Posch M, Walczak R, Swedlow JR, Mattaj IW (2009) The Nup107-160 nucleoporin complex promotes mitotic events via control of the localization state of the chromosome passenger complex. Mol Biol Cell 20(24):5260–5275. https://doi.org/10.1091/mbc.e09-05-0377

Polioudaki H, Kourmouli N, Drosou V, Bakou A, Theodoropoulos PA, Singh PB, Giannakouros T, Georgatos SD (2001) Histones H3/H4 form a tight complex with the inner nuclear membrane protein LBR and heterochromatin protein 1. EMBO Rep 2(10):920–925. https://doi.org/10.1093/embo-reports/kve199

Pratt CH, Curtain M, Donahue LR, Shopland LS (2011) Mitotic defects lead to pervasive aneuploidy and accompany loss of RB1 activity in mouse LmnaDhe dermal fibroblasts. PLoS One 6(3):e18065. https://doi.org/10.1371/journal.pone.0018065

Qi R, Xu N, Wang G, Ren H, Li S, Lei J, Lin Q, Wang L, Gu X, Zhang H, Jiang Q, Zhang C (2015) The lamin-A/C-LAP2alpha-BAF1 protein complex regulates mitotic spindle assembly and positioning. J Cell Sci 128(15):2830–2841. https://doi.org/10.1242/jcs.164566

Ranade D, Koul S, Thompson J, Prasad KB, Sengupta K (2017) Chromosomal aneuploidies induced upon lamin B2 depletion are mislocalized in the interphase nucleus. Chromosoma 126(2):223–244. https://doi.org/10.1007/s00412-016-0580-y

Rasala BA, Orjalo AV, Shen Z, Briggs S, Forbes DJ (2006) ELYS is a dual nucleoporin/kinetochore protein required for nuclear pore assembly and proper cell division. Proc Natl Acad Sci USA 103(47):17801–17806. https://doi.org/10.1073/pnas.0608484103

Ribeiro SA, Gatlin JC, Dong Y, Joglekar A, Cameron L, Hudson DF, Farr CJ, McEwen BF, Salmon ED, Earnshaw WC, Vagnarelli P (2009) Condensin regulates the stiffness of vertebrate centromeres. Mol Biol Cell 20(9):2371–2380. https://doi.org/10.1091/mbc.e08-11-1127

Rodriguez-Bravo V, Maciejowski J, Corona J, Buch HK, Collin P, Kanemaki MT, Shah JV, Jallepalli PV (2014) Nuclear pores protect genome integrity by assembling a premitotic and Mad1-dependent anaphase inhibitor. Cell 156(5):1017–1031. https://doi.org/10.1016/j.cell.2014.01.010

Rowat AC, Lammerding J, Ipsen JH (2006) Mechanical properties of the cell nucleus and the effect of emerin deficiency. Biophys J 91(12):4649–4664. https://doi.org/10.1529/biophysj.106.086454

Rowat AC, Jaalouk DE, Zwerger M, Ung WL, Eydelnant IA, Olins DE, Olins AL, Herrmann H, Weitz DA, Lammerding J (2013) Nuclear envelope composition determines the ability of neutrophil-type cells to passage through micron-scale constrictions. J Biol Chem 288(12):8610–8618. https://doi.org/10.1074/jbc.M112.441535

Rowley MJ, Corces VG (2018) Organizational principles of 3D genome architecture. Nat Rev Genet 19(12):789–800. https://doi.org/10.1038/s41576-018-0060-8

Ruben BS, Brahmachari S, Contessoto VG, Cheng RR, Oliveira Junior AB, Di Pierro M, Onuchic JN (2023) Structural reorganization and relaxation dynamics of axially stressed chromosomes. Biophys J 122(9):1633–1645. https://doi.org/10.1016/j.bpj.2023.03.029

Salina D, Bodoor K, Eckley DM, Schroer TA, Rattner JB, Burke B (2002) Cytoplasmic dynein as a facilitator of nuclear envelope breakdown. Cell 108(1):97–107

Salina D, Enarson P, Rattner JB, Burke B (2003) Nup358 integrates nuclear envelope breakdown with kinetochore assembly. J Cell Biol 162(6):991–1001. https://doi.org/10.1083/jcb.200304080

Samejima K, Samejima I, Vagnarelli P, Ogawa H, Vargiu G, Kelly DA, de Lima Alves F, Kerr A, Green LC, Hudson DF, Ohta S, Cooke CA, Farr CJ, Rappsilber J, Earnshaw WC (2012) Mitotic chromosomes are compacted laterally by KIF4 and condensin and axially by topoisomerase IIα. J Cell Biol 199(5):755–770. https://doi.org/10.1083/jcb.201202155

Schmitt J, Benavente R, Hodzic D, Hoog C, Stewart CL, Alsheimer M (2007) Transmembrane protein Sun2 is involved in tethering mammalian meiotic telomeres to the nuclear envelope. Proc Natl Acad Sci USA 104(18):7426–7431. https://doi.org/10.1073/pnas.0609198104

Schneider MWG, Gibson BA, Otsuka S, Spicer MFD, Petrovic M, Blaukopf C, Langer CCH, Batty P, Nagaraju T, Doolittle LK, Rosen MK, Gerlich DW (2022) A mitotic chromatin phase transition prevents perforation by microtubules. Nature 609(7925):183–190. https://doi.org/10.1038/s41586-022-05027-y

Schreiner SM, Koo PK, Zhao Y, Mochrie SG, King MC (2015) The tethering of chromatin to the nuclear envelope supports nuclear mechanics. Nat Commun 6:7159. https://doi.org/10.1038/ncomms8159

Schuller AP, Wojtynek M, Mankus D, Tatli M, Kronenberg-Tenga R, Regmi SG, Dip PV, Lytton-Jean AKR, Brignole EJ, Dasso M, Weis K, Medalia O, Schwartz TU (2021) The cellular environment shapes the nuclear pore complex architecture. Nature 598(7882):667–671. https://doi.org/10.1038/s41586-021-03985-3

Schweizer N, Ferras C, Kern DM, Logarinho E, Cheeseman IM, Maiato H (2013) Spindle assembly checkpoint robustness requires Tpr-mediated regulation of Mad1/Mad2 proteostasis. J Cell Biol 203(6):883–893. https://doi.org/10.1083/jcb.201309076

Shin JW, Spinler KR, Swift J, Chasis JA, Mohandas N, Discher DE (2013) Lamins regulate cell trafficking and lineage maturation of adult human hematopoietic cells. Proc Natl Acad Sci USA 110(47):18892–18897. https://doi.org/10.1073/pnas.1304996110

Shintomi K, Takahashi TS, Hirano T (2015) Reconstitution of mitotic chromatids with a minimum set of purified factors. Nat Cell Biol 17(8):1014–1023. https://doi.org/10.1038/ncb3187

Shogren-Knaak M, Ishii H, Sun JM, Pazin MJ, Davie JR, Peterson CL (2006) Histone H4-K16 acetylation controls chromatin structure and protein interactions. Science 311(5762):844–847. https://doi.org/10.1126/science.1124000

Singh PB, Miller JR, Pearce J, Kothary R, Burton RD, Paro R, James TC, Gaunt SJ (1991) A sequence motif found in a Drosophila heterochromatin protein is conserved in animals and plants. Nucleic Acids Res 19(4):789–794. https://doi.org/10.1093/nar/19.4.789

Solovei I, Wang AS, Thanisch K, Schmidt CS, Krebs S, Zwerger M, Cohen TV, Devys D, Foisner R, Peichl L, Herrmann H, Blum H, Engelkamp D, Stewart CL, Leonhardt H, Joffe B (2013) LBR and lamin A/C sequentially tether peripheral heterochromatin and inversely regulate differentiation. Cell 152(3):584–598. https://doi.org/10.1016/j.cell.2013.01.009

Splinter D, Tanenbaum ME, Lindqvist A, Jaarsma D, Flotho A, Yu KL, Grigoriev I, Engelsma D, Haasdijk ED, Keijzer N, Demmers J, Fornerod M, Melchior F, Hoogenraad CC, Medema RH, Akhmanova A (2010) Bicaudal D2, dynein, and kinesin-1 associate with nuclear pore complexes and regulate centrosome and nuclear positioning during mitotic entry. PLoS Biol 8(4):e1000350. https://doi.org/10.1371/journal.pbio.1000350

Starr DA, Fridolfsson HN (2010) Interactions between nuclei and the cytoskeleton are mediated by SUN-KASH nuclear-envelope bridges. Annu Rev Cell Dev Biol 26:421–444. https://doi.org/10.1146/annurev-cellbio-100109-104037

Stephens AD, Banigan EJ, Adam SA, Goldman RD, Marko JF (2017) Chromatin and lamin A determine two different mechanical response regimes of the cell nucleus. Mol Biol Cell 28(14):1984–1996. https://doi.org/10.1091/mbc.E16-09-0653

Stephens AD, Liu PZ, Banigan EJ, Almassalha LM, Backman V, Adam SA, Goldman RD, Marko JF (2018) Chromatin histone modifications and rigidity affect nuclear morphology independent of lamins. Mol Biol Cell 29(2):220–233. https://doi.org/10.1091/mbc.E17-06-0410

Stephens AD, Banigan EJ, Marko JF (2019a) Chromatin's physical properties shape the nucleus and its functions. Curr Opin Cell Biol 58:76–84. https://doi.org/10.1016/j.ceb.2019.02.006

Stephens AD, Liu PZ, Kandula V, Chen H, Almassalha LM, Herman C, Backman V, O'Halloran T, Adam SA, Goldman RD, Banigan EJ, Marko JF (2019b) Physicochemical mechanotransduction alters nuclear shape and mechanics via heterochromatin formation. Mol Biol Cell 30(17):2320–2330. https://doi.org/10.1091/mbc.E19-05-0286

Stewart-Hutchinson PJ, Hale CM, Wirtz D, Hodzic D (2008) Structural requirements for the assembly of LINC complexes and their function in cellular mechanical stiffness. Exp Cell Res 314(8):1892–1905. https://doi.org/10.1016/j.yexcr.2008.02.022

Stiff T, Echegaray-Iturra FR, Pink HJ, Herbert A, Reyes-Aldasoro CC, Hochegger H (2020) Prophase-specific perinuclear actin coordinates centrosome separation and positioning to ensure accurate chromosome segregation. Cell Rep 31(8):107681. https://doi.org/10.1016/j.celrep.2020.107681

Strom AR, Biggs RJ, Banigan EJ, Wang X, Chiu K, Herman C, Collado J, Yue F, Ritland Politz JC, Tait LJ, Scalzo D, Telling A, Groudine M, Brangwynne CP, Marko JF, Stephens AD (2021) HP1alpha is a chromatin crosslinker that controls nuclear and mitotic chromosome mechanics. elife 10. https://doi.org/10.7554/eLife.63972

Sun M, Biggs R, Hornick J, Marko JF (2018) Condensin controls mitotic chromosome stiffness and stability without forming a structurally contiguous scaffold. Chromosom Res 26(4):277–295. https://doi.org/10.1007/s10577-018-9584-1

Swift J, Ivanovska IL, Buxboim A, Harada T, Dingal PC, Pinter J, Pajerowski JD, Spinler KR, Shin JW, Tewari M, Rehfeldt F, Speicher DW, Discher DE (2013) Nuclear lamin-A scales with tissue stiffness and enhances matrix-directed differentiation. Science 341(6149):1240104. https://doi.org/10.1126/science.1240104

Tang W, Chen X, Wang X, Zhu M, Shan G, Wang T, Dou W, Wang J, Law J, Gong Z, Hopyan S, Huang X, Sun Y (2023) Indentation induces instantaneous nuclear stiffening and unfolding of nuclear envelope wrinkles. Proc Natl Acad Sci USA 120(36):e2307356120. https://doi.org/10.1073/pnas.2307356120

Turgay Y, Champion L, Balazs C, Held M, Toso A, Gerlich DW, Meraldi P, Kutay U (2014) SUN proteins facilitate the removal of membranes from chromatin during nuclear envelope breakdown. J Cell Biol 204(7):1099–1109. https://doi.org/10.1083/jcb.201310116

Turkmen AM, Saik NO, Ullman KS (2023) The dynamic nuclear envelope: resilience in health and dysfunction in disease. Curr Opin Cell Biol 85:102230. https://doi.org/10.1016/j.ceb.2023.102230

Uemura T, Ohkura H, Adachi Y, Morino K, Shiozaki K, Yanagida M (1987) DNA topoisomerase II is required for condensation and separation of mitotic chromosomes in S. pombe. Cell 50(6):917–925. https://doi.org/10.1016/0092-8674(87)90518-6

Ungricht R, Kutay U (2017) Mechanisms and functions of nuclear envelope remodelling. Nat Rev Mol Cell Biol 18(4):229–245. https://doi.org/10.1038/nrm.2016.153

Uroz M, Wistorf S, Serra-Picamal X, Conte V, Sales-Pardo M, Roca-Cusachs P, Guimera R, Trepat X (2018) Regulation of cell cycle progression by cell-cell and cell-matrix forces. Nat Cell Biol 20(6):646–654. https://doi.org/10.1038/s41556-018-0107-2

Van de Vosse DW, Wan Y, Lapetina DL, Chen WM, Chiang JH, Aitchison JD, Wozniak RW (2013) A role for the nucleoporin Nup170p in chromatin structure and gene silencing. Cell 152(5):969–983. https://doi.org/10.1016/j.cell.2013.01.049

van Schaik T, Vos M, Peric-Hupkes D, Hn Celie P, van Steensel B (2020) Cell cycle dynamics of lamina-associated DNA. EMBO Rep 21(11):e50636. https://doi.org/10.15252/embr.202050636

van Steensel B, Belmont AS (2017) Lamina-associated domains: links with chromosome architecture, heterochromatin, and gene repression. Cell 169(5):780–791. https://doi.org/10.1016/j.cell.2017.04.022

Venturini V, Pezzano F, Catala Castro F, Hakkinen HM, Jimenez-Delgado S, Colomer-Rosell M, Marro M, Tolosa-Ramon Q, Paz-Lopez S, Valverde MA, Weghuber J, Loza-Alvarez P, Krieg M, Wieser S, Ruprecht V (2020) The nucleus measures shape changes for cellular proprioception to control dynamic cell behavior. Science 370(6514). https://doi.org/10.1126/science.aba2644

Vitiello E, Moreau P, Nunes V, Mettouchi A, Maiato H, Ferreira JG, Wang I, Balland M (2019) Acto-myosin force organization modulates centriole separation and PLK4 recruitment to ensure centriole fidelity. Nat Commun 10(1):52. https://doi.org/10.1038/s41467-018-07965-6

Walther N, Ellenberg J (2018) Quantitative live and super-resolution microscopy of mitotic chromosomes. Methods Cell Biol 145:65–90. https://doi.org/10.1016/bs.mcb.2018.03.014

Ward GE, Kirschner MW (1990) Identification of cell cycle-regulated phosphorylation sites on nuclear lamin C. Cell 61(4):561–577. https://doi.org/10.1016/0092-8674(90)90469-u

Wen B, Wu H, Shinkai Y, Irizarry RA, Feinberg AP (2009) Large histone H3 lysine 9 dimethylated chromatin blocks distinguish differentiated from embryonic stem cells. Nat Genet 41(2):246–250. https://doi.org/10.1038/ng.297

Wilkins BJ, Rall NA, Ostwal Y, Kruitwagen T, Hiragami-Hamada K, Winkler M, Barral Y, Fischle W, Neumann H (2014) A cascade of histone modifications induces chromatin condensation in mitosis. Science 343(6166):77–80. https://doi.org/10.1126/science.1244508

Williams JF, Surovtsev IV, Schreiner SM, Chen Z, Raiymbek G, Nguyen H, Hu Y, Biteen JS, Mochrie SGJ, Ragunathan K, King MC (2024) The condensation of HP1alpha/Swi6 imparts nuclear stiffness. Cell Rep 43(7):114373. https://doi.org/10.1016/j.celrep.2024.114373

Witt H, Harju J, Chameau EMJ, Bruinsma CMA, Clement TVM, Nielsen CF, Hickson ID, Peterman EJG, Broedersz CP, Wuite GJL (2024) Ion-mediated condensation controls the mechanics of mitotic chromosomes. Nat Mater 23(11):1556–1562. https://doi.org/10.1038/s41563-024-01975-0

Wreggett KA, Hill F, James PS, Hutchings A, Butcher GW, Singh PB (1994) A mammalian homologue of Drosophila heterochromatin protein 1 (HP1) is a component of constitutive heterochromatin. Cytogenet Cell Genet 66(2):99–103. https://doi.org/10.1159/000133676

Wu W, McHugh T, Kelly DA, Pidoux AL, Allshire RC (2022) Establishment of centromere identity is dependent on nuclear spatial organization. Curr Biol 32(14):3121–3136 e3126. https://doi.org/10.1016/j.cub.2022.06.048

Xie W, Burke B (2016) Lamins. Curr Biol 26(9):R348–R350. https://doi.org/10.1016/j.cub.2016.01.055

Ye Q, Worman HJ (1996) Interaction between an integral protein of the nuclear envelope inner membrane and human chromodomain proteins homologous to Drosophila HP1. J Biol Chem 271(25):14653–14656. https://doi.org/10.1074/jbc.271.25.14653

Ye Q, Callebaut I, Pezhman A, Courvalin JC, Worman HJ (1997) Domain-specific interactions of human HP1-type chromodomain proteins and inner nuclear membrane protein LBR. J Biol Chem 272(23):14983–14989. https://doi.org/10.1074/jbc.272.23.14983

Zhang YL, Frangos JA, Chachisvilis M (2006) Laurdan fluorescence senses mechanical strain in the lipid bilayer membrane. Biochem Biophys Res Commun 347(3):838–841. https://doi.org/10.1016/j.bbrc.2006.06.152

Zhiteneva A, Bonfiglio JJ, Makarov A, Colby T, Vagnarelli P, Schirmer EC, Matic I, Earnshaw WC (2017) Mitotic post-translational modifications of histones promote chromatin compaction in vitro. Open Biol 7(9). https://doi.org/10.1098/rsob.170076

Zimmerli CE, Allegretti M, Rantos V, Goetz SK, Obarska-Kosinska A, Zagoriy I, Halavatyi A, Hummer G, Mahamid J, Kosinski J, Beck M (2021) Nuclear pores dilate and constrict in cellulo. Science 374(6573):eabd9776. https://doi.org/10.1126/science.abd9776

Zuccolo M, Alves A, Galy V, Bolhy S, Formstecher E, Racine V, Sibarita JB, Fukagawa T, Shiekhattar R, Yen T, Doye V (2007) The human Nup107-160 nuclear pore subcomplex contributes to proper kinetochore functions. EMBO J 26(7):1853–1864. https://doi.org/10.1038/sj.emboj.7601642

Chapter 10
Forceful Beginnings: Mechanotransduction and Mechanosensation at the Meiotic Cell Nucleus During Gamete Development

Rachel Nenstiel and Chenshu Liu

Abstract For most eukaryotes, sexual reproduction depends on meiosis, a specialized cell division and differentiation program that halves the parental genome to produce gametes such as eggs, sperm, and pollen. Accurate segregation of homologous chromosomes during meiosis and the subsequent maturation of gametes are essential for reproductive success. Growing evidence suggests that mechanical forces acting on meiotic nuclei play essential roles in the development of healthy gametes. Here, we highlight how mechanical forces participate in key events through gamete development by focusing on three distinct yet related processes over space and time during the protracted meiotic prophase: (1) dynamic chromosome reorganization in early prophase, (2) control of gamete quality, and (3) gamete maturation. We center this chapter on *Caenorhabditis elegans*, in which the molecular logic of nuclear mechanotransduction during gamete development is especially well resolved. Where relevant, we summarize key findings in other model organisms to underscore conserved and divergent mechanisms. In addition to synthesizing current knowledge, this chapter outlines challenges and opportunities in studying the mechanisms and roles of force transmission and sensation at the nuclear envelope during gamete development—an emerging area with the potential to inform strategies that would improve reproductive health.

10.1 Introduction

For sexually reproducing organisms, the continuity of a species and the diversity among its populations depend on meiosis and the successful maturation of gametes, such as sperm, eggs, and pollen. Meiosis includes one round of DNA replication

R. Nenstiel · C. Liu (✉)
Department of Biological Sciences, Lehigh University, Bethlehem, PA, USA
e-mail: chenshu.liu@lehigh.edu

M. Kloc, J. Wosik (eds.), *Mechanotransduction and Mechanosensing*, Results and Problems in Cell Differentiation 78, https://doi.org/10.1007/978-3-032-19509-8_10

followed by two successive rounds of cell division, first segregating homologs, then separating sister chromatids, eventually creating gametes that have only half the somatic genome content (Bhalla and Dernburg 2008; Zickler and Kleckner 2023). During meiotic prophase I, chromosomes undergo dramatic reorganization: Homologs must find each other, align along their lengths, and repair DNA double-strand breaks to establish genetic crossovers that ensure accurate segregation (Bhalla and Dernburg 2008). Errors in any of these steps could cause aneuploidy, a hallmark of infertility, miscarriages, and congenital developmental disorders (Charalambous et al. 2023; Heffner 2004; Hassold and Hunt 2001). Meanwhile, the successful generation of gametes also requires that meiotic nuclei maintain structural integrity and functional competence across extended timescales, ranging from days in many invertebrates to decades in humans (Scharf et al. 2021; Charalambous et al. 2023; Hoffman et al. 2021; Huan Lu et al. 2022; Liu et al. 2023; Huelgas-Morales et al. 2020; Tsatskis et al. 2020). These requirements are increasingly understood to be not only biochemical but also mechanical (Kim et al. 2022; Link and Jantsch 2019; Vasse et al. 2025).

The nuclear envelope (NE)—composed of the outer and inner nuclear membranes, nuclear lamina, and associated protein complexes—sits at the interface between chromosomes and the cytoplasm. The NE not only serves as the physical barrier to protect a cell's genome but also provides a dynamic platform that transmits chemical and mechanical signals originating from the cytoplasm to dynamically regulate nuclear movement, chromosome organization, and gene expression (Maurer and Lammerding 2019; Zhu et al. 2018; Miroshnikova and Wickström 2022). Key to the mechanical coupling between nucleus and the cytoplasm are LInker of Nucleoskeleton and Cytoskeleton (LINC) complexes, which span the NE via SUN (Sad1/UNC-84)-domain proteins at the inner nuclear membrane and KASH-domain proteins at the outer nuclear membrane (Chang et al. 2015; Tapley and Starr 2013). These complexes connect the cell nucleus or individual chromosomes to the cytoskeleton, such as microtubules and actin, enabling mechanical forces generated in the cytoplasm to drive chromosome movements, reposition nuclei, and deform the NE (Chang et al. 2015; Burke 2018; Méjat and Misteli 2010; Kalukula et al. 2022). At the same time, the NE and its underlying lamina, together with inner nuclear membrane (INM) proteins, confer mechanical stability on the nucleus and are increasingly implicated in the maintenance of genome integrity and cellular response to mechanical stimuli in diverse cellular and developmental contexts (Venturini et al. 2020; Lomakin et al. 2020; Kovacs et al. 2023; Earle et al. 2020; Denais et al. 2016).

Work in the nematode *Caenorhabditis elegans* has been especially instrumental in revealing how mechanical forces and nuclear architecture intersect during meiosis and gamete development. The discovery of chromosome-specific pairing centers and their associated zinc finger proteins, the identification of meiotic LINC complexes and their cytoskeletal partners, and the characterization of NE-based checkpoints and stabilization mechanisms have collectively provided a framework in which chromosome dynamics, force transmission, and nuclear integrity can be dissected at the molecular level (Villeneuve 1994; MacQueen et al. 2005; Phillips et al. 2005, 2009b; Phillips and Dernburg 2006; Sato et al. 2009; Harper et al. 2011; Wynne et al. 2012; Bhalla and Dernburg 2005; Liu et al. 2023; Liu and Dernburg 2024; Kim et al. 2023). These advances, together with the optical transparency and genetic tractability of *C. elegans* (Brenner 1974; Phillips et al. 2009a), make it a powerful system for understanding how mechanical forces contribute to gamete development.

In this chapter, we focus on three mechanistically linked questions about how forces acting on the meiotic nucleus shape gamete development over space and time. First, we examine how mechanical forces regulate chromosome dynamics during meiotic prophase, emphasizing the role of NE-associated factors in regulating homolog pairing, synapsis, and recombination. Second, we discuss emerging evidence that mechanical inputs and mechanosensitive pathways at the NE contribute to quality control removing defective gametes via apoptosis. Third, we explore how mechanical forces acting on the nucleus regulate gamete maturation, including nuclear positioning, NE stabilization, and force-dependent nuclear rotation and NE fluctuations during prolonged prophase I arrest (Fig. 10.1). We center our discussion on *C. elegans*, where nuclear mechanotransduction in meiosis has been studied in great detail, and, where relevant, we draw on findings from other model systems to highlight conserved principles and organism-specific solutions.

By integrating historical genetic insights with recent advances in imaging, biophysics, and mechanobiology, we aim to provide a cohesive view of how the meiotic nucleus functions as a force-sensitive organelle. This perspective opens new avenues for understanding how the mechanical properties of the NE and its associated machineries contribute to reproductive robustness, how their failure leads to germ cell loss or aneuploidy, and how these mechanisms might be leveraged to improve reproductive health.

10.2 How Mechanical Forces Regulate Chromosome Dynamics During Meiotic Prophase

A defining feature of meiosis is the dramatic reorganization of the genome at the onset of prophase. Homologous chromosomes must find one another, align along their length, and assemble a proteinaceous synaptonemal complex (SC) between their axes (Page and Hawley 2004; Adams and Davies 2023). Homolog pairing and synapsis are prerequisites for successful reductional segregation: They bring homologs into close proximity to ensure the formation of crossovers, enabling segregation to opposite spindle poles in anaphase I and providing the basis of genetic diversity (Zickler and Kleckner 2023; Charalambous et al. 2023). Over the past three decades, genetic and cytological studies have revealed a network of factors that link chromosome ends to the nuclear envelope (NE) and to the cytoskeleton, thereby harnessing mechanical forces to drive this reorganization.

10.2.1 *C. elegans* Pairing Centers as Cis-Acting Factors Essential for Homolog Pairing

Early genetics in *Caenorhabditis elegans* provided the first clues that specific chromosomal sites promote meiotic pairing. Herman and Kari showed that recombination can occur between the X chromosome and duplications that include sequences near the left end of X (Herman and Kari 1989). McKim and colleagues further demonstrated that translocations that disrupt or relocate specific chromosomal regions strongly suppress pairing and recombination, implicating dedicated "pairing sites" (McKim et al. 1988, 1993).

Leveraging these observations and the fact that the "pairing center" activity is concentrated at a single region on each chromosome, Villeneuve identified the first *cis*-acting pairing locus in an otherwise normal X chromosome: *me8*, mapped to the extreme left end of X (Villeneuve 1994). Because *C. elegans* is predominantly self-fertilizing hermaphrodite (XX), wild-type hermaphrodites produce only ~0.2% male (XO) progeny, and any defect in X-chromosome pairing would yield a "high incidence of males" (*him*) phenotype (Villeneuve 1994; Hodgkin et al. 1979). Homozygous deletions spanning me8 (*Df/Df*) or *Df/me8* heterozygotes produce very high frequencies of male progeny correlated with severely reduced X recombination. Meanwhile, 7 DAPI-staining bodies at diakinesis (2 X univalents and 5 autosomal bivalents), rather than the 6 bivalents seen in wild type (which has 6 pairs of homologous chromosomes) or 12 univalents, should pairing and synapsis of all chromosomes fail, were observed, supporting X-chromosome specific defect. Duplications of the *me8* region in *trans* fail to rescue these defects, indicating that the *me8* region acts in *cis* as a dedicated X-pairing center (PC) (Villeneuve 1994).

The discovery of pairing centers (PCs) established *C. elegans* as a premier system for dissecting the molecular basis of homolog pairing, synapsis (the assembly of synaptonemal complex), and recombination, due to its genetic tractability and optical transparency, which enable high-resolution cytology and live imaging of meiotic nuclei (Rog and Dernburg 2013).

10.2.2 Pairing Center Proteins Act in Trans to Promote Homolog Pairing and Synapsis

The development of fluorescence in situ hybridization (FISH) and antibody staining against axis and SC components allowed visualization of homolog pairing and synapsis (Dernburg et al. 1998; MacQueen et al. 2002). In mutants lacking the X-PC on both homologs, FISH revealed diminished pairing both near and distal to the PC, and most pachytene nuclei contain chromosome axes lacking overlapping SC staining, indicating extensive synapsis defects (MacQueen et al. 2005).

A forward genetic screen for *him* mutants identified *him-8* (Hodgkin et al. 1979), which specifically disrupts X pairing and synapsis without affecting autosomes

(Phillips et al. 2005). Because *him-8* maps to chromosome IV rather than X, it must act in *trans*. HIM-8 encodes a C2H2 zinc finger protein that forms distinct nuclear foci during meiosis. Combined FISH and immunostaining showed that HIM-8 foci colocalize with the X-PC and associate with the NE (Phillips et al. 2005).

Phillips and Dernburg then showed that *him-8* is one of a small family of related zinc finger proteins encoded in a tandem array on chromosome IV (Phillips and Dernburg 2006). Whereas HIM-8 promotes X pairing, ZIM-1, ZIM-2, and ZIM-3 collectively promote pairing of the five autosomes, each recognizing sequence motifs enriched in the cognate PC (Phillips et al. 2009b). HIM-8 (chromosome X), ZIM-1 (chromosomes II and III), ZIM-2 (chromosome V), and ZIM-3 (chromosomes I and IV) foci all localize to the NE at meiotic entry (leptotene/zygotene), concomitant with a polarized clustering of chromosomes in the nucleus (Phillips and Dernburg 2006; Phillips et al. 2009b). These findings established that PCs near chromosome ends are bound by specific zinc finger proteins, which then connect to NE-based machinery.

10.2.3 LINC Complexes, Cytoskeletal Forces, and Prophase Chromosome Motion

The polarized configuration of chromosomes in *C. elegans* meiotic nuclei resembles the "bouquet" in other organisms, where telomeres cluster at one side of the nucleus via NE attachment (Rubin et al. 2020; Kim et al. 2022). In fission yeast, for example, telomeres attach to the NE through the SUN-domain protein Sad1 at the inner nuclear membrane and a KASH partner at the outer membrane, which connects to the spindle pole body and microtubules (Hagan and Yanagida 1995; Miki et al. 2004; Shimanuki et al. 1997). Dynein-driven "horsetail" movements are required for homolog alignment (Ding et al. 2004; Yamamoto et al. 1999; Chikashige et al. 1994).

In *C. elegans*, SUN-1 at the inner nuclear membrane and its outer nuclear membrane partner ZYG-12 form the meiotic LINC complex (Malone et al. 2003; Tzur et al. 2006; Minn et al. 2009; Penkner et al. 2007). A *sun-1* mutant exhibits extensive synapsis between nonhomologous chromosomes, suggesting that LINC-mediated forces help ensure homology-dependent synapsis (Penkner et al. 2007, 2009). Using GFP-tagged ZYG-12, Sato et al. showed that LINC complexes form dynamic patches that colocalize with PCs on all chromosomes. Cytoplasmic dynein also concentrates at these patches (Sato et al. 2009). Disrupting microtubules or depleting dynein impairs pairing and delays synapsis, while loss of SUN-1 or ZYG-12 leads to promiscuous SC assembly between non-homologs (Penkner et al. 2009; Sato et al. 2009; Liu et al. 2023; Daryabeigi et al. 2016). These observations indicate that force transmission across the NE via LINC complexes is essential for both promoting homolog pairing and for regulating synapsis.

Live imaging further revealed that PCs and LINC complexes mediate large-scale, processive chromosome movements along the NE that depend on dynein and microtubules but not F-actin (Baudrimont et al. 2010; Wynne et al. 2012). Dynein-driven motion also promotes the extension of SC along paired chromosomes (Rog and Dernburg 2015), which regulates the number and distribution of crossovers between homologs for genetic recombination (Rog et al. 2017; Zhang et al. 2018, 2025; Zickler and Kleckner 2023). The timing of these force-generating connections is tightly controlled: At meiotic entry, the kinase CHK-2 phosphorylates PC proteins to recruit Polo-like kinase PLK-2, which then phosphorylates NE proteins to promote pairing and synapsis (Harper et al. 2011; Kim et al. 2015; Zhang et al. 2023; Labella et al. 2011). Together, work in *C. elegans* shows how *cis*-acting PCs and *trans*-acting zinc finger proteins, LINC complexes, motors, and accessory/regulatory proteins are integrated at the NE to use mechanical force for chromosome reorganization (Kim et al. 2022, 2023; Link and Jantsch 2019).

10.2.4 Lessons from Other Organisms

Beyond *C. elegans*, NE-based chromosomal attachments and telomere bouquets are essential for accurate meiotic chromosome segregation in a wide range of organisms. In fission and budding yeasts, plants, flies, and mammals, SUN/KASH LINC complexes link chromosome ends to the cytoskeleton, supporting chromosome movements that are key to synapsis and recombination, despite differences in attachment sites (pairing centers versus telomeres or centromeres), cytoskeletal organization (microtubules versus actin filaments), or the order of events between synapsis and recombination.

In fission yeast *Schizosaccharomyces pombe*, a network of telomere-associated proteins tether telomeres to the Sad1/Kms1 LINC complex, which uses dynein and microtubules for the "horsetail" movement of chromosomes, promoting homolog pairing and recombination (Chikashige et al. 1994, 2006; Hagan and Yanagida 1995; Shimanuki et al. 1997; Yamamoto et al. 1999; Ding et al. 2004). In budding yeast *S. cerevisiae*, telomeres move along F-actin cables, mediated by LINC complexes composed of the SUN-protein Mps3 and KASH-domain-like proteins Csm4 and Mps2, and a myosin motor (Myo2) (Conrad et al. 2007, 2008; Wanat et al. 2008; Lee et al. 2020; Koszul et al. 2008; Fan et al. 2020). In plants such as *Arabidopsis thaliana*, SUN-domain proteins SUN1/2 and the KASH-domain-containing partner SINE3 connect telomeres to the NE and drive kinesin-dependent chromosome motion (Cromer et al. 2024; Cai et al. 2025). In *Drosophila melanogaster*, centromere clustering and pairing require SUN/KASH (Klaroid/Klarsicht)-mediated NE attachments and the microtubule motor dynein (Christophorou et al. 2015). In zebrafish meiotic cells, adjacent to where telomeres cluster at the NE, there is a high density of microtubules in the cytoplasm (Saito et al. 2014), which are important to bouquet formation (Elkouby et al. 2016). Interestingly, a "zygotene cilium" anchors the bouquet centrosome to resist telomere rotation and pulling, promoting synapsis and oogenesis (Mytlis et al. 2022). Although LINC complex components for zebrafish meiotic NE remain to be identified, several key meiotic telomere binding proteins (TERB1, TERB2, and MAJIN) are conserved between zebrafish and mice (da Cruz et al. 2020). In mice, TERB1 and TERB2 connect the telomere protein TRF1 to the SUN1/KASH5-LINC complexes in the presence of MAJIN (Shibuya et al. 2014, 2015; Ding et al. 2007; Morimoto et al. 2012), enabling dynein-/microtubule-mediated concomitant nuclear rotation and chromosome movement during prophase I of spermatogenesis (Horn et al. 2013; Lee et al. 2015).

In *C. elegans* meiosis and *Drosophila* female meiosis, homologous chromosome synapsis does not require recombination (Dernburg et al. 1998; McKim et al. 1998). Budding yeast, *Arabidopsis*, zebrafish, and mammals all have recombination-dependent synapsis (Rubin et al. 2020; Higgins et al. 2005; Lambing et al. 2022). While molecular details vary, a common theme emerges: Dedicated chromosomal regions connect to the NE and, likely through LINC complexes, use cytoskeletal motors and mechanical forces to promote homolog pairing, synapsis, and crossover formation.

Despite this progress, many molecular players and the precise mechanisms by which forces are integrated at the NE to control chromosome behavior remain to be defined. *C. elegans* continues to provide a powerful framework for mechanistic dissection, while cross-species comparisons reveal conserved principles and organism-specific innovations.

10.3 How Mechanical Forces Act on the Cell Nucleus to Regulate Gamete Quality Control

NE proteins that transduce mechanical forces to chromosomes in early prophase remain at the NE long after pairing and synapsis are established (Morimoto et al. 2012; Horn et al. 2013; Viera et al. 2015; Penkner et al. 2007, 2009; Sato et al. 2009; Wynne et al. 2012; Liu et al. 2023). This persistence indicates that NE-based mechanotransduction may also play additional roles during gametogenesis, including quality control of developing gametes. Here, we discuss emerging evidence suggesting important roles of mechanotransduction and mechanosensation in gamete quality control.

10.3.1 An NE-Based Synapsis Checkpoint in C. elegans

The concept of an NE-based quality control mechanism traces back to the observation that synapsis defects trigger apoptosis during *C. elegans* oogenesis (Villeneuve 1994; Bhalla and Dernburg 2005). Villeneuve's original analysis of X-PC mutants provided a key clue: When both X chromosomes lack their PCs, defective X pairing and synapsis produce very high frequencies of male progeny (>30% XO versus 0.2% in wild type). By contrast, when only one X lacks its PC, interactions between X chromosomes remain defective, but only 5–7% of progeny are male. This implies that many defective meiotic nuclei are eliminated before contributing to the gamete pool.

Up to ~50% of oocytes normally undergo apoptosis in the *C. elegans* germline, suggesting that germline apoptosis could serve as a quality-control "filter" (Lettre and Hengartner 2006; Raiders et al. 2018). Indeed, blocking germline apoptosis with a *ced-4* null mutation dramatically increases the frequency of male progeny in worms with a single X-PC, demonstrating that apoptosis selectively culls oocytes with synapsis defects (Bhalla and Dernburg 2005).

Genetic analysis, combined with counting apoptotic nuclei in the germline, revealed that this "synapsis checkpoint" acts independently of DNA damage checkpoints but requires unsynapsed PCs, the X-PC protein HIM-8, the AAA+ ATPase PCH-2/TRIP13, the Polo-like kinase PLK-2, and additional chromosomal and NE proteins (e.g., SC subunits and Mad1/2) (Bhalla and Dernburg 2005; Harper et al. 2011; Bohr et al. 2015, 2016; Devigne and Bhalla 2021). However, how meiotic nuclei detect and respond to synapsis failure at the NE has remained unclear.

10.3.2 PLK-2, Lamin, and Checkpoint Activation-Induced NE Remodeling

At meiotic onset, PCs recruit PLK-2 via CHK-2-dependent phosphorylation (Harper et al. 2011; Kim et al. 2015). Once recruited to the NE, PLK-2 phosphorylates the nuclear lamin protein LMN-1 to induce transient and partial solubilization of the nuclear lamina (Link et al. 2018; Velez-Aguilera et al. 2020; Liu and Dernburg 2024). These remodeling events facilitate LINC-mediated chromosome movement during pairing. Once synapsis is complete, PLK-2 dissociates from PCs and relocalizes to the SC (Harper et al. 2011; Labella et al. 2011; Brandt et al. 2020). In mutants with persistent synapsis defects, however, PLK-2 remains associated with PCs at the NE (Castellano-Pozo et al. 2020; Bohr et al. 2016; Sato-Carlton et al. 2017).

This raised the possibility that persistent PLK-2 at PCs might signal synapsis failure. A key challenge was distinguishing whether PLK-2's NE localization drives the checkpoint response or is merely a consequence of it. To address this, Liu and Dernburg developed an in vivo chemically induced proximity (CIP) system in *C. elegans* to recruit PLK-2 artificially to PCs via induced dimerization with HIM-8, which binds to X-PCs even after synapsis is complete (Liu and Dernburg 2024). Forced recruitment of PLK-2 to X-PCs in otherwise normal nuclei is sufficient to trigger apoptosis, and this response requires PLK-2-dependent phosphorylation of LMN-1. These results suggest that lamin remodeling at the NE is a key effector of the synapsis checkpoint.

Multiple synapsis mutants show persistent NE-associated foci of SUN-1, ZYG-12, PLK-2, and PCs that undergo rapid movements across the NE late in pachytene, implicating a mechanical response (Wynne et al. 2012; Castellano-Pozo et al. 2020; Sato-Carlton et al. 2017; Bohr et al. 2016; Zhang et al. 2012). Liu and Dernburg further showed that synapsis checkpoint activation correlates with increased whole-NE fluctuations—protrusions and invaginations of the NE over time (Liu and Dernburg 2024). While quantitative measurements of NE mechanics in this context are still lacking, the association between NE fluctuations, persistent LINC complexes foci at the NE, and apoptosis points to a role for force transmission through the NE in quality control.

10.3.3 LINC Complexes, Mechanosensitive Channels, and Checkpoint Execution

If force transmission is important for the synapsis checkpoint, compromising LINC-mediated mechanics should affect apoptosis. Indeed, acute depletion of SUN-1 or a point mutation in *sun-1(jf18)* that weakens mechanotransduction across the SUN-KASH bridge abrogates asynapsis-induced apoptosis (Penkner et al. 2007; Baudrimont et al. 2010; Liu and Dernburg 2024). This demonstrates that mechanical coupling through LINC complexes is required for the checkpoint response.

Recent work revealed an additional layer of mechanosensation in synapsis checkpoint response. The mechanosensitive ion channel Piezo1/PEZO-1, which localizes to the meiotic NE, is required for apoptosis triggered by synapsis defects (Bai et al. 2020; Liu and Dernburg 2024). In response to asynapsis, PEZO-1 redistributes within the NE to co-cluster with LINC complexes (Liu and Dernburg 2024). Together, these findings suggest that NE-localized mechanotransducers (LINC) work with mechanosensors (Piezo) to generate and detect abnormal forces at the NE and trigger apoptosis.

How force-transducing NE proteins originally deployed for early prophase chromosome dynamics are repurposed at later stages remains an open question. Possibilities include recruitment of new binding partners or altered posttranslational modifications of NE proteins by kinases such as CHK-2, PLK-2, MPK-1/ERK, and CDKs, which dominate distinct meiotic stages (Arur et al. 2009, 2011; MacQueen and Villeneuve 2001; Zhang et al. 2023, 2025; Lee et al. 2007; Martinez-Perez and Villeneuve 2005; Roelens et al. 2019; Haversat et al. 2022; Harper et al. 2011; Labella et al. 2011; Brandt et al. 2020). Another major question is how NE-based mechanotransduction interfaces with canonical DNA damage and recombination checkpoints (Huang and Roig 2023). Additionally, how mechanosensitive ion channels act in oocyte quality control remains an important open question.

10.3.4 Implications for Mammalian Meiotic Quality Control

Although the NE-based synapsis checkpoint has been most clearly defined in *C. elegans*, similar principles may operate in mammalian meiosis. In mice and humans, recombination precedes and is required for synapsis, and defects in recombination can arrest meiosis and trigger apoptosis (Huang and Roig 2023; Di Giacomo et al. 2005; Baudat et al. 2000; Jan et al. 2018). During mouse spermatogenesis, CDK2 localizes to both crossover sites and telomeres engaged with the NE. When crossover formation is perturbed, CDK2 persists at telomeres, suggesting a telomere/NE-linked surveillance mechanism (Ashley et al. 2001; Viera et al. 2009, 2015; Palmer et al. 2020; Reynolds et al. 2013; Holloway et al. 2014; Tu et al. 2017). Indeed, mammalian CDKs phosphorylate lamins to disassemble the nuclear lamina during mitosis (Khan et al. 2025; Liu et al. 2025; Chi et al. 2020). Thus, a quality control mechanism at the NE that monitors synapsis could potentially help ensure crossover, a prerequisite for accurate chromosome segregation in mammals.

Whether mammals employ a synapsis checkpoint that directly senses mechanical inputs at the NE remains unresolved. Nonetheless, the *C. elegans* framework provides a conceptual model in which persistent force-bearing NE domains, specialized kinases, and mechanosensitive channels work together to monitor chromosome status and eliminate defective gametes. Mechanistic insights from model organisms could shed light on the sexually dimorphic yet poorly understood quality control in mammals (Hunt and Hassold 2002; Huang and Roig 2023; Baudat et al. 2000; Jan et al. 2018; Charalambous et al. 2023).

10.4 How Mechanical Forces Act on the Cell Nucleus to Regulate Gamete Maturation

Throughout gamete development and maturation (Zhang 2018), the nucleus experiences a variety of mechanical forces (Wolke et al. 2007; Lee et al. 2015; Wynne et al. 2012; Liu et al. 2023; Nagamatsu et al. 2019; Wang and Yang 2025; Almonacid et al. 2018). The same forces that are essential for chromosome dynamics and likely promote gamete quality control could also jeopardize fertility if NE integrity is compromised (Tsatskis et al. 2020; Liu et al. 2023). Consequently, mechanotransduction at the NE during gamete maturation must be tightly regulated. Here, we examine NE-associated forces and responses from three angles: nuclear positioning, NE stabilization, and force-dependent nuclear rotation and NE fluctuations during prophase I arrest.

10.4.1 Mechanical Forces and Nuclear Positioning During Gametogenesis

For mechanical forces to act on specific NE substrates or be transmitted to chromatin, meiotic nuclei must be correctly positioned within germ cells. In the syncytial *C. elegans* gonad, meiotic nuclei are initially tethered near the plasma membrane facing the outer surface of the gonad (Hirsh et al. 1976; Kimble and Crittenden 2005; Albert Hubbard 2007) and then repositioned toward the center of maturing oocytes (Green et al. 2011). This tethering and repositioning require a network of proteins involved in cytoskeletal dynamics and associated signaling.

A central requirement for proper nuclear positioning is the interaction between LINC complexes and the cytoskeleton. The SUN-1/ZYG-12 LINC complex is present at the NE throughout oogenesis, and depletion of ZYG-12 causes dramatic nuclear mispositioning: Meiotic nuclei detach from the gonad surface and fall into the central rachis, including in regions where oocytes typically mature (Zhou et al. 2009; Liu et al. 2023). A similar phenotype was observed when depleting various components of the cytoplasmic dynein motor, which also localizes to the NE (Zhang et al. 2015; Liu et al. 2023). These observations highlight the importance of mechanical tethering in maintaining ordered nuclear positioning along the germline axis, which may help align nuclei with spatial gradients of regulatory factors such as CHK-2 and MPK-1 in the syncytial *C. elegans* gonad (Zhang et al. 2023; Arur et al. 2009; Kim et al. 2015).

Hydrodynamic forces from cytoplasmic flow also vary along the germline (Chartier et al. 2021; Wolke et al. 2007), raising the possibility that the tethering strength between the NE and the cytoskeleton adapts to changing mechanical environments. How LINC complexes and the cytoskeleton tune their mechanical properties across regions of different flow remains to be determined.

Nuclear position is also actively regulated during mammalian gametogenesis. In maturing mouse oocytes, which lack centrosomes, the formin-2 (Fmn2)-dependent actin network generates active diffusion that centers the nucleus. Actin-coated vesicles, driven by MyosinVb, create a pressure gradient that propels the nucleus toward the center of the oocyte (Almonacid et al. 2015). Nuclear position correlates with maturation competence: Centrally located germinal vesicles (GVs) are more likely to support successful meiotic progression and subsequent embryo development (Brunet and Maro 2007; Levi et al. 2013). Thus, force-generating mechanisms that center the nucleus may promote oocyte competence, in part by homogenizing the cytoplasm and ensuring uniform distribution of key factors (Almonacid et al. 2015, 2018).

10.4.2 Stabilization of the Oocyte NE to Resist External Mechanical Stress and Maintain the Ovarian Reserve

For oocytes to mature, their nuclei must be mechanically stable during the growth and long arrest phases before meiotic resumption. Unlike the nuclei of somatic cells, the nucleus of oocytes can sensitively respond to mechanical forces acting on the cell surface. When a whole oocyte is mechanically deformed by external indentation, the nucleus shows little resistance and will exhibit a similar deformation—a phenomenon named "nuclear shape mimicry" (Vos et al. 2026). Growing evidence suggests that compromised NE stability of oocytes correlates with increased death of maturing oocytes (Link et al. 2013; Hakim et al. 2025; Tsatskis et al. 2020; Liu et al. 2023), and multiple INM proteins likely confer mechanical stability to the maturing oocyte nucleus.

The nuclear lamina is a major determinant of NE stiffness in many cell types. In mammals, the ratio of A/C- and B-type lamins often scales with tissue stiffness (Swift et al. 2013), and numerous LMNA mutations cause laminopathies with systemic or reproductive phenotypes: Women with LMNA mutations show increased infertility risk (Worman 2012; Vantyghem et al. 2008), implicating lamin function in gamete development. However, the presence of multiple lamin isoforms in mammalian germ cells and the fact that lamins are widely expressed in both somatic and germ cells have made it difficult to examine germline-specific functions (Link et al. 2013).

C. elegans, with a single lamin gene, *lmn-1*, offers a simpler system (Liu et al. 2000; Green et al. 2011; Huelgas-Morales et al. 2020; Haithcock et al. 2005; Gregory et al. 2023; Link et al. 2018). Using germline-specific auxin-inducible degradation (Zhang et al. 2015), LMN-1 can be depleted in germ cells without affecting somatic tissues (Liu et al. 2023). Upon LMN-1 depletion, nuclei of maturing oocytes collapse in late prophase I (diplotene stage). Although collapsed nuclei display hypercondensed chromatin reminiscent of apoptosis, nuclear collapse does not require CED-4-dependent apoptosis, and live imaging shows that collapse completes within minutes—far faster than germline apoptotic execution (Gartner et al. 2000). This indicates that nuclear collapse reflects acute mechanical failure of the NE rather than programmed cell death.

Live imaging also reveals that LMN-1-depleted oocyte nuclei exhibit polarized accumulation of the LINC complex at the NE prior to collapse, a phenotype suppressed by co-depletion of dynein components. Co-depleting dynein or LINC complexes rescues diplotene collapse, whereas co-depleting additional INM proteins such as EMR-1/Emerin and LEM-2, or SAMP-1, together with LMN-1, exacerbates the phenotype, leading to earlier LINC polarization and collapse soon after meiotic entry. Importantly, these exacerbated collapses are also rescued when SUN-1 is co-depleted, demonstrating that EMR-1/LEM-2 and SAMP-1 provide LMN-1-independent resistance to LINC-mediated forces. Measurements of nuclear mechanics confirm that LMN-1-depleted oocyte nuclei are softer. Together, these findings support a model in which force transmitted via LINC complexes can compromise NE integrity when the lamina and INM scaffolds are weakened, and in which a cooperative network of INM proteins stabilizes meiotic nuclei by distributing LINC-mediated forces (Liu et al. 2023).

In addition to lamin, NEMP1 provides another conserved NE component implicated in fertility. Human genome-wide association studies identified multiple single-nucleotide polymorphisms near NEMP1 associated with age at menopause, including a rare nonsynonymous coding variant linked to earlier menopause by almost 4 years (Tsatskis et al. 2020). Nemp1 encodes a multi-pass INM protein required for fertility in flies, worms, zebrafish, and mice. Loss of Nemp1 leads to abnormal nuclear morphology, reduced nuclear stiffness in oocytes, and reduced oocyte numbers (Hakim et al. 2025; Tsatskis et al. 2020). Nemp1 associates with Emerin in cell lysates, and its loss in female mice results in defects in DNA repair, synapsis, and recombination, leading to CHK2/ATM-dependent apoptosis and premature oocyte loss. Pharmacological or genetic inhibition of CHK2/ATM partially rescues DNA repair defects, apoptosis, and oocyte numbers (Tsatskis et al. 2020; Hakim et al. 2025). These data indicate that Nemp1 is important for maintaining the ovarian reserve, potentially by stabilizing the NE and/or promoting efficient DNA repair.

Similarly, LMN-1 depletion in *C. elegans* triggers SPO-11-independent DNA damage and apoptosis in late pachytene, concomitant with diplotene nuclear collapse (Liu et al. 2023). How softer NEs in the absence of Nemp1 or lamin translate into increased DNA damage remains unresolved but may involve altered force transmission, impaired chromatin organization, or defective repair factor recruitment.

Collectively, recent work on lamin and Nemp1 supports a picture in which INM proteins act together to stabilize the oocyte NE, protecting germ cell nuclei from mechanical failure and contributing to a viable ovarian reserve. Further work is needed to disentangle the relationship between meiotic defects-triggered apoptosis and global oocyte nuclear collapse following NE destabilization.

10.4.3 Force-Dependent Nuclear Rotation and NE Fluctuations During Prophase I Arrest

In mammals, the entire oocyte reserve is established before birth (Charalambous et al. 2023; Zhang 2018). Primary oocytes are arrested in the dictyate (diplotene) stage of prophase I following the completion of recombination/crossover, within primordial follicles in the ovarian cortex (Zhang 2018; Pei et al. 2023; Tsatskis et al. 2020). In humans, this arrest can last up to several decades, during which a small subset of oocytes periodically resumes meiosis to produce mature eggs (Charalambous et al. 2023; Pei et al. 2023). The ability of primary oocytes to remain dormant yet competent for resumption is central to sustained fertility, and emerging evidence suggests that nuclear mechanics plays an important role during this extended arrest (Biswas et al. 2022).

In mice, compressive forces exerted by granulosa cells and the cortical extracellular matrix (ECM) are critical for maintaining oocyte dormancy. Using cultured neonatal ovaries containing primordial follicles, Nagamatsu and colleagues showed that enzymatic digestion of ECM relaxes mechanical stress, leading to FOXO3 export from the nucleus and oocyte activation (Nagamatsu et al. 2019). Reapplying exogenous pressure restores nuclear FOXO3 and dormancy. Live imaging revealed that this compressive stress is also required for nuclear rotation in dormant oocytes; nuclear rotation depends on dynein, and dynein inhibition halts rotation and promotes FOXO3 nuclear export (Nagamatsu et al. 2019). These findings suggest that ECM-mediated compression and cytoskeleton-driven nuclear rotation help maintain dormancy and reproductive longevity.

This raises a broader "Goldilocks" problem for ovarian mechanics. On one hand, abnormally soft cortical environments could potentially trigger premature exit from dormancy and depletion of the primordial follicle pool (Nagamatsu 2021). On the other hand, aging is associated with ovarian fibrosis and global stiffening (Balough et al. 2024; Amargant et al. 2020); thus, it is possible that age-dependent changes to overall ovarian tissue stiffness could cause defective oocyte growth, inhibiting activation of primordial follicles in a stiffer ovarian cortex, or both, jeopardizing fertility (Anderson et al. 2025; Wang and Yang 2025; Biswas et al. 2022; Vasse et al. 2025). Experimental softening of fibrotic ovaries through collagen removal or antifibrotic drugs partially restores tissue mechanics and ovulatory function in aged mice, extending reproductive lifespan (Umehara et al. 2022). Together, these studies suggest that primordial follicles are maintained within a physiological range of ECM stiffness and compressive stress that supports FOXO3-dependent dormancy, whereas age-associated fibrosis shifts the ovary outside this range, compromising follicle activation and ovulation (Nagamatsu 2021; Vasse et al. 2025).

From the NE perspective, several questions arise. For example, is nuclear rotation itself required for dormancy, or is it a readout of appropriate mechanical coupling to the ECM and cytoskeleton? Furthermore, how is nuclear rotation linked to nuclear-cytoplasmic transport of FOXO3? Notably, nuclear rotation has been observed in various cell types and often requires LINC complexes (Christophorou et al. 2015; Lee et al. 2015; Wu et al. 2018). Given that lamins help distribute LINC-generated forces and prevent nuclear collapse in *C. elegans* oocytes (Liu et al. 2023), lamins may also be important for maintaining dormancy in mammalian oocytes. Interestingly, primary oocytes in mice show relatively low levels of lamin A/C but high levels of Nemp1 (Tsatskis et al. 2020), suggesting that Nemp1 may help anchor LINC complexes or otherwise contribute to nuclear mechanics in the dormant state.

As follicles grow and exit the cortex, oocytes remain in prophase I arrest until germinal vesicle (GV) breakdown, when the NE disassembles to allow chromosome segregation (Pei et al. 2023; Zhang 2018). During this phase, fully grown GV-stage mouse oocytes exhibit NE fluctuations driven by Fmn2-dependent cytoplasmic F-actin (Almonacid et al. 2019). In the absence of Fmn2 or actin polymerization, both NE fluctuations and chromatin motion are reduced. Importantly, nuclear viscosity appears unchanged, implying that changes in chromatin motion reflect altered external forces transmitted through the NE.

These NE fluctuations have functional consequences. Contrary to earlier views of global transcriptional shutdown in fully grown oocytes (Bouniol-Baly et al. 1999; De La Fuente and Eppig 2001), Almonacid and colleagues showed that a subset of loci remains transcriptionally active prior to meiotic resumption. In *Fmn2* mutants, gene expression profiles are altered, with many genes downregulated across multiple chromosomes. This suggests that cytoplasmic forces transmitted through the NE, by modulating chromatin motion, fine-tune transcriptional programs that build competence during prolonged arrest. Disruption of these mechanical inputs may contribute to poor developmental outcomes (Almonacid et al. 2019).

Taken together, these studies suggest that mechanical forces acting on the oocyte nucleus—from ECM-mediated compression and nuclear rotation in primordial follicles to actin-driven NE fluctuations in fully grown oocytes—are integral to maintaining dormancy, building competence, and supporting reproductive success. How these forces are integrated with NE composition, chromatin organization, and signaling pathways across the oocyte's lifetime remain a rich area for future work.

10.5 Conclusions

In summary, emerging work across sexually reproducing eukaryotes reveals that the meiotic nucleus is far from a passive genome container: It functions as an active hub for force transmission, mechanosensation, and cellular decision-making throughout gamete development. When considered across the three areas highlighted in this chapter—early chromosome dynamics, gamete quality control, and gamete maturation—a coherent framework emerges: Mechanical inputs received at the NE are integrated through signaling networks and ultimately translated into changes in chromosome organization, nuclear integrity, or cell fate (Fig. 10.1).

First, studies in *C. elegans* demonstrate how *cis*-acting chromosome regions and their cognate zinc finger proteins engage LINC complexes and cytoskeletal motors to generate large-scale chromosome motions that promote homolog pairing and regulate synapsis. This framework provides a conceptual template for understanding telomere-based bouquets and NE-associated chromosome dynamics in yeast, flies, plants, and mammals, highlighting both conserved LINC-mediated force transmission and the diversity of attachment sites and cytoskeletal elements across species. Second, work in *C. elegans* also defines an NE-based synapsis checkpoint in which Polo-like kinases, LINC complex clusters, and mechanosensitive ion channels implement a "stress test" by mechanically destabilizing the NE to detect synapsis defects and trigger apoptosis. Whether analogous NE-based quality control mechanisms operate in mammals and how they intersect with recombination and DNA-damage surveillance remain important open questions. Third, recent studies of oocyte maturation reveal that forces acting on the nucleus during growth and prolonged prophase I arrest—from ovarian cortex compression and dynein-driven

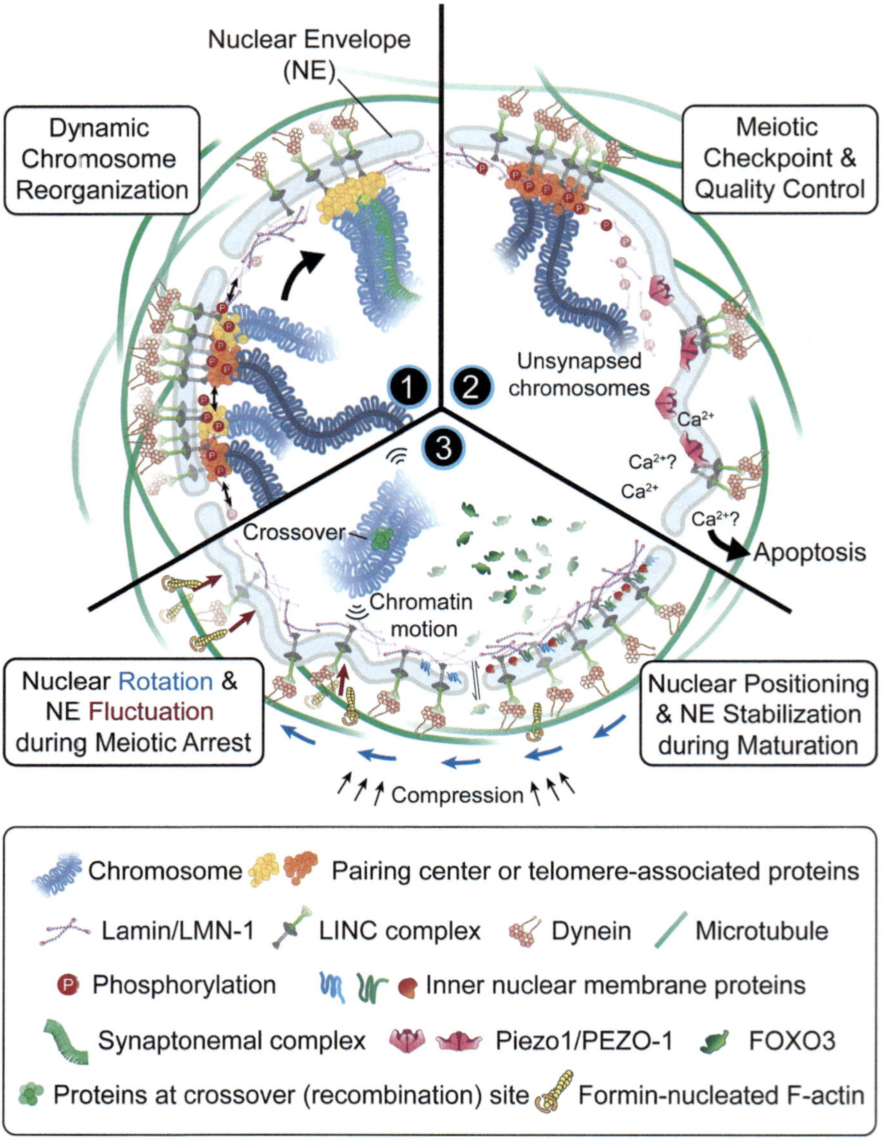

Fig. 10.1 Mechanotransduction and mechanosensation at the germ cell nuclear envelope during gamete development. The three distinct yet related processes, over space and time during gametogenesis, discussed in this chapter, are illustrated in ①, ②, and ③. We discuss how mechanical forces regulate chromosome dynamics, including homolog pairing and synapsis in early prophase (①), how mechanical forces act on the cell nucleus to regulate gamete quality control (②), and how they regulate gamete maturation (③). All processes discussed here are during meiotic prophase I. Piezo1/PEZO-1 is depicted as localizing to the INM for illustrative purposes only, as its exact localization and function remain unclear. Several modes of movement are highlighted, including pairing center (or telomere, depending on the organism)-driven chromosome movement along the NE (double-headed arrows in ①), dynein-dependent nuclear rotation (blue arrows in ③), and F-actin-dependent NE fluctuation (dark red arrows in ③) and related chromatin motion (triple arch in ③). The direction of nuclear rotation is arbitrary and for illustrative purposes only. The change of nuclear volume typically associated with meiotic prophase is not shown

rotation to F-actin-dependent NE fluctuations—shape dormancy, competence, and the ovarian reserve. Together with the roles of lamins and Nemp1 in preserving NE integrity, these findings support a "Goldilocks" model in which an optimal range of ECM stiffness and NE mechanics protects germ cell nuclei from collapse, DNA damage, and premature loss, whereas both insufficient and excessive force exposure can jeopardize fertility.

Looking ahead, several directions seem particularly promising. First, quantitative biophysical measurements of nuclear and NE mechanics in intact germlines—combined with targeted perturbations of specific NE proteins and cytoskeletal elements—will be crucial for linking force and deformation to specific molecular perturbation in a causal manner. Second, directly probing how germ cells respond to mechanical forces applied to the NE will be essential for understanding how varying magnitudes and modes of force—for example, pushing versus pulling—are interpreted to support accurate yet robust gamete production. Third, cross-species comparisons, informed by mechanistic studies in *C. elegans*, will be essential for distinguishing deeply conserved pathways from lineage-specific adaptations in gametogenesis.

The meiotic nucleus functions as a mechanically active organelle whose ability to transmit, interpret, and withstand forces is essential for chromosome reorganization, gamete quality control, and maturation. Work in *C. elegans* and other model organisms has continued to provide a mechanistic foundation for understanding NE-based force transmission, and the recent identification of SUN and KASH mutations in human infertility patients underscores the broader relevance of this machinery to reproductive health (Meng et al. 2023; Xiang et al. 2022; Bentebbal et al. 2021; Yang et al. 2022; Wu et al. 2022; Zhang et al. 2022; Hou et al. 2023). As tools for molecular perturbation, mechanical manipulation, and quantitative measurement continue to evolve, it may become possible to test whether targeted modulation of nuclear mechanics can improve oocyte quality, extend reproductive longevity, or mitigate infertility linked to NE defects.

Acknowledgments We thank members of the Liu Lab for critically reading the manuscript.

Author Contributions Writing, reviewing, editing, R.N. and C.L.; supervision, C.L. All authors have read and agreed to the published version of the manuscript.

Conflicts of Interest The authors declare no conflict of interest.

Funding Work in our lab during the preparation of this manuscript has been supported by the National Institutes of Health (1R35GM160292) and the Ralph E. Powe Junior Faculty Enhancement Award from the Oak Ridge Associated Universities to C.L.

References

Adams IR, Davies OR (2023) Meiotic chromosome structure, the synaptonemal complex, and infertility. Annu Rev Genomics Hum Genet 24:35–61. https://doi.org/10.1146/annurev-genom-110122-090239

Albert Hubbard EJ (2007) Caenorhabditis elegans germ line: a model for stem cell biology. Dev Dyn 236(12):3343–3357. https://doi.org/10.1002/dvdy.21335

Almonacid M, Ahmed WW, Bussonnier M, Mailly P, Betz T, Voituriez R, Gov NS, Verlhac M-H (2015) Active diffusion positions the nucleus in mouse oocytes. Nat Cell Biol 17(4):470–479. https://doi.org/10.1038/ncb3131

Almonacid M, Terret M-E, Verlhac M-H (2018) Control of nucleus positioning in mouse oocytes. Semin Cell Dev Biol 82:34–40. https://doi.org/10.1016/j.semcdb.2017.08.010

Almonacid M, Al Jord A, El-Hayek S, Othmani A, Coulpier F, Lemoine S, Miyamoto K, Grosse R, Klein C, Piolot T, Mailly P, Voituriez R, Genovesio A, Verlhac M-H (2019) Active fluctuations of the nuclear envelope shape the transcriptional dynamics in oocytes. Dev Cell 51(2):145–157.e110. https://doi.org/10.1016/j.devcel.2019.09.010

Amargant F, Manuel SL, Tu Q, Parkes WS, Rivas F, Zhou LT, Rowley JE, Villanueva CE, Hornick JE, Shekhawat GS, Wei J-J, Pavone ME, Hall AR, Pritchard MT, Duncan FE (2020) Ovarian stiffness increases with age in the mammalian ovary and depends on collagen and hyaluronan matrices. Aging Cell 19(11):e13259. https://doi.org/10.1111/acel.13259

Anderson RA, Marston AL, Telfer EE (2025) Oocyte development: it's all about quality. Reprod Biomed Online 50(4). https://doi.org/10.1016/j.rbmo.2025.104804

Arur S, Ohmachi M, Nayak S, Hayes M, Miranda A, Hay A, Golden A, Schedl T (2009) Multiple ERK substrates execute single biological processes in Caenorhabditis elegans germ-line development. Proc Natl Acad Sci 106(12):4776–4781. https://doi.org/10.1073/pnas.0812285106

Arur S, Ohmachi M, Berkseth M, Nayak S, Hansen D, Zarkower D, Schedl T (2011) MPK-1 ERK controls membrane organization in C. elegans oogenesis via a sex-determination module. Dev Cell 20(5):677–688. https://doi.org/10.1016/j.devcel.2011.04.009

Ashley T, Walpita D, Rooij DG (2001) Localization of two mammalian cyclin dependent kinases during mammalian meiosis. J Cell Sci 114(4):685–693. https://doi.org/10.1242/jcs.114.4.685

Bai X, Bouffard J, Lord A, Brugman K, Sternberg PW, Cram EJ, Golden A (2020) Caenorhabditis elegans PIEZO channel coordinates multiple reproductive tissues to govern ovulation. elife 9:e53603. https://doi.org/10.7554/eLife.53603

Balough JL, Dipali SS, Velez K, Kumar TR, Duncan FE (2024) Hallmarks of female reproductive aging in physiologic aging mice. Nat Aging 4(12):1711–1730. https://doi.org/10.1038/s43587-024-00769-y

Baudat F, Manova K, Yuen JP, Jasin M, Keeney S (2000) Chromosome synapsis defects and sexually dimorphic meiotic progression in mice lacking Spo11. Mol Cell 6(5):989–998. https://doi.org/10.1016/S1097-2765(00)00098-8

Baudrimont A, Penkner A, Woglar A, Machacek T, Wegrostek C, Gloggnitzer J, Fridkin A, Klein F, Gruenbaum Y, Pasierbek P, Jantsch V (2010) Leptotene/zygotene chromosome movement via the SUN/KASH protein bridge in Caenorhabditis elegans. PLoS Genet 6(11):e1001219. https://doi.org/10.1371/journal.pgen.1001219

Bentebbal SA, Meqbel BR, Salter A, Allan V, Burke B, Horn HF (2021) A human infertility-associated KASH5 variant promotes mitochondrial localization. Sci Rep 11(1):10133. https://doi.org/10.1038/s41598-021-89439-2

Bhalla N, Dernburg AF (2005) A conserved checkpoint monitors meiotic chromosome synapsis in Caenorhabditis elegans. Science 310(5754):1683–1686

Bhalla N, Dernburg AF (2008) Prelude to a division. Annu Rev Cell Dev Biol 24(1):397–424. https://doi.org/10.1146/annurev.cellbio.23.090506.123245

Biswas A, Ng BH, Prabhakaran VSO, Chan CJ (2022) Squeezing the eggs to grow: the mechanobiology of mammalian folliculogenesis. Front Cell Dev Biol 10:1038107

Bohr T, Nelson CR, Klee E, Bhalla N (2015) Spindle assembly checkpoint proteins regulate and monitor meiotic synapsis in C. elegans. J Cell Biol 211(2):233–242. https://doi.org/10.1083/jcb.201409035

Bohr T, Ashley G, Eggleston E, Firestone K, Bhalla N (2016) Synaptonemal complex components are required for meiotic checkpoint function in Caenorhabditis elegans. Genetics 204(3):987–997. https://doi.org/10.1534/genetics.116.191494

Bouniol-Baly C, Hamraoui L, Guibert J, Beaujean N, Szöllösi MS, Debey P (1999) Differential transcriptional activity associated with chromatin configuration in fully grown mouse germinal vesicle oocytes. Biol Reprod 60(3):580–587. https://doi.org/10.1095/biolreprod60.3.580

Brandt JN, Hussey KA, Kim Y (2020) Spatial and temporal control of targeting Polo-like kinase during meiotic prophase. J Cell Biol 219(11). https://doi.org/10.1083/jcb.202006094

Brenner S (1974) The genetics of Caenorhabditis elegans. Genetics 77(1):71–94

Brunet S, Maro B (2007) Germinal vesicle position and meiotic maturation in mouse oocyte. Reproduction 133(6):1069–1072. https://doi.org/10.1530/REP-07-0036

Burke B (2018) LINC complexes as regulators of meiosis. Curr Opin Cell Biol 52:22–29. https://doi.org/10.1016/j.ceb.2018.01.005

Cai B, Tiscareno-Andrade M, Luo Y, Lefranc S, Cao F, Chambon A, Yuan X, Peuch M, Zhang Y, Hurel A, Guérin J, Vrielynck N, Mézard C, Andrey P, Cromer L, Yang C, Grelon M (2025) Identification of the cytoplasmic motor–LINC complex involved in rapid chromosome movements during meiotic prophase in Arabidopsis thaliana. Nat Plants 11(8):1608–1627. https://doi.org/10.1038/s41477-025-02043-4

Castellano-Pozo M, Pacheco S, Sioutas G, Jaso-Tamame AL, Dore MH, Karimi MM, Martinez-Perez E (2020) Surveillance of cohesin-supported chromosome structure controls meiotic progression. Nat Commun 11(1):4345. https://doi.org/10.1038/s41467-020-18219-9

Chang W, Worman HJ, Gundersen GG (2015) Accessorizing and anchoring the LINC complex for multifunctionality. J Cell Biol 208(1):11–22. https://doi.org/10.1083/jcb.201409047

Charalambous C, Webster A, Schuh M (2023) Aneuploidy in mammalian oocytes and the impact of maternal ageing. Nat Rev Mol Cell Biol 24(1):27–44. https://doi.org/10.1038/s41580-022-00517-3

Chartier NT, Mukherjee A, Pfanzelter J, Fürthauer S, Larson BT, Fritsch AW, Amini R, Kreysing M, Jülicher F, Grill SW (2021) A hydraulic instability drives the cell death decision in the nematode germline. Nat Phys 17(8):920–925. https://doi.org/10.1038/s41567-021-01235-x

Chi Y, Carter JH, Swanger J, Mazin AV, Moritz RL, Clurman BE (2020) A novel landscape of nuclear human CDK2 substrates revealed by in situ phosphorylation. Sci Adv 6(16):eaaz9899. https://doi.org/10.1126/sciadv.aaz9899

Chikashige Y, Ding D-Q, Funabiki H, Haraguchi T, Mashiko S, Yanagida M, Hiraoka Y (1994) Telomere-led premeiotic chromosome movement in fission yeast. Science 264(5156):270–273. https://doi.org/10.1126/science.8146661

Chikashige Y, Tsutsumi C, Yamane M, Okamasa K, Haraguchi T, Hiraoka Y (2006) Meiotic proteins Bqt1 and Bqt2 tether telomeres to form the bouquet arrangement of chromosomes. Cell 125(1):59–69. https://doi.org/10.1016/j.cell.2006.01.048

Christophorou N, Rubin T, Bonnet I, Piolot T, Arnaud M, Huynh J-R (2015) Microtubule-driven nuclear rotations promote meiotic chromosome dynamics. Nat Cell Biol 17(11):1388–1400. https://doi.org/10.1038/ncb3249

Conrad MN, Lee C-Y, Wilkerson JL, Dresser ME (2007) MPS3 mediates meiotic bouquet formation in Saccharomyces cerevisiae. Proc Natl Acad Sci 104(21):8863–8868. https://doi.org/10.1073/pnas.0606165104

Conrad MN, Lee C-Y, Chao G, Shinohara M, Kosaka H, Shinohara A, Conchello JA, Dresser ME (2008) Rapid telomere movement in meiotic prophase is promoted by *NDJ1*, *MPS3* and *CSM4* and is modulated by recombination. Cell 133(7):1175–1187. https://doi.org/10.1016/j.cell.2008.04.047

Cromer L, Tiscareno-Andrade M, Lefranc S, Chambon A, Hurel A, Brogniez M, Guérin J, Le Masson I, Adam G, Charif D, Andrey P, Grelon M (2024) Rapid meiotic prophase chromosome movements in Arabidopsis thaliana are linked to essential reorganization at the nuclear envelope. Nat Commun 15(1):5964. https://doi.org/10.1038/s41467-024-50169-4

da Cruz I, Brochier-Armanet C, Benavente R (2020) The TERB1-TERB2-MAJIN complex of mouse meiotic telomeres dates back to the common ancestor of metazoans. BMC Evol Biol 20(1):55. https://doi.org/10.1186/s12862-020-01612-9

Daryabeigi A, Woglar A, Baudrimont A, Silva N, Paouneskou D, Vesely C, Rauter M, Penkner A, Jantsch M, Jantsch V (2016) Nuclear envelope retention of LINC complexes Is promoted by SUN-1 oligomerization in the *Caenorhabditis elegans* germ line. Genetics 203(2):733

De La Fuente R, Eppig JJ (2001) Transcriptional activity of the mouse oocyte genome: companion granulosa cells modulate transcription and chromatin remodeling. Dev Biol 229(1):224–236. https://doi.org/10.1006/dbio.2000.9947

Denais CM, Gilbert RM, Isermann P, McGregor AL, te Lindert M, Weigelin B, Davidson PM, Friedl P, Wolf K, Lammerding J (2016) Nuclear envelope rupture and repair during cancer cell migration. Science 352(6283):353. https://doi.org/10.1126/science.aad7297

Dernburg AF, McDonald K, Moulder G, Barstead R, Dresser M, Villeneuve AM (1998) Meiotic recombination in C. elegans initiates by a conserved mechanism and Is dispensable for homologous chromosome synapsis. Cell 94(3):387–398. https://doi.org/10.1016/S0092-8674(00)81481-6

Devigne A, Bhalla N (2021) Mad1's ability to interact with Mad2 is essential to regulate and monitor meiotic synapsis in C. elegans. PLoS Genet 17(11):e1009598. https://doi.org/10.1371/journal.pgen.1009598

Di Giacomo M, Barchi M, Baudat F, Edelmann W, Keeney S, Jasin M (2005) Distinct DNA-damage-dependent and -independent responses drive the loss of oocytes in recombination-defective mouse mutants. Proc Natl Acad Sci 102(3):737–742. https://doi.org/10.1073/pnas.0406212102

Ding D-Q, Yamamoto A, Haraguchi T, Hiraoka Y (2004) Dynamics of homologous chromosome pairing during meiotic prophase in fission yeast. Dev Cell 6(3):329–341. https://doi.org/10.1016/S1534-5807(04)00059-0

Ding X, Xu R, Yu J, Xu T, Zhuang Y, Han M (2007) SUN1 Is required for telomere attachment to nuclear envelope and gametogenesis in mice. Dev Cell 12(6):863–872. https://doi.org/10.1016/j.devcel.2007.03.018

Earle AJ, Kirby TJ, Fedorchak GR, Isermann P, Patel J, Iruvanti S, Moore SA, Bonne G, Wallrath LL, Lammerding J (2020) Mutant lamins cause nuclear envelope rupture and DNA damage in skeletal muscle cells. Nat Mater 19(4):464–473. https://doi.org/10.1038/s41563-019-0563-5

Elkouby YM, Jamieson-Lucy A, Mullins MC (2016) Oocyte polarization Is coupled to the chromosomal bouquet, a conserved polarized nuclear configuration in meiosis. PLoS Biol 14(1):e1002335. https://doi.org/10.1371/journal.pbio.1002335

Fan J, Jin H, Koch BA, Yu H-G (2020) Mps2 links Csm4 and Mps3 to form a telomere-associated LINC complex in budding yeast. Life Sci Alliance 3(12):e202000824. https://doi.org/10.26508/lsa.202000824

Gartner A, Milstein S, Ahmed S, Hodgkin J, Hengartner MO (2000) A conserved checkpoint pathway mediates DNA damage-induced apoptosis and cell cycle arrest in *C. elegans*. Mol Cell 5(3):435–443. https://doi.org/10.1016/S1097-2765(00)80438-4

Green RA, Kao H-L, Audhya A, Arur S, Mayers Jonathan R, Fridolfsson HN, Schulman M, Schloissnig S, Niessen S, Laband K, Wang S, Starr DA, Hyman AA, Schedl T, Desai A, Piano F, Gunsalus Kristin C, Oegema K (2011) A high-resolution *C. elegans* essential gene network based on phenotypic profiling of a complex tissue. Cell 145(3):470–482. https://doi.org/10.1016/j.cell.2011.03.037

Gregory EF, Ragle JM, Ward JD, Starr DA (2023) Split-GFP lamin as a tool for studying C. elegans LMN-1 dynamics in vivo. MicroPubl Biol. https://doi.org/10.17912/MICROPUB.BIOLOGY.001022

Hagan I, Yanagida M (1995) The product of the spindle formation gene sad1+ associates with the fission yeast spindle pole body and is essential for viability. J Cell Biol 129(4):1033–1047. https://doi.org/10.1083/jcb.129.4.1033

Haithcock E, Dayani Y, Neufeld E, Zahand AJ, Feinstein N, Mattout A, Gruenbaum Y, Liu J (2005) Age-related changes of nuclear architecture in *Caenorhabditis elegans*. P Natl Acad Sci USA 102(46):16690. https://doi.org/10.1073/pnas.0506955102

Hakim BA-O, Tsatskis YA-O, Zhang L, Choi E, Zhang Y, Hodzic D, Wu Q, Zhang M, Pashei M, Ha K, Brill JA-O, Brieño-Enríquez MA, Jurisicova A, McNeill HA-O (2025) Nuclear Envelope

Membrane Protein 1 plays crucial and conserved roles in female meiosis. Res Sq. https://doi.org/10.21203/rs.3.rs-7159889/v1. LID – rs.3.rs-7159889 [pii] (2693–5015 (Electronic))

Harper NC, Rillo R, Jover-Gil S, Assaf ZJ, Bhalla N, Dernburg Abby F (2011) Pairing centers recruit a polo-like kinase to orchestrate meiotic chromosome dynamics in C. elegans. Dev Cell 21(5):934–947

Hassold T, Hunt P (2001) To err (meiotically) is human: the genesis of human aneuploidy. Nat Rev Genet 2(4):280–291. https://doi.org/10.1038/35066065

Haversat J, Woglar A, Klatt K, Akerib CC, Roberts V, Chen S-Y, Arur S, Villeneuve AM, Kim Y (2022) Robust designation of meiotic crossover sites by CDK-2 through phosphorylation of the MutSγ complex. Proc Natl Acad Sci U S A 119(21):e2117865119. https://doi.org/10.1073/pnas.2117865119

Heffner LJ (2004) Advanced maternal age—how old Is too old? N Engl J Med 351(19):1927–1929. https://doi.org/10.1056/NEJMp048087

Herman RK, Kari CK (1989) Recombination between small X chromosome duplications and the X chromosome in Caenorhabditis elegans. Genetics 121(4):723–737. https://doi.org/10.1093/genetics/121.4.723

Higgins JD, Sanchez-Moran E, Armstrong SJ, Jones GH, Franklin FCH (2005) The Arabidopsis synaptonemal complex protein ZYP1 is required for chromosome synapsis and normal fidelity of crossing over. Genes Dev 19(20):2488–2500

Hirsh D, Oppenheim D, Klass M (1976) Development of the reproductive system of Caenorhabditis elegans. Dev Biol 49(1):200–219. https://doi.org/10.1016/0012-1606(76)90267-0

Hodgkin J, Horvitz HR, Brenner S (1979) Nondisjunction mutants of the nematode Caenorhabditis elegans. Genetics 91(1):67–94. https://doi.org/10.1093/genetics/91.1.67

Hoffman JM, Dudeck SK, Patterson HK, Austad SN (2021) Sex, mating and repeatability of Drosophila melanogaster longevity. R Soc Open Sci 8(8):210273. https://doi.org/10.1098/rsos.210273

Holloway JK, Sun X, Yokoo R, Villeneuve AM, Cohen PE (2014) Mammalian CNTD1 is critical for meiotic crossover maturation and deselection of excess precrossover sites. J Cell Biol 205(5):633–641. https://doi.org/10.1083/jcb.201401122

Horn HF, Kim DI, Wright GD, Wong ESM, Stewart CL, Burke B, Roux KJ (2013) A mammalian KASH domain protein coupling meiotic chromosomes to the cytoskeleton. J Cell Biol 202(7):1023–1039. https://doi.org/10.1083/jcb.201304004

Hou X, Zeb A, Dil S, Zhou J, Zhang H, Shi B, Muhammad Z, Khan I, Zaman Q, Shah WA, Jiang X, Wu L, Ma H, Shi Q (2023) A homozygous KASH5 frameshift mutation causes diminished ovarian reserve, recurrent miscarriage, and non-obstructive azoospermia in humans. Front Endocrinol 14:1128362

Huan Lu LM, Zhang Y, Feng Y, Zhang J, Wang S (2022) Current animal model systems for ovarian aging research. Aging Dis 13(4):1183–1195

Huang Y, Roig I (2023) Genetic control of meiosis surveillance mechanisms in mammals. Front Cell Dev Biol 11:1127440

Huelgas-Morales G, Sanders M, Mekonnen G, Tsukamoto T, Greenstein D (2020) Decreased mechanotransduction prevents nuclear collapse in a *Caenorhabditis elegans* laminopathy. Proc Natl Acad Sci 117(49):31301. https://doi.org/10.1073/pnas.2015050117

Hunt PA, Hassold TJ (2002) Sex matters in meiosis. Science 296(5576):2181–2183. https://doi.org/10.1126/science.1071907

Jan SZ, Jongejan A, Korver CM, van Daalen SKM, van Pelt AMM, Repping S, Hamer G (2018) Distinct prophase arrest mechanisms in human male meiosis. Development 145(16):dev160614. https://doi.org/10.1242/dev.160614

Kalukula Y, Stephens AD, Lammerding J, Gabriele S (2022) Mechanics and functional consequences of nuclear deformations. Nat Rev Mol Cell Biol 23(9):583–602. https://doi.org/10.1038/s41580-022-00480-z

Khan A, Zhang C, Nguyen PG, Metts JM, Collins LC, Jain K, Mills CA, Vlach L, Li K, Brademeyer AL, Bowman BM, Major MB, Aubé J, Herring LE, Rathmell WK, Mason FM, Davis IJ, Zhang Q, Strahl BD (2025) A SETD2–CDK1–lamin axis maintains nuclear morphology and genome stability. Nat Cell Biol 27(8):1327–1341. https://doi.org/10.1038/s41556-025-01723-9

Kim Y, Kostow N, Dernburg AF (2015) The chromosome axis mediates feedback control of CHK-2 to ensure crossover formation in C. elegans. Dev Cell 35(2):247–261. https://doi.org/10.1016/j.devcel.2015.09.021

Kim HJ, Liu C, Dernburg AF (2022) How and why chromosomes interact with the cytoskeleton during meiosis. Genes 13(5). https://doi.org/10.3390/genes13050901

Kim HJ, Liu C, Zhang L, Dernburg AF (2023) MJL-1 is a nuclear envelope protein required for homologous chromosome pairing and regulation of synapsis during meiosis in C. elegans. Sci Adv 9(6):eadd1453. https://doi.org/10.1126/sciadv.add1453

Kimble J, Crittenden SL (2005) Germline proliferation and its control. In: WormBook: the online review of C. elegans biology [Internet]. WormBook, Pasadeena. 2005-2018. Available from: https://www.ncbi.nlm.nih.gov/books/NBK19769/

Koszul R, Kim KP, Prentiss M, Kleckner N, Kameoka S (2008) Meiotic chromosomes move by linkage to dynamic actin cables with transduction of force through the nuclear envelope. Cell 133(7):1188–1201. https://doi.org/10.1016/j.cell.2008.04.050

Kovacs MT, Vallette M, Wiertsema P, Dingli F, Loew D, Nader GPF, Piel M, Ceccaldi R (2023) DNA damage induces nuclear envelope rupture through ATR-mediated phosphorylation of lamin A/C. Mol Cell 83(20):3659–3668.e3610. https://doi.org/10.1016/j.molcel.2023.09.023

Labella S, Woglar A, Jantsch V, Zetka M (2011) Polo kinases establish links between meiotic chromosomes and cytoskeletal forces essential for homolog pairing. Dev Cell 21(5):948–958

Lambing C, Kuo P, Kim J, Osman K, Whitbread AL, Yang J, Choi K, Franklin FCH, Henderson IR (2022) Differentiated function and localisation of SPO11-1 and PRD3 on the chromosome axis during meiotic DSB formation in Arabidopsis thaliana. PLoS Genet 18(7):e1010298. https://doi.org/10.1371/journal.pgen.1010298

Lee M-H, Ohmachi M, Arur S, Nayak S, Francis R, Church D, Lambie E, Schedl T (2007) Multiple functions and dynamic activation of MPK-1 extracellular signal-regulated kinase signaling in Caenorhabditis elegans germline development. Genetics 177(4):2039–2062. https://doi.org/10.1534/genetics.107.081356

Lee C-Y, Horn Henning F, Stewart Colin L, Burke B, Bolcun-Filas E, Schimenti John C, Dresser Michael E, Pezza Roberto J (2015) Mechanism and regulation of rapid telomere prophase movements in mouse meiotic chromosomes. Cell Rep 11(4):551–563. https://doi.org/10.1016/j.celrep.2015.03.045

Lee C-Y, Bisig CG, Conrad MM, Ditamo Y, Previato de Almeida L, Dresser ME, Pezza RJ (2020) Extranuclear structural components that mediate dynamic chromosome movements in yeast meiosis. Curr Biol 30(7):1207–1216.e1204. https://doi.org/10.1016/j.cub.2020.01.054

Lettre G, Hengartner MO (2006) Developmental apoptosis in C. elegans: a complex CEDnario. Nat Rev Mol Cell Biol 7(2):97–108. https://doi.org/10.1038/nrm1836

Levi M, Ghetler Y, Shulman A, Shalgi R (2013) Morphological and molecular markers are correlated with maturation-competence of human oocytes. Hum Reprod 28(9):2482–2489. https://doi.org/10.1093/humrep/det261

Link J, Jantsch V (2019) Meiotic chromosomes in motion: a perspective from Mus musculus and Caenorhabditis elegans. Chromosoma 128(3):317–330. https://doi.org/10.1007/s00412-019-00698-5

Link J, Jahn D, Schmitt J, Göb E, Baar J, Ortega S, Benavente R, Alsheimer M (2013) The meiotic nuclear lamina regulates chromosome dynamics and promotes efficient homologous recombination in the mouse. PLoS Genet 9(1):e1003261. https://doi.org/10.1371/journal.pgen.1003261

Link J, Paouneskou D, Velkova M, Daryabeigi A, Laos T, Labella S, Barroso C, Pacheco Piñol S, Montoya A, Kramer H, Woglar A, Baudrimont A, Markert SM, Stigloher C, Martinez-Perez E, Dammermann A, Alsheimer M, Zetka M, Jantsch V (2018) Transient and partial nuclear lamina disruption promotes chromosome movement in early meiotic prophase. Dev Cell 45(2):212–225.e217. https://doi.org/10.1016/j.devcel.2018.03.018

Liu C, Dernburg AF (2024) Chemically induced proximity reveals a Piezo-dependent meiotic checkpoint at the oocyte nuclear envelope. Science 386(6724):eadm7969. https://doi.org/10.1126/science.adm7969

Liu J, Ben-Shahar TR, Riemer D, Treinin M, Spann P, Weber K, Fire A, Gruenbaum Y (2000) Essential roles for Caenorhabditis elegans lamin gene in nuclear organization, cell cycle progression, and spatial organization of nuclear pore complexes. Mol Biol Cell 11(11):3937–3947. https://doi.org/10.1091/mbc.11.11.3937

Liu C, Rex R, Lung Z, Wang JS, Wu F, Kim HJ, Zhang L, Sohn LL, Dernburg AF (2023) A cooperative network at the nuclear envelope counteracts LINC-mediated forces during oogenesis in C. elegans. Sci Adv 9(28):eabn5709. https://doi.org/10.1126/sciadv.abn5709

Liu S, Xiong F, Dou Z, Chu L, Yao Y, Wang M, Yao X, Liu X, Wang Z (2025) Phosphorylation of lamin A/C regulates the structural integrity of the nuclear envelope. J Biol Chem 301(1). https://doi.org/10.1016/j.jbc.2024.108033

Lomakin AJ, Cattin CJ, Cuvelier D, Alraies Z, Molina M, Nader GPF, Srivastava N, Sáez PJ, Garcia-Arcos JM, Zhitnyak IY, Bhargava A, Driscoll MK, Welf ES, Fiolka R, Petrie RJ, De Silva NS, González-Granado JM, Manel N, Lennon-Duménil AM, Müller DJ, Piel M (2020) The nucleus acts as a ruler tailoring cell responses to spatial constraints. Science 370(6514):eaba2894. https://doi.org/10.1126/science.aba2894

MacQueen AJ, Villeneuve AM (2001) Nuclear reorganization and homologous chromosome pairing during meiotic prophase require C. elegans chk-2. Genes Dev 15(13):1674–1687

MacQueen AJ, Colaiácovo MP, McDonald K, Villeneuve AM (2002) Synapsis-dependent and -independent mechanisms stabilize homolog pairing during meiotic prophase in C. elegans. Genes Dev 16(18):2428–2442

MacQueen AJ, Phillips CM, Bhalla N, Weiser P, Villeneuve AM, Dernburg AF (2005) Chromosome sites play dual roles to establish homologous synapsis during meiosis in C. elegans. Cell 123(6):1037–1050. https://doi.org/10.1016/j.cell.2005.09.034

Malone CJ, Misner L, Le Bot N, Tsai M-C, Campbell JM, Ahringer J, White JG (2003) The C. elegans hook protein, ZYG-12, mediates the essential attachment between the centrosome and nucleus. Cell 115(7):825–836. https://doi.org/10.1016/S0092-8674(03)00985-1

Martinez-Perez E, Villeneuve AM (2005) HTP-1-dependent constraints coordinate homolog pairing and synapsis and promote chiasma formation during C. elegans meiosis. Genes Dev 19(22):2727–2743

Maurer M, Lammerding J (2019) The driving force: nuclear mechanotransduction in cellular function, fate, and disease. Annu Rev Biomed Eng 21:443–468. https://doi.org/10.1146/annurev-bioeng-060418-052139

McKim KS, Howell AM, Rose AM (1988) The effects of translocations on recombination frequency in Caenorhabditis elegans. Genetics 120(4):987–1001. https://doi.org/10.1093/genetics/120.4.987

McKim KS, Peters K, Rose AM (1993) Two types of sites required for meiotic chromosome pairing in Caenorhabditis elegans. Genetics 134(3):749–768. https://doi.org/10.1093/genetics/134.3.749

McKim KS, Green-Marroquin BL, Sekelsky JJ, Chin G, Steinberg C, Khodosh R, Hawley RS (1998) Meiotic synapsis in the absence of recombination. Science 279(5352):876–878. https://doi.org/10.1126/science.279.5352.876

Méjat A, Misteli T (2010) LINC complexes in health and disease. Nucleus 1(1):40–52. https://doi.org/10.4161/nucl.1.1.10530

Meng Q, Shao B, Zhao D, Fu X, Wang J, Li H, Zhou Q, Gao T (2023) Loss of SUN1 function in spermatocytes disrupts the attachment of telomeres to the nuclear envelope and contributes to non-obstructive azoospermia in humans. Hum Genet 142(4):531–541. https://doi.org/10.1007/s00439-022-02515-z

Miki F, Kurabayashi A, Tange Y, Okazaki K, Shimanuki M, Niwa O (2004) Two-hybrid search for proteins that interact with Sad1 and Kms1, two membrane-bound components of the spindle pole body in fission yeast. Mol Gen Genomics 270(6):449–461. https://doi.org/10.1007/s00438-003-0938-8

Minn IL, Rolls MM, Hanna-Rose W, Malone CJ (2009) SUN-1 and ZYG-12, mediators of centrosome–nucleus attachment, are a functional SUN/KASH pair in Caenorhabditis elegans. Mol Biol Cell 20(21):4586–4595. https://doi.org/10.1091/mbc.e08-10-1034

Miroshnikova YA, Wickström SA (2022) Mechanical forces in nuclear organization. Cold Spring Harb Perspect Biol 14(1):a039685. https://doi.org/10.1101/cshperspect.a039685

Morimoto A, Shibuya H, Zhu X, Kim J, Ishiguro K-i, Han M, Watanabe Y (2012) A conserved KASH domain protein associates with telomeres, SUN1, and dynactin during mammalian meiosis. J Cell Biol 198(2):165–172. https://doi.org/10.1083/jcb.201204085

Mytlis A, Kumar V, Qiu T, Deis R, Hart N, Levy K, Masek M, Shawahny A, Ahmad A, Eitan H, Nather F, Adar-Levor S, Birnbaum RY, Elia N, Bachmann-Gagescu R, Roy S, Elkouby YM (2022) Control of meiotic chromosomal bouquet and germ cell morphogenesis by the zygotene cilium. Science 376(6599):eabh3104. https://doi.org/10.1126/science.abh3104

Nagamatsu G (2021) Regulation of primordial follicle formation, dormancy, and activation in mice. J Reprod Dev 67(3):189–195. https://doi.org/10.1262/jrd.2021-040

Nagamatsu G, Shimamoto S, Hamazaki N, Nishimura Y, Hayashi K (2019) Mechanical stress accompanied with nuclear rotation is involved in the dormant state of mouse oocytes. Sci Adv 5(6):eaav9960. https://doi.org/10.1126/sciadv.aav9960

Page SL, Hawley RS (2004) The genetics and molecular biology of the synaptonemal complex. Annu Rev Cell Dev Biol 20:525–558. https://doi.org/10.1146/annurev.cellbio.19.111301.155141. (1081-0706 (Print))

Palmer N, Talib SZA, Singh P, Goh CMF, Liu K, Schimenti JC, Kaldis P (2020) A novel function for CDK2 activity at meiotic crossover sites. PLoS Biol 18(10):e3000903. https://doi.org/10.1371/journal.pbio.3000903

Pei Z, Deng K, Xu C, Zhang S (2023) The molecular regulatory mechanisms of meiotic arrest and resumption in oocyte development and maturation. Reprod Biol Endocrinol 21(1):90. https://doi.org/10.1186/s12958-023-01143-0

Penkner A, Tang L, Novatchkova M, Ladurner M, Fridkin A, Gruenbaum Y, Schweizer D, Loidl J, Jantsch V (2007) The nuclear envelope protein Matefin/SUN-1 Is required for homologous pairing in C. elegans meiosis. Dev Cell 12(6):873–885. https://doi.org/10.1016/j.devcel.2007.05.004

Penkner AM, Fridkin A, Gloggnitzer J, Baudrimont A, Machacek T, Woglar A, Csaszar E, Pasierbek P, Ammerer G, Gruenbaum Y, Jantsch V (2009) Meiotic chromosome homology search involves modifications of the nuclear envelope protein Matefin/SUN-1. Cell 139(5):920–933. https://doi.org/10.1016/j.cell.2009.10.045

Phillips CM, Dernburg AF (2006) A family of zinc-finger proteins Is required for chromosome-specific pairing and synapsis during meiosis in C. elegans. Dev Cell 11(6):817–829. https://doi.org/10.1016/j.devcel.2006.09.020

Phillips CM, Wong C, Bhalla N, Carlton PM, Weiser P, Meneely PM, Dernburg AF (2005) HIM-8 binds to the X chromosome pairing center and mediates chromosome-specific meiotic synapsis. Cell 123(6):1051–1063. https://doi.org/10.1016/j.cell.2005.09.035

Phillips CM, McDonald KL, Dernburg AF (2009a) Cytological analysis of meiosis in Caenorhabditis elegans. In: Keeney S (ed) Meiosis: volume 2, cytological methods. Humana Press, Totowa, pp 171–195. https://doi.org/10.1007/978-1-60761-103-5_11

Phillips CM, Meng X, Zhang L, Chretien JH, Urnov FD, Dernburg AF (2009b) Identification of chromosome sequence motifs that mediate meiotic pairing and synapsis in C. elegans. Nat Cell Biol 11(8):934–942. http://www.nature.com/ncb/journal/v11/n8/suppinfo/ncb1904_S1.html

Raiders SA, Eastwood MD, Bacher M, Priess JR (2018) Binucleate germ cells in Caenorhabditis elegans are removed by physiological apoptosis. PLoS Genet 14(7):e1007417. https://doi.org/10.1371/journal.pgen.1007417

Reynolds A, Qiao H, Yang Y, Chen JK, Jackson N, Biswas K, Holloway JK, Baudat F, de Massy B, Wang J, Höög C, Cohen PE, Hunter N (2013) RNF212 is a dosage-sensitive regulator of crossing-over during mammalian meiosis. Nat Genet 45(3):269–278. https://doi.org/10.1038/ng.2541

Roelens B, Barroso C, Montoya A, Cutillas P, Zhang W, Woglar A, Girard C, Martinez-Perez E, Villeneuve AM (2019) Spatial regulation of polo-like kinase activity during Caenorhabditis elegans meiosis by the nucleoplasmic HAL-2/HAL-3 complex. Genetics 213(1):79–96. https://doi.org/10.1534/genetics.119.302479

Rog O, Dernburg AF (2013) Chromosome pairing and synapsis during Caenorhabditis elegans meiosis. Curr Opin Cell Biol 25(3):349–356. https://doi.org/10.1016/j.ceb.2013.03.003

Rog O, Dernburg AF (2015) Direct visualization reveals kinetics of meiotic chromosome synapsis. Cell Rep 10(10):1639–1645. https://doi.org/10.1016/j.celrep.2015.02.032

Rog O, Köhler S, Dernburg AF (2017) The synaptonemal complex has liquid crystalline properties and spatially regulates meiotic recombination factors. elife 6:e21455. https://doi.org/10.7554/eLife.21455

Rubin T, Macaisne N, Huynh J-R (2020) Mixing and matching chromosomes during female meiosis. Cells 9(3):696. https://doi.org/10.3390/cells9030696

Saito K, Sakai C, Kawasaki T, Sakai N (2014) Telomere distribution pattern and synapsis initiation during spermatogenesis in zebrafish. Dev Dyn 243(11):1448–1456. https://doi.org/10.1002/dvdy.24166

Sato A, Isaac B, Phillips CM, Rillo R, Carlton PM, Wynne DJ, Kasad RA, Dernburg AF (2009) Cytoskeletal forces span the nuclear envelope to coordinate meiotic chromosome pairing and synapsis. Cell 139(5):907–919. https://doi.org/10.1016/j.cell.2009.10.039

Sato-Carlton A, Nakamura-Tabuchi C, Chartrand SK, Uchino T, Carlton PM (2017) Phosphorylation of the synaptonemal complex protein SYP-1 promotes meiotic chromosome segregation. J Cell Biol 217(2):555–570. https://doi.org/10.1083/jcb.201707161

Scharf A, Pohl F, Egan BM, Kocsisova Z, Kornfeld K (2021) Reproductive aging in Caenorhabditis elegans: from molecules to ecology. Front Cell Dev Biol 9:718522

Shibuya H, Ishiguro K-i, Watanabe Y (2014) The TRF1-binding protein TERB1 promotes chromosome movement and telomere rigidity in meiosis. Nat Cell Biol 16(2):145–156. https://doi.org/10.1038/ncb2896

Shibuya H, Hernández-Hernández A, Morimoto A, Negishi L, Höög C, Watanabe Y (2015) MAJIN links telomeric DNA to the nuclear membrane by exchanging telomere cap. Cell 163(5):1252–1266. https://doi.org/10.1016/j.cell.2015.10.030

Shimanuki M, Miki F, Ding DQ, Chikashige Y, Hiraoka Y, Horio T, Niwa O (1997) A novel fission yeast gene, kms1 +, is required for the formation of meiotic prophase-specific nuclear architecture. Mol Gen Genet MGG 254(3):238–249. https://doi.org/10.1007/s004380050412

Swift J, Ivanovska IL, Buxboim A, Harada T, Dingal PCDP, Pinter J, Pajerowski JD, Spinler KR, Shin J-W, Tewari M, Rehfeldt F, Speicher DW, Discher DE (2013) Nuclear lamin-A scales with tissue stiffness and enhances matrix-directed differentiation. Science 341(6149):1240104

Tapley EC, Starr DA (2013) Connecting the nucleus to the cytoskeleton by SUN–KASH bridges across the nuclear envelope. Curr Opin Cell Biol 25(1):57–62. https://doi.org/10.1016/j.ceb.2012.10.014

Tsatskis Y, Rosenfeld R, Pearson JD, Boswell C, Qu Y, Kim K, Fabian L, Mohammad A, Wang X, Robson MI, Krchma K, Wu J, Gonçalves J, Hodzic D, Wu S, Potter D, Pelletier L, Dunham WH, Gingras A-C, Sun Y, Meng J, Godt D, Schedl T, Ciruna B, Choi K, Perry JRB, Bremner R, Schirmer EC, Brill JA, Jurisicova A, McNeill H (2020) The NEMP family supports metazoan fertility and nuclear envelope stiffness. Sci Adv 6(35):eabb4591. https://doi.org/10.1126/sciadv.abb4591

Tu Z, Bayazit MB, Liu H, Zhang J, Busayavalasa K, Risal S, Shao J, Satyanarayana A, Coppola V, Tessarollo L, Singh M, Zheng C, Han C, Chen Z, Kaldis P, Gustafsson J-Å, Liu K (2017) Speedy A–Cdk2 binding mediates initial telomere–nuclear envelope attachment during meiotic prophase I independent of Cdk2 activation. Proc Natl Acad Sci U S A 114(3):592–597. https://doi.org/10.1073/pnas.1618465114

Tzur YB, Margalit A, Melamed-Book N, Gruenbaum Y (2006) Matefin/SUN-1 is a nuclear envelope receptor for CED-4 during *Caenorhabditis elegans* apoptosis. Proc Natl Acad Sci 103(36):13397. https://doi.org/10.1073/pnas.0604224103

Umehara T, Winstanley YE, Andreas E, Morimoto A, Williams EJ, Smith KM, Carroll J, Febbraio MA, Shimada M, Russell DL, Robker RL (2022) Female reproductive life span is extended by targeted removal of fibrotic collagen from the mouse ovary. Sci Adv 8(24):eabn4564. https://doi.org/10.1126/sciadv.abn4564

Vantyghem MC, Vincent-Desplanques D, Defrance-Faivre F, Capeau J, Fermon C, Valat AS, Lascols O, Hecart AC, Pigny P, Delemer B, Vigouroux C, Wemeau JL (2008) Fertility and obstetrical complications in women with LMNA-related familial partial lipodystrophy. J Clin Endocrinol Metabol 93(6):2223–2229. https://doi.org/10.1210/jc.2007-2521

Vasse J, Fiscus J, Fraison E, Salle B, David L, Labrune E (2025) Biomechanical properties of ovarian tissue and their impact on the activation of follicular growth: a narrative review. Reprod Biomed Online 50:3. https://doi.org/10.1016/j.rbmo.2024.104450

Velez-Aguilera G, Nkombo Nkoula S, Ossareh-Nazari B, Link J, Paouneskou D, Van Hove L, Joly N, Tavernier N, Verbavatz J-M, Jantsch V, Pintard L (2020) PLK-1 promotes the merger of the parental genome into a single nucleus by triggering lamina disassembly. elife 9:e59510. https://doi.org/10.7554/eLife.59510

Venturini V, Pezzano F, Català Castro F, Häkkinen H-M, Jiménez-Delgado S, Colomer-Rosell M, Marro M, Tolosa-Ramon Q, Paz-López S, Valverde Miguel A, Weghuber J, Loza-Alvarez P, Krieg M, Wieser S, Ruprecht V (2020) The nucleus measures shape changes for cellular proprioception to control dynamic cell behavior. Science 370(6514):eaba2644. https://doi.org/10.1126/science.aba2644

Viera A, Rufas JS, Martínez I, Barbero JL, Ortega S, Suja JA (2009) CDK2 is required for proper homologous pairing, recombination and sex-body formation during male mouse meiosis. J Cell Sci 122(12):2149–2159. https://doi.org/10.1242/jcs.046706

Viera A, Alsheimer M, Gómez R, Berenguer I, Ortega S, Symonds CE, Santamaría D, Benavente R, Suja JA (2015) CDK2 regulates nuclear envelope protein dynamics and telomere attachment in mouse meiotic prophase. J Cell Sci 128(1):88–99. https://doi.org/10.1242/jcs.154922

Villeneuve AM (1994) A cis-acting locus that promotes crossing over between X chromosomes in Caenorhabditis elegans. Genetics 136(3):887–902. https://doi.org/10.1093/genetics/136.3.887

Vos BE, Vadapalli Y, Muenker T, Jentoft IMA, Todisco E, Eskandari MA, Schuh M, Lenart P, Betz T (2026) Direct mechanical communication of cellular to nuclear shapes in oocytes. Biophys J 125:1–13. https://doi.org/10.1016/j.bpj.2025.08.012

Wanat JJ, Kim KP, Koszul R, Zanders S, Weiner B, Kleckner N, Alani E (2008) Csm4, in collaboration with Ndj1, mediates telomere-led chromosome dynamics and recombination during yeast meiosis. PLoS Genet 4(9):e1000188. https://doi.org/10.1371/journal.pgen.1000188

Wang H, Yang L (2025) Ovarian mechanobiology: understanding the interplay between mechanics and follicular development. Cells 14(5):355. https://doi.org/10.3390/cells14050355

Wolke U, Jezuit EA, Priess JR (2007) Actin-dependent cytoplasmic streaming in *C. elegans* oogenesis. Development 134(12):2227. https://doi.org/10.1242/dev.004952

Worman HJ (2012) Nuclear lamins and laminopathies. J Pathol 226(2):316–325. https://doi.org/10.1002/path.2999

Wu YK, Umeshima H, Kurisu J, Kengaku M (2018) Nesprins and opposing microtubule motors generate a point force that drives directional nuclear motion in migrating neurons. Development 145(5). https://doi.org/10.1242/dev.158782

Wu H, Zhang X, Hua R, Li Y, Cheng L, Li K, Liu Y, Gao Y, Shen Q, Wang G, Lv M, Xu Y, He X, Cao Y, Liu M (2022) Homozygous missense mutation in CCDC155 disrupts the transmembrane distribution of CCDC155 and SUN1, resulting in non-obstructive azoospermia and premature ovarian insufficiency in humans. Hum Genet 141(11):1795–1809. https://doi.org/10.1007/s00439-022-02459-4

Wynne DJ, Rog O, Carlton PM, Dernburg AF (2012) Dynein-dependent processive chromosome motions promote homologous pairing in C. elegans meiosis. J Cell Biol 196(1):47–64

Xiang M, Wang Y, Wang K, Kong S, Lu M, Zhang J, Duan Z, Zha X, Shi X, Wang F, Cao Y, Zhu F (2022) Novel mutation and deletion in SUN5 cause male infertility with acephalic spermatozoa syndrome. Reprod Sci 29(2):646–651. https://doi.org/10.1007/s43032-021-00665-5

Yamamoto A, West RR, McIntosh JR, Hiraoka Y (1999) A cytoplasmic dynein heavy chain Is required for oscillatory nuclear movement of meiotic prophase and efficient meiotic recombination in fission yeast. J Cell Biol 145(6):1233–1250. https://doi.org/10.1083/jcb.145.6.1233

Yang C, Lin X, Ji Z, Huang Y, Zhang L, Luo J, Chen H, Li P, Tian R, Zhi E, Hong Y, Zhou Z, Zhang F, Li Z, Yao C (2022) Novel bi-allelic variants in KASH5 are associated with meiotic arrest and non-obstructive azoospermia. Mol Hum Reprod 28(7):gaac021. https://doi.org/10.1093/molehr/gaac021

Zhang M (2018) Oocyte meiotic arrest. In: Skinner MK (ed) Encyclopedia of reproduction, 2nd edn. Academic Press, Oxford, pp 153–158. https://doi.org/10.1016/B978-0-12-801238-3.64443-4

Zhang W, Miley N, Zastrow MS, MacQueen AJ, Sato A, Nabeshima K, Martinez-Perez E, Mlynarczyk-Evans S, Carlton PM, Villeneuve AM (2012) HAL-2 promotes homologous pairing during Caenorhabditis elegans meiosis by antagonizing inhibitory effects of Synaptonemal complex precursors. PLoS Genet 8(8):e1002880. https://doi.org/10.1371/journal.pgen.1002880

Zhang L, Ward JD, Cheng Z, Dernburg AF (2015) The auxin-inducible degradation (AID) system enables versatile conditional protein depletion in C. elegans. Development 142(24):4374–4384

Zhang L, Köhler S, Rillo-Bohn R, Dernburg AF (2018) A compartmentalized signaling network mediates crossover control in meiosis. elife 7:e30789. https://doi.org/10.7554/eLife.30789

Zhang Q, Tao C, Gao S, Li S, Xu B, Ke H, Wang Y, Zhang F, Qin Y, Zhang L, Guo T (2022) Homozygous variant in KASH5 causes premature ovarian insufficiency by disordered meiotic homologous pairing. J Clin Endocrinol Metabol 107(9):2589–2597. https://doi.org/10.1210/clinem/dgac368

Zhang L, Stauffer WT, Wang JS, Wu F, Yu Z, Liu C, Kim HJ, Dernburg AF (2023) Recruitment of polo-like kinase couples synapsis to meiotic progression via inactivation of CHK-2. elife 12:e84492. https://doi.org/10.7554/eLife.84492

Zhang L, Stauffer W, Liu C, Shao H, Abuzahriyeh N, Jiang R, Zwicker D, Liu X, Yao X, Dernburg AF (2025) Crossover patterning through condensation and coarsening of pro-crossover factors. Nat Cell Biol 27(7):1161–1174. https://doi.org/10.1038/s41556-025-01688-9

Zhou K, Rolls MM, Hall DH, Malone CJ, Hanna-Rose W (2009) A ZYG-12–dynein interaction at the nuclear envelope defines cytoskeletal architecture in the C. elegans gonad. J Cell Biol 186(2):229–241. https://doi.org/10.1083/jcb.200902101

Zhu R, Liu C, Gundersen GG (2018) Nuclear positioning in migrating fibroblasts. Semin Cell Dev Biol 82:41–50. https://doi.org/10.1016/j.semcdb.2017.11.006

Zickler D, Kleckner N (2023) Meiosis: dances between homologs. Annu Rev Genet 57:1–63. https://doi.org/10.1146/annurev-genet-061323-044915

Chapter 11
Modulation of the Mechanosensory Matrisome by Serotonergic Signaling

Iván P. Uray, Sussy Cervantes Ramirez, and Karen L. Uray

Abstract The role of, perhaps, the least understood biogenic amine, serotonin, goes beyond transmitting information within the neuroendocrine system. Serotonin has emerged as a critical factor in diverse signaling events across various organs and tissues. In the central nervous system, serotonin, also referred to by its chemical name 5-hydroxytryptamine, regulates behavior, mood, anxiety, sleep, and learning. In the periphery, serotonin signaling is a critical factor in gastrointestinal homeostasis, platelet activation, mammary gland development, lactation, and involution. Various aspects of mammary and gut physiology depend on related functions, where mechanical and biochemical signals converge. In this chapter, we explore some of the overlapping mechanosensory functions modulated by serotonergic signaling that characterize the relatively static mammary gland and the mechanical active gastrointestinal system in their physiologic and disease states.

11.1 Mechanotransduction in the Mammary Gland

11.1.1 Matrix Rigidity and Stiffness as Mechanosensory Signals in the Breast

The relationship between tissue and extracellular matrix (ECM) stiffness and cancer progression has been known for decades (Paszek and Weaver 2004). The

I. P. Uray (✉) · S. C. Ramirez
Faculty of Medicine, Department of Biochemistry and Molecular Biology, University of Debrecen, Debrecen, Hungary

Faculty of Medicine, Department of Clinical Oncology, University of Debrecen, Debrecen, Hungary
e-mail: uray.ivan@med.unideb.hu

K. L. Uray
Faculty of Medicine, Department of Medical Chemistry, University of Debrecen, Debrecen, Hungary

M. Kloc, J. Wosik (eds.), *Mechanotransduction and Mechanosensing*, Results and Problems in Cell Differentiation 78, https://doi.org/10.1007/978-3-032-19509-8_11

mammary gland is a biological system that must adapt to radical changes in tensile and compressive forces under physiologic and pathologic conditions. Malignant transformation of breast cells coincides with significant changes in mechanical forces, including elevated compression forces, high tensional stress, and increased extracellular matrix stiffness (Pathak and Kumar 2012). Although not sufficient to induce carcinogenesis alone, increased ECM stiffness is a significant enabling factor driving tumor progression by creating favorable physical and biochemical conditions for proliferation, cell mobility and invasiveness, neovessel formation, and even therapy resistance (Nia et al. 2020).

As a typical mechanosensing mechanism, excess collagen or fibronectin in the extracellular space allosterically changes integrin conformation. Integrin β subunits recruit the cytoskeletal linker proteins talin, paxillin, kindlin, and vinculin, which form a force-dependent catch bond with F-actin, leading to the assembly and reorganization of the actin cytoskeleton (del Rio et al. 2009; Huang et al. 2017). Talin and vinculin are part of a multicomponent system involved in mechanosensing in cell-matrix adhesions. Both exist in autoinhibited forms, and activation of vinculin requires binding to mechanically activated talin (Wang et al. 2021). Importantly, the interactions of talin with integrins, microtubules, and the actin cytoskeleton are heavily influenced by cyclin-dependent kinase 1, indicating a strong coordination with cell cycle events (Gough et al. 2021).

The assembly of integrin-activated focal adhesions with the actomyosin network is essential for both mechanosensing and the transmission of force to extracellular contacts. Mechanosensory signals of substrate rigidity transmitted through talin control the acetylation of the microtubular system at focal adhesions. Microtubule acetylation tunes cellular mechanosensitivity by inducing translocation of the yes-associated protein and the release of the guanine nucleotide exchange factor GET-H1 from microtubules, which activates RhoA, actomyosin contractility, and traction forces (Seetharaman et al. 2022). Changes in serotonin levels could potentially modulate this process through the activation of the HTR2A serotonin receptor, which interferes with alpha-tubulin acetylation, as shown in neurons (Ohtani et al. 2014). Similarly, the serotonylation of RhoA may alter the activation of the actomyosin network in a serotonin-dependent manner (Guilluy et al. 2009). Comparison of these mechanisms in mammary and colonic epithelial cells should provide further insights into new therapeutic options for breast and colorectal cancers.

Upon mechanical cues, focal adhesion points activate Src family kinases and focal adhesion kinase (FAK), which enhances PI3K activity (Levental et al. 2009). Increased cytoskeletal tension enables complex formation by the serum response factor (SRF) and myocardin-related transcription factor (MRTF) as a coactivator to express cytoskeletal and muscle development genes. In addition, allosteric changes in integrins activate other critical pathways by inhibiting myosin phosphatase target (MYPT) 1 through phosphorylation by the integrin-linked kinase (ILK). Dephosphorylation by myosin light chain phosphatase is essential for the activity of mammalian Ste20-like kinase 1/2 and large tumor suppressor 1/2 (LATS1/2). In the absence of mechanotransductive signals, these kinases block the nuclear translocation of yes-associated transcriptional regulator/tafazzin (YAP/TAZ) (Sabra et al. 2017). Tissue density, mechanical stretching, and the activation of select G-protein-coupled receptors reduce the proteasomal turnover of the coactivators YAP and TAZ in the Hippo pathway. In turn, YAP/TAZ coactivates the transcription factor TEAD, whose target genes include the key cell cycle regulators cyclin D1 and forkhead box M1. Protein tyrosine phosphatase 14 (PTPN14) reduces nuclear YAP/TAZ levels by binding to these coactivators, promoting LATS-mediated phosphorylation, and inducing their cytoplasmic retention and ubiquitin-mediated degradation (Blakely et al. 2024).

11.1.2 The Relevance of Mechanotransduction in the Mammary Gland

Mechanotransduction influences normal mammary gland physiology and pathology, particularly during lactation and involution, a complex process of regression and tissue remodeling after weaning (Biet et al. 2016; Carneiro et al. 2008; Keating and Spencer 2019; Matsuda et al. 2004). Mechanical forces arising from milk accumulation can trigger responses in mammary epithelial cells, regulating their function and survival. For example, mechanical stretching during milk engorgement is thought to induce a cascade of signal transduction pathways that alter cell shape, tight junction integrity, and apoptotic signaling mechanisms (Quaglino et al. 2009; Singh et al. 2008, 2015). Distension of the mammary epithelium due to milk accumulation can activate mechanotransduction pathways, potentially leading to increased tight junction permeability and subsequent loss of cell survival signals (Singh et al. 2008).

Beyond its physical properties, density, stiffness, and interstitial fluid balance, ECM composition dictates the response to mechanical cues in the mammary gland by modulating the activity of mechanosensitive ion channels, integrin receptors, and focal adhesions. Enhanced expression of mechanosensitive channels, such as Piezo 1, during lactation supports cellular responsiveness to mechanical stimuli (Paszek and Weaver 2004; Stewart et al. 2021). Furthermore, interactions with the ECM and other cells can lead to mechanoregulation of gene expression and signaling dynamics, affecting processes such as cell migration and differentiation (Afshar et al. 2023; Provenzano et al. 2009). Mechanotransduction and serotonin signaling are intrinsically linked to mechanical forces, influencing the activation of signaling pathways responsible for cellular responses that determine lactation persistence, involution, and the risk of neoplastic transformation in the mammary gland. Hence, the pharmacologic modulation of serotonin pathways has a significant impact on mammary gland homeostasis and development and marked effects on sustaining the secretory phenotype, proliferation, adhesion, and motility of breast cells (Matsuda et al. 2004; Lengyel et al. 2026). In the gut, Piezo ion channels mediate serotonin release from enterochromaffin (EC) cells, and the loss of Piezo 1 results in intestinal dyskinesia and constipation (Yan et al. 2025). Inhibition of cationic mechanosensitive channels increases tryptophan hydroxylase (TPH) 1 expression and serotonin levels and activates the p38 signaling pathway, suggesting the functional pressure-sensitive regulation of serotonin availability, linked to MAPK activation (Zhu et al. 2023).

11.1.3 Correlations Between Stromal Stiffness and Cancer Pathogenesis

The stroma in any solid organ contains cellular and noncellular elements of connective tissue, blood vessels, and inflammatory cells. In solid tumors, stroma formation is dictated by cancer cell-host interactions and provides mechanical and nutritional support for the neoplastic cells in the tumor parenchyma. The mechanical properties of the stroma largely depend on the composition of the ECM and cellular density. Stromal responses to altered mechanosensory cues can be observed in disease-free breast tissues in individuals at high risk for developing cancer. Concurrently, mammographic density is a major risk factor for the development of breast cancer, reflecting modifications in the composition of breast tissue, both in the stromal and glandular compartments (Provenzano et al. 2009; Zdanowski et al. 2023). In addition to increased and altered ECM content, this high-density phenotype is associated with reduced adipocytes, repressed fatty acid translocase CD36, increased DNA damage signaling, telomere malfunction, and activin A expression (DeFilippis et al. 2014). Activin A belongs to the transforming growth factor-β (TGFβ) superfamily and can activate canonical and noncanonical TGFβ pathways, affecting bone resorption and formation, cytoskeletal F-actin organization, MAPK activation, and cell cycle regulation. Interestingly, even fetal activin A serum concentrations are affected by the maternal use of selective serotonin reuptake inhibitors (SSRIs) (Bellissima et al. 2011). The SSRI fluoxetine increases fatty acid uptake and the expression of prostaglandin biosynthetic enzymes (Ayyash and Holloway 2022), highlighting another link between systemic serotonin homeostasis and ECM density.

ECM rigidity modulates cell growth in glioma cell lines (Ulrich et al. 2009), hepatocytes (Semler et al. 2000), fibroblasts (Wang et al. 2000), and neural stem cells (Saha et al. 2008). Of note, stiffness is not uniform but often heterogeneous across tumors. In orthotopic murine glioblastoma multiforme, tumors are stiffer than peritumoral brain tissues; tumor stiffness increases from the tumor edges to the tumor core (Sohrabi et al. 2023).

11.1.4 Integrins Are Central Players in Mechanosensory Functions

Integrins are central players in mechanosensing and the translation of external mechanical cues to biochemical signals. Coded by 27 distinct genes, 18 alpha and 9 beta subunits can form 24 distinct heterodimeric integrin receptors. All integrins undergo posttranslational modifications, including N-glycosylation, an essential modification for the intact function of these proteins (Gu and Taniguchi 2004). Ligand binding to these receptors includes various components of the extracellular matrix, primarily fibronectin, collagens, and laminins; binding of integrins to cadherins and immunoglobulin-like proteins (ICAMs) mediates connections to adjacent cells (Hynes 2002).

Integrins act as both mechanosensors and mechanotransducers to regulate cell fate. Both functions are bidirectional. Cells can adapt to an altered physical environment by changing the integrin heterodimer composition, cytoskeletal coupling, membrane properties, or the properties of the ECM with which integrins interact. Mechanical signals exerted by the extracellular matrix or other cells induce marked allosteric changes in the integrin ectodomains, which may occur on a timescale of <1 s (Takagi et al. 2002; Wang 2017). These "outside-in" signals are transduced to the cytoskeletal actin network by linker proteins, such as talin, vinculin, and kindlin (Ziegler et al. 2008). However, many heterodimers can also mediate retrograde signaling via their extracellular regions and reciprocally impact ECM organization or cell-cell interactions. Myosin II-generated cytoskeletal forces can cause the $\alpha5\beta1$ integrin to switch between relaxed and tensioned states, thereby exerting force on the ECM through fibronectin bonds (Friedland et al. 2009). Ligand-binding affinity is enhanced in response to "inside-out" signals derived from activated G-proteins or via coupling with other receptors (Takagi et al. 2002). Similarly, the adhesion of circulating metastatic cancer cells is promoted by "inside-out" signaling via the focal adhesion kinase FAK-integrin axis (Downey-Biechler et al. 2019). Stabilization of the high-affinity conformation induces integrin clustering and strengthens adhesions (Gahmberg et al. 2009) through actin cytoskeleton organization and activation of FAK and the Src family of kinases. These impulses initiate major pro-survival and mitogenic signaling events via the activation of the PI3K/Akt, Ras/Erk, and YAP/TAZ pathways, demonstrating the interconnected nature of integrins and receptor tyrosine kinases (Hynes 2009).

Integrin clustering precipitated by collagen fibers promotes the contraction and motility of cancer cells. This process is dependent on the recruitment of numerous players in the focal adhesion signaling pathway, including FAK, Src, paxillin, Rac, Rho, and Ras. In addition, the activation of PI3K and AP-1 helps promote the migration and invasion of cancer cells (Friedland et al. 2009; Rubashkin et al. 2014). In contrast, endothelial cells exposed to permanent shear stress utilize a mechanical sensor complex composed of platelet and endothelial cell adhesion molecule 1 (PECAM1), vascular endothelial cadherin, and the VEGF receptor (Lakshmikanthan et al. 2015).

11.1.5 The Role of Serotonin in Mammary Gland Homeostasis and Disease

Serotonin (also known as 5-hydroxytryptamine or 5-HT) is produced by hydroxylation and decarboxylation of tryptophan, with the rate-limiting step catalyzed by TPH-1 and TPH-2 in peripheral tissues and neurons, respectively. Approximately 90% of serotonin production occurs in the EC cells of the gut, and significant amounts are stored and transported in dense granules of circulating platelets, serving as a signal for aggregation and immune cell recruitment upon release (Schoenichen et al. 2019). Serotonin maintains the homeostasis of the mammary gland throughout all developmental stages, including lactation and cell turnover during involution after milk stasis. Yet, the impact of serotonin is best viewed in conjunction with key hormonal factors, particularly estrogen (Field et al. 2025). Acting through both autocrine and paracrine mechanisms, many of the serotonin-related responses are mediated by seven subfamilies of serotonin receptors (HTRs) (Paszek and Weaver 2004; Pathak and Kumar 2012; Nia et al. 2020; del Rio et al. 2009; Huang et al. 2017; Wang et al. 2021; Gough et al. 2021) that couple with the G_s-, G_i-, or G_q-type alpha chains of trimeric G proteins, eliciting potentially antagonistic signals (Lengyel et al. 2026). However, the amount of available extracellular serotonin is modulated by the widely expressed serotonin transporter (SERT), enabling intracellular breakdown by the monoamine oxidase enzyme or transamidation of select glutamine residues in target proteins by transglutaminases (Bader 2019).

11.1.6 Implications for Serotonin and Cancers

Tumor tissue may be differentially affected by serotonin, depending on its concentration. Low levels of serotonin may cause vasoconstriction in the tumor vasculature and reduce blood supply, leading to a growth inhibitory effect (Huhnt and Lubbe 1995). At higher concentrations, serotonin may enhance tumor progression by activating HTR2A and HTR2B receptors and promoting cell growth and survival (Balakrishna et al. 2021). Another key function of serotonin was demonstrated by the decreased circulating calcium and estrogen levels induced by fluoxetine treatment before birth (Domingues et al. 2023). Nevertheless, serotonin is often thought to have positive effects on cancer progression, partly due to the altered HTR profile expressed by cancer cells, favoring pro-tumorigenic signals. For example, gastric cancer progression is associated with overexpression of HTR2A, HTR2B, HTR3A, and HTR7 receptors, which stimulate the MAPK and PI3K/Akt pathways and promote proliferation, survival, and angiogenesis (Abedini et al. 2023). Breast cancer cells frequently upregulate HTR1A and HTR2A to support amplified tumor growth (Jayachandran et al. 2023). In addition, breast cancer cells, particularly triple-negative lines, express significantly higher levels of serotonin than non-tumor epithelial cells (Xie et al. 2021). Colorectal tumor tissue seems to rely on the HTR2B receptor to trigger Akt/mTOR signaling, activate S6 kinase B and CREB1, and induce epithelial-mesenchymal transition (Li et al. 2024). However, some conflicting results call for a more nuanced evaluation of the role of serotonin in tumor pathogenesis (Chen et al. 2024). For example, in colorectal cancer cells, serotonin promotes expansion. However, impaired serotonin synthesis leads to extensive DNA damage in enterocytes, inducing tumor-promoting inflammatory reactions, indicating a dual nature of serotonin effects on tumor progression (Kannen et al. 2020). Apart from the classical pathways regulating proliferation and viability, serotonin exposure induces characteristic changes in the production of mechanosensory proteins.

11.1.7 Involvement of Serotonin Signaling in Mechanotransduction

Notably, although changes in ECM rigidity activate YAP/TAZ and the Hippo pathway via multiple routes, including MYPT1/LATS1/2, the FAK/Rac/PAK pathway, and directly through Src, serotonin levels can also modulate YAP/TAZ activity (Shu et al. 2018). In colorectal cancer cells, SERT mediates the influx of serotonin, upregulating YAP expression via the activation of ROCK1/2 by serotonylated RhoA and inducing proliferation (Yu et al. 2023). Starlinger et al. demonstrated increased proliferation in a posthepatectomy liver regeneration model; YAP was activated via serotonin-induced tyrosine phosphorylation (Starlinger et al. 2023). These studies indicate a serotonin-dependent mechanism for activating the Hippo pathway that mimics mechanosensory events in the absence of the mechanical cue. Tumors forming a stiff stroma exhibit elevated Rho-dependent cytoskeletal tension, enhanced ERK activation, and increased ROCK-generated contractility and focal adhesions (Paszek et al. 2005). The cells in developing tumors with a stiffening microenvironment alleviate the mechanical challenge and reestablish force equilibrium in several ways. The activation of the Rho/ROCK pathway by mechanosensory signals stabilizes β-catenin and enhances collagen production, resulting in increased actomyosin-mediated cellular tension (Samuel et al. 2011). In inflammatory bowel disease, this process represents a feed-forward entry point for serotoninergic signaling, as elevated serotonin stimulates the Wnt/β-catenin signaling pathway, inducing an increase in intestinal EC cells and further promoting serotonin secretion (Yu et al. 2025). Reinforcing the relevance of the serotonin-beta-catenin axis, another study demonstrated the ability of the HTR7 receptor to induce the phosphorylation of dishevelled by CK1ε, leading to β-catenin accumulation and, ultimately, the activation of Wnt/β-catenin-dependent tumor growth and dissemination. Not surprisingly, overexpression of HTR7 in colorectal cancer tissue correlated with tumor burden, pathological grade, and mortality rate in CRC patients (Lv et al. 2025).

11.1.8 Connection Between Serotonin Homeostasis and Integrins

Genetic interactions were identified between the SERT (SLC6A4) gene and ITGB3, encoding for the β3 subunit that forms the αIIbβ3 and αvβ3 integrin receptor complexes. The loss of a single allele of ITGB3 attenuated SERT function via an αvβ3-mediated mechanism (Mazalouskas et al. 2015). Serotonin uptake was reduced because decreased integrin β3 subunit expression limited the pool of active SERT molecules, leading to a reduction in the effective SSRI dose. This interaction of protein complexes containing integrin β3 and SERT is not restricted to neural cells; integrin β3 and SERT complexes also affect platelet function, impacting serotonin blood levels (Carneiro et al. 2008; Gabriele et al. 2019). Furthermore, fibrinogen, an activator of integrin αIIbβ3, may increase SERT activity in human platelets. As an example of the reciprocal effect, SSRIs inhibit platelet activation by attenuating collagen-induced platelet aggregation, resulting in cardioprotective effects (Tseng et al. 2010).

Reactome pathway profiles in GSEA analyses of human mammary epithelial cells treated with the SSRI paroxetine revealed a marked enrichment of genes related to integrin-cell surface interactions, collagen formation, the Rho GTPase cycle, and $G_{q/12/12}$ signaling events.(Lengyel et al. 2026) Specific transcripts overrepresented included integrins β8, β6, and α2; ICAM 1 and 2; PECAM1; collagens 8, 2, 9, and 5; fibronectin; and thrombospondin. The spectrum of co-regulated genes suggests a strong influence of paroxetine on the expression of mechanosensory proteins. The impact of paroxetine on the uptake of serotonin indicates an increased exposure of serotonin receptors to the transmitter via a paracrine mechanism and reduced intracellular serotonin levels. Less intracellular serotonin is associated with decreased free radical production due to serotonin breakdown by monoamine oxidases, less DNA damage, and reduced modification of serotonylation targets (Lengyel et al. 2026; I. Uray, unpublished observation).

Selective SSRIs are the first-line treatment for major depressive disorder. Because of the alterations in serotonin levels in cancer patients, the potential impact of SSRIs on breast carcinogenesis has been debated but deemed unsubstantiated (Ashbury et al. 2012). Interestingly, despite the marked expression of the SERT in platelets, there is no strong evidence to support potential SSRI effects on platelet functions (Mokhtarian et al. 2025). However, serotonin uptake via the SERT triggered by the activation of platelet-derived growth factor receptor β and mitogen-activated protein kinases influences the proliferation and migration of pulmonary vascular smooth muscle cells (Liu et al. 2007).

11.1.9 Integrin Function Controlling Cell Proliferation, Invasion, and Tumorigenesis

Select integrins, including integrin β1, were identified as core components of tumor progression signatures associated with the induction of neoplasia. Antibody-mediated blockade of β1 integrins inhibited tumor progression in transplanted skin cancer (Reuter et al. 2009). Furthermore, in a chemical carcinogenesis model, Ferreira et al. showed that the expression of a constitutively active β1 integrin subunit induced the neoplastic conversion of papillomas to squamous cell carcinomas in the basal layer of the mouse epidermis (Ferreira et al. 2009). The heterodimers, consisting of α5 and β1 integrin subunits, may stimulate cell proliferation and enable cell migration, angiogenesis, and immune evasion. Thus, this heterodimeric integrin complex is generally considered pro-tumorigenic (Hou et al. 2020).

A more direct mechanism for sustaining proliferative signaling involves integrin crosstalk with IGF1 or EGF receptors. By forming a complex with the receptor tyrosine kinase Her-2 and activating STAT3 and c-Jun, α6β4 integrin induces tumor and metastasis formation in breast cancer (Guo et al. 2006). Moreover, through this interaction, integrins can control EGF receptor trafficking from the plasma membrane to the endosomes and control receptor activity (Caswell et al. 2009). IGF1 is a growth factor critical for cancer cell proliferation and migration. The αvβ3 and α6β4 integrins may be instrumental for recruiting IGF1 to the IGF1-IGF1R complex, significantly enhancing the phosphorylation and activation of both IGF1R and β3 integrin (Takada et al. 2017).

Several lines of evidence link the ITGB3 gene, which encodes the integrin beta3 subunit, with the serotonin system, likely via the modulation of the SERT. The ITGB3 gene polymorphism coding Leu33Pro (rs5918, Pl(A2)) produces hyperactive αvβ3 receptors that alter whole-blood serotonin levels; thus, this polymorphism was proposed as a risk modifier for autism spectrum disorder (Dohn et al. 2017). Recent studies have revealed that integrin signaling drives multiple stem cell functions, including tumor initiation, epithelial plasticity, metastatic reactivation, and resistance to oncogene- and immune-targeted therapies (Cooper and Giancotti 2019).

The integrin heterodimers most commonly associated with breast cancers include αvβ6, αvβ3, α11β1, α6β1, α5β1, α4β1, and α3β1. Integrins α2 and α9 were overrepresented in the triple-negative subgroup, and α6 and α11 conferred stem cell properties and drug resistance (Chaudhary et al. 2024; Goel et al. 2013; Luo et al. 2018; Mierke et al. 2011; Moore et al. 2014; Sharma et al. 2017; Subbaram et al. 2014; Zeltz et al. 2019).

11.1.10 Fibronectin and Collagen in the Healthy and Transformed States

Fibronectin is a dimeric glycoprotein with two subunits of approximately 220 kDa. Fibronectin contains several globular domains that unfold when exposed to physiological levels of force (Baneyx et al. 2002). Fibronectin is one of the main constituents of the complex tumor stroma microenvironment. Conformational flexibility is a key characteristic that makes fibronectin a suitable mechanoregulator of the extracellular matrix. Fibronectin matrix assembly requires mechanical stimulation, which originates from cellular traction forces that induce a conformational change and expose cryptic binding sites, allowing polymerization (Carraher and Schwarzbauer 2013). As a result, fibronectin molecules align within fibers with alternating overlapping regions of the N- and C-termini of fibronectin (Fruh et al. 2015). Assembly of the fibronectin network is a prerequisite for the deposition of collagen (Kubow et al. 2015). Conversely, collagen deposition augments fibronectin remodeling with the help of matrix metalloproteinases, stabilizing the ECM, facilitating cell migration, and promoting proliferation. A multimodular collagen binding site within the fibronectin protein mediates the synergistic relationship between collagen and fibronectin in the matrix (Hahn and Yamada 1979). The structural support for the organ and the tensile strength of the connective tissue is provided by fibrillar collagen types I, II, and III. Type IV is a non-fibrillar collagen that constitutes the basal lamina, along with laminins, entactin, and proteoglycans (Insua-Rodriguez and Oskarsson 2016).

Cells can sense their environment and respond to altered mechanical resistance due to increased cytoskeletal tension. The translocation of $\alpha5\beta1$ integrins is associated with increased cytoskeletal tension along the actin cytoskeletal bundles, which elongate fibronectin molecules and induce their polymerization. In the tumor stroma, fibronectin is frequently upregulated due to enhanced synthesis by cancer-associated fibroblasts (Kalluri and Zeisberg 2006) or tumor cells themselves. In breast tumors, cancer-associated fibroblasts produce a stiff matrix of unfolded fibronectin, which initiates the switch from a predominantly $\alpha5\beta1$ integrin heterodimer to an $\alpha v\beta3$ binding receptor, generating elevated pro-angiogenic signaling through VEGF.

In addition to expression levels, fibronectin conformation is critical in tumorigenesis. A strong connection between $\alpha5\beta1$ integrin and the ECM is dependent on the native conformation of fibronectin. Once fibronectin conformation is altered during cellular transformation, some matrix interactions become dysregulated, resulting in changes to downstream signaling pathways. Furthermore, fibronectin is subject to posttranslational modification by serotonin (Wei et al. 2012). Serotonin is covalently attached to glutamine residues on proteins in pulmonary artery smooth muscle cells, stimulating their proliferation and migration (Liu et al. 2011). The transglutaminase-mediated transamidation of serotonin modulates fibronectin function, triggering additional processes, including platelet activation and extracellular matrix accumulation.

Stiffening of the extracellular matrix induces fibronectin unfolding, which induces a more pro-angiogenic impact of integrins and enhances TGFβ activity. TGFβ is released from mesenchymal cells in response to force and alters cell morphology and ECM production (Subramanian et al. 2018). Tumor cell invasion preferentially directs cells toward a stiffer ECM—a phenomenon known as durotaxis (Lo et al. 2000). These changes are associated with αvβ3 integrin recruitment, resulting in cytoskeleton remodeling at the leading edge of cells, directing the migration of tumor cells toward the surrounding stroma (Kostic and Sheetz 2006).

Integrin-mediated binding of cells to fibronectin leads to clustering of growth factor receptors (Yamada and Weston 1974). Thus, the fibronectin matrix is not only a mechanotransducing network but also a chemical reservoir of signaling molecules for cells. ECM components such as heparin or heparan sulfate facilitate an extended conformation of fibronectin to enhance VEGF binding (Mitsi et al. 2006) in a pH-dependent manner. Low pH in the tumor stroma may contribute to altered fibronectin conformation. The impact of serotonin on mechanotransduction is, in many cases, reciprocal, as is the serotonin production of EC cells, and may be modulated by mechanosensory signals (Yu et al. 2025).

11.1.11 Tuning the Mechanosensory Matrix by Neurohormonal Regulation of Chromatin Remodeling Factors

In addition to the physical responses to external forces, mechanotransduction events elicit changes in gene expression via the transcription factors MRTF, YAP/TAZ, TEAD, SRF, AP-1, and HIF1α. Pharmacological blockade of the mechanosensitive transient receptor potential cation channel TRPM7 curbed MRTF transcriptional and tumorigenic activity by Mg^{2+}-dependent inhibition of RhoA activity and actin polymerization in hepatocellular carcinoma cells (Voringer et al. 2020). Several transcription factors are influenced by nuclear actin binding (Guettler et al. 2008; Vartiainen et al. 2007). Moreover, mechanosensory stimuli in the form of increased extracellular stiffness can lead to genome-wide alterations. Cumulative tension on the nuclear membrane and the associated increase in histone deacetylase activity transform transiently activated fibroblasts into myofibroblasts (Walker et al. 2021).

Less direct but longer-term regulation of the mechanosensory proteins and the matrisome was revealed more recently in the context of higher-order chromatin structure modulation. Silencing of the ATPase subunit of the mammalian SWI/SNF chromatin remodeling complex in nonmalignant breast epithelial cells resulted in marked changes in genes related to calcium and lipid metabolism, and a suppression of ECM components (Barutcu et al. 2016). ARID1A is a SWI/SNF component tumor suppressor frequently lost in various cancers. The cellular effects of ARID1A on proliferation and tumorigenesis is dependent on YAP/TAZ, and the ARID1A-containing SWI/SNF complex acts as a mechanoregulated inhibitor of YAP and TAZ (Chang et al. 2018; Hillmer and Link 2019).

Strikingly, serotonylation may function as an epigenetic marker regulating gene expression (Bader 2019). Transamidation of serotonin to the glutamine 5 residue on histone H3, known as H3Q5Ser, acts as a permissive posttranslational modification in conjunction with neighboring lysine 4 trimethylation (H3Kme3) events. H3Q5Ser potentiates the function of H3K4me3 by stabilizing H3K4me3, preventing dynamic turnover, or improving its recognition by downstream effectors (Zhao et al. 2021).

We previously reported the upregulation of ARID1A protein and chromatin-wide redistribution of SWI/SNF by the nonselective adrenergic inhibitor carvedilol combined with bexarotene in normal and malignant breast epithelial cells (Jdeed et al. 2022a, b). Comparisons of ChIP-Seq profiles for combination-treated MCF7 cells showed an overrepresentation of genes related to epithelial-mesenchymal transition of the cellular phenotype, an E- to N-cadherin isoform switch, and suppression of the IGF-1 signaling pathway (Jdeed et al. 2022a). In normal breast epithelial cells, the blockade of the adrenergic receptors combined with bexarotene (Carv+Bex) increased the recruitment of ARID1A and BRG1 to enhancers of the FOXQ1 and KLF4 genes and suppressed the expression of genes mediating TGFβ signaling, indicating mechanosensory reprogramming (Jdeed et al. 2022b).

A gene regulatory region whose accessibility was reduced by the SWI/SNF complex by both carvedilol and paroxetine belongs to the ZG16B (PAUF) gene, which codes for a small, secreted protein. Proliferation, motility, and clonogenic growth was increased in normal breast cells exposed to this serotonin-responsive protein, and the exposed cells exhibited a marked shift in matrix properties toward increased density and cell-matrix interactions, predominantly upregulating fibronectin, talin, vinculin, basement membrane forming collagen IV, laminins, and the integrin isoforms β4, α5, and α2 (Lengyel et al. 2025).

Of note, some of the chromatin remodeling events leading to concerted shifts in mechanosensory proteins and matrix proteins were induced by both serotonergic and adrenergic interventions, possibly due to the overlapping G-protein couplings of serotonin and adrenergic receptors. The central role of ARID1A in the maintenance of epithelial cell identity indicates a strong potential for the therapeutic targeting of epigenetic mechanisms and chromatin accessibility for the reduction of cancer risk and tumorigenic transformation in the mammary gland.

11.2 Mechanotransduction in the Gastrointestinal Tract

11.2.1 Mechanotransduction in Enterochromaffin Cells in the Gastrointestinal Tract

The gastrointestinal (GI) system is a mechanically active organ that relies on a wide variety of mechanical signals for normal function. Multiple cell types in the gut are mechanosensitive, including enterochromaffin (EC) cells, smooth muscle cells, interstitial cells of Cajal (ICCs), and intrinsic and extrinsic neurons. These cells respond to a variety of mechanical cues, including shear stress, hydrostatic and intraluminal pressure, and cyclical stretch. The response of mechanosensitive cells to mechanical cues can elicit a variety of secretory and motility responses in the gut. Thus, the subject of mechanotransduction in the GI tract is expansive. Herein, we will focus on mechanotransduction in EC cells resulting in the production of serotonin as an example of mechanotransduction in the GI tract. Approximately 90% of serotonin is synthesized in the EC cells of the gut from tryptophan (Bellono et al. 2017; Chugani 2011); the rate-limiting step in serotonin synthesis is tryptophan hydroxylase, and most serotonin is stored in platelets (Schoenichen et al. 2019). The effects of serotonin are widespread, especially in the gut and nervous system.

11.2.2 Mechanosensitive Enterochromaffin Cells

EC cells can sense a variety of mechanical cues, including flow shear stress, intraluminal pressure, centrifugal forces, stretch, compression, membrane distortion, and cell volume changes (Linan-Rico et al. 2016). These mechanical cues modulate gastrointestinal motility. EC cells respond to mechanical cues by releasing serotonin, and serotonin receptors located on smooth muscle cells, ICCs, and enteric neurons mediate changes in gastrointestinal motility. Although serotonin release from EC cells was originally thought to elicit peristalsis, real-time recordings of serotonin release from EC cells suggest that the compression of EC cells during peristalsis triggers serotonin release rather than the other way around (Keating and Spencer 2010, 2019; Spencer et al. 2011). Thus, serotonin may modulate rather than elicit peristalsis (Huang et al. 2017).

Many of the responses to mechanical forces in EC cells are triggered by mechanical stimulation of Piezo2 channels (Wang et al. 2017). Of note, piezo channels are also expressed in sensory neurons and are necessary for normal gastrointestinal motility (Servin-Vences et al. 2023). Mechanical stimulation of EC cells activates Piezo2 channels, causing an inward ion current, including calcium influx (Alcaino et al. 2018). Increased intracellular calcium triggers the fusion of serotonin-containing vesicles with the cell membrane to release serotonin. Blocking Piezo channels slows intestinal transit, demonstrating the importance of Piezo2 channels in gastrointestinal motility (Zhu et al. 2023).

11.2.3 Piezo2 Structure and Mechanosensing

The structure of Piezo2 is complex. Piezo2 channels are homotrimers resembling a three-bladed propeller; each monomer forms 38 transmembrane regions (Wang et al. 2019). The three subunits form propeller-like structures (Saotome et al. 2018). The pore of the channel, which is formed at the center of the three propellers, conducts nonselective cations, slightly favoring Ca^{2+} (Coste et al. 2010). Other structures relevant to the mechanical gating of Piezo channels include the intracellular beam and clasp and an extracellular cap (Wang et al. 2019). In cell-free experiments, lateral stretch of PIEZO1 straightens the propeller blades and opens the channel, suggesting the mechanical mechanism for opening PIEZO channels (Lin et al. 2019). Compared with Piezo1, Piezo2 channels are less sensitive to membrane stretch and more sensitive to plasma membrane indentation (Shin et al. 2019) and cytoskeletal forces (Romero et al. 2020).

11.2.4 Regulation of Piezo2 Activity

A number of cellular and extracellular changes can alter Piezo2 activity. While most of these effects have been demonstrated in other cell types outside the GI tract, these changes may also affect Piezo2 activity, and therefore mechanotransduction, in the GI tract. Changes in lipids, cytoskeletal proteins, and extracellular matrix may affect Piezo2 responsiveness to mechanical signals.

Many ion channels are affected by the mechanical properties of the cell membranes (Cordero-Morales and Vasquez 2018). Thus, it is not surprising that the fatty acid composition of the cell membrane affects Piezo channels (Romero et al. 2020). Fatty acids may alter the rigidity of the lipid bilayer to affect mechanosensitive channels. For example, the fatty acid margaric acid inhibited Piezo2 sensitivity to mechanical stimuli in neuro-20 cells (Romero et al. 2020). Stomatin-like protein 3 (STROML3) and cholesterol deficiency also attenuated the mechanosensitivity of Piezo2 in sensory neurons, demonstrating that Piezo2 interaction with STROML3 facilitates force transduction (Qi et al. 2015). Depleting $PI(4,5)P_2$ and OI(4)P inhibits Piezo2 activation in DRG neurons (Borbiro et al. 2015). These results are supported by experiments showing that Mtmr2 and tetonin3, which regulate phospholipids, also modulate Piezo2 activity in sensory neurons (Anderson et al. 2018; Narayanan et al. 2018). In addition, TRPV1 channel activations of phospholipase C can deplete $PI(4,5)P_2$ and, thus, inhibit Piezo2, indicating an interaction between TRPV1 channels and Piezo channels (Borbiro et al. 2015). Although these interactions were mainly shown in neuronal cells, similar changes in cell membranes may also affect the mechanosensitivity and function of Piezo2 channels in the gut.

Unlike Piezo1, Piezo2 requires an intact cytoskeleton, indicating that cytoskeletal forces can modulate Piezo2 activity (Eijkelkamp et al. 2013). In the force-filament model of Piezo2 mechanosensitivity, force is transmitted to Piezo2 via the cytoskeleton (Chuang and Chen 2022). Although direct evidence is lacking, this model of Piezo2 activation is supported by several lines of evidence. Piezo1, but not Piezo2, can be activated by membrane tension in excised membrane patches, providing evidence that Piezo2 requires an intact cytoskeleton (Wang et al. 2019; Moroni et al. 2018). In addition, treatment with latrunculin A, an inhibitor of actin polymerization, strongly reduced the mechanosensitivity of Piezo2 (Eijkelkamp et al. 2013). In a possible feedback loop, Ca^{2+} influx mediated by Piezo2 activates RhoA and the formation of stress fibers and focal adhesions in MDA-MB-BrM2 cells (Pardo-Pastor et al. 2018).

Several endogenous regulators of Piezo2 have been identified recently. Coexpression of TMEM120A (also known as TACAN) and Piezo2 inhibited the mechanical activation of Piezo2 in HEK293 and N2A cells (Del Rosario et al. 2022). TMEM120A and Piezo2 are both expressed in DRG neurons. Subsequent studies showed that TMEM120A elevates phosphatidic acid and lysophosphatidic acid levels, which inhibit the mechanical activation of Piezo2 (but not Piezo1) in N2A cells (Gabrielle et al. 2024).

11.2.5 EC Cells, Serotonin, and Piezo2 in Gastrointestinal Diseases

Enterochromaffin cells and serotonin secretion play critical roles in normal physiological function in the GI tract. Therefore, it is not surprising that dysregulation of EC cells and serotonin contributes to a number of GI diseases. While the contribution of EC cells and serotonin to irritable bowel syndrome (IBS) is unclear, several lines of evidence support a role. A systematic review of clinical biopsy results revealed that EC cells and serotonin are increased in patients with IBS (Martin-Vinas and Quigley 2016). Increased numbers of EC cells and increased expression of tryptophan hydroxylase, an enzyme involved in serotonin synthesis, were detected in the cecum and colon of IBS-D patients compared with healthy controls (Yang et al. 2021). EC cell hyperplasia was also detected in a rodent maternal separation model of early life stress-induced IBS (Tao et al. 2022). Finally, serotonin antagonists can be used to treat diarrhea and abdominal pain in patients with IBS, and a meta-analysis showed that serotonin antagonists relieve the symptoms of patients with non-constipated or diarrhea-dominant IBS (Andresen et al. 2008; Spiller 2007; Vanuytsel et al. 2014).

EC cells are increased, and expression of the SERT is decreased in patients with inflammatory bowel disease (IBD) (El-Salhy et al. 1997; Tada et al. 2016). In several clinical studies comparing patients with Crohn's disease to healthy controls, tryptophan hydroxylase 1 was elevated, and the SERT was downregulated (Chojnacki et al. 2013; Shajib et al. 2019). In another clinical study, serotonin levels were increased in active Crohn's disease compared with patients in remission, and the SERT decreased in the ileum and colon of patients with Crohn's disease (Manzella et al. 2020). Several other studies also demonstrated that the SERT was reduced in the epithelium of patients with active Crohn's disease and ulcerative colitis (Jorandli et al. 2020; Qasem et al. 2021). Although the exact role of serotonin signaling in IBD is still unclear, serotonin is associated with inflammation and likely contributes to the development and maintenance of IBD.

Piezo channels are essential for the normal function of the GI tract. The secretion of serotonin by EC cells regulates a variety of physiological secretory and motility processes and may also be involved in the pathology of IBD (Linan-Rico et al. 2013), IBS (Tao et al. 2022), and bacterial-induced diarrhea. Knockdown of Piezo2 in mice with functional constipation reduced serotonin release from EC cells and exacerbated the slowed GI motility (Yan et al. 2025). Interestingly, knockdown of Piezo2 decreased tryptophan hydroxylase 1 levels, suggesting that Piezo channels regulate both serotonin production and release (Yan et al. 2025).

The importance of Piezo2 channels in regulating gastrointestinal motility is highlighted by a study of patients deficient in Piezo2 (Servin-Vences et al. 2023). Although this was a small cohort, the patients reported a wide array of GI symptoms, including urgency, difficulty defecating, watery or lumpy stools, and urgency (Servin-Vences et al. 2023). Interestingly, several subjects reported difficulties sensing bowel movement (Servin-Vences et al. 2023). The authors also showed that Piezo2 in the dorsal root is important for sensing gut contents and colonic distension (Servin-Vences et al. 2023).

As a mechanically active organ, the GI tract must distinguish and respond to a variety of important mechanical cues to modulate GI motility for the optimization of nutrient absorption. To detect mechanical cues, the GI tract has a variety of mechanosensitive cells, including EC cells, smooth muscle cells, ICCs, and intrinsic and extrinsic neurons. The EC cells in the GI mucosal lining help detect mechanical and chemical cues and modulate gastrointestinal motility. Piezo2 channels are key mechanosensors mediating both physiological and pathological responses of EC cells, including serotonin release. Although Piezo2 regulation has mainly been studied in neurons, similar regulatory mechanisms may also occur in EC cells. The role of EC cells and Piezo2 channels in various diseases, including IBD and IBS, are not fully understood but may represent targets for treating these diseases. Thus, understanding the regulation of Piezo2 channels in EC cells is important for targeting EC cells and serotonin release for the treatment of IBD and IBS.

11.3 Summary and Conclusions

The examples presented here confirm that changes in ECM composition correlate with cell fate transitions at the crossroads of differentiation, proliferation, or apoptosis. A great wealth of new evidence speaks for the integral regulatory role of serotonin, and possibly other monoaminergic neuroendocrine factors, in determining and tuning of the physical characteristics of breast tissue structure and gastrointestinal function in distinct but overlapping ways. Monoaminergic regulation emerges as a therapeutic target in conditions involving an imbalance of mechanosensory and mechanotransductive processes.

References

Abedini F, Amjadi O, Hedayatizadeh-Omran A, Lira SA, Ahangari G (2023) Serotonin receptors and acetylcholinesterase gene expression alternations: the potential value on tumor microenvironment of gastric cancer. Oncology 101(7):415–424

Afshar K, Sanaei MJ, Ravari MS, Pourbagheri-Sigaroodi A, Bashash D (2023) An overview of extracellular matrix and its remodeling in the development of cancer and metastasis with a glance at therapeutic approaches. Cell Biochem Funct 41(8):930–952

Alcaino C, Knutson KR, Treichel AJ, Yildiz G, Strege PR, Linden DR et al (2018) A population of gut epithelial enterochromaffin cells is mechanosensitive and requires Piezo2 to convert force into serotonin release. Proc Natl Acad Sci USA 115(32):E7632–E7641

Anderson EO, Schneider ER, Matson JD, Gracheva EO, Bagriantsev SN (2018) TMEM150C/Tentonin3 is a regulator of mechano-gated ion channels. Cell Rep 23(3):701–708

Andresen V, Montori VM, Keller J, West CP, Layer P, Camilleri M (2008) Effects of 5-hydroxytryptamine (serotonin) type 3 antagonists on symptom relief and constipation in nonconstipated irritable bowel syndrome: a systematic review and meta-analysis of randomized controlled trials. Clin Gastroenterol Hepatol 6(5):545–555

Ashbury JE, Levesque LE, Beck PA, Aronson KJ (2012) Selective serotonin reuptake inhibitor (SSRI) antidepressants, Prolactin and breast cancer. Front Oncol 2:177

Ayyash A, Holloway AC (2022) Fluoxetine-induced hepatic lipid accumulation is mediated by prostaglandin endoperoxide synthase 1 and is linked to elevated 15-deoxy-Delta(12,14) PGJ(2). J Appl Toxicol 42(6):1004–1015

Bader M (2019) Serotonylation: serotonin signaling and epigenetics. Front Mol Neurosci 12:288

Balakrishna P, George S, Hatoum H, Mukherjee S (2021) Serotonin pathway in cancer. Int J Mol Sci 22(3):1268

Baneyx G, Baugh L, Vogel V (2002) Fibronectin extension and unfolding within cell matrix fibrils controlled by cytoskeletal tension. Proc Natl Acad Sci USA 99(8):5139–5143

Barutcu AR, Lajoie BR, Fritz AJ, McCord RP, Nickerson JA, van Wijnen AJ et al (2016) SMARCA4 regulates gene expression and higher-order chromatin structure in proliferating mammary epithelial cells. Genome Res 26(9):1188–1201

Bellissima V, Visser GH, Ververs TF, van Bel F, Termote JU, van der Heide M et al (2011) Antenatal maternal antidepressants drugs affect Activin A concentrations in maternal blood, in amniotic fluid and in fetal cord blood. J Matern Fetal Neonatal Med 24(Suppl 2):31–34

Bellono NW, Bayrer JR, Leitch DB, Castro J, Zhang C, O'Donnell TA et al (2017) Enterochromaffin cells are gut chemosensors that couple to sensory neural pathways. Cell 170(1):185–98 e16

Biet J, Poole CA, Stelwagen K, Margerison JK, Singh K (2016) Primary cilia distribution and orientation during involution of the bovine mammary gland. J Dairy Sci 99(5):3966–3978

Blakely WJ, Hatterschide J, White EA (2024) HPV18 E7 inhibits LATS1 kinase and activates YAP1 by degrading PTPN14. MBio 15(10):e0181124

Borbiro I, Badheka D, Rohacs T (2015) Activation of TRPV1 channels inhibits mechanosensitive Piezo channel activity by depleting membrane phosphoinositides. Sci Signal 8(363):ra15

Carneiro AM, Cook EH, Murphy DL, Blakely RD (2008) Interactions between integrin alphaIIb-beta3 and the serotonin transporter regulate serotonin transport and platelet aggregation in mice and humans. J Clin Invest 118(4):1544–1552

Carraher CL, Schwarzbauer JE (2013) Regulation of matrix assembly through rigidity-dependent fibronectin conformational changes. J Biol Chem 288(21):14805–14814

Caswell PT, Vadrevu S, Norman JC (2009) Integrins: masters and slaves of endocytic transport. Nat Rev Mol Cell Biol 10(12):843–853

Chang L, Azzolin L, Di Biagio D, Zanconato F, Battilana G, Lucon Xiccato R et al (2018) The SWI/SNF complex is a mechanoregulated inhibitor of YAP and TAZ. Nature 563(7730):265–269

Chaudhary P, Yadav K, Lee HJ, Kang KW, Mo J, Kim JA (2024) siRNA treatment targeting integrin alpha11 overexpressed via EZH2-driven axis inhibits drug-resistant breast cancer progression. Breast Cancer Res 26(1):72
Chen L, Huang S, Wu X, He W, Song M (2024) Serotonin signalling in cancer: emerging mechanisms and therapeutic opportunities. Clin Transl Med 14(7):e1750
Chojnacki C, Wisniewska-Jarosinska M, Kulig G, Majsterek I, Reiter RJ, Chojnacki J (2013) Evaluation of enterochromaffin cells and melatonin secretion exponents in ulcerative colitis. World J Gastroenterol 19(23):3602–3607
Chuang YC, Chen CC (2022) Force from filaments: the role of the cytoskeleton and extracellular matrix in the gating of mechanosensitive channels. Front Cell Dev Biol 10:886048
Chugani DC (2011) alpha-methyl-L-tryptophan: mechanisms for tracer localization of epileptogenic brain regions. Biomark Med 5(5):567–575
Cooper J, Giancotti FG (2019) Integrin signaling in cancer: mechanotransduction, stemness, epithelial plasticity, and therapeutic resistance. Cancer Cell 35(3):347–367
Cordero-Morales JF, Vasquez V (2018) How lipids contribute to ion channel function, a fat perspective on direct and indirect interactions. Curr Opin Struct Biol 51:92–98
Coste B, Mathur J, Schmidt M, Earley TJ, Ranade S, Petrus MJ et al (2010) Piezo1 and Piezo2 are essential components of distinct mechanically activated cation channels. Science 330(6000):55–60
DeFilippis RA, Fordyce C, Patten K, Chang H, Zhao J, Fontenay GV et al (2014) Stress signaling from human mammary epithelial cells contributes to phenotypes of mammographic density. Cancer Res 74(18):5032–5044
del Rio A, Perez-Jimenez R, Liu R, Roca-Cusachs P, Fernandez JM, Sheetz MP (2009) Stretching single talin rod molecules activates vinculin binding. Science 323(5914):638–641
Del Rosario JS, Gabrielle M, Yudin Y, Rohacs T (2022) TMEM120A/TACAN inhibits mechanically activated PIEZO2 channels. J Gen Physiol 154(8):e202213164
Dohn MR, Kooker CG, Bastarache L, Jessen T, Rinaldi C, Varney S et al (2017) The gain-of-function integrin beta3 Pro33 variant alters the serotonin system in the mouse brain. J Neurosci 37(46):11271–11284
Domingues RR, Teixeira NN, Frizzarini WS, Beard AD, Connelly MK, Vang A et al (2023) The antidepressant fluoxetine (Prozac(R)) modulates serotonin signaling to alter maternal peripartum calcium homeostasis. Sci Rep 13(1):21832
Downey-Biechler C, Craig DH, More SK, Basson MD (2019) Inside-out signaling through FAK-integrin axis may regulate circulating cancer cell metastatic adhesion. Proc Natl Acad Sci USA 116(40):19795–19796
Eijkelkamp N, Linley JE, Torres JM, Bee L, Dickenson AH, Gringhuis M et al (2013) A role for Piezo2 in EPAC1-dependent mechanical allodynia. Nat Commun 4:1682
El-Salhy M, Danielsson A, Stenling R, Grimelius L (1997) Colonic endocrine cells in inflammatory bowel disease. J Intern Med 242(5):413–419
Ferreira M, Fujiwara H, Morita K, Watt FM (2009) An activating beta1 integrin mutation increases the conversion of benign to malignant skin tumors. Cancer Res 69(4):1334–1342
Field SL, Arendt LM, Hernandez LL, Laporta J (2025) Characterizing serotonin expression throughout bovine mammary gland developmental stages and its relationship with 17beta-estradiol at puberty. PLoS One 20(3):e0319914
Friedland JC, Lee MH, Boettiger D (2009) Mechanically activated integrin switch controls alpha5beta1 function. Science 323(5914):642–644
Fruh SM, Schoen I, Ries J, Vogel V (2015) Molecular architecture of native fibronectin fibrils. Nat Commun 6:7275
Gabriele S, Canali M, Lintas C, Sacco R, Tirindelli MC, Ricciardello A et al (2019) Evidence that ITGB3 promoter variants increase serotonin blood levels by regulating platelet serotonin transporter trafficking. Hum Mol Genet 28(7):1153–1161
Gabrielle M, Yudin Y, Wang Y, Su X, Rohacs T (2024) Phosphatidic acid is an endogenous negative regulator of PIEZO2 channels and mechanical sensitivity. Nat Commun 15(1):7020

Gahmberg CG, Fagerholm SC, Nurmi SM, Chavakis T, Marchesan S, Gronholm M (2009) Regulation of integrin activity and signalling. Biochim Biophys Acta 1790(6):431–444

Goel HL, Pursell B, Chang C, Shaw LM, Mao J, Simin K et al (2013) GLI1 regulates a novel neuropilin-2/alpha6beta1 integrin based autocrine pathway that contributes to breast cancer initiation. EMBO Mol Med 5(4):488–508

Gough RE, Jones MC, Zacharchenko T, Le S, Yu M, Jacquemet G et al (2021) Talin mechanosensitivity is modulated by a direct interaction with cyclin-dependent kinase-1. J Biol Chem 297(1):100837

Gu J, Taniguchi N (2004) Regulation of integrin functions by N-glycans. Glycoconj J 21(1–2):9–15

Guettler S, Vartiainen MK, Miralles F, Larijani B, Treisman R (2008) RPEL motifs link the serum response factor cofactor MAL but not myocardin to rho signaling via actin binding. Mol Cell Biol 28(2):732–742

Guilluy C, Eddahibi S, Agard C, Guignabert C, Izikki M, Tu L et al (2009) RhoA and Rho kinase activation in human pulmonary hypertension: role of 5-HT signaling. Am J Respir Crit Care Med 179(12):1151–1158

Guo W, Pylayeva Y, Pepe A, Yoshioka T, Muller WJ, Inghirami G et al (2006) Beta 4 integrin amplifies ErbB2 signaling to promote mammary tumorigenesis. Cell 126(3):489–502

Hahn LH, Yamada KM (1979) Identification and isolation of a collagen-binding fragment of the adhesive glycoprotein fibronectin. Proc Natl Acad Sci USA 76(3):1160–1163

Hillmer RE, Link BA (2019) The roles of hippo signaling transducers Yap and Taz in chromatin remodeling. Cells 8(5):502

Hou J, Yan D, Liu Y, Huang P, Cui H (2020) The roles of integrin alpha5beta1 in human cancer. Onco Targets Ther 13:13329–13344

Huang DL, Bax NA, Buckley CD, Weis WI, Dunn AR (2017) Vinculin forms a directionally asymmetric catch bond with F-actin. Science 357(6352):703–706

Huhnt W, Lubbe AS (1995) Venules and arterioles in xenotransplanted human colon adenocarcinoma critically constrict with hyperthermia and serotonin. J Cancer Res Clin Oncol 121(5):267–274

Hynes RO (2002) Integrins: bidirectional, allosteric signaling machines. Cell 110(6):673–687

Hynes RO (2009) The extracellular matrix: not just pretty fibrils. Science 326(5957):1216–1219

Insua-Rodriguez J, Oskarsson T (2016) The extracellular matrix in breast cancer. Adv Drug Deliv Rev 97:41–55

Jayachandran P, Battaglin F, Strelez C, Lenz A, Algaze S, Soni S et al (2023) Breast cancer and neurotransmitters: emerging insights on mechanisms and therapeutic directions. Oncogene 42(9):627–637

Jdeed S, Erdos E, Balint BL, Uray IP (2022a) The role of ARID1A in the nonestrogenic modulation of IGF-1 signaling. Mol Cancer Res 20(7):1071–1082

Jdeed S, Lengyel M, Uray IP (2022b) Redistribution of the SWI/SNF complex dictates coordinated transcriptional control over epithelial-mesenchymal transition of normal breast cells through TGF-beta signaling. Cells 11(17):2633

Jorandli JW, Thorsvik S, Skovdahl HK, Kornfeld B, Saeterstad S, Gustafsson BI et al (2020) The serotonin reuptake transporter is reduced in the epithelium of active Crohn's disease and ulcerative colitis. Am J Physiol Gastrointest Liver Physiol 319(6):G761–G7G8

Kalluri R, Zeisberg M (2006) Fibroblasts in cancer. Nat Rev Cancer 6(5):392–401

Kannen V, Bader M, Sakita JY, Uyemura SA, Squire JA (2020) The dual role of serotonin in colorectal cancer. Trends Endocrinol Metab 31(8):611–625

Keating DJ, Spencer NJ (2010) Release of 5-hydroxytryptamine from the mucosa is not required for the generation or propagation of colonic migrating motor complexes. Gastroenterology 138(2):659-70–670 e1–2

Keating DJ, Spencer NJ (2019) What is the role of endogenous gut serotonin in the control of gastrointestinal motility? Pharmacol Res 140:50–55

Kostic A, Sheetz MP (2006) Fibronectin rigidity response through Fyn and p130Cas recruitment to the leading edge. Mol Biol Cell 17(6):2684–2695

Kubow KE, Vukmirovic R, Zhe L, Klotzsch E, Smith ML, Gourdon D et al (2015) Mechanical forces regulate the interactions of fibronectin and collagen I in extracellular matrix. Nat Commun 6:8026

Lakshmikanthan S, Zheng X, Nishijima Y, Sobczak M, Szabo A, Vasquez-Vivar J et al (2015) Rap1 promotes endothelial mechanosensing complex formation, NO release and normal endothelial function. EMBO Rep 16(5):628–637

Lengyel M, Molnar A, Nagy T, Jdeed S, Garai I, Horvath Z et al (2025) Zymogen granule protein 16B (ZG16B) is a druggable epigenetic target to modulate the mammary extracellular matrix. Cancer Sci 116(1):81–94

Lengyel M, Arkosy P, Uray IP (2026) G protein subtype preference dictates paroxetine-enhanced serotonin receptor response in non-malignant breast epithelial cells. Biochem Pharmacol 244:117584

Levental KR, Yu H, Kass L, Lakins JN, Egeblad M, Erler JT et al (2009) Matrix crosslinking forces tumor progression by enhancing integrin signaling. Cell 139(5):891–906

Li T, Wei L, Zhang X, Fu B, Zhou Y, Yang M et al (2024) Serotonin receptor HTR2B facilitates colorectal cancer metastasis via CREB1-ZEB1 Axis-mediated epithelial-mesenchymal transition. Mol Cancer Res 22(6):538–554

Lin YC, Guo YR, Miyagi A, Levring J, MacKinnon R, Scheuring S (2019) Force-induced conformational changes in PIEZO1. Nature 573(7773):230–234

Linan-Rico A, Wunderlich JE, Grants IS, Frankel WL, Xue J, Williams KC et al (2013) Purinergic autocrine regulation of mechanosensitivity and serotonin release in a human EC model: ATP-gated P2X3 channels in EC are downregulated in ulcerative colitis. Inflamm Bowel Dis 19(11):2366–2379

Linan-Rico A, Ochoa-Cortes F, Beyder A, Soghomonyan S, Zuleta-Alarcon A, Coppola V et al (2016) Mechanosensory signaling in enterochromaffin cells and 5-HT release: potential implications for gut inflammation. Front Neurosci 10:564

Liu Y, Li M, Warburton RR, Hill NS, Fanburg BL (2007) The 5-HT transporter transactivates the PDGFbeta receptor in pulmonary artery smooth muscle cells. FASEB J 21(11):2725–2734

Liu Y, Wei L, Laskin DL, Fanburg BL (2011) Role of protein transamidation in serotonin-induced proliferation and migration of pulmonary artery smooth muscle cells. Am J Respir Cell Mol Biol 44(4):548–555

Lo CM, Wang HB, Dembo M, Wang YL (2000) Cell movement is guided by the rigidity of the substrate. Biophys J 79(1):144–152

Luo J, Yao JF, Deng XF, Zheng XD, Jia M, Wang YQ et al (2018) 14, 15-EET induces breast cancer cell EMT and cisplatin resistance by up-regulating integrin alphavbeta3 and activating FAK/PI3K/AKT signaling. J Exp Clin Cancer Res 37(1):23

Lv M, Hu C, Li J, Zhang Y, Lu C, Zhang S et al (2025) Serotonin receptor type 7 (5-HT7) promotes colorectal cancer progression by activating Wnt/beta-catenin signaling via interacting with CK1epsilon. Oncogene. https://doi.org/10.1038/s41388-025-03645

Manzella CR, Jayawardena D, Pagani W, Li Y, Alrefai WA, Bauer J et al (2020) Serum serotonin differentiates between disease activity states in Crohn's patients. Inflamm Bowel Dis 26(10):1607–1618

Martin-Vinas JJ, Quigley EM (2016) Immune response in irritable bowel syndrome: a systematic review of systemic and mucosal inflammatory mediators. J Dig Dis 17(9):572–581

Matsuda M, Imaoka T, Vomachka AJ, Gudelsky GA, Hou Z, Mistry M et al (2004) Serotonin regulates mammary gland development via an autocrine-paracrine loop. Dev Cell 6(2):193–203

Mazalouskas M, Jessen T, Varney S, Sutcliffe JS, Veenstra-VanderWeele J, Cook EH Jr et al (2015) Integrin beta3 haploinsufficiency modulates serotonin transport and antidepressant-sensitive behavior in mice. Neuropsychopharmacology 40(8):2015–2024

Mierke CT, Frey B, Fellner M, Herrmann M, Fabry B (2011) Integrin alpha5beta1 facilitates cancer cell invasion through enhanced contractile forces. J Cell Sci 124(Pt 3):369–383

Mitsi M, Hong Z, Costello CE, Nugent MA (2006) Heparin-mediated conformational changes in fibronectin expose vascular endothelial growth factor binding sites. Biochemistry 45(34):10319–10328

Mokhtarian A, Siguret V, Jourdi G (2025) Effects of selective serotonin reuptake inhibitors on platelet functions: a literature review. Curr Opin Hematol 32(1):22–33

Moore KM, Thomas GJ, Duffy SW, Warwick J, Gabe R, Chou P et al (2014) Therapeutic targeting of integrin alphavbeta6 in breast cancer. J Natl Cancer Inst 106(8):dju169

Moroni M, Servin-Vences MR, Fleischer R, Sanchez-Carranza O, Lewin GR (2018) Voltage gating of mechanosensitive PIEZO channels. Nat Commun 9(1):1096

Narayanan P, Hutte M, Kudryasheva G, Taberner FJ, Lechner SG, Rehfeldt F et al (2018) Myotubularin related protein-2 and its phospholipid substrate PIP(2) control Piezo2-mediated mechanotransduction in peripheral sensory neurons. Elife 7:e32346

Nia HT, Munn LL, Jain RK (2020) Physical traits of cancer. Science 370(6516):eaaz0868

Ohtani A, Kozono N, Senzaki K, Shiga T (2014) Serotonin 2A receptor regulates microtubule assembly and induces dynamics of dendritic growth cones in rat cortical neurons in vitro. Neurosci Res 81–82:11–20

Pardo-Pastor C, Rubio-Moscardo F, Vogel-Gonzalez M, Serra SA, Afthinos A, Mrkonjic S et al (2018) Piezo2 channel regulates RhoA and actin cytoskeleton to promote cell mechanobiological responses. Proc Natl Acad Sci USA 115(8):1925–1930

Paszek MJ, Weaver VM (2004) The tension mounts: mechanics meets morphogenesis and malignancy. J Mammary Gland Biol Neoplasia 9(4):325–342

Paszek MJ, Zahir N, Johnson KR, Lakins JN, Rozenberg GI, Gefen A et al (2005) Tensional homeostasis and the malignant phenotype. Cancer Cell 8(3):241–254

Pathak A, Kumar S (2012) Independent regulation of tumor cell migration by matrix stiffness and confinement. Proc Natl Acad Sci USA 109(26):10334–10339

Provenzano PP, Inman DR, Eliceiri KW, Keely PJ (2009) Matrix density-induced mechanoregulation of breast cell phenotype, signaling and gene expression through a FAK-ERK linkage. Oncogene 28(49):4326–4343

Qasem A, Naser AE, Naser SA (2021) Enteropathogenic infections modulate intestinal serotonin transporter (SERT) function by activating Toll-like receptor 2 (TLR-2) in Crohn's disease. Sci Rep 11(1):22624

Qi Y, Andolfi L, Frattini F, Mayer F, Lazzarino M, Hu J (2015) Membrane stiffening by STOML3 facilitates mechanosensation in sensory neurons. Nat Commun 6:8512

Quaglino A, Salierno M, Pellegrotti J, Rubinstein N, Kordon EC (2009) Mechanical strain induces involution-associated events in mammary epithelial cells. BMC Cell Biol 10:55

Reuter JA, Ortiz-Urda S, Kretz M, Garcia J, Scholl FA, Pasmooij AM et al (2009) Modeling inducible human tissue neoplasia identifies an extracellular matrix interaction network involved in cancer progression. Cancer Cell 15(6):477–488

Romero LO, Caires R, Nickolls AR, Chesler AT, Cordero-Morales JF, Vasquez V (2020) A dietary fatty acid counteracts neuronal mechanical sensitization. Nat Commun 11(1):2997

Rubashkin MG, Cassereau L, Bainer R, DuFort CC, Yui Y, Ou G et al (2014) Force engages vinculin and promotes tumor progression by enhancing PI3K activation of phosphatidylinositol (3,4,5)-triphosphate. Cancer Res 74(17):4597–4611

Sabra H, Brunner M, Mandati V, Wehrle-Haller B, Lallemand D, Ribba AS et al (2017) beta1 integrin-dependent Rac/group I PAK signaling mediates YAP activation of Yes-associated protein 1 (YAP1) via NF2/merlin. J Biol Chem 292(47):19179–19197

Saha K, Keung AJ, Irwin EF, Li Y, Little L, Schaffer DV et al (2008) Substrate modulus directs neural stem cell behavior. Biophys J 95(9):4426–4438

Samuel MS, Lopez JI, McGhee EJ, Croft DR, Strachan D, Timpson P et al (2011) Actomyosin-mediated cellular tension drives increased tissue stiffness and beta-catenin activation to induce epidermal hyperplasia and tumor growth. Cancer Cell 19(6):776–791

Saotome K, Murthy SE, Kefauver JM, Whitwam T, Patapoutian A, Ward AB (2018) Structure of the mechanically activated ion channel Piezo1. Nature 554(7693):481–486

Schoenichen C, Bode C, Duerschmied D (2019) Role of platelet serotonin in innate immune cell recruitment. Front Biosci (Landmark Ed) 24(3):514–526
Seetharaman S, Vianay B, Roca V, Farrugia AJ, De Pascalis C, Boeda B et al (2022) Microtubules tune mechanosensitive cell responses. Nat Mater 21(3):366–377
Semler EJ, Ranucci CS, Moghe PV (2000) Mechanochemical manipulation of hepatocyte aggregation can selectively induce or repress liver-specific function. Biotechnol Bioeng 69(4):359–369
Servin-Vences MR, Lam RM, Koolen A, Wang Y, Saade DN, Loud M et al (2023) PIEZO2 in somatosensory neurons controls gastrointestinal transit. Cell 186(16):3386–3399 e15
Shajib MS, Chauhan U, Adeeb S, Chetty Y, Armstrong D, Halder SLS et al (2019) Characterization of serotonin signaling components in patients with inflammatory bowel disease. J Can Assoc Gastroenterol 2(3):132–140
Sharma R, Sharma R, Khaket TP, Dutta C, Chakraborty B, Mukherjee TK (2017) Breast cancer metastasis: putative therapeutic role of vascular cell adhesion molecule-1. Cell Oncol (Dordr) 40(3):199–208
Shin KC, Park HJ, Kim JG, Lee IH, Cho H, Park C et al (2019) The Piezo2 ion channel is mechanically activated by low-threshold positive pressure. Sci Rep 9(1):6446
Shu B, Zhai M, Miao X, He C, Deng C, Fang Y et al (2018) Serotonin and YAP/VGLL4 balance correlated with progression and poor prognosis of hepatocellular carcinoma. Sci Rep 8(1):9739
Singh K, Davis SR, Dobson JM, Molenaar AJ, Wheeler TT, Prosser CG et al (2008) cDNA microarray analysis reveals that antioxidant and immune genes are upregulated during involution of the bovine mammary gland. J Dairy Sci 91(6):2236–2246
Singh K, Swanson KM, Henderson HV, Erdman RA, Stelwagen K (2015) The effect of milking reinitiation following extended nonmilking periods on lactation in primiparous dairy cows. J Dairy Sci 98(11):7666–7674
Sohrabi A, Lefebvre A, Harrison MJ, Condro MC, Sanazzaro TM, Safarians G et al (2023) Microenvironmental stiffness induces metabolic reprogramming in glioblastoma. Cell Rep 42(10):113175
Spencer NJ, Nicholas SJ, Robinson L, Kyloh M, Flack N, Brookes SJ et al (2011) Mechanisms underlying distension-evoked peristalsis in Guinea pig distal colon: is there a role for enterochromaffin cells? Am J Physiol Gastrointest Liver Physiol 301(3):G519–G527
Spiller R (2007) Recent advances in understanding the role of serotonin in gastrointestinal motility in functional bowel disorders: alterations in 5-HT signalling and metabolism in human disease. Neurogastroenterol Motil 19(Suppl 2):25–31
Starlinger P, Brunnthaler L, Watkins R, Pereyra D, Stift J, Finsterbusch M et al (2023) Tyrosine phosphorylation of YAP-1 in biliary epithelial cells mediates posthepatectomy liver regeneration and is affected by serotonin. J Cell Biochem 124(5):687–700
Stewart TA, Hughes K, Stevenson AJ, Marino N, Ju AL, Morehead M et al (2021) Mammary mechanobiology – investigating roles for mechanically activated ion channels in lactation and involution. J Cell Sci 134(1):jcs248849
Subbaram S, Lyons SP, Svenson KB, Hammond SL, McCabe LG, Chittur SV et al (2014) Integrin alpha3beta1 controls mRNA splicing that determines Cox-2 mRNA stability in breast cancer cells. J Cell Sci 127(Pt 6):1179–1189
Subramanian A, Kanzaki LF, Galloway JL, Schilling TF (2018) Mechanical force regulates tendon extracellular matrix organization and tenocyte morphogenesis through TGFbeta signaling. elife 7:7
Tada Y, Ishihara S, Kawashima K, Fukuba N, Sonoyama H, Kusunoki R et al (2016) Downregulation of serotonin reuptake transporter gene expression in healing colonic mucosa in presence of remaining low-grade inflammation in ulcerative colitis. J Gastroenterol Hepatol 31(8):1443–1452
Takada Y, Takada YK, Fujita M (2017) Crosstalk between insulin-like growth factor (IGF) receptor and integrins through direct integrin binding to IGF1. Cytokine Growth Factor Rev 34:67–72
Takagi J, Petre BM, Walz T, Springer TA (2002) Global conformational rearrangements in integrin extracellular domains in outside-in and inside-out signaling. Cell 110(5):599–511

Tao E, Zhu Z, Hu C, Long G, Chen B, Guo R et al (2022) Potential roles of enterochromaffin cells in early life stress-induced irritable bowel syndrome. Front Cell Neurosci 16:837166

Tseng YL, Chiang ML, Huang TF, Su KP, Lane HY, Lai YC (2010) A selective serotonin reuptake inhibitor, citalopram, inhibits collagen-induced platelet aggregation and activation. Thromb Res 126(6):517–523

Ulrich TA, de Juan Pardo EM, Kumar S (2009) The mechanical rigidity of the extracellular matrix regulates the structure, motility, and proliferation of glioma cells. Cancer Res 69(10):4167–4174

Vanuytsel T, Tack JF, Boeckxstaens GE (2014) Treatment of abdominal pain in irritable bowel syndrome. J Gastroenterol 49(8):1193–1205

Vartiainen MK, Guettler S, Larijani B, Treisman R (2007) Nuclear actin regulates dynamic subcellular localization and activity of the SRF cofactor MAL. Science 316(5832):1749–1752

Voringer S, Schreyer L, Nadolni W, Meier MA, Woerther K, Mittermeier C et al (2020) Inhibition of TRPM7 blocks MRTF/SRF-dependent transcriptional and tumorigenic activity. Oncogene 39(11):2328–2344

Walker CJ, Crocini C, Ramirez D, Killaars AR, Grim JC, Aguado BA et al (2021) Nuclear mechanosensing drives chromatin remodelling in persistently activated fibroblasts. Nat Biomed Eng 5(12):1485–1499

Wang N (2017) Cellular adhesion: instant integrin mechanosensing. Nat Mater 16(12):1173–1174

Wang HB, Dembo M, Wang YL (2000) Substrate flexibility regulates growth and apoptosis of normal but not transformed cells. Am J Physiol Cell Physiol 279(5):C1345–C1350

Wang F, Knutson K, Alcaino C, Linden DR, Gibbons SJ, Kashyap P et al (2017) Mechanosensitive ion channel Piezo2 is important for enterochromaffin cell response to mechanical forces. J Physiol 595(1):79–91

Wang L, Zhou H, Zhang M, Liu W, Deng T, Zhao Q et al (2019) Structure and mechanogating of the mammalian tactile channel PIEZO2. Nature 573(7773):225–229

Wang Y, Yao M, Baker KB, Gough RE, Le S, Goult BT et al (2021) Force-dependent interactions between talin and full-length vinculin. J Am Chem Soc 143(36):14726–14737

Wei L, Warburton RR, Preston IR, Roberts KE, Comhair SA, Erzurum SC et al (2012) Serotonylated fibronectin is elevated in pulmonary hypertension. Am J Physiol Lung Cell Mol Physiol 302(12):L1273–L1279

Xie QE, Du X, Wang M, Xie F, Zhang Z, Cao Y et al (2021) Identification of serotonin as a predictive marker for breast cancer patients. Int J Gen Med 14:1939–1948

Yamada KM, Weston JA (1974) Isolation of a major cell surface glycoprotein from fibroblasts. Proc Natl Acad Sci USA 71(9):3492–3496

Yan X, Ma P, Wang W, Zeng W, Li Y, Hou Y et al (2025) Piezo knockdown reduces 5-hydroxytryptamine release from enterochromaffin cells and exacerbates intestinal dyskinesia in mice with functional constipation. Int J Mol Med 56(6):178

Yang J, Wang P, Liu T, Lin L, Li L, Kou G et al (2021) Involvement of mucosal flora and enterochromaffin cells of the caecum and descending colon in diarrhoea-predominant irritable bowel syndrome. BMC Microbiol 21(1):316

Yu H, Qu T, Yang J, Dai Q (2023) Serotonin acts through YAP to promote cell proliferation: mechanism and implication in colorectal cancer progression. Cell Commun Signal 21(1):75

Yu J, Wang Z, Chen Y, Dong Y (2025) Restraint stress disrupted intestinal homeostasis via 5-HT/HTR7/Wnt/beta-catenin/NF-kB signaling. Int J Mol Sci 26(9):4021

Zdanowski A, Sartor H, Feldt M, Skarping I (2023) Mammographic density in relation to breast cancer recurrence and survival in women receiving neoadjuvant chemotherapy. Front Oncol 13:1177310

Zeltz C, Alam J, Liu H, Erusappan PM, Hoschuetzky H, Molven A et al (2019) alpha11beta1 integrin is induced in a subset of cancer-associated fibroblasts in desmoplastic tumor stroma and mediates in vitro cell migration. Cancers (Basel) 11(6):765

Zhao S, Chuh KN, Zhang B, Dul BE, Thompson RE, Farrelly LA et al (2021) Histone H3Q5 serotonylation stabilizes H3K4 methylation and potentiates its readout. Proc Natl Acad Sci USA 118(6):e2016742118

Zhu Z, Chen X, Chen S, Hu C, Guo R, Wu Y et al (2023) Examination of the mechanism of Piezo ion channel in 5-HT synthesis in the enterochromaffin cell and its association with gut motility. Front Endocrinol (Lausanne) 14:1193556
Ziegler WH, Gingras AR, Critchley DR, Emsley J (2008) Integrin connections to the cytoskeleton through talin and vinculin. Biochem Soc Trans 36(Pt 2):235–239

Part III
Tissue- and Organ-Specific Mechanotransduction in Health and Disease

Chapter 12
Ciliary-Mediated Mechanotransduction in Skeletal Development and Diseases

Maryam Faisal, Lauren Tran, Yolanda Gutierrez, Jiarui Bi, Hiroyuki Yamaguchi, and Yoshihiro Komatsu

Abstract Cells have a mechanism that converts physical forces into biochemical signals, a critical process known as "mechanotransduction." Mechanotransduction plays a vital role in regulating multiple developmental and homeostatic events, including skeletal formation. Although underappreciated for decades, accumulated evidence now suggests that primary cilia play an important role in transducing mechanical stimuli. Since defects in the cilia result in multiple skeletal diseases, the importance of understanding the mechanisms between the mechanotransductive

M. Faisal
Department of Pediatrics, McGovern Medical School, The University of Texas Health Science Center at Houston, Houston, TX, USA

Department of Bioengineering, Rice University George R. Brown School of Engineering, Houston, TX, USA

L. Tran · H. Yamaguchi
Department of Pediatrics, McGovern Medical School, The University of Texas Health Science Center at Houston, Houston, TX, USA

Y. Gutierrez
Department of Pediatrics, McGovern Medical School, The University of Texas Health Science Center at Houston, Houston, TX, USA

University of Texas Rio Grande Valley School of Medicine, Edinburg, TX, USA

J. Bi
Department of Pediatrics, McGovern Medical School, The University of Texas Health Science Center at Houston, Houston, TX, USA

Department of Oral Biological and Medical Sciences, University of British Columbia, Vancouver, BC, Canada

Y. Komatsu (✉)
Department of Pediatrics, McGovern Medical School, The University of Texas Health Science Center at Houston, Houston, TX, USA

Graduate Program in Genetics and Epigenetics, The University of Texas Graduate School of Biomedical Sciences at Houston, Houston, TX, USA
e-mail: Yoshihiro.Komatsu@uth.tmc.edu

M. Kloc, J. Wosik (eds.), *Mechanotransduction and Mechanosensing*, Results and Problems in Cell Differentiation 78, https://doi.org/10.1007/978-3-032-19509-8_12

role of the cilia and skeletal health is paramount. However, the cellular mechanism of how cilia function as mechanosensors to control skeletal growth and homeostasis remains unclear. In addition, little is known about whether the manipulation of ciliary signaling related to defective mechanotransduction improves the skeletal disease conditions seen in osteoarthritis and osteoporosis. In this chapter, we provide a comprehensive overview of cilia and their role in mechanotransduction during skeletal development, with a focus on the function of key ciliary genes and intracellular signaling pathways in different skeletal cell types. We also highlight the broad impact of ciliary defects on mechanical stress-related skeletal diseases.

12.1 Introduction

Mechanotransduction is the process by which cells convert mechanical stimuli and stress into biochemical and transcriptional responses (Li et al. 1997; Chalfie 2009; Kloc and Wosik 2025). The main biomechanical cues that cells respond to are hydrostatic pressure, tensile force, fluid shear stress, extracellular matrix stiffness, and extracellular fluid viscosity, which are important for tissue homeostasis in various organs and cell types (Wilson et al. 1995; Mascharak et al. 2024; Hinz 2007).

In the body, bone is the tissue designed to sustain the greatest mechanical loads, and mechanical loading and signaling heavily regulate bone development and remodeling (Lecka–Czernik et al. 2025; Pal 2014). Developmentally, most bone arises from the mesoderm, but some craniofacial and middle ear bones arise from neural crest cells (Berendsen and Olsen 2015; Bronner and LeDouarin 2012; Noden and Trainor 2005; Neben and Merrill 2015; Mishina and Snider 2014). The two main components of mature bone are cortical and trabecular bone, which are comprised of four primary cell types: osteoblasts, osteoclasts and osteocytes. (Wang et al. 2022a; Ozcivici et al. 2010; Baldridge et al. 2010). Osteoblasts are fibroblast-like cells that originate from multipotent mesenchymal stem cells (Kovacs et al. 2021; Hojo et al. 2024; Mizoguchi and Ono 2021). The primary function of osteoblasts is the synthesis of the extracellular matrix of bone by the secretion of type I collagen (Yamauchi and Sricholpech 2012; Clews et al. 2025; Unlu et al. 2014). Mineralization transforms osteoblasts into osteocytes, post-proliferative cells residing within bone, representing the most differentiated state of osteoblast lineage (Kovacs et al. 2021; Ornitz and Marie 2015). Osteoclasts are cells instrumental in the processes of bone resorption and reformation necessary for bone remodeling (Nakashima and Takayanagi 2011; Rauner and Bozec 2026; Zhou et al. 2022) and differentiate from myeloid cells to remove old osteocytes and initiate bone resorption (Takahashi et al. 1988; Takayanagi 2007).

The primary cilium, a small microtubule-based organelle present on almost every cell type in the body, is critically important (Pazour and Rosenbaum 2002; Rosenbaum and Witman 2002). This organelle senses and transduces essential extracellular signals, regulating many cellular behaviors (Mitchison and Valente 2017; Yamaguchi et al. 2024). It is present in several skeletal cell types, and it has been implicated in prevalent skeletal diseases such as osteoarthritis (Rux et al. 2022; McGlashan et al. 2008) and congenital craniofacial bone abnormalities (Li et al. 2022; Schock and Brugmann 2017; Yuan and Yang 2016; Adel Al-Lami et al. 2016). In bone, the cilium's function has primarily been studied in the context of embryonic bone development; however, recent studies suggest that primary cilia also act as mechanosensors in adult bone homeostasis (Nguyen and Jacobs 2013; Ti et al. 2025). In this chapter, we explore the role of primary cilia in mechanotransduction during skeletal development and maintenance. We examine how cilia mediate cellular responses to mechanical stimuli, their involvement in skeletal disease, and the potential for targeting ciliary signaling through therapeutic strategies.

12.2 The Role of Cilium as a Chemosensor

Cilia exhibit a range of structural and functional diversity (Derderian et al. 2023; Ishikawa and Marshall 2011; Satir and Christensen 2007). Most cilia are formed during the interphase stage of the cell cycle (Marshall and Rosenbaum 1999; Plotnikova et al. 2009). They extend from a basal body, derived from an older centriole known as the mother centriole, that acts as an anchor for an axoneme, a core of nine microtubule pairs arranged in a ring (Fig. 12.1). This axoneme is surrounded by a lipid bilayer ciliary membrane, a specialized extension of the cellular membrane (Wood and Rosenbaum 2015; Sorokin 1962; Gilula and Satir 1972; Mill et al. 2023). Whether or not the axoneme contains a central microtubule pair determines if the cilia are categorized as $9 + 2$ (present) or $9 + 0$ (absent). Both $9 + 2$ and $9 + 0$ cilia can be further designated as motile or immotile. Research on ciliary architecture suggests that the presence of certain sub-ciliary structures, such as the central pair, dynein arms, and radial spokes, is responsible for motility (Brown and Witman 2014; Horani et al. 2014; Teves et al. 2016). Primary cilia are recognized as $9 + 0$ immotile cilia. In contrast to motile cilia that are present in only a few specific cell types in humans, primary cilia are found on virtually all mammalian cells (Satir and Christensen 2007; Venkatesh 2017). However, they are not a constant presence on these cells. Primary cilia are assembled and disassembled in a dynamic process that is cell cycle dependent (Wheatley et al. 1996; Quarmby and Parker 2005; Nguyen and Jacobs 2013).

Despite their immotile nature, primary cilia are incredibly important. Rather than facilitating movement, they act as cellular antennae, detecting a wide range of external signals and transmitting them to the cell (Berbari et al. 2009; Nishimura et al. 2019; Singla and Reiter 2006; Nauli et al. 2003). This sensory function enables the cell to respond to its environment, playing a crucial role in processes such as development and tissue homeostasis (Mill et al. 2023; Huangfu and Anderson 2005). Proteins required for the structure and function of the primary cilium are transported to their designated locations via intraflagellar transport (IFT). IFT is a bidirectional process that utilizes motor proteins to move cargo along the microtubule axoneme

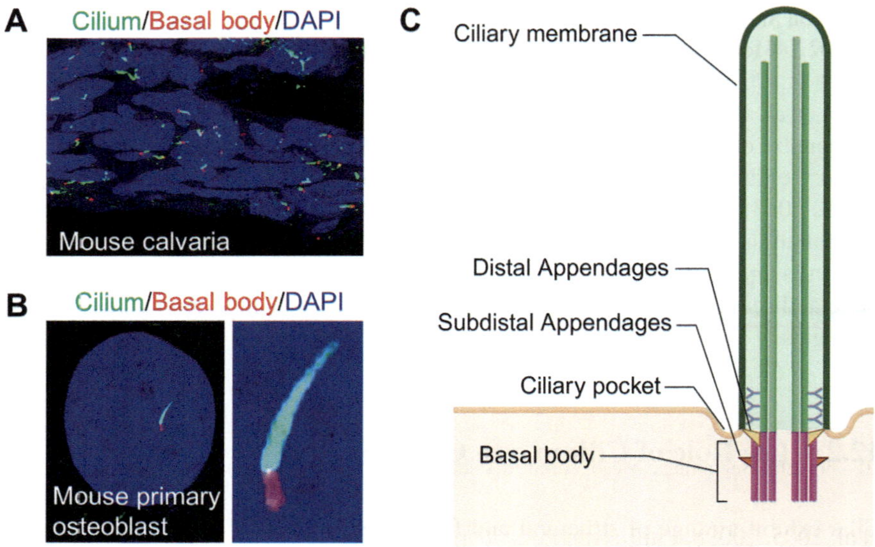

Fig. 12.1 Primary cilium; antenna-like organelle. (**a**, **b**) Immunofluorescent images of primary cilium in mouse calvaria (**a**) and primary osteoblast (**b**). Ciliary membrane was stained with anti-acetylated-tubulin antibody (green), and basal body was stained with anti-γ-tubulin antibody (magenta). (**c**) Schematic illustration of primary cilium. The primary cilium is a nonmotile antenna-like organelle that extends as a solitary unit from the centrosomal mother centriole (basal body). The cilium comprises a microtubule-based axoneme containing a ring of nice outer microtubule doublets. Between the basal body and cilium is the ciliary transition zone, which contains specialized gating structures such as Y-links that, along with the basal body transition fibers, are often referred to as the distal appendages. The distal appendages and subdistal appendages are known as a crucial regulator of ciliogenesis. The ciliary pocket, the specific membrane domain which locates at the base of primary cilia in most cell types, acts as a platform for cilia-associated vesicular trafficking and as an interface with the actin cytoskeleton

(Kozminski et al. 1993; Huangfu et al. 2003; Louvi and Grove 2011). IFT-B complexes, driven by kinesin motors, move cargo toward the tip of the cilium (anterograde transport), while IFT-A complexes, driven by dynein motors, move cargo back toward the base (retrograde transport) (Jing et al. 2023; Pigino et al. 2009). This constant transport ensures the cilium's structure is maintained and that it can properly sense and relay signals. The following section provides a summary of the cilia-mediated signal transduction pathways.

12.2.1 Ciliary Hedgehog Signaling

The Hedgehog (Hh) signaling pathway is arguably the most well-studied and critical pathway involving primary cilia (Fig. 12.2). In the absence of an Hh ligand, the key signaling protein, smoothened (SMO), is sequestered within vesicles and prevented

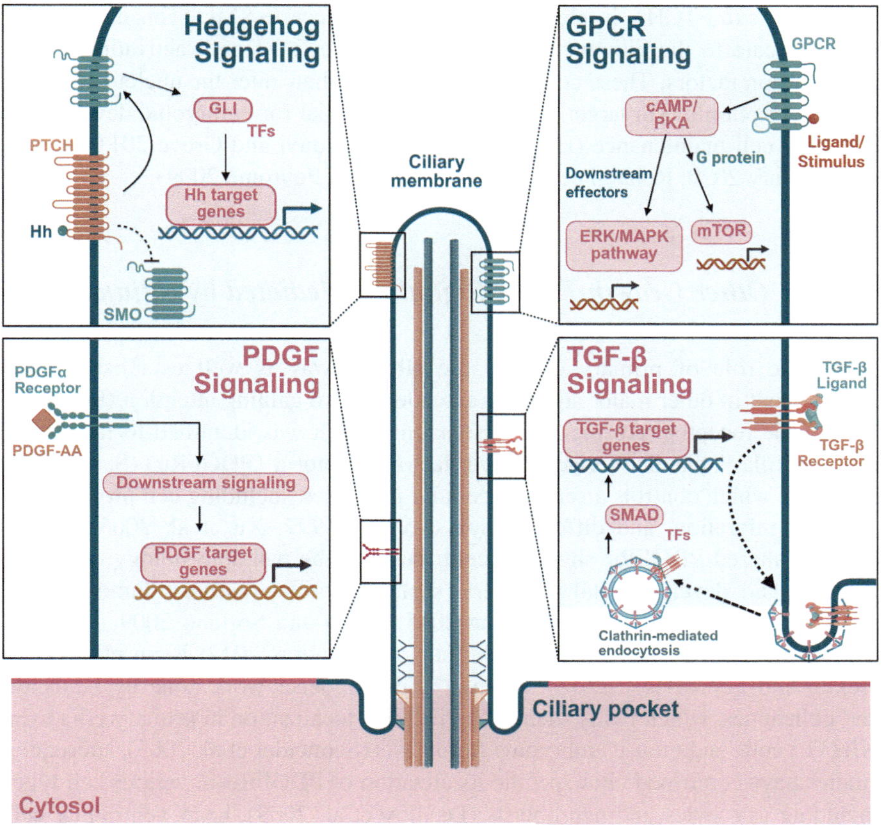

Fig. 12.2 Primary cilium as a chemosensor. With binding of Hh to PTCH1 extracellular domains, PTCH1 exits from the surface of the cilium and internalizes at the ciliary base. Removal of PTCH1 causes concomitant enrichment and activation of smoothened (SMO) in cilia. Hh stimulation results in dissociation of suppressor of FUSED (SUFU) from GLI transcription factors, formation of full-length activator forms of GLIs (GLIA), which translocate from the cilium into the nucleus to induce transcriptional activation of Hh target genes. Upon binding cilia-located receptor, PDGFRα, with its specific ligand, PDGFAA, downstream pathways, including P13K-AKT and MEK1/2-ERK1/2-RSK pathways, are activated to control cell migration, proliferation, and differentiation in multiple tissues. Receptors of transforming growth factor-β (TGF-β) are recruited to the primary cilium to activate the receptor-regulated SMADs (SMAD2/3) pathway via internalization of active receptors by clathrin-mediated endocytosis at the ciliary pocket. Binding of agonists to various G protein-coupled receptors (GPCRs) expressed on cilia modulates a variety of signaling pathways on different cell types

from reaching the primary cilium (Louvi and Grove 2011; Corbit et al. 2005; Gigante and Caspary 2020). A transmembrane protein called Patched1 (PTCH1), localized to the cilium, actively inhibits SMO. This inhibition leads to the cleavage of the Gli family of transcription factors (GLI2 and GLI3) into their repressor forms, which then suppress the transcription of Hh target genes. However, when an Hh ligand binds to PTCH1, it releases the inhibitory effect on SMO. This allows SMO to translocate to the cilium, where it promotes the full-length activation of GLI transcription factors. These activated GLI proteins then enter the nucleus to induce the transcription of Hh target genes, which are crucial for embryonic development and stem cell maintenance (Huangfu et al. 2003; Louvi and Grove 2011; Gigante and Caspary 2020; Rohatgi et al. 2007; Aulehla and Pourquie 2009).

12.2.2 Other Growth Factor Signaling Mediated by Cilium

While the role of primary cilia in the Hh pathway is well established, their involvement in other major signaling cascades is also gaining attention (Fig. 12.2). One of the receptor tyrosine kinase signaling which was identified to localize on primary cilia is platelet-derived growth factor receptor α (PDGFRα) (Schneider et al. 2005), which controls a series of cellular processes, including cell proliferation, survival, migration, and differentiation (Soriano 1997; Xu et al. 2005). Consequently, altered PDGFRα signaling contributes to the pathophysiology of various diseases and developmental disorders, such as fibrotic diseases, tumorigenesis, and cancer (Farahani and Xaymardan 2015; Olson and Soriano 2009; Demoulin and Essaghir 2014; Demoulin and Montano-Almendras 2012; Ryan et al. 2006; Heldin and Lennartsson 2013). Starting with a pioneer work done by Schneider and colleagues, which firstly identified PDGFRα localization in primary cilia using NIH3T3 cells and mouse embryonic fibroblasts (Schneider et al. 2005), subsequent studies have confirmed cilia-specific localization of PDGFRα in various cell types, including astrocytes and neuroblasts (Danilov et al. 2009), heart ventricular cells (Gerhardt et al. 2013), osteoblasts (Noda et al. 2016), human embryonic stem cells (Awan et al. 2010), human ovarian surface epithelium cells (Egeberg et al. 2012), and human placental choriocarcinoma JAR cells (Sheu et al. 2022). When a PDGF-AA binds to PDGFRα located in primary cilia, it causes the phosphorylation of PDGFRα, which activates downstream cascades like the RAS-MAPK and P13K pathways (Nishimura et al. 2019; Christensen et al. 2017; Pandey et al. 2023). Also, the localization of PDGFRα in primary cilia is regulated by the interactions with infraglabellar transport protein 20 (IFT20) and E3 ubiquitin ligases, c-Cbl and Cbl-b, which is required for Cbl-mediated ubiquitination and internalization of PDGFRα for feedback inhibition of receptor signaling (Schmid et al. 2017).

The Wnt signaling pathway is a complex network that regulates cell fate, proliferation, and polarity (Louvi and Grove 2011; Vuong and Mlodzik 2023; Anvarian et al. 2019; Nusse and Varmus 1982). In certain cell types, primary cilia appear to serve as signaling platforms to transduce Wnt Signaling. For example, some studies suggest that the cilium can sequester key signaling components, modulating the response to Wnt ligands (Lancaster et al. 2011; May-Simera and Kelley 2012; Gerdes et al. 2009). However, this relationship is highly context dependent and may differ across tissues (Ocbina et al. 2009; Corbit et al. 2008). In a more recent finding, primary cilia were shown to promote the differentiation of human neurons by shifting Wnt signaling toward the noncanonical pathway (Coschiera et al. 2024). Similarly, the TGF-β signaling pathway, which controls cell growth and differentiation, is also influenced by primary cilia (Anvarian et al. 2019; Clement et al. 2013; Goldgof et al. 2016). Recent research indicates that the ciliary pocket, the invagination at the base of the cilium (Molla-Herman et al. 2010), is a site for endocytosis that helps regulate TGF-β signaling. This regulation is crucial for processes like brain and heart development, and defects can lead to diseases such as microcephaly and congenital heart disease (Clement et al. 2013; Shaikh Qureshi and Hentges 2024).

12.2.3 G Protein-Coupled Receptors and ERK/MAPK Signaling

Beyond developmental pathways, cilium also functions as a sensory hub for a vast array of other extracellular signals. The ciliary membrane is enriched with a wide variety of G protein-coupled receptors (GPCRs), which are activated by various stimuli (Marinissen and Gutkind 2001; Schou et al. 2015). Once activated, these GPCRs initiate downstream signaling cascades. For instance, in the context of polycystic kidney disease, GPCRs in the cilium regulate downstream pathways such as mammalian target of rapamycin (mTOR) and cyclic adenosine monophosphate (cAMP), which influence cell proliferation and cyst formation (Walker et al. 2019; Sussman et al. 2020; Buqaileh et al. 2025). While these events can occur in multiple cellular compartments, localized ciliary cAMP/PKA activity has been shown to regulate ciliary length and stability in several contexts (Kim et al. 2011; Basten and Giles 2013). In parallel, elevated cAMP/PKA signaling can activate ERK/MAPK in epithelial cells, such as those affected in polycystic kidney disease (Yamaguchi et al. 2003). Although these processes are not always directly linked, together they underscore how cAMP-mediated signaling can coordinate structural and proliferative responses across both ciliary and cytoplasmic domains, highlighting the intricate connection between ciliary structure, signaling, and overall cellular physiology.

12.3 Ciliary Defects and Skeletal Diseases

In recent years, dysfunctions of primary cilia, referred to as "ciliopathies," have increasingly been recognized as key contributors to a variety of congenital skeletal disorders (Mitchison and Valente 2017; Yamaguchi et al. 2024; Zhang et al. 2018; Badano et al. 2006; Kaku and Komatsu 2017). The phenotypic link between cilia defects and skeletal diseases in humans is further detailed in Table 12.1.

Table 12.1 Mutations of ciliary genes cause skeletal ciliopathies

Syndrome	OMIM ID	Skeletal phenotypes	Identified genes
Bardet-Biedl syndrome	#209900	Polydactyly, lower articular joint thickness	BBS1 (Muller et al. 2010), BBS10 (Muller et al. 2010), MKS1 (Brunetti-Pierri et al. 2021), IFT172 (Bujakowska et al. 2015), TTC21B (Davis et al. 2011)
Cranioectodermal dysplasia/Sensenbrenner syndrome	#218330	Dolichocephaly, sagittal syndrome, short thorax and limbs, brachydactyly, ectodermal defects, congenital heart diseases	IFT122 (WDR10) (Walczak-Sztulpa et al. 2010), IFT140 (Bayat et al. 2017), IFT144 (WDR19) (Ishida et al. 2021), SPAG17 (Córdova-Fletes et al. 2018), WDR35 (Gilissen et al. 2010), IFT43 (Arts et al. 2011)
Ellis-van Creveld syndrome	#225500	Short rib and limbs, dysplastic teeth and nail, postaxial polydactyly	EVC1 (Ruiz-Perez et al. 2000), EVC2 (Galdzicka et al. 2002)
Jeune syndrome/asphyxiating thoracic dystrophy	#208500	Short ribs and limbs, small rigid thoracic cage, nail dysplasia, poor development of the iliac	DYNC2H1 (Dagoneau et al. 2009), WDR34 (Schmidts et al. 2013a), IFT140 (Miller et al. 2013), IFT80 (Beales et al. 2007)
Joubert syndrome	#213300	Polydactyly, scoliosis, bones shortening or bowing	JBTS17 (Damerla et al. 2015), FAM149B1 (Shaheen et al. 2019), SUFU (De Mori et al. 2017), IFT172 (Halbritter et al. 2013), KIAA053 (Hammarsjö et al. 2017), INPP5E (Travaglini et al. 2013), TMEM67 (Komatsu et al. 2016), TMEM216 (Valente et al. 2010), CPLANE1 (Vilboux et al. 2017; Bachmann-Gagescu et al. 2015), ARL13B (Cantagrel et al. 2008), KIF7 (Dafinger et al. 2011), MKS3 (Baala et al. 2007; Smith et al. 2006), CC2D2A (MKS6) (Gorden et al. 2008), AHI1 (Vilboux et al. 2017; Parisi et al. 2006)
Mainzer-Saldino syndrome	#266920	Short-rib thoracic dysplasia, polydactyly, cone-shaped epiphyses, small thoraces	IFT140 (Perrault et al. 2012), IFT172 (Halbritter et al. 2013; Helm et al. 2017), WDR19 (Ryan et al. 2018), TEKT1 (Ryan et al. 2018)

Meckel syndrome	#249000	Polydactyly, bowing of the long bones of the limbs	MKS1 (Bialas et al. 2009), TMEM216 (MKS2) (Valente et al. 2010; Edvardson et al. 2010), TMEM67 (MKS3) (Vilboux et al. 2017), B9D1 (Romani et al. 2014), TMEM67 (Leightner et al. 2013), TTC21B (Davis et al. 2011), CC2D2A (MKS6) (Tallila et al. 2008), CEP290 (MKS4) (Frank et al. 2008)
Meckel syndrome	#249000	Polydactyly, bowing of the long bones of the limbs	MKS1 (Bialas et al. 2009), TMEM216 (MKS2) (Valente et al. 2010; Edvardson et al. 2010), TMEM67 (MKS3) (Vilboux et al. 2017), B9D1 (Romani et al. 2014), TMEM67 (Leightner et al. 2013), TTC21B (Davis et al. 2011), CC2D2A (MKS6) (Tallila et al. 2008), CEP290 (MKS4) (Frank et al. 2008)
Sensenbrenner syndrome	#218330	Sagittal craniosynostosis, brachydactyly, narrow thorax, short long bones, joint laxity	IFT122 (Walczak-Sztulpa et al. 2010; Takahara et al. 2018), SPAG17 (Córdova-Fletes et al. 2018), IFT52 (Girisha et al. 2016), IFT140 (Walczak-Sztulpa et al. 2022), IFT144 (WDR19) (Ishida et al. 2021)
Short-rib polydactyly syndrome	#269860	Short ribs, polydactyly, narrow thorax, shortened long bones	WDR60 (McInerney-Leo et al. 2013), DYNC2H1 (Badiner et al. 2017; Zhang et al. 2020)
Oral-facial-digital syndrome	#311200	Cleft of the hard or soft palate, hypodontia, dental abnormalities, widely spaced eyes, micrognathia, brachydactyly, syndactyly, clinodactyly of the fifth finger, duplicated great toe	OFD1 (Ferrante et al. 2001; Thauvin-Robinet et al. 2009), C2CD3 (Thauvin-Robinet et al. 2014), CEP164 (Strong et al. 2021), RAB34 (Bruel et al. 2023)

The relationship between mutations of ciliary genes and skeletal ciliopathies with Online Mendelian Inheritance in Man (OMIM) ID

12.3.1 Ellis-Van Creveld Syndrome

Ellis-van Creveld (EVC) syndrome is an autosomal recessive skeletal disease, occurring in approximately 1 in 140,000, characterized by multiple clinical features, such as short ribs, limbs, dysplastic teeth and nails, and postaxial polydactyly (Wang et al. 2023; Aubert-Mucca et al. 2023). Two causative genes of EVC, EVC and EVC2, have been identified as associated with this syndrome (Ruiz-Perez et al. 2000; Galdzicka et al. 2002; Da Silva et al. 2023). Elevated fibroblast growth factor (FGF) signaling and a disrupted Hh-Parathyroid hormone-related protein (PTHrP) feedback loop in growth plates result in the limb dwarfism phenotype of EVC (Zhang et al. 2016).

12.3.2 Jeune Syndrome

Jeune syndrome (JS) is a rare but severe skeletal ciliopathy characterized by short ribs and limbs; a small, rigid thoracic cage; nail dysplasia; and poor development of the iliac (Keppler-Noreuil et al. 2011). JS shows significant genetic heterogeneity, and mutations in DYNC2H1 and WDR34 have been widely demonstrated to cause the disease (Schmidts et al. 2013a; Dagoneau et al. 2009). In JS patients with severe skeletal phenotypes, IFT80 and IFT140 mutations have also been identified (Miller et al. 2013; Beales et al. 2007). Because of these mutations, it has been proposed that the phenotypes result from disrupted ciliogenesis and chondral differentiation with altered Hh and Wnt signaling pathways (Emiralioglu et al. 2018; Schmidts et al. 2013b). Likewise, the loss of GRK2 in chondrocytes disrupts chondrogenesis and impairs Hh and canonical Wnt signaling and has been associated with JS (Bosakova et al. 2020). GRK2 functions in multiple tissues throughout the body as a G protein-coupled receptor kinase, and further investigations may therefore help identify specific targets that contribute to JS (Bosakova et al. 2020).

12.3.3 Cranioectodermal Dysplasia

Individuals with cranioectodermal dysplasia (CED), also known as Sensenbrenner syndrome, exhibit craniosynostosis, short limbs, and a narrow thorax (Konstantinidou et al. 2009; Lin et al. 2013). Ectodermal defects such as sparse hair, abnormal nails, and dental abnormalities are also common symptoms of CED (Lin et al. 2013). Disruption of SPAG17, IFT52, IFT122, IFT140, or IFT144 causes CED (Walczak-Sztulpa et al. 2010, 2020; Bayat et al. 2017; Córdova-Fletes et al. 2018; Gilissen et al. 2010).

12.3.4 Bardet-Biedl Syndrome

Bardet-Biedl syndrome (BBS) is an inherited genetic disorder that affects multiple systems in the body. BBS is characterized by retinal dystrophy, obesity, polydactyly, and joint abnormalities (Klein and Ammann 1969; Melluso et al. 2023). At least 24 different genes have been identified as causative genes of BBS (Niederlova et al. 2019; Dollfus et al. 2024). Among them, BBS1 and BBS10 are commonly mutated in BBS (Muller et al. 2010).

12.3.5 Meckel Syndrome

Meckel syndrome (MS) is a rare genetic disorder characterized by abnormalities such as cystic kidney dysplasia, polydactyly, and altered limb formation (Salonen and Paavola 1998; Hartill et al. 2017). MKS1, B9D1, TMEM67, TMEM216, TTC21B, and others have been identified as causative genes of MS (Davis et al. 2011; Valente et al. 2010; Vilboux et al. 2017; Bialas et al. 2009; Romani et al. 2014), and this mutation induces defects in primary cilia (Barker et al. 2014), thereby altering multiple signaling pathways, including Wnt signaling (Ahmed et al. 2025).

12.4 The Role of Cilium as a Mechanotransduction Regulator

Mechanical forces are crucial in various biological processes, such as cell fate switching and development, prompting growing interest in their roles (Hoffman et al. 2011; Ingber 2006; Mammoto and Ingber 2010). Mechanotransduction refers to the conversion of mechanical stimuli into biochemical signals through distinct mechanisms, thereby activating downstream signaling pathways that affect cell behavior in specific ways.

12.4.1 Variety of Mechanosensitive Channels

Calcium signaling is central to mechanotransduction and includes mechanosensitive components like Piezo, transient receptor potential (TRP), polycystin, and others (Coste et al. 2010; Clapham 2003; Retailleau and Duprat 2014; Sjaastad and Nelson 1997; Arcangeli and Becchetti 2006). Piezo is a mechanosensitive ion channel protein consisting of Piezo1 and Piezo2, both of which are nonselective cationic channels that are permeable to Ca^{2+}. Their gating can be stimulated by various mechanical stimuli such as membrane stretching, compression, poking, and stress (Lin et al. 2022). Piezo is widely expressed in various organs in humans and mice, and prior studies have demonstrated its active roles in physiological processes and development. For example, Piezo1 is expressed in embryonic vascular endothelial cells, and its inactivation in mice leads to vascular remodeling defects (Li et al. 2014). The role of Piezo has also been studied in the skeletal system. Bone-specific *Piezo1* knockout mice display impaired skeletal development and remodeling and defective bone structure (Clapham 2003; Sun et al. 2019; Li et al. 2019; Zhou et al. 2020; Wang et al. 2020). Similarly, TRP channels are nonselective cation channels that allow Ca^{2+} influx in response to various stimuli, including temperature, mechanical stress, and ligands (Clapham 2003). TRP ion channels influence skeletal stability by regulating both extracellular and intracellular Ca^{2+} balance processes (Lieben and Carmeliet 2012). Compared to Piezo1 and Piezo2, which sense high-intensity mechanical input, transient receptor potential vanilloid (TRPV) 4 mediates low-intensity mechanical stress (Du et al. 2020). Mutations in the TRPV4 gene in humans have been reported in several skeletal dysplasias (Kang et al. 2012; Rock et al. 2008; Krakow et al. 2009). Previous studies using Trp-related gene knockout mice led to impaired bone development and homeostasis (Lieben and Carmeliet 2012; Aliprantis et al. 2008; Van Cromphaut et al. 2001; Hoenderop et al. 2003). In particular, a study of *Trpv4* knockout mice demonstrated insensitivity to hind limb unloading and suppression of bone loss during the unloading (Mizoguchi et al. 2008). Further evidence showed that, as expressed in osteoclasts, TRPV4 regulates osteoclast differentiation and activity through NFATc1 signaling and autophagy (Masuyama et al. 2008). Lastly, the polycystin channel is a protein complex, primarily formed by polycystin 1 and polycystin 2. It functions as a mechanosensitive calcium channel that senses fluid flow and regulates calcium signaling, with polycystin 1 reported to be involved in bone development and formation via Runx2 regulation (Xiao et al. 2023; Merrick et al. 2019; Gargalionis et al. 2024).

12.4.2 Integrins as Mechanosensory Molecules

Integrins are another well-studied key mechanosensory molecule. Integrins, as transmembrane proteins, mediate adhesion between the cell actin cytoskeleton and the extracellular matrix (ECM) and promote bidirectional signaling to regulate multiple biological functions in cells (Campbell and Humphries 2011; Hynes 2002; Miranti and Brugge 2002; Takada et al. 2007). In mammals, at least 24 distinct receptor heterodimers are assembled as integrins from 18 α and 8 β subunits (Hynes 2002; Kadry and Calderwood 2020). Certain integrins are reported to be restricted to specific cell types: αIIbβ3 in platelets, αEβ7 in T cells and dendritic cells, α4β7 in memory T cells, α4β1 and α4β2 in leukocytes, and αvβ6 in epithelial cells (Bi et al. 2017; Koivisto et al. 2018). By binding to the ECM and connecting to the actin cytoskeleton, integrins sense mechanical tension through distinct adhesion types, including focal adhesions, focal complexes, nascent adhesions, podosome-type adhesions, and fibrillar adhesions (Zaidel-Bar et al. 2004; Geiger and Yamada 2011; Collin et al. 2008). Focal adhesions are large structures, approximately 2 μm wide and 3–10 μm long. They link integrins with extracellular ECM fibrils and intracellular contractile actomyosin fibers, through Rho GTPase activity (Schwartz 2010). In contrast, focal complexes are small circular structures, about 2 μm in diameter, that rely on the activation of Cdc42 or Rac (Schwartz 2010). Other adhesions, including nascent adhesions, podosome-type adhesions, and fibrillar adhesions, have also been shown to play a role in mechanotransduction (Collin et al. 2008; Bruinsma 2005). In response to mechanical forces in the ECM, tension or compression of adhesions can enlarge or recruit new cytoskeletal proteins for matrix assembly, which helps resist the force. Integrins transduce these biological signals intracellularly to regulate gene expression and cell behavior. For example, mechanical stressors trigger fibroblasts to express genes encoding collagens, fibronectin, and metalloproteinases and their inhibitors, leading to the assembly of a dense ECM (Chiquet et al. 2003). Fibronectin, upon binding via integrin, senses tension from actomyosin, opening binding sites and transforming into fibrils during matrix assembly (Baneyx et al. 2002; Zhong et al. 1998). Furthermore, these regulatory actions involve signaling pathways that are downstream of integrins (Schwartz 2010; Katsumi et al. 2004). In skeletal development and remodeling, integrins also play an important role in mediating mechanical stress (Liang et al. 2015; Zeng et al. 2015; Vlaikou et al. 2017; Yang et al. 2024). Studies have shown that cyclic tensile stress increases osteoblast proliferation through the ERK1/2, a pathway that was suppressed by integrin β1 inhibition (Zeng et al. 2015; Yang et al. 2024; Yan et al. 2012). Similarly, the stretching force in osteoblasts upregulated integrin β1 and osteogenesis-related gene expression, an effect reduced by integrin β1 knockdown (Yan et al. 2016). Further study revealed that osteocyte-specific β1 knockout mice had severely low bone mass, along with impaired bone formation and collagen I fiber integrity under tibial loading (Qin et al. 2022). Similar to β1, β3 also plays a crucial role in bone metabolism under mechanostress. However, β3 regulates osteoclast precursor cell differentiation and osteoclast adhesion to the ECM (Li et al. 1995; Hong et al. 2014; Kameyama et al. 2013; Wang et al. 2022b; Qin et al. 2023). For instance, osteocyte-specific β3 knockout mice subjected to mechanical stress exhibited impaired long bone formation, resulting from abnormal cell morphology characterized by reduced numbers and lengths of dendritic processes in osteocytes, alongside decreased focal adhesion-related molecules (Qin et al. 2023).

Primary cilia and integrins are thought to be involved in mechanotransduction that enables cells to integrate extracellular mechanical cues into intracellular biochemical responses (Geoghegan et al. 2021; Ong et al. 2020; Pittman and Solecki 2023). In a renal cystogenetic study, deletion of β1 integrins prevented cystogenesis caused by ciliary defects (via cilia assembly gene Ift88 ablation), reducing the pathogenic process, including increased inflammatory infiltrate and fibrosis in mice (Yoo et al. 2020). Similarly, knockdown of ITGB1 in Hs578T cells by shRNAs significantly increased the number of ciliated cells (Failler et al. 2021). In skeletal development and maintenance, integrins and primary cilia, components of mechanosensing systems, form complex networks that enable the sensing and regulation of mechanical signals. In osteoblasts and osteocytes, integrins and primary cilia can be directly stimulated by a fluid flow stimulus, with cilia amplification and the activation of integrin at the ciliary base (Vaughan et al. 2015). In the estrogen-withdrawal postmenopausal osteoporosis model, antagonism of αvβ3 integrins reduces actin contractility and promotes cilium elongation, thereby activating the Hh signaling pathway and osteoclastogenesis, leading to bone loss (Geoghegan et al. 2021). Another study discovered that α2, α3, and β1 integrins are found on primary cilia in chondrocytes (McGlashan et al. 2006). However, since the number of integrin-positive cilia varies, not all cilia display integrins (McGlashan et al. 2006). The function of these integrins present on cilia remains elusive, and further investigation is required to determine whether their function is distinct from that of those not present on cilia.

12.4.3 The Role of Cilium as a Mechanosensor in Skeletal Development

Primary cilia act as mechanosensors, detecting mechanical stress, such as shear stress, fluid flow, and function as Ca^{2+} signaling mediators (Fig. 12.3). Several studies have shown that deflection of primary cilia in kidney epithelium increases intracellular Ca^{2+} entry via polycystin (Praetorius and Spring 2001; Nauli et al. 2003, 2008; AbouAlaiwi et al. 2009; Masyuk et al. 2006). The bending of the primary cilium causes deformation of the cytoskeleton and stretching of the membrane, facilitating the influx of extracellular Ca^{2+} through polycystin 2 to enter the intracellular space, as well as the release of Ca^{2+} from intracellular stores via the activation of ryanodine receptors (Jin et al. 2014). This process indicates the link between primary cilia and stretch-activated ion channels in mediating intracellular Ca^{2+} level.

Primary cilia in osteoblasts and osteocytes respond to mechanical stimulation, facilitating mechanotransduction at the transcriptional level and changing the cellular activity (Malone et al. 2007). Interestingly, while primary cilia deflection in kidney cells relies on polycystin channels for Ca^{2+} flux, the same Ca^{2+} dependence was not observed in bone cells. Instead, other channels appear relevant in bone: TRPV4 has been shown to localize to regions of the primary cilium (Corrigan et al. 2018). In vitro, primary cilia increased flow-induced Ca^{2+} flux in the cell plasma through TRPV4, but not polycystin 2 or Piezo1, to induce an osteogenic response in osteocytes (Lee et al. 2015). A study using mesenchymal stem cell (MSC) osteogenesis showed that MSCs lacking functional primary cilia exhibit the inhibition of an osteogenic response to TRPV4 activation (Corrigan et al. 2018). In the chondrocytes of the mandibular condyle, IFT88-mediated Piezo1 regulates the early differentiation of the chondrocytes under cyclic tensile strain (Zhang et al. 2022). However, evidence of the molecular link between Piezo and cilia in skeletal

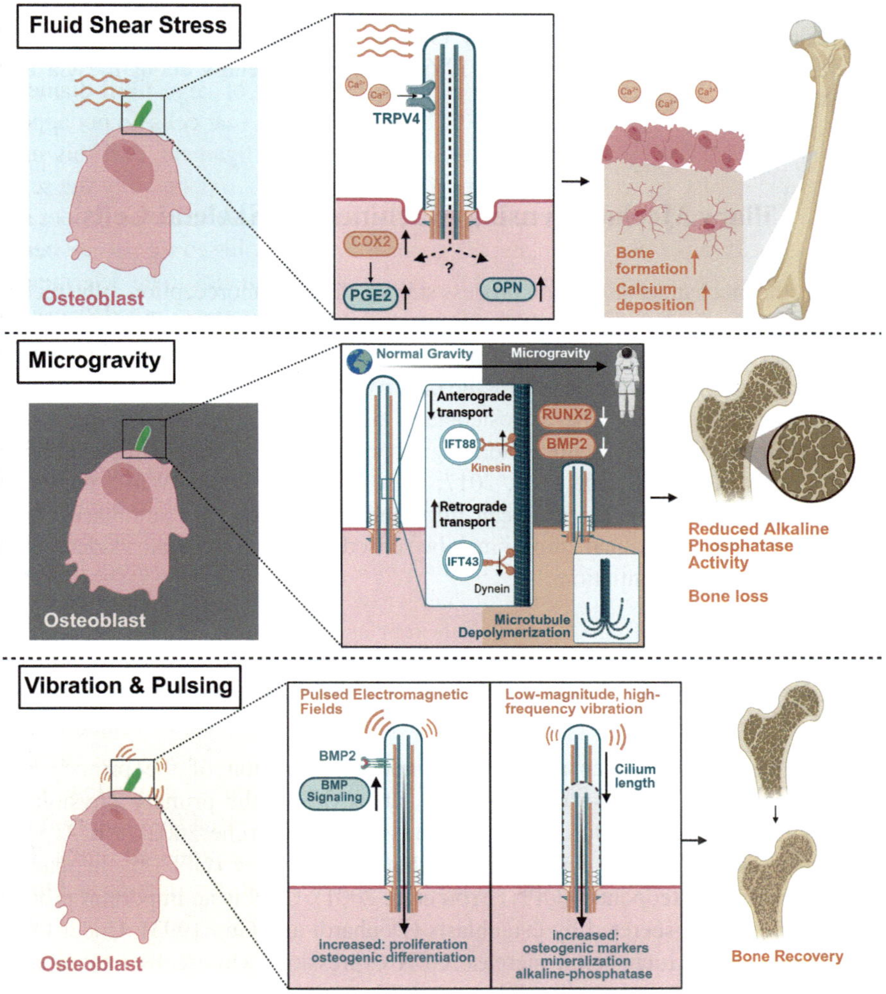

Fig. 12.3 Primary cilium as a mechanosensor in osteoblasts. Primary cilia in osteoblasts and osteocytes respond to mechanical stimulation, facilitating mechanotransduction at the transcriptional level and changing the cellular activity. TRPV4 was shown to localize to areas of the primary cilium, and primary cilia increased flow-induced Ca^{2+} flux in the cell plasma through TRPV4 to induce osteogenic response in osteocytes. Under oscillatory fluid flow (OFF), osteoblasts with primary cilia increase osteopontin (OPN) expression, cyclooxygenase 2 (COX2), and prostaglandin E2 (PGE2). In situations of microgravity, alkaline phosphatase activity was reduced along with protein levels of Runx2, BMP2, and other osteogenic markers, which might be regulated by primary cilia-dependent microtubule depolymerization. Pulsed electromagnetic fields (PEMFs) treatment stimulates osteogenic differentiation and maturation by primary cilium-mediated upregulation of BMP signaling. Low-magnitude, high-frequency vibration (LMHFV) stimulation results in shorter and fewer primary cilia in osteoblasts, which increases the expression of osteogenic markers, mineralization, and alkaline-phosphatase activity

development and function remains elusive. Further investigations are necessary to determine the precise role of primary cilia in regulating Ca^{2+} channels during mechanical stress. Please see "Sect. 12.6.1" for more details about the Ca and TRPV4 signaling in Chondrocytes.

12.5 Ciliary Mechanosensing Machinery in Skeletal Cells

In bone, the mechanotransduction process starts with mechanoreception—the reception of a mechanical signal—followed by its coupling of the mechanical signal and the cellular response (Duncan and Turner 1995). Studies on mechanotransduction in the skeletal system were driven by the observation that during increased loading, osteoblasts are activated, whereas osteoclasts are partially suppressed (Frost 1987, 2003). Conversely, during reduced loading, bone formation is suppressed, and bone resorption is increased (Uda et al. 2017; Wise and King 2008). While the exact mechanisms behind this are not been fully understood, here, we summarize how various types of mechanical stimuli may be sensed by cilia to regulate skeletogenic cell growth and differentiation.

12.5.1 Fluid Shear Stress (FSS) in Osteoblasts and Osteocytes

In osteoblasts, mechanosensation is an important function of the primary cilium, especially in response to fluid shear stress (FSS), the primary physiologically present mechanical force on bone (Basso and Heersche 2002). MC3T3-E1 osteoblasts were found to respond to oscillatory fluid flow (OFF) by increasing mRNA levels of osteopontin (OPN) (You et al. 2001). OPN is an important protein of the bone matrix secreted by osteoblasts (Denhardt and Guo 1993). Under OFF, osteoblasts with primary cilia increase OPN expression, while cells treated with chloral hydrate to abrogate cilia did not result in increased OPN levels (Malone et al. 2007).

Interestingly, in murine sagittal sutures subjected to expansion mechanical stress by orthopedic spring wires, OPN protein expression increased under mechanical stress but also subsequently resulted in increased bone formation at the edge of the parietal bone in contact with the suture gap. This effect was not seen in OPN-deficient mice subjected to the same tensile stress (Morinobu et al. 2009), highlighting the mechanical links between OPN expression and bone formation under mechanical stress. It should be noted that the expression of the cytokine prostaglandin E_2 (PGE_2) and the expression of its upstream enzyme cyclooxygenase 2 (COX2), both important for osteoblast proliferation and metabolism, were increased in response to dynamic fluid flow, but this effect was not seen without the primary cilium in both MC3T3 osteoblasts and in MLO-Y4 osteocytes (Malone et al. 2007). Using MLO-A5 osteoblasts, another group demonstrated that fluid flow reduces the number of cells expressing primary cilia (Delaine-Smith et al. 2014). They found that PGE_2 levels increase after OFF application, and this effect is reversed by chloral hydrate treatment to remove cilia. Chloral hydrate exposure also reduced the amount of calcium deposited by the cells in response to OFF, showing that primary cilia are important for OFF-induced calcium deposition.

The mechanotransductive role of primary cilia in osteoblasts and osteocytes has also been demonstrated in vivo. Interestingly, osteoblast-specific deletion of *Kif3a* did not display overt skeletal phenotypes (Temiyasathit et al. 2012). However, the wild-type mice displayed significantly increased rMS/BS (mineralizing surface/bone surface), rMAR (mineral apposition rate), and rBFR/BS (bone formation rate/bone surface) after being subjected to cyclic axial compressive ulnar loading. This increase was significantly less in the *Kif3a* mutant mice. These findings indicate that cilia deletion in osteoblasts and osteocytes reduces mechanical loading-induced bone formation.

Subjecting MC3T3 osteoblasts to fluid flow stress also induces Ca^{2+} mobilization (You et al. 2001), triggering intracellular Ca^{2+} release (Chen et al. 2000). Notably, kidney epithelial cells, in which the primary cilia's mechanotransductive role has been well studied, show a similar response to FSS as osteoblasts (Praetorius and Spring 2001). In kidney cells, this effect occurs through the stretch-sensitive Ca^{2+} channel polycystin 2 (Nauli et al. 2003). Application of a constant FSS in $Kif3a^{flox/+}$, Osteocalcin (*Oc*)-*Cre:*$Kif3a^{flox/+}$, $Kif3a^{flox/null}$, and *Oc-Cre:*$Kif3a^{flox/null}$ primary osteoblasts resulted in a gene dose-dependent reduction in flow-induced intracellular calcium response. A gene dose-dependent reduction in cilium length was also observed between these groups, but a much stronger relationship was observed in the flow-induced intracellular calcium response (Qiu et al. 2012). These findings indicate that FSS may be a more sensitive measure of primary cilia dysfunction than cilia length.

Additional studies suggest that the Ca^{2+} environment of the primary cilium may be distinct from that of the cytosol (Lee et al. 2015). Therefore, despite flow-induced Ca^{2+} increases in MLO-Y4 osteocytes being independent of primary cilia, the osteocyte primary cilium may mediate mechanotransduction by forming a Ca^{2+} microdomain. A fluorescence resonance energy transfer (FRET)-based Ca^{2+} biosensor fused to the sequence of the primary cilium-specific protein *Arl13b* in MLO-Y4 osteocytes was used, allowing detection of ciliary and cytosolic Ca^{2+} increases. Under OFF, a majority of detected Ca^{2+} peaks occurred in the primary cilium prior to the cytosol. Additionally, with thapsigargin treatment, which depletes intracellular Ca^{2+} stores, the peak mean signals were significantly reduced in the cytosol, but not in the primary cilium. These results indicate that Ca^{2+}-permeable channels on the primary cilium also mediate flow-induced extracellular Ca^{2+} entry. It should be noted that TRPV4 knockdown blocks flow-induced increase in *Cox2* mRNA expression, but Piezo1 knockdown did not, indicating that TRPV4 knockdown may impair transcriptional effects of osteocyte mechanotransduction. Together, these findings indicate that cilia-mediated mechanotransduction in kidney epithelial cells, which depends on PC2, follows a mechanism distinct from that in osteocytes and osteoblasts (Lee et al. 2015).

12.5.2 Microgravity in Osteoblasts and Osteocytes

Studies on bone mechanotransduction have often focused on how increased loading increases bone formation, but, importantly, the opposite also occurs. In situations of microgravity, astronauts and cosmonauts often experience a reduction in the mass of load-bearing bone. This phenomenon has also been linked to the primary cilium (Man et al. 2022). For example, in primary rat calvarial osteoblasts, primary cilia shrank and disappeared almost completely in a time-dependent manner after exposure to stimulated microgravity induced by a random positioning machine (Shi et al. 2017). In these osteoblasts, alkaline phosphatase activity was reduced along with protein levels of Runx2, BMP2, and other osteogenic markers. Treatment with cytochalasin D, an inhibitor of actin polymerization, elongates the primary cilium (Kim et al. 2010), prevents the abrogation of primary cilia in microgravity conditions, and attenuates the effects of microgravity on osteogenesis and mineralization (Shi et al. 2017).

Beyond signaling mechanisms, other mechanistic insights have been made, specifically regarding effects on ciliary structure and function. For example, microgravity reversibly depolymerizes the microtubule network in rat calvarial osteoblasts, resulting in decrease in the maximum branch length of microtubules (Shi et al. 2020). Interestingly, microtubule stabilization to prevent depolymerization does not inhibit the previously seen disappearance of primary cilia in microgravity conditions. However, preventing ciliogenesis by chloral hydrate or a siRNA targeting *Ift88* hinders microgravity-induced microtubule depolymerization (Shi et al. 2020), indicating that microtubule depolymerization in microgravity occurs via a ciliary mechanism. A separate study investigating MLO-Y4 osteocytes found that ciliary length is also reduced in microgravity. Under microgravity in MLO-Y4 osteocytes, the ratio of anterograde/retrograde particle number and the ratio of protein expression of IFT88/IFT43 were increased. However, the size of anterograde IFT particles was smaller under microgravity, possibly explaining the smaller ciliary size (Ding et al. 2020). More work is needed to better elucidate the mechanisms behind decreased cilia size in microgravity conditions.

12.5.3 Other Mechanical Forces in Osteoblasts and Osteocytes

Beyond FSS and microgravity, ciliary mechanotransduction in osteoblasts and osteocytes under a few other mechanical forces has also been investigated, particularly related to mechanical treatments for bone-related diseases. For instance, pulsed electromagnetic fields (PEMFs) have been considered as a potential treatment for osteoporosis (Funk et al. 2009). PEMFs applied to rat calvarial osteoblasts resulted in increased proliferation and osteogenic differentiation, but this effect was abrogated when siRNA targeting *Ift88* was used (Yan et al. 2015). Mechanistically, PEMF treatment upregulates BMP signaling, and BMP receptor II was also upregulated after PEMF treatment and localized at the base of the primary cilium. After *Ift88* siRNA treatment, these effects were abolished. Furthermore, knockdown of BMP receptor II reversed the osteogenic effects of PEMF treatment, indicating that PEMFs stimulate osteogenic differentiation and maturation of osteoblasts by primary cilium-mediated upregulation of BMP receptor II expression and subsequent activation of BMP signaling (Xie et al. 2016). Low-magnitude, high-frequency vibration (LMHFV) is another mechanical treatment being considered for osteoporosis (Wysocki et al. 2011). LMHFV stimulation results in shorter primary cilia and fewer ciliated cells. However, it also results in increased osteogenic differentiation which is contrary to what is typically expected to occur when the cilium shortens (Li et al. 2020). All of these effects were reversed after chloral hydrate treatment, highlighting the role of the primary cilium in LMHFV-induced osteogenesis (Li et al. 2020).

12.5.4 Ciliary Mechanosensing Machinery in Chondrocytes

It has been widely known that primary cilia in chondrocytes are crucial for skeletal development and maintain homeostasis due to their involvement in the propagation of various signaling pathways (Chinipardaz et al. 2022; Kitami et al. 2019; Haycraft et al. 2007; Koyama et al. 2007; Song et al. 2007; Liu et al. 2020; Yamaguchi et al. 2022). In addition to the role of primary cilia as chemosensors, articular and growth plate cartilages experience various mechanical loads, including compressive and tensile strain, fluid shear stress, hydrostatic pressure, and osmotic stress. For example, axonemal orientation in the cilia of equine articular chondrocytes is more pronounced in load-bearing than in non-load-bearing areas (Farnum and Wilsman 2011). Additionally, application of compressive strain to chondrocytes alters the length and incidence of cilia over time (McGlashan et al. 2010).

As described earlier, mechanical stimulations typically activate calcium signaling pathways in primary cilia, and this also occurs in chondrocytes (Fig. 12.4). For example, after application of cyclic compression to wild-type chondrocytes, mRNA expression of Aggrecan was increased, but not in cilia-deleted ORPK cells (Wann et al. 2012). Aggrecan is a structural protein important for the function of articular cartilage and its load-bearing properties (Kiani et al. 2002). Sulfated glycosaminoglycan synthesis increased after load application in wild-type chondrocytes but not in ORPK cells, highlighting the role of primary cilia in mechanosensitive proteoglycan synthesis. Mechanosensitive Ca^{2+} signaling is present in chondrocytes, just as it is in osteocytes and osteoblasts (Roberts et al. 2001). Specifically, after mechanical loading, the percentage of cells exhibiting an intracellular Ca^{2+} transient increased in wild-type chondrocytes but not in ORPK cells. Taken together, these findings indicate that the primary cilium is important for mechanotransduction in chondrocytes (Wann et al. 2012).

An additional study reveals the role of ciliary localization of polycystins in chondrocytes (Thompson et al. 2021). Application of uniaxial cyclic tensile

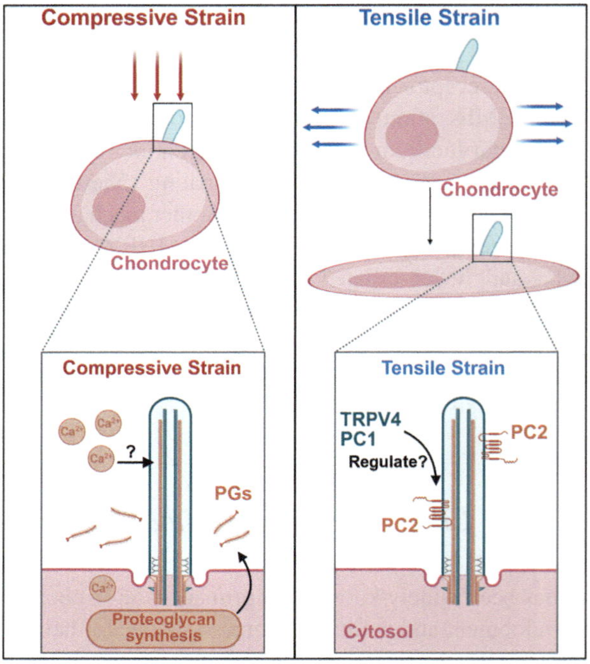

Fig. 12.4 Primary cilium as mechanosensor in chondrocyte. Application of compressive strain on chondrocytes affects the length and presence of cilia to change based on the time of compressive strain application. Primary cilia in chondrocytes have a critical role in mechanosensitive proteoglycans (PGs) synthesis under compressive strain. With uniaxial cyclic tensile strain on chondrocytes, colocalization of polycystin 2 (PC2) with cilia is increased. PC2 colocalization with cilia may be regulated by polycystin 1 (PC1) and transient receptor potential vanilloid (TRPV) 4

strain on bovine articular chondrocytes decreased cilia prevalence and length after 24 h and increased colocalization of PC2 with cilia. Interestingly, this was PC1-dependent: Knockdown of PC1 using siRNA reduced overall colocalization of PC2 with the primary cilium ion response to strain, compared to the control. Conversely, knockdown of PC2 resulted in a significant decrease in cilia length and prevalence, which was further exacerbated strain application (Thompson et al. 2021). PC2 colocalization with the cilium remained increased after strain with or without ATP, but treatment with EGTA, a Ca^{2+} chelator, or inhibition of TRPV4 reduced PC2 colocalization. This further highlights the importance of the ciliary mechanotransduction in chondrocytes (Thompson et al. 2021).

12.5.5 *Ciliary Mechanosensing Machinery in Osteoclasts*

To date, there is no clear evidence of the presence of cilia in osteoclasts. However, osteoclast function can be influenced by other skeletal cell types. Some studies have shown that this can occur in a cilia-dependent manner in response to mechanical stress (Malone et al. 2007; Temiyasathit and Jacobs 2010; Maycas et al. 2015; Tirado-Cabrera et al. 2022). In preosteoclasts, receptor activator of NF-κB ligand (RANKL) binds to receptor activator of NF-κB (RANK) and promotes osteoclast differentiation, which is inhibited by osteoprotegerin (OPG), a decoy receptor of

RANKL (Yasuda et al. 1999). Both osteocytes and osteoblasts produce OPG and RANKL (Zhao et al. 2002). In MLO-Y4 osteocytes, exposure to dynamic fluid flow increased the ratio of OPG/RANKL, but this effect was not seen in cells in which primary cilia were abrogated (Malone et al. 2007). This indicates that the primary cilium may indirectly regulate bone resorption in response to mechanical stress. In fact, in heterozygous *Pkd1* mutant mice, there was similarly a decrease in the OPG/RANKL ratio, highlighting the possible link between bone resorption and cilia loss in osteoclasts (Temiyasathit and Jacobs 2010).

12.6 Ciliary Dysfunction and Skeletal Diseases

The human skeleton is composed of diverse connective tissues that not only provide structural support and facilitate movement but also play a vital role in withstanding mechanical stress during daily functions (Quadri and Upadhyai 2023). Here, we particularly focus on possible involvement of cilia in osteoarthritis and osteoporosis, diseases in which the bone's ability to respond to mechanical cues is compromised (Yuan and Yang 2016).

12.6.1 Ciliary Dysfunction in Osteoarthritis

Osteoarthritis (OA) is a chronic degenerative joint disease most often observed in older adults, individuals with obesity, and those with prior joint injuries (Yue and Berman 2022). It is characterized by progressive breakdown of articular cartilage, leading to synovial inflammation and chronic joint damage (Abramoff and Caldera 2020). Common clinical features include joint pain, stiffness, swelling, and reduced range of motion, while imaging may show joint space narrowing and osteophyte formation (Abramoff and Caldera 2020). Although classified as a musculoskeletal disorder, OA can significantly affect overall health by disrupting sleep, contributing to depressed mood, and reducing independence (Hawker 2019). Management is multifactorial, ranging from lifestyle changes, physical therapy, and medications to more invasive options such as total joint arthroplasty (Salman et al. 2023). Recent research has identified new potential contributors to OA development, including the role of primary cilia in joint health, highlighting opportunities for future therapeutic approaches.

Prior studies highlight the critical role of chondrocyte primary cilia in both mechanotransduction and OA pathogenesis (Irianto et al. 2014; Wann et al. 2012; Ruhlen and Marberry 2014). Using a chondrocyte-specific *Ift88* deletion mouse model in which cilia were depleted from chondrocytes, cartilage was observed to be thicker yet mechanically weaker (Irianto et al. 2014). Specifically, there was a significant reduction in both instantaneous and equilibrium compressive moduli at higher indentation strains, indicating a compromise in the cartilage's load-bearing capacity (Irianto et al. 2014). In addition, cilia-deficient mice exhibited increased expression of OA markers (e.g., MMP13 and ADAMTS5) and abnormal bone morphology (Irianto et al. 2014). Similar findings were reported showing increased expression of *Mmp13*, *Adamts5*, *ColX*, and *Runx2* in chondrocyte-specific *Ift88* knockout mice (Chang et al. 2012). Another study using a joint-specific deletion of *Ift88* mouse model also showed OA-like phenotype with reduced articular cartilage stiffness and disorganized ECM (Rux et al. 2022). These findings support the concept that primary cilia dysfunction can accelerate osteoarthritic changes by disrupting both cartilage integrity and joint structure. It has been found that IFT88 protein expression is decreased in human OA chondrocytes compared to normal chondrocytes (Sun et al. 2025). In mice, the status of the primary cilium in OA is currently a matter of debate in the literature. Studies examining the cilium's prevalence and dimensions have yielded disparate results: Certain reports document reduced presence of cilia, but at a longer length (Sun et al. 2025; Wu et al. 2024), whereas contrasting findings indicate both an increased presence and length of cilia (McGlashan et al. 2008). Therefore, additional rigorous studies are necessary to elucidate the precise role of the primary cilium in OA pathology.

Inflammatory cytokines such as IL-1 promote cilia elongation via PKA, but IL-1 has also been found to reduce ciliary presence in mouse chondrocytes (Sun et al. 2025), similar to the general phenotype seen in OA. The mechanism behind IL-1-mediated cilia elongation was also investigated and it was found that integrin $\alpha 1\beta 1$ and focal adhesions are required for this (Stam and Clark 2023). Other studies have explored the role of calcium-mediated mechanical signaling in OA via primary cilia and TRPV4. One study found that the decrease in ciliary presence seen in OA correlated with increased TRPV-4 expression, but the channel was dysfunctional (Sun et al. 2025). In normal mouse chondrocytes, TRPV4 inhibition reduced Ca^{2+}-dependent NF-κB activation and increased COL2, but in *Ift88* deleted cKO chondrocytes or in chondrocytes with IL-1 induced cilia disruption (to model OA), this effect was no longer seen despite TRPV4 expression increase, indicating primary cilia integrity in chondrocytes governs the TRPV4/Ca2+/NF-κB/COL2 signaling axis (Sun et al. 2025). The dual disruption of mechanical and inflammatory signaling demonstrates the multifaceted role of primary cilia in maintaining cartilage homeostasis and preventing OA progression (Fig. 12.5).

12.6.2 Ciliary Dysfunction in Osteoporosis

Osteoporosis is an ongoing chronic condition that has many etiologies. It is commonly observed in older adults, particularly those with certain risk factors such as hormonal imbalances, a diet low in calcium or vitamin D, smoking, specific medications, high alcohol consumption, and various comorbidities (Johnston and Dagar 2020). Another etiology of osteoporosis is postmenopausal osteoporosis, which results from decreased estrogen production and leads to brittle and weakened

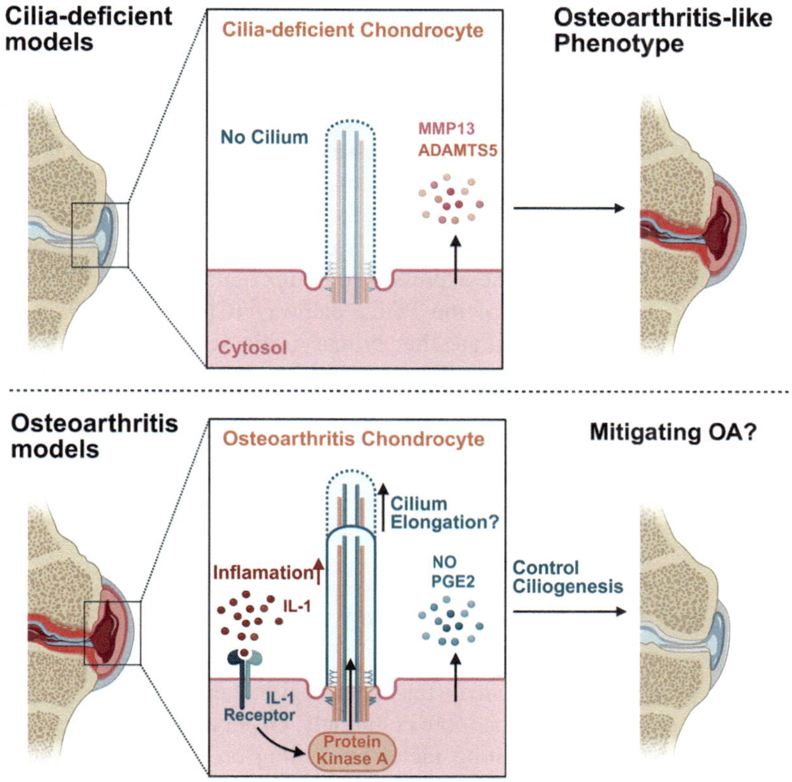

Fig. 12.5 Ciliary dysfunction and skeletal diseases. Primary cilia in chondrocytes function as mechano-inflammatory transducers, which may function in pathological conditions. Inflammatory cytokines that are present in osteoarthritis (OA), such as IL-1, promote ciliary elongation via protein kinase A (PKA) and integrins. This modulates the release of inflammatory mediators such as PGE2 and NO. In OA, there is also an observed decrease in ciliary incidence or number which is often modeled by cilia-deficient chondrocyte models. In cilia deficiency models, the absence of cilia increases the expression of osteoarthritis markers, such as MMP13 and ADAMTS5 and increases Hh pathway markers such as Ptch1 and Gli1. In IL-1-induced cilia deficiency and cilia-deficient mouse models, TRPV4 is nonfunctional and a reduction in COL2 expression is observed

bones (Cosman et al. 2024). Current guidelines for osteoporosis include preventative, nonpharmacologic, and pharmacologic management. Starting with preventive measures, especially for women, it is recommended to increase calcium intake and weight-bearing exercise (Srivastava and Deal 2002; Yong and Logan 2021). If preventive measures fail, nonpharmacologic treatment is the first line of treatment, such as increasing oral intake of calcium and vitamin D, followed by pharmacologic management with bisphosphonates, teprotide, or denosumab (Arceo-Mendoza and Camacho 2021).

Although elderly and postmenopausal osteoporosis is the most common etiology, its pathological mechanisms are still poorly understood. Therefore, recent research has begun to explore the role of primary cilia in bone health and osteoporosis. This relationship was explored by using osteoblast- and/or osteocyte-specific deletion of *Kif3a* in mice. While cilia mutant mice exhibited normal skeletal morphology at baseline (Temiyasathit et al. 2012), when subjected to cyclical axial ulnar unloading, they showed a significant decrease in anabolic response compared to controls (Temiyasathit et al. 2012). Complementing these findings, another study demonstrated that primary cilia are essential for the differentiation of mesenchymal stem cells under mechanical stimulation. Mesenchymal stem cells exposed to mechanical loading increased the expression of Runx2 and Ocn via cilia-mediated calcium influx and activation of the PGE2 pathway (Chen et al. 2016). Taken together, these studies demonstrate that primary cilia serve as mechanosensory organelles in both bone-forming cells and progenitor cells, converting mechanical forces into signals that can increase bone formation. Ciliary dysfunction can therefore be detrimental to skeletal adaptation to loading. This mechanism may provide an alternative etiology for osteoporosis in at-risk individuals and offer new therapeutic guidance and targets.

12.6.3 Oxidative Stress and Primary Cilia in Aging

Although OA and osteoporosis are distinct pathologies, they are alike in the fact that they affect the elderly population more commonly. Given this shared demographic, we wanted to explore the limiting factor of gaining on primary cilia; however, evidence is scarce.

Since bone mechanosensitivity decreases with age, it is plausible that age-related skeletal changes are partly driven by ciliary dysfunction (Verbruggen et al. 2023). Some studies have demonstrated that osteocyte primary cilia possess adenylyl cyclase 3 and GPCR signaling receptors. When cilia are disrupted, these signaling mechanisms are diminished, which parallels the reduced cAMP signaling observed in aged bone (Lian et al. 2023; Chen et al. 2025). In terms of cartilage and primary cilia, it is inferred, via the mechanisms discussed in the OA section, that age-related degeneration of primary cilia demonstrates similar findings observed in primary cilia-deficient mice (Chinipardaz et al. 2022; Yuan and Yang 2016; Zhang et al. 2023).

Further evidence of aging and primary cilia has been observed via the effects of oxidative stress. However, currently, the available evidence on the effects of oxidative stress, primary cilia, and bone is limited. The available evidence focuses on other tissues. For example, in the respiratory epithelium, it has been observed that as one ages, the oxidant tone increases activation of PLCe (Bailey et al. 2018). Thus, suppression of PKCe recovers ciliary beat frequency and mucociliary transport (Bailey et al. 2018). Supporting this notion, the use of antioxidants reduces PKCe activity and restores primary cilia function. As expected, ciliary signaling, including Hh, Wnt, PDGF, GPCR, and cAMP, is disrupted via changes in reactive oxygen species (Silva and Cavadas 2023). Along with signaling pathway disruption, there are also physical changes, such as altered tissue length, which impact ligand and mechanosensory signaling (Silva and Cavadas 2023). Taken together, these findings suggest that aging exerts a multifactorial impact on primary cilia, altering their structure and signaling. While much of the current understanding comes from animal models or nonskeletal tissues, parallels across systems suggest a potential role for ciliary dysfunction in the pathogenesis of age-related bone and cartilage degeneration. This demonstrates a gap in current research and suggests that future studies could guide the development of further treatment options for the population.

12.7 Therapeutic Implications of Utilizing Cilia Agonists and Antagonists

Pathway-specific pharmacologic modulation offers targeted approaches for treating skeletal ciliopathies. For instance, Hh pathway agonists or antagonists, particularly those modulating SMO, may normalize defective chondrocyte signaling in ciliopathies where ciliary function is partially retained (Huangfu et al. 2003; Wu et al. 2012). Mechanical signaling pathways, including possible involvement of phosphorylated FAK, are also under investigation (Antoniades et al. 2017; Leucht et al. 2007; Liu et al. 2022; Castillo et al. 2012).

It has been reported that HDAC6 inhibitor (Gradilone et al. 2013) and Forskolin (Besschetnova et al. 2010) can improve ciliary formation or stability. Therefore, these pharmacological agents may be used to restore ciliogenesis, which directly or indirectly contributes to recovering the defective cilia-mediated mechanotransduction. On the other hand, ciliogenesis inhibitors, including Ciliobrevin A/D (Firestone et al. 2012) and Roscovitine (Husson et al. 2016), have also demonstrated the potential to suppress cilia development. Additional possible ciliogenic agonists and antagonists are listed (Table 12.2).

12.8 Future Direction in Ciliary Mechanotransduction Research

In this book chapter, we emphasize the importance of cilia's role as a cellular antenna that transduces mechanical signals to regulate skeletal development and homeostasis. However, there are several outstanding questions that still need to be answered. Mechanistically, there is still more to be understood on the signaling pathways that mediate ciliary mechanotransduction in osteoblasts, osteocytes, chondrocytes, and osteoclasts. Many recent studies have broadened this field, exploring ciliary mechanical signaling in other skeletal-related cell types, including mesenchymal stem cells (Corrigan et al. 2018; Johnson et al. 2018, 2021; Hoey et al. 2011), periodontal ligament cells (Moore and Konermann 2024), periosteal

Table 12.2 Potential cilia agonists and antagonists

Small molecule	Medication category	Target diseases/symptoms	Effect for ciliogenesis	Experimental model (s)	References
Tubastatin-A	Histon-deacetylase 6 (HADC6) inhibitor	N/A (under investigation)	Induce ciliogenesis	Cholangiocarcinoma (CCA) cells, CCA model in rats	Gradilone et al. (2013)
Vorinostat	Nonspecific HADC inhibitor	Cutaneous T-cell lymphoma	Aligned cilia	Olfactory cilia in *Xenopus laevis*	Novak et al. (2025)
Clofibrate	Fibric acid derivative	Hypertriglyceridemia	Induce ciliogenesis	Human cancer cell line	Khan et al. (2016)
Gefitinib	Tyrosine kinase inhibitor	Non-small cell lung carcinoma (NSCLC)	Induce ciliogenesis	Human cancer cell line	Khan et al. (2016)
Sirolimus	Immunosuppressant Kinase inhibitor mTOR inhibitor	Organ transplant rejections, lymphangioleiomyomatosis, perivascular epithelioid cell tumors	Induce ciliogenesis	Human cancer cell line	Khan et al. (2016)
Imexon	Cyanoaziridine derivatives	N/A (under investigation)	Induce ciliogenesis	Human cancer cell line	Khan et al. (2016)
Dexamethasone	Glucocorticoid	Bronchial asthma, endocrine, rheumatic disorders	Induce ciliogenesis	Human cancer cell line	Khan et al. (2016)
Simvastatin	HMG-CoA reductase inhibitor	Hyperlipidemia, cardiovascular diseases	Induce ciliogenesis	Osteoblast	Iwaya et al. (2024)
Jasplakinolide	Actin polymerization inducer	N/A (under investigation)	Induce ciliogenesis	Human retinal pigmented epithelial cells	Nagai and Mizuno (2017)
Mitotic inhibitor-181	M-phase inhibitor	N/A (under investigation)	Elongate cilia	Retinal pigment epithelial-1 cells	Gholkar et al. (2024)

Forskolin	Intracellular cAMP promoter	N/A under investigation, dietary supplement is available for obese (Abbasi et al. (2025))	Elongate cilia	Immortalized kidney collecting duct line (IMCD), embryonically derived kidney epithelia (MEK)	Besschetnova et al. (2010)
Lithium chloride (LiCl)	Ionic compound	N/A (under investigation)	Elongate cilia	*Chlamydomonas reinhardtii*	Bigge et al. (2023)
Interleukin-1 beta (IL-1β) inhibitor	Inflammatory cytokine inhibitor	Rheumatoid arthritis, autoimmune diseases, multiple inflammatory diseases	Elongate cilia	Chondrocytes	Wann and Knight (2012)
Tumor necrosis factor-alpha (TNF-α) inhibitor	Inflammatory cytokine inhibitor	Rheumatoid arthritis, ankylosing spondylitis, multiple inflammatory diseases	Elongate cilia	Human retinal pigmented epithelial cells	Kumari et al. (2024)
Cytochalasin D	Cell-permeable mycotoxin, inhibitor of actin polymerization	N/A (under investigation)	Elongate cilia	Human retinal pigmented epithelial cells	Kim et al. (2015)
Fibroblast growth factor 2 (FGF2)	Growth factor	N/A (under investigation; several drugs that target FGFR pathway have been approved by FDA for cholangiocarcinoma and bladder cancer)	Elongate cilia	NIH3T3 cells	Kunova Bosakova et al. (2019)

(continued)

Table 12.2 (continued)

Small molecule	Medication category	Target diseases/symptoms	Effect for ciliogenesis	Experimental model (s)	References
Rapamycin	mTOR inhibitor, immunosuppressant	Cancer, organ transplant rejection, tuberous sclerosis complex, facial angiofibroma	Elongate cilia	Porcine renal epithelial cells from proximal tubule (LLCPK), mouse vascular endothelial (ET) cells	Sherpa et al. (2016)
			Shorten cilia	Zebrafish	Yuan et al. (2012)
Ciliobrevin A	Cytoplasmic dynein inhibitor	N/A (under investigation)	Inhibit ciliogenesis	NIH3T3 cells	Firestone et al. (2012)
Ciliobrevin D	Cytoplasmic dynein inhibitor	N/A (under investigation)	Inhibit ciliogenesis	NIH3T3 cells	Firestone et al. (2012)
MLN4924	NEDD8-activating enzyme (NAE) inhibitor	N/A (under investigation)	Inhibit ciliogenesis	BEAS2B cells	Mao et al. (2019)
S-CR8	Cyclin-dependent kinases (CDKs) inhibitor	N/A (under investigation)	Shorter cilia	Kidney epithelial cells	Husson et al. (2016)
Roscovitine	Cyclin-dependent kinases (CDKs) inhibitor	N/A (under investigation)	Shorter cilia	Kidney epithelial cells	Husson et al. (2016)

Potential ciliogenesis inducers and/or inhibitors are listed based on the studies conducted in various cell types

progenitors (Moore et al. 2018), and bone marrow cells (Coughlin et al. 2016). Understanding ciliary mechanotransduction in these cell types may be key in further translational studies, especially for tissue regeneration. Indeed, mechanical signaling in cilia has been implicated in bone healing after orthopedic implantation (Leucht et al. 2013) and in post-injury cartilage regeneration (Tao et al. 2023). Obtaining a more complete picture of ciliary mechanotransduction in both progenitor and differentiated skeletal cell types is needed to expand on these findings and develop therapeutics for skeletal diseases and injury. Translationally, there are still questions about how cilia function in skeletogenic cells in aging and diseases. For example, while structural and/or functional deficiency of cilia has been implicated in idiopathic scoliosis (Oliazadeh et al. 2017, 2022) and craniosynostosis (Tiberio et al. 2021), the pathological link between cilia-mediated mechanotransduction and attenuated or augmented mesenchymal cell differentiation remains elusive. This underscores the importance of cilia as mechanical sensors across all stages of life, from development to aging, and the need to further expand on their dynamic roles.

The story of the primary cilium began not in humans but in the single-celled green alga *Chlamydomonas*. Nearly three decades ago, the publication of this organism's ciliary proteome laid the foundation for much of the research discussed in this chapter (Kozminski et al. 1993). At that time, nobody expected that discoveries made from a microscopic alga would enable us to determine the physiological and pathological mechanisms of the largest and most complex organ system—the skeleton. As our understanding of cilia-mediated mechanotransduction's mechanisms continues to grow, this small cellular projection will undoubtedly remain a key in novel, far-reaching discoveries.

Acknowledgments We apologize for the authors we could not cite their significant work due to space limitations. This work was partially supported by the McGovern Medical School at UTHealth Houston Retention Fund (Y.K.), a research grant NIDCR/NIH R01DE025897 (Y.K.), and by the Bone Disease Program of Texas Rolanette and Berdon Lawrence Research Award (H.Y.).

References

Abbasi M et al (2025) Anti-obesity and metabolic effects of forskolin in obese C57BL/6J mice. Int J Mol Sci 26(14):6607

AbouAlaiwi WA et al (2009) Ciliary polycystin-2 is a mechanosensitive calcium channel involved in nitric oxide signaling cascades. Circ Res 104(7):860–869

Abramoff B, Caldera FE (2020) Osteoarthritis: pathology, diagnosis, and treatment options. Med Clin North Am 104(2):293–311

Adel Al-Lami H, Barrell WB, Liu KJ (2016) Micrognathia in mouse models of ciliopathies. Biochem Soc Trans 44(6):1753–1759. https://doi.org/10.1042/bst20160241./p

Ahmed M et al (2025) Cleavage of the Meckel-Gruber syndrome protein TMEM67 by ADAMTS9 uncouples Wnt signaling and ciliogenesis. Nat Commun 16(1):4946

Aliprantis AO et al (2008) NFATc1 in mice represses osteoprotegerin during osteoclastogenesis and dissociates systemic osteopenia from inflammation in cherubism. J Clin Invest 118(11):3775–3789

Antoniades I et al (2017) Addressing the functional determinants of FAK during ciliogenesis in multiciliated cells. J Biol Chem 292(2):488–504
Anvarian Z et al (2019) Cellular signalling by primary cilia in development, organ function and disease. Nat Rev Nephrol 15(4):199–219
Arcangeli A, Becchetti A (2006) Complex functional interaction between integrin receptors and ion channels. Trends Cell Biol 16(12):631–639
Arceo-Mendoza RM, Camacho PM (2021) Postmenopausal osteoporosis: latest guidelines. Endocrinol Metab Clin N Am 50(2):167–178
Arts HH et al (2011) *C14ORF179* encoding IFT43 is mutated in Sensenbrenner syndrome. J Med Genet 48(6):390–395
Aubert-Mucca M et al (2023) Ellis-Van Creveld syndrome: clinical and molecular analysis of 50 individuals. J Med Genet 60(4):337–345
Aulehla A, Pourquie O (2009) More than patterning—hox genes and the control of posterior axial elongation. Dev Cell 17(4):439–440
Awan A et al (2010) Immunoflourescence and mRNA analysis of human embryonic stem cells (hESCs) grown under feeder-free conditions. Methods Mol Biol 584:195–210
Baala L et al (2007) The Meckel-Gruber syndrome gene, MKS3, is mutated in Joubert syndrome. Am J Hum Genet 80(1):186–194
Bachmann-Gagescu R et al (2015) Joubert syndrome: a model for untangling recessive disorders with extreme genetic heterogeneity. J Med Genet 52(8):514–522
Badano JL et al (2006) The ciliopathies: an emerging class of human genetic disorders. Annu Rev Genomics Hum Genet 7:125–148
Badiner N et al (2017) Mutations in DYNC2H1, the cytoplasmic dynein 2, heavy chain 1 motor protein gene, cause short-rib polydactyly type I, Saldino-Noonan type. Clin Genet 92(2):158–165
Bailey KL et al (2018) Oxidative stress associated with aging activates protein kinase Cε, leading to cilia slowing. Am J Physiol Lung Cell Mol Physiol 315(5):L882–l890
Baneyx G, Baugh L, Vogel V (2002) Fibronectin extension and unfolding within cell matrix fibrils controlled by cytoskeletal tension. Proc Natl Acad Sci 99(8):5139–5143
Barker AR, Thomas R, Dawe HR (2014) Meckel-Gruber syndrome and the role of primary cilia in kidney, skeleton, and central nervous system development. Organogenesis 10(1):96–107
Basten SG, Giles RH (2013) Functional aspects of primary cilia in signaling, cell cycle and tumorigenesis. Cilia 2(1):6
Basso N, Heersche JNM (2002) Characteristics of in vitro osteoblastic cell loading models. Bone 30(2):347–351. https://doi.org/10.1016/S8756-3282(01)00678-0
Bayat A, Kerr B, Douzgou S (2017) The evolving craniofacial phenotype of a patient with Sensenbrenner syndrome caused by IFT140 compound heterozygous mutations. Clin Dysmorphol 26(4):247–251
Beales PL et al (2007) IFT80, which encodes a conserved intraflagellar transport protein, is mutated in Jeune asphyxiating thoracic dystrophy. Nat Genet 39(6):727–729
Berbari NF et al (2009) The primary cilium as a complex signaling center. Curr Biol 19(13):R526–R535
Berendsen AD, Olsen BR (2015) Bone development. Bone 80:14–18. https://doi.org/10.1016/j.bone.2015.04.035
Besschetnova TY et al (2010) Identification of signaling pathways regulating primary cilium length and flow-mediated adaptation. Curr Biol 20(2):182–187
Baldridge D et al (2010) Signaling pathways in human skeletal dysplasias. Annu Rev Genomics Hum Genet 11:189–217. https://doi.org/10.1146/annurev-genom-082908-150158./p
Bi J et al (2017) Suppression of $\alpha v\beta 6$ integrin expression by polymicrobial oral biofilms in gingival epithelial cells. Sci Rep 7(1):4411
Bialas NJ et al (2009) Functional interactions between the ciliopathy-associated Meckel syndrome 1 (MKS1) protein and two novel MKS1-related (MKSR) proteins. J Cell Sci 122(Pt 5):611–624
Bigge BM, Dougherty LL, Avasthi P (2023) Lithium-induced ciliary lengthening sparks Arp2/3 complex–dependent endocytosis. Mol Biol Cell 34(4):ar26

Bosakova M et al (2020) Mutations in GRK2 cause Jeune syndrome by impairing Hedgehog and canonical Wnt signaling. EMBO Mol Med 12(11):e11739
Bronner ME, LeDouarin NM (2012) Development and evolution of the neural crest: an overview. Dev Biol 366(1):2–9. https://doi.org/10.1016/j.ydbio.2011.12.042./pp11
Brown JM, Witman GB (2014) Cilia and diseases. Bioscience 64(12):1126–1137
Bruel AL et al (2023) Pathogenic RAB34 variants impair primary cilium assembly and cause a novel oral-facial-digital syndrome. Hum Mol Genet 32(18):2822–2831
Bruinsma R (2005) Theory of force regulation by nascent adhesion sites. Biophys J 89(1):87–94
Brunetti-Pierri R et al (2021) Mild clinical presentation of Joubert syndrome in a male adult carrying biallelic MKS1 truncating variants. Diagnostics (Basel) 11(7):1218
Bujakowska KM et al (2015) Mutations in IFT172 cause isolated retinal degeneration and Bardet-Biedl syndrome. Hum Mol Genet 24(1):230–242
Buqaileh R, Alshriem LA, AbouAlaiwi W (2025) Ciliary G-protein coupled receptor signaling in polycystic kidney disease. Int J Mol Sci 26(11):4971
Campbell ID, Humphries MJ (2011) Integrin structure, activation, and interactions. Cold Spring Harb Perspect Biol 3(3):a004994
Cantagrel V et al (2008) Mutations in the cilia gene ARL13B lead to the classical form of Joubert syndrome. Am J Hum Genet 83(2):170–179
Castillo AB et al (2012) Focal adhesion kinase plays a role in osteoblast mechanotransduction in vitro but does not affect load-induced bone formation in vivo. PLoS One 7(9):e43291
Chang CF, Ramaswamy G, Serra R (2012) Depletion of primary cilia in articular chondrocytes results in reduced Gli3 repressor to activator ratio, increased Hedgehog signaling, and symptoms of early osteoarthritis. Osteoarthr Cartil 20(2):152–161
Chalfie M (2009) Neurosensory mechanotransduction. Nat Rev Mol Cell Biol 10(1):44–52. https://doi.org/10.1038/nrm2595
Chen NX et al (2000) Ca(2+) regulates fluid shear-induced cytoskeletal reorganization and gene expression in osteoblasts. Am J Physiol Cell Physiol 278(5):C989–C997
Chen JC et al (2016) Mechanical signals promote osteogenic fate through a primary cilia-mediated mechanism. FASEB J 30(4):1504–1511
Chen N et al (2025) Mechanosignaling in osteoporosis: when cells feel the force. Int J Mol Sci 26(9):4007
Chinipardaz Z et al (2022) Role of primary cilia in bone and cartilage. J Dent Res 101(3):253–260
Chiquet M et al (2003) How do fibroblasts translate mechanical signals into changes in extracellular matrix production? Matrix Biol 22(1):73–80
Christensen ST et al (2017) Primary cilia and coordination of receptor tyrosine kinase (RTK) and transforming growth factor β (TGF-β) signaling. Cold Spring Harb Perspect Biol 9(6):a028167
Clapham DE (2003) TRP channels as cellular sensors. Nature 426(6966):517–524
Clement CA et al (2013) TGF-β signaling is associated with endocytosis at the pocket region of the primary cilium. Cell Rep 3(6):1806–1814
Clews C et al (2025) Taking a closer look at matrix vesicle biogenesis. J Bone Miner Res 40(8):931–945. https://doi.org/10.1093/jbmr/zjaf076./pp22
Collin O et al (2008) Self-organized podosomes are dynamic mechanosensors. Curr Biol 18(17):1288–1294
Corbit KC et al (2005) Vertebrate smoothened functions at the primary cilium. Nature 437(7061):1018–1021
Corbit KC et al (2008) Kif3a constrains beta-catenin-dependent Wnt signalling through dual ciliary and non-ciliary mechanisms. Nat Cell Biol 10(1):70–76. https://doi.org/10.1038/ncb1670
Córdova-Fletes C et al (2018) Uncommon runs of homozygosity disclose homozygous missense mutations in two ciliopathy-related genes (SPAG17 and WDR35) in a patient with multiple brain and skeletal anomalies. Eur J Med Genet 61(3):161–167
Corrigan MA et al (2018) TRPV4-mediates oscillatory fluid shear mechanotransduction in mesenchymal stem cells in part via the primary cilium. Sci Rep 8(1):3824
Coschiera A et al (2024) Primary cilia promote the differentiation of human neurons through the WNT signaling pathway. BMC Biol 22(1):48

Cosman F et al (2024) Goal-directed osteoporosis treatment: ASBMR/BHOF task force position statement 2024. J Bone Miner Res 39(10):1393–1405. https://doi.org/10.1093/jbmr/zjae119

Coste B et al (2010) Piezo1 and Piezo2 are essential components of distinct mechanically activated cation channels. Science 330(6000):55–60

Coughlin TR et al (2016) Primary cilia expression in bone marrow in response to mechanical stimulation in explant bioreactor culture. Eur Cell Mater 32:111–122

Da Silva JD et al (2023) Establishing an objective clinical spectrum, genotype-phenotype correlations, and CRMP1 as a modifier in the Ellis-van Creveld syndrome: the first systematic review of EVC- and EVC2-associated conditions. Genet Med Open 1(1):100781

Dafinger C et al (2011) Mutations in KIF7 link Joubert syndrome with Sonic Hedgehog signaling and microtubule dynamics. J Clin Invest 121(7):2662–2667

Dagoneau N et al (2009) DYNC2H1 mutations cause asphyxiating thoracic dystrophy and short rib-polydactyly syndrome, type III. Am J Hum Genet 84(5):706–711

Damerla RR et al (2015) Novel Jbts17 mutant mouse model of Joubert syndrome with cilia transition zone defects and cerebellar and other ciliopathy related anomalies. Hum Mol Genet 24(14):3994–4005

Danilov AI et al (2009) Ultrastructural and antigenic properties of neural stem cells and their progeny in adult rat subventricular zone. Glia 57(2):136–152

Davis EE et al (2011) TTC21B contributes both causal and modifying alleles across the ciliopathy spectrum. Nat Genet 43(3):189–196

De Mori R et al (2017) Hypomorphic recessive variants in SUFU impair the sonic hedgehog pathway and cause Joubert syndrome with cranio-facial and skeletal defects. Am J Hum Genet 101(4):552–563

Delaine-Smith RM, Sittichokechaiwut A, Reilly GC (2014) Primary cilia respond to fluid shear stress and mediate flow-induced calcium deposition in osteoblasts. FASEB J 28(1):430–439

Demoulin JB, Essaghir A (2014) PDGF receptor signaling networks in normal and cancer cells. Cytokine Growth Factor Rev 25(3):273–283

Demoulin JB, Montano-Almendras CP (2012) Platelet-derived growth factors and their receptors in normal and malignant hematopoiesis. Am J Blood Res 2(1):44–56

Denhardt DT, Guo X (1993) Osteopontin: a protein with diverse functions. FASEB J 7(15):1475–1482

Derderian C, Canales GI, Reiter JF (2023) Seriously cilia: a tiny organelle illuminates evolution, disease, and intercellular communication. Dev Cell 58(15):1333–1349. https://doi.org/10.1016/j.devcel.2023.06.013./pp41

Ding D et al (2020) The microgravity induces the ciliary shortening and an increased ratio of anterograde/retrograde intraflagellar transport of osteocytes. Biochem Biophys Res Commun 530(1):167–172

Dollfus H et al (2024) Bardet-Biedl syndrome improved diagnosis criteria and management: inter European reference networks consensus statement and recommendations. Eur J Hum Genet 32(11):1347–1360

Du G et al (2020) Roles of TRPV4 and piezo channels in stretch-evoked Ca(2+) response in chondrocytes. Exp Biol Med (Maywood) 245(3):180–189

Duncan RL, Turner CH (1995) Mechanotransduction and the functional response of bone to mechanical strain. Calcif Tissue Int 57(5):344–358

Edvardson S et al (2010) Joubert syndrome 2 (JBTS2) in Ashkenazi Jews is associated with a TMEM216 mutation. Am J Hum Genet 86(1):93–97

Egeberg DL et al (2012) Primary cilia and aberrant cell signaling in epithelial ovarian cancer. Cilia 1(1):15

Emiralioglu N et al (2018) DYNC2H1 mutation causes Jeune syndrome and recurrent lung infections associated with ciliopathy. Clin Respir J 12(3):1017–1020

Failler M et al (2021) Whole-genome screen identifies diverse pathways that negatively regulate ciliogenesis. Mol Biol Cell 32(2):169–185

Farahani RM, Xaymardan M (2015) Platelet-derived growth factor receptor alpha as a marker of mesenchymal stem cells in development and stem cell biology. Stem Cells Int 2015:362753

Farnum CE, Wilsman NJ (2011) Orientation of primary cilia of articular chondrocytes in three-dimensional space. Anat Rec 294(3):533–549
Ferrante MI et al (2001) Identification of the gene for oral-facial-digital type I syndrome. Am J Hum Genet 68(3):569–576
Firestone AJ et al (2012) Small-molecule inhibitors of the AAA+ ATPase motor cytoplasmic dynein. Nature 484(7392):125–129
Frank V et al (2008) Mutations of the CEP290 gene encoding a centrosomal protein cause Meckel-Gruber syndrome. Hum Mutat 29(1):45–52
Frost HM (1987) Bone "mass" and the "mechanostat": a proposal. Anat Rec 219(1):1–9
Frost HM (2003) Bone's mechanostat: a 2003 update. Anat Rec A Discov Mol Cell Evol Biol 275(2):1081–1101
Funk RHW, Monsees T, Özkucur N (2009) Electromagnetic effects—from cell biology to medicine. Prog Histochem Cytochem 43(4):177–264
Galdzicka M et al (2002) A new gene, EVC2, is mutated in Ellis-van Creveld syndrome. Mol Genet Metab 77(4):291–295
Gargalionis AN et al (2024) Runx2 and polycystins in bone mechanotransduction: challenges for therapeutic opportunities. Int J Mol Sci 25(10):5291
Geiger B, Yamada KM (2011) Molecular architecture and function of matrix adhesions. Cold Spring Harb Perspect Biol 3(5):a005033
Geoghegan IP, McNamara LM, Hoey DA (2021) Estrogen withdrawal alters cytoskeletal and primary ciliary dynamics resulting in increased Hedgehog and osteoclastogenic paracrine signalling inosteocytes. Sci Rep 11(1):9272. https://doi.org/10.1038/s41598-021-88633-6./pp206
Gerdes JM, Davis EE, Katsanis N (2009) The vertebrate primary cilium in development, homeostasis, and disease. Cell 137(1):32–45
Gerhardt C et al (2013) The ciliary protein Ftm is required for ventricular wall and septal development. PLoS One 8(2):e57545
Gholkar AA et al (2024) MI-181 modulates cilia length and restores cilia length in cells with defective shortened cilia. ACS Chem Biol 19(8):1733–1742
Gigante ED, Caspary T (2020) Signaling in the primary cilium through the lens of the hedgehog pathway. WIREs Dev Biol 9(6):e377
Gilissen C et al (2010) Exome sequencing identifies *WDR35* variants involved in Sensenbrenner syndrome. Am J Hum Genet 87(3):418–423
Gilula NB, Satir P (1972) The ciliary necklace. A ciliary membrane specialization. J Cell Biol 53(2):494–509. https://doi.org/10.1083/jcb.53.2.494./p
Girisha KM et al (2016) A homozygous nonsense variant in IFT52 is associated with a human skeletal ciliopathy. Clin Genet 90(6):536–539
Goldgof GM et al (2016) Comparative chemical genomics reveal that the spiroindolone antimalarial KAE609 (Cipargamin) is a P-type ATPase inhibitor. Sci Rep 6(1):27806
Gorden NT et al (2008) CC2D2A is mutated in Joubert syndrome and interacts with the ciliopathy-associated basal body protein CEP290. Am J Hum Genet 83(5):559–571
Gradilone SA et al (2013) HDAC6 inhibition restores ciliary expression and decreases tumor growth. Cancer Res 73(7):2259–2270
Halbritter J et al (2013) Defects in the IFT-B component IFT172 cause Jeune and Mainzer-Saldino syndromes in humans. Am J Hum Genet 93(5):915–925
Hammarsjö A et al (2017) Novel KIAA0753 mutations extend the phenotype of skeletal ciliopathies. Sci Rep 7(1):15585
Hartill V et al (2017) Meckel-Gruber syndrome: an update on diagnosis, clinical management, and research advances. Front Pediatr 5:244
Hawker GA (2019) Osteoarthritis is a serious disease. Clin Exp Rheumatol 37 Suppl 120(5):3–6
Haycraft CJ et al (2007) Intraflagellar transport is essential for endochondral bone formation. Development 134(2):307–316
Heldin CH, Lennartsson J (2013) Structural and functional properties of platelet-derived growth factor and stem cell factor receptors. Cold Spring Harb Perspect Biol 5(8):a009100

Helm BM et al (2017) Partial uniparental isodisomy of chromosome 16 unmasks a deleterious biallelic mutation in IFT140 that causes Mainzer-Saldino syndrome. Hum Genomics 11(1):16

Hoenderop JG et al (2003) Renal Ca^{2+} wasting, hyperabsorption, and reduced bone thickness in mice lacking TRPV5. J Clin Invest 112(12):1906–1914

Hoey DA, Kelly DJ, Jacobs CR (2011) A role for the primary cilium in paracrine signaling between mechanically stimulated osteocytes and mesenchymal stem cells. Biochem Biophys Res Commun 412(1):182–187

Hoffman BD, Grashoff C, Schwartz MA (2011) Dynamic molecular processes mediate cellular mechanotransduction. Nature 475(7356):316–323. https://doi.org/10.1038/nature10316./pp138

Hojo H, Tani S, Ohba S (2024) Modeling of skeletal development and diseases using human pluripotent stem cells. J Bone Miner Res 40(1):5–19. https://doi.org/10.1093/jbmr/zjae178./pp19

Hong JM et al (2014) Electromagnetically controllable osteoclast activity. Bone 62:99–107

Horani A, Brody SL, Ferkol TW (2014) Picking up speed: advances in the genetics of primary ciliary dyskinesia. Pediatr Res 75(1):158–164

Hinz B (2007) Formation and function of the myofibroblast during tissue repair. J Invest Dermatol 127(3):526–537. https://doi.org/10.1038/sj.jid.5700613

Huangfu D, Anderson KV (2005) Cilia and hedgehog responsiveness in the mouse. Proc Natl Acad Sci 102(32):11325–11330

Huangfu D et al (2003) Hedgehog signalling in the mouse requires intraflagellar transport proteins. Nature 426(6962):83–87

Husson H et al (2016) Reduction of ciliary length through pharmacologic or genetic inhibition of CDK5 attenuates polycystic kidney disease in a model of nephronophthisis. Hum Mol Genet 25(11):2245–2255

Hynes RO (2002) Integrins: bidirectional, allosteric signaling machines. Cell 110(6):673–687. https://doi.org/10.1016/s0092-8674(02)00971-6./pp176

Ingber DE (2006) Cellular mechanotransduction: putting all the pieces together again. Faseb J 20(7):811–827. https://doi.org/10.1096/fj.05-5424rev./pp139

Irianto J et al (2014) Depletion of chondrocyte primary cilia reduces the compressivemodulus of articular cartilage. J Biomech 47(2):579–582. https://doi.org/10.1016/j.jbiomech.2013.11.040./pp272

Ishida Y et al (2021) Molecular basis of ciliary defects caused by compound heterozygous IFT144/WDR19 mutations found in cranioectodermal dysplasia. Hum Mol Genet 30(3–4):213–225

Ishikawa H, Marshall WF (2011) Ciliogenesis: building the cell's antenna. Nat Rev Mol Cell Biol 12(4):222–234. https://doi.org/10.1038/nrm3085./pp42

Iwaya C, Suzuki A, Iwata J (2024) Loss of Sc5d results in micrognathia due to a failure in osteoblast differentiation. J Adv Res 65:153–165

Jin X et al (2014) Cilioplasm is a cellular compartment for calcium signaling in response to mechanical and chemical stimuli. Cell Mol Life Sci 71(11):2165–2178

Jing J et al (2023) Hedgehog signaling in tissue homeostasis, cancers and targeted therapies. Signal Transduct Target Ther 8(1):315

Johnson GP et al (2018) Mesenchymal stem cell mechanotransduction is cAMP dependent and regulated by adenylyl cyclase 6 and the primary cilium. J Cell Sci 131(21):jcs222737

Johnson GP, Fair S, Hoey DA (2021) Primary cilium-mediated MSC mechanotransduction is dependent on Gpr161 regulation of hedgehog signalling. Bone 145:115846

Johnston CB, Dagar M (2020) Osteoporosis in older adults. Med Clin North Am 104(5):873–884

Kadry YA, Calderwood DA (2020) Chapter 22: Structural and signaling functions of integrins. Biochim Biophys Acta Biomembr 1862(5):183206. https://doi.org/10.1016/j.bbamem.2020.183206./p

Kaku M, Komatsu Y (2017) Functional diversity of ciliary proteins in bone development and disease. Curr Osteoporos Rep 15(2):96–102

Kameyama S et al (2013) Short-term mechanical stress inhibits osteoclastogenesis via suppression of DC-STAMP in RAW264.7 cells. Int J Mol Med 31(2):292–298

Kang SS et al (2012) Human skeletal dysplasia caused by a constitutive activated transient receptor potential vanilloid 4 (TRPV4) cation channel mutation. Exp Mol Med 44(12):707–722
Katsumi A et al (2004) Integrins in mechanotransduction. J Biol Chem 279(13):12001–12004
Keppler-Noreuil KM et al (2011) Clinical insights gained from eight new cases and review of reported cases with Jeune syndrome (asphyxiating thoracic dystrophy). Am J Med Genet A 155a(5):1021–1032
Khan NA et al (2016) Identification of drugs that restore primary cilium expression in cancer cells. Oncotarget 7(9):9975
Kiani C et al (2002) Structure and function of aggrecan. Cell Res 12(1):19–32
Kim J et al (2010) Functional genomic screen for modulators of ciliogenesis and cilium length. Nature 464(7291):1048–1051
Kim S et al (2011) Nde1-mediated inhibition of ciliogenesis affects cell cycle re-entry. Nat Cell Biol 13(4):351–360
Kim J et al (2015) Actin remodelling factors control ciliogenesis by regulating YAP/TAZ activity and vesicle trafficking. Nat Commun 6(1):6781
Kitami M et al (2019) IFT20 is required for the maintenance of cartilaginous matrix in condylar cartilage. Biochem Biophys Res Commun 509(1):222–226
Klein D, Ammann F (1969) The syndrome of Laurence-Moon-Bardet-Biedl and allied diseases in Switzerland. Clinical, genetic and epidemiological studies. J Neurol Sci 9(3):479–513
Kloc M, Wosik J (2025) Mechanical forces, nucleus, chromosomes, and chromatin. Biomolecules 15(3):354. https://doi.org/10.3390/biom15030354
Koivisto L et al (2018) Integrin αvβ6: structure, function and role in health and disease. Int J Biochem Cell Biol 99:186–196
Komatsu Y et al (2016) TMEM67 mutations found in a case of Joubert syndrome with renal hypodysplasia. CEN Case Rep 5(2):137–140
Konstantinidou AE et al (2009) Cranioectodermal dysplasia: a probable ciliopathy. Am J Med Genet A 149a(10):2206–2211
Kovacs CS et al (2021) The role of biomineralization in disorders of skeletal development and tooth formation. Nat Rev Endocrinol 17(6):336–349. https://doi.org/10.1038/s41574-021-00488-z
Koyama E et al (2007) Conditional Kif3a ablation causes abnormal hedgehog signaling topography, growth plate dysfunction, and excessive bone and cartilage formation during mouse skeletogenesis. Development 134(11):2159–2169
Kozminski KG et al (1993) A motility in the eukaryotic flagellum unrelated to flagellar beating. Proc Natl Acad Sci 90(12):5519–5523
Krakow D et al (2009) Mutations in the gene encoding the calcium-permeable ion channel TRPV4 produce spondylometaphyseal dysplasia, Kozlowski type and metatropic dysplasia. Am J Hum Genet 84(3):307–315
Kumari A et al (2024) TNF-alpha promotes cilia elongation via mixed lineage kinases signaling in mouse fibroblasts and human RPE-1 cells. Cytoskeleton 81(11):639–647
Kunova Bosakova M et al (2019) Fibroblast growth factor receptor influences primary cilium length through an interaction with intestinal cell kinase. Proc Natl Acad Sci 116(10):4316–4325
Lancaster MA, Schroth J, Gleeson JG (2011) Subcellular spatial regulation of canonical Wnt signalling at the primary cilium. Nat Cell Biol 13(6):700–707
Lee KL et al (2015) The primary cilium functions as a mechanical and calcium signaling nexus. Cilia 4(1):7
Lecka-Czernik B, Rosen CJ, Napoli N (2025) The role of bone in whole-body energy metabolism. Nat Rev Endocrinol 21(12):743–756. https://doi.org/10.1038/s41574-025-01162-4
Leightner AC et al (2013) The Meckel syndrome protein meckelin (TMEM67) is a key regulator of cilia function but is not required for tissue planar polarity. Hum Mol Genet 22(10):2024–2040
Leucht P et al (2007) FAK-mediated mechanotransduction in skeletal regeneration. PLoS One 2(4):e390
Leucht P et al (2013) Primary cilia act as mechanosensors during bone healing around an implant. Med Eng Phys 35(3):392–402

Li CF et al (1995) Estrogen enhances alpha v beta 3 integrin expression by avian osteoclast precursors via stabilization of beta 3 integrin mRNA. Mol Endocrinol 9(7):805–813

Li S et al (1997) Fluid shear stress activation of focal adhesion kinase. linking to mitogen-activated protein kinases. J Biol Chem 272(48):30455–30462. https://doi.org/10.1074/jbc.272.48.30455

Li J et al (2014) Piezo1 integration of vascular architecture with physiological force. Nature 515(7526):279–282

Li X et al (2019) Stimulation of Piezo1 by mechanical signals promotes bone anabolism. elife 8:e49631

Li Y-H et al (2020) Primary cilia respond to intermittent low-magnitude, high-frequency vibration and mediate vibration-induced effects in osteoblasts. Am J Phys Cell Phys 318(1):C73–C82

Li X et al (2022) Role of primary cilia in skeletal disorders. Stem Cells Int 2022:6063423

Lian F et al (2023) Recent advances in primary cilia in bone metabolism. Front Endocrinol (Lausanne) 14:1259650

Liang W et al (2015) Integrin β1 gene therapy enhances in vitro creation of tissue-engineered cartilage under periodic mechanical stress. Cell Physiol Biochem 37(4):1301–1314

Lieben L, Carmeliet G (2012) The involvement of TRP channels in bone homeostasis. Front Endocrinol (Lausanne) 3:99

Lin AE et al (2013) Sensenbrenner syndrome (Cranioectodermal dysplasia): clinical and molecular analyses of 39 patients including two new patients. Am J Med Genet A 161a(11):2762–2776

Lin Y, Ren J, McGrath C (2022) Mechanosensitive Piezo1 and Piezo2 ion channels in craniofacial development and dentistry: recent advances and prospects. Front Physiol 13:1039714

Liu M et al (2020) IFT80 is required for fracture healing through controlling the regulation of TGF-β signaling in chondrocyte differentiation and function. J Bone Miner Res 35(3):571–582

Liu P et al (2022) Effects of mechanical stress stimulation on function and expression mechanism of osteoblasts. Front Bioeng Biotechnol 10:830722

Louvi A, Grove EA (2011) Cilia in the CNS: the quiet organelle claims center stage. Neuron 69(6):1046–1060

Malone AM et al (2007) Primary cilia mediate mechanosensing in bone cells by a calcium-independent mechanism. Proc Natl Acad Sci USA 104(33):13325–13330

Mammoto T, Ingber DE (2010) Mechanical control of tissue and organ development. Development 137(9):1407–1420. https://doi.org/10.1242/dev.024166./p

Man J et al (2022) The effects of microgravity on bone structure and function. NPJ Microgravity 8(1):9

Mao H et al (2019) Neddylation inhibitor MLN4924 suppresses cilia formation by modulating AKT1. Protein Cell 10(10):726–744

Mascharak S et al (2024) Modelling and targeting mechanical forces in organ fibrosis. Nat Rev Bioeng 2(4):305–323. https://doi.org/10.1038/s44222-023-00144-3

Marinissen MJ, Gutkind JS (2001) G-protein-coupled receptors and signaling networks: emerging paradigms. Trends Pharmacol Sci 22(7):368–376

Marshall WF, Rosenbaum JL (1999) Cell division: the renaissance of the centriole. Curr Biol 9(6):R218–R220. https://doi.org/10.1016/s0960-9822(99)80133-x./pp44

Masuyama R et al (2008) TRPV4-mediated calcium influx regulates terminal differentiation of osteoclasts. Cell Metab 8(3):257–265

Masyuk AI et al (2006) Cholangiocyte cilia detect changes in luminal fluid flow and transmit them into intracellular Ca^{2+} and cAMP signaling. Gastroenterology 131(3):911–920

Maycas M et al (2015) Role of the parathyroid hormone type 1 receptor (PTH1R) as a mechanosensor in osteocyte survival. J Bone Miner Res 30(7):1231–1244

May-Simera HL, Kelley MW (2012) Cilia, Wnt signaling, and the cytoskeleton. Cilia 1(1):7

McGlashan SR, Jensen CG, Poole CA (2006) Localization of extracellular matrix receptors on the chondrocyte primary cilium. J Histochem Cytochem 54(9):1005–1014

McGlashan SR et al (2008) Primary cilia in osteoarthritic chondrocytes: from chondrons to clusters. Dev Dyn 237(8):2013–2020. https://doi.org/10.1002/dvdy.21501./p

McGlashan SR et al (2010) Mechanical loading modulates chondrocyte primary cilia incidence and length. Cell Biol Int 34(5):441–446

McInerney-Leo AM et al (2013) Short-rib polydactyly and Jeune syndromes are caused by mutations in WDR60. Am J Hum Genet 93(3):515–523
Melluso A et al (2023) Bardet-Biedl syndrome: current perspectives and clinical outlook. Ther Clin Risk Manag 19:115–132
Merrick D et al (2019) Polycystin-1 regulates bone development through an interaction with the transcriptional coactivator TAZ. Hum Mol Genet 28(1):16–30
Mill P, Christensen ST, Pedersen LB (2023) Primary cilia as dynamic and diverse signalling hubs in development and disease. Nat Rev Genet 24(7):421–441. https://doi.org/10.1038/s41576-023-00587-9
Miller KA et al (2013) Cauli: a mouse strain with an Ift140 mutation that results in a skeletal ciliopathy modelling Jeune syndrome. PLoS Genet 9(8):e1003746
Miranti CK, Brugge JS (2002) Sensing the environment: a historical perspective on integrin signal transduction. Nat Cell Biol 4(4):E83–E90
Mitchison HM, Valente EM (2017) Motile and non-motile cilia in human pathology: from function to phenotypes. J Pathol 241(2):294–309. https://doi.org/10.1002/path.4843./p
Mishina Y, Snider TN (2014) Neural crest cell signaling pathways critical to cranial bone development and pathology. Exp Cell Res 325(2):138–147. https://doi.org/10.1016/j.yexcr.2014.01.019./p
Mizoguchi F et al (2008) Transient receptor potential vanilloid 4 deficiency suppresses unloading-induced bone loss. J Cell Physiol 216(1):47–53
Molla-Herman A et al (2010) The ciliary pocket: an endocytic membrane domain at the base of primary and motile cilia. J Cell Sci 123(10):1785–1795. https://doi.org/10.1242/jcs.059519
Moore ER, Konermann A (2024) Investigating the role of primary cilia and bone morphogenetic protein signaling in periodontal ligament response to orthodontic strain in vivo and in vitro: a pilot study. Int J Mol Sci 25(23):12648
Moore ER et al (2018) Periosteal progenitors contribute to load-induced bone formation in adult mice and require primary cilia to sense mechanical stimulation. Stem Cell Res Ther 9(1):190
Morinobu M et al (2009) Osteopontin expression in osteoblasts and osteocytes during bone formation under mechanical stress in the calvarial suture in vivo. J Bone Miner Res 18(9):1706–1715
Muller J et al (2010) Identification of 28 novel mutations in the Bardet-Biedl syndrome genes: the burden of private mutations in an extensively heterogeneous disease. Hum Genet 127(5):583–593
Mizoguchi T, Ono N (2021) The diverse origin of bone-forming osteoblasts. J Bone Miner Res 36(8):1432–1447. https://doi.org/10.1002/jbmr.4410./p
Nagai T, Mizuno K (2017) Jasplakinolide induces primary cilium formation through cell rounding and YAP inactivation. PLoS One 12(8):e0183030
Nakashima T, Takayanagi H (2011) New regulation mechanisms of osteoclast differentiation. Ann N Y Acad Sci 1240:E13–E18. https://doi.org/10.1111/j.1749-6632.2011.06373.x./pp25
Nauli SM et al (2003) Polycystins 1 and 2 mediate mechanosensation in the primary cilium of kidney cells. Nat Genet 33(2):129–137
Nauli SM et al (2008) Endothelial cilia are fluid shear sensors that regulate calcium signaling and nitric oxide production through polycystin-1. Circulation 117(9):1161–1171
Noden DM, Trainor PA (2005) Relations and interactions between cranial mesoderm and neural crest populations. J Anat 207(5):575–601. https://doi.org/10.1111/j.1469-7580.2005.00473.x./pp12
Neben CL, Merrill AE (2015) Signaling pathways in craniofacial development: insights from rare skeletal disorders. Curr Top Dev Biol 115:493–542. https://doi.org/10.1016/bs.ctdb.2015.09.005./pp13
Nguyen AM, Jacobs CR (2013) Emerging role of primary cilia as mechanosensors in osteocytes. Bone 54(2):196–204. https://doi.org/10.1016/j.bone.2012.11.016./pp39
Niederlova V et al (2019) Meta-analysis of genotype-phenotype associations in Bardet-Biedl syndrome uncovers differences among causative genes. Hum Mutat 40(11):2068–2087

Nishimura Y et al (2019) Primary cilia as signaling hubs in health and disease. Adv Sci 6(1):1801138

Noda K et al (2016) Canonical and noncanonical intraflagellar transport regulates craniofacial skeletal development. Proc Natl Acad Sci 113(19):E2589–E2597

Novak R et al (2025) AI-enabled drug prediction and gene network analysis reveal therapeutic use of vorinostat for Rett syndrome in preclinical models. Commun Med 5(1):249

Nusse R, Varmus HE (1982) Many tumors induced by the mouse mammary tumor virus contain a provirus integrated in the same region of the host genome. Cell 31(1):99–109

Ocbina PJ, Tuson M, Anderson KV (2009) Primary cilia are not required for normal canonical Wnt signaling in the mouse embryo. PLoS One 4(8):e6839. https://doi.org/10.1371/journal.pone.0006839

Oliazadeh N et al (2017) Identification of elongated primary cilia with impaired mechanotransduction in idiopathic scoliosis patients. Sci Rep 7(1):44260

Oliazadeh N et al (2022) Altered mechanotransduction in adolescent idiopathic scoliosis osteoblasts: an exploratory in vitro study. Sci Rep 12(1):1846

Olson LE, Soriano P (2009) Increased PDGFRalpha activation disrupts connective tissue development and drives systemic fibrosis. Dev Cell 16(2):303–313

Ong T et al (2020) Siah2 integrates mitogenic and extracellular matrix signals linking neuronal progenitor ciliogenesis with germinal zone occupancy. Nat Commun 11(1):5312. https://doi.org/10.1038/s41467-020-19063-7./pp207

Ornitz DM, Marie PJ (2015) Fibroblast growth factor signaling in skeletal development and disease. Genes Dev 29(14):1463–1486. https://doi.org/10.1101/gad.266551.115

Ozcivici E et al (2010) Mechanical signals as anabolic agents in bone. Nat Rev Rheumatol 6(1):50–59. https://doi.org/10.1038/nrrheum.2009.239./pp16

Pal S (2014) Mechanical properties of biological materials. In: Pal S (ed) Design of Artificial Human Joints & Organs. Springer US, Boston, MA, pp 23–40

Pandey P et al (2023) New insights about the PDGF/PDGFR signaling pathway as a promising target to develop cancer therapeutic strategies. Biomed Pharmacother 161:114491

Parisi MA et al (2006) AHI1 mutations cause both retinal dystrophy and renal cystic disease in Joubert syndrome. J Med Genet 43(4):334–339

Pazour GJ, Rosenbaum JL (2002) Intraflagellar transport and cilia-dependent diseases. Trends Cell Biol 12(12):551–555. https://doi.org/10.1016/s0962-8924(02)02410-8./pp30

Perrault I et al (2012) Mainzer-Saldino syndrome is a ciliopathy caused by IFT140 mutations. Am J Hum Genet 90(5):864–870

Pigino G et al (2009) Electron-tomographic analysis of intraflagellar transport particle trains in situ. J Cell Biol 187(1):135–148

Pittman AE, Solecki DJ (2023) Cooperation between primary cilia signaling and integrin receptor extracellular matrix engagement regulates progenitor proliferationand neuronal differentiation in the developing cerebellum. Front Cell Dev Biol 11:1127638. https://doi.org/10.3389/fcell.2023.1127638./p

Plotnikova OV, Pugacheva EN, Golemis EA (2009) Chapter 7 – Primary cilia and the cell cycle. In: Sloboda RD (ed) Methods in cell biology. Academic Press, pp 137–160

Praetorius HA, Spring KR (2001) Bending the MDCK cell primary cilium increases intracellular calcium. J Membr Biol 184(1):71–79

Qin L et al (2022) Osteocyte β1 integrin loss causes low bone mass and impairs bone mechanotransduction in mice. J Orthop Translat 34:60–72

Qin L et al (2023) Osteocyte β3 integrin promotes bone mass accrual and force-induced bone formation in mice. J Orthop Translat 40:58–71

Qiu N et al (2012) Disruption of Kif3a in osteoblasts results in defective bone formation and osteopenia. J Cell Sci 125(Pt 8):1945–1957

Quadri N, Upadhyai P (2023) Primary cilia in skeletal development and disease. Exp Cell Res 431(1):113751

Quarmby LM, Parker JDK (2005) Cilia and the cell cycle? J Cell Biol 169(5):707–710

Rauner M, Bozec A (2026) Mechanisms of osteoclast activation in inflammatory bone loss in rheumatoid arthritis. Nat Rev Rheumatol 22(1):42–461. https://doi.org/10.1038/s41584-025-01323-9./pp26
Retailleau K, Duprat F (2014) Polycystins and partners: proposed role in mechanosensitivity. J Physiol 592(12):2453–2471
Roberts SR et al (2001) Mechanical compression influences intracellular Ca^{2+} signaling in chondrocytes seeded in agarose constructs. J Appl Physiol (1985) 90(4):1385–1391
Rock MJ et al (2008) Gain-of-function mutations in TRPV4 cause autosomal dominant brachyolmia. Nat Genet 40(8):999–1003
Rohatgi R, Milenkovic L, Scott MP (2007) Patched1 regulates hedgehog signaling at the primary cilium. Science 317(5836):372–376
Romani M et al (2014) Mutations in B9D1 and MKS1 cause mild Joubert syndrome: expanding the genetic overlap with the lethal ciliopathy Meckel syndrome. Orphanet J Rare Dis 9:72
Rosenbaum JL, Witman GB (2002) Intraflagellar transport. Nat Rev Mol Cell Biol 3(11):813–825. https://doi.org/10.1038/nrm952./p
Ruhlen R, Marberry K (2014) The chondrocyte primary cilium. Osteoarthritis Cartilage 22(8):1071–1076. https://doi.org/10.1016/j.joca.2014.05.011./p
Ruiz-Perez VL et al (2000) Mutations in a new gene in Ellis-van Creveld syndrome and Weyers acrodental dysostosis. Nat Genet 24(3):283–286
Rux D et al (2022) Primary cilia direct murine articular cartilage tidemark patterning through hedgehog signaling and ambulatory load. J Bone Miner Res 37(6):1097–1116. https://doi.org/10.1002/jbmr.4506./pp34
Ryan PE et al (2006) Regulating the regulator: negative regulation of Cbl ubiquitin ligases. Trends Biochem Sci 31(2):79–88
Ryan R et al (2018) Functional characterization of tektin-1 in motile cilia and evidence for TEKT1 as a new candidate gene for motile ciliopathies. Hum Mol Genet 27(2):266–282
Salman LA et al (2023) Osteoarthritis: a narrative review of molecular approaches to disease management. Arthritis Res Ther 25(1):27
Salonen R, Paavola P (1998) Meckel syndrome. J Med Genet 35(6):497–501
Satir P, Christensen ST (2007) Overview of structure and function of mammalian cilia. Annu Rev Physiol 69:377–400. https://doi.org/10.1146/annurev.physiol.69.040705.141236./p
Schmid FM et al (2017) IFT20 modulates ciliary PDGFRα signaling by regulating the stability of Cbl E3 ubiquitin ligases. J Cell Biol 217(1):151–161
Schmidts M et al (2013a) Mutations in the gene encoding IFT dynein complex component WDR34 cause Jeune asphyxiating thoracic dystrophy. Am J Hum Genet 93(5):932–944
Schmidts M et al (2013b) Exome sequencing identifies DYNC2H1 mutations as a common cause of asphyxiating thoracic dystrophy (Jeune syndrome) without major polydactyly, renal or retinal involvement. J Med Genet 50(5):309–323
Schneider L et al (2005) PDGFRalphaalpha signaling is regulated through the primary cilium in fibroblasts. Curr Biol 15(20):1861–1866
Schou KB, Pedersen LB, Christensen ST (2015) Ins and outs of GPCR signaling in primary cilia. EMBO Rep 16(9):1099–1113
Schock EN, Brugmann SA (2017) Discovery, diagnosis, and etiology of craniofacial ciliopathies. Cold Spring Harb Perspect Biol 9(9). https://doi.org/10.1101/cshperspect.a028258./pp36
Schwartz MA (2010) Integrins and extracellular matrix in mechanotransduction. Cold Spring Harb Perspect Biol 2(12):a005066
Shaheen R et al (2019) Bi-allelic mutations in FAM149B1 cause abnormal primary cilium and a range of ciliopathy phenotypes in humans. Am J Hum Genet 104(4):731–737
Shaikh Qureshi WM, Hentges KE (2024) Functions of cilia in cardiac development and disease. Ann Hum Genet 88(1):4–26
Sherpa RT et al (2016) Rapamycin increases length and mechanosensory function of primary cilia in renal epithelial and vascular endothelial cells. Int Educ Res J 2(12):91–97
Sheu S-Y et al (2022) Platelet-derived growth factor-AA promotes placental choriocarcinoma JAR cell proliferation via primary cilia. Taiwan J Obstet Gynecol 61(2):299–305

Shi W et al (2017) Microgravity induces inhibition of osteoblastic differentiation and mineralization through abrogating primary cilia. Sci Rep 7(1):1866

Shi W et al (2020) Primary cilia act as microgravity sensors by depolymerizing microtubules to inhibit osteoblastic differentiation and mineralization. Bone 136:115346

Silva DF, Cavadas C (2023) Primary cilia shape hallmarks of health and aging. Trends Mol Med 29(7):567–579

Singla V, Reiter JF (2006) The primary cilium as the cell's antenna: signaling at a sensory organelle. Science 313(5787):629–633

Sjaastad MD, Nelson WJ (1997) Integrin-mediated calcium signaling and regulation of cell adhesion by intracellular calcium. BioEssays 19(1):47–55

Smith UM et al (2006) The transmembrane protein meckelin (MKS3) is mutated in Meckel-Gruber syndrome and the wpk rat. Nat Genet 38(2):191–196

Song B et al (2007) Development of the post-natal growth plate requires intraflagellar transport proteins. Dev Biol 305(1):202–216

Soriano P (1997) The PDGF alpha receptor is required for neural crest cell development and for normal patterning of the somites. Development 124(14):2691–2700

Sorokin S (1962) Centrioles and the formation of rudimentary cilia by fibroblasts and smooth muscle cells. J Cell Biol 15(2):363–377. https://doi.org/10.1083/jcb.15.2.363./pp47

Srivastava M, Deal C (2002) Osteoporosis in elderly: prevention and treatment. Clin Geriatr Med 18(3):529–555

Stam LB, Clark AL (2023) Chondrocyte primary cilia lengthening and shortening in response to mediators of osteoarthritis; a role for integrin $\alpha 1\beta 1$ and focal adhesions. Osteoarthr Cartil Open 5(2):100357

Strong A et al (2021) Expanding the genetic landscape of oral-facial-digital syndrome with two novel genes. Am J Med Genet A 185(8):2409–2416

Sun W et al (2019) The mechanosensitive Piezo1 channel is required for bone formation. elife 8:e47454

Sun Y et al (2025) Primary cilia integrity governs TRPV4-mediated NF-κB/COL2 signaling in osteoarthritis pathogenesis. Cell Signal 136:112141

Sussman CR et al (2020) Modulation of polycystic kidney disease by G-protein coupled receptors and cyclic AMP signaling. Cell Signal 72:109649

Takada Y, Ye X, Simon S (2007) The integrins. Genome Biol 8(5):215

Takahashi N et al (1988) Osteoblastic cells are involved in osteoclast formation. Endocrinology 123(5):2600–2602. https://doi.org/10.1210/endo-123-5-2600./pp28

Takayanagi H (2007) Osteoimmunology: shared mechanisms and crosstalk between the immune and bone systems. Nat Rev Immunol 7(4):292–304. https://doi.org/10.1038/nri2062./p

Takahara M et al (2018) Ciliopathy-associated mutations of IFT122 impair ciliary protein trafficking but not ciliogenesis. Hum Mol Genet 27(3):516–528

Tallila J et al (2008) Identification of *CC2D2A* as a Meckel syndrome gene adds an important piece to the ciliopathy puzzle. Am J Hum Genet 82(6):1361–1367

Tao D et al (2023) Primary cilia support cartilage regeneration after injury. Int J Oral Sci 15(1):22

Temiyasathit S, Jacobs CR (2010) Osteocyte primary cilium and its role in bone mechanotransduction. Ann N Y Acad Sci 1192(1):422–428

Temiyasathit S et al (2012) Mechanosensing by the primary cilium: deletion of Kif3A reduces bone formation due to loading. PLoS One 7(3):e33368

Teves ME et al (2016) Mammalian axoneme central pair complex proteins: broader roles revealed by gene knockout phenotypes. Cytoskeleton 73(1):3–22

Thauvin-Robinet C et al (2009) Genomic deletions of OFD1 account for 23% of oral-facial-digital type 1 syndrome after negative DNA sequencing. Hum Mutat 30(2):E320–E329

Thauvin-Robinet C et al (2014) The oral-facial-digital syndrome gene C2CD3 encodes a positive regulator of centriole elongation. Nat Genet 46(8):905–911

Thompson CL et al (2021) Polycystin-2 is required for chondrocyte mechanotransduction and traffics to the primary cilium in response to mechanical stimulation. Int J Mol Sci 22(9):4313

Ti H et al (2025) Primary cilia as mechanosensors in musculoskeletal homeostasis and disease. Pharmacol Res 219:107887. https://doi.org/10.1016/j.phrs.2025.107887./p
Tiberio F, Parolini O, Lattanzi W (2021) Ciliary signalling and mechanotransduction in the pathophysiology of craniosynostosis. Genes 12(7):1073
Tirado-Cabrera I et al (2022) PTH1R translocation to primary cilia in mechanically-stimulated osteocytes prevents osteoclast formation via regulation of CXCL5 and IL-6 secretion. J Cell Physiol 237(10):3927–3943
Travaglini L et al (2013) Phenotypic spectrum and prevalence of INPP5E mutations in Joubert syndrome and related disorders. Eur J Hum Genet 21(10):1074–1078
Uda Y et al (2017) Osteocyte mechanobiology. Curr Osteoporos Rep 15(4):318–325
Unlu G et al (2014) Trafficking mechanisms of extracellular matrix macromolecules: insights from vertebrate development and human diseases. Int J Biochem Cell Biol 47:57–67. https://doi.org/10.1016/j.biocel.2013.11.005./p
Valente EM et al (2010) Mutations in TMEM216 perturb ciliogenesis and cause Joubert, Meckel and related syndromes. Nat Genet 42(7):619–625
Van Cromphaut SJ et al (2001) Duodenal calcium absorption in vitamin D receptor-knockout mice: functional and molecular aspects. Proc Natl Acad Sci USA 98(23):13324–13329
Vaughan TJ et al (2015) Bone cell mechanosensation of fluid flow stimulation: a fluid-structure interaction model characterising the role integrin attachments and primary cilia. Biomech Model Mechanobiol 14(4):703–718
Venkatesh D (2017) Primary cilia. J Oral Maxillofac Pathol 21(1):8–10
Verbruggen SW, Sittichokechaiwut A, Reilly GC (2023) Osteocytes and primary cilia. Curr Osteoporos Rep 21(6):719–730
Vilboux T et al (2017) Molecular genetic findings and clinical correlations in 100 patients with Joubert syndrome and related disorders prospectively evaluated at a single center. Genet Med 19(8):875–882
Vlaikou AM et al (2017) Mechanical stress affects methylation pattern of GNAS isoforms and osteogenic differentiation of hAT-MSCs. Biochim Biophys Acta, Mol Cell Res 1864(8):1371–1381
Vuong LT, Mlodzik M (2023) Chapter Three – The complex relationship of Wnt-signaling pathways and cilia. In: Iomini C, Sun Y (eds) Current topics in developmental biology. Academic Press, New York, pp 95–125
Walczak-Sztulpa J et al (2010) Cranioectodermal dysplasia, Sensenbrenner syndrome, is a ciliopathy caused by mutations in the IFT122 gene. Am J Hum Genet 86(6):949–956
Walczak-Sztulpa J et al (2020) Compound heterozygous IFT140 variants in two Polish families with Sensenbrenner syndrome and early onset end-stage renal disease. Orphanet J Rare Dis 15(1):36
Walczak-Sztulpa J et al (2022) Identical IFT140 variants cause variable skeletal ciliopathy phenotypes-challenges for the accurate diagnosis. Front Genet 13:931822
Walker RV et al (2019) Ciliary exclusion of Polycystin-2 promotes kidney cystogenesis in an autosomal dominant polycystic kidney disease model. Nat Commun 10(1):4072
Wang L et al (2020) Mechanical sensing protein PIEZO1 regulates bone homeostasis via osteoblast-osteoclast crosstalk. Nat Commun 11(1):282
Wang L et al (2022a) Mechanical regulation of bone remodeling. Bone Res 10(1):16. https://doi.org/10.1038/s41413-022-00190-4./pp15
Wang Q et al (2022b) Substrate stiffness regulates the differentiation profile and functions of osteoclasts via cytoskeletal arrangement. Cell Prolif 55(1):e13172
Wang J et al (2023) A homozygous EVC mutation in a prenatal fetus with Ellis-van Creveld syndrome. Mol Genet Genomic Med 11(8):e2183
Wann AK et al (2012) Primary cilia mediate mechanotransduction through control of ATP-induced Ca^{2+} signaling in compressed chondrocytes. Faseb J 26(4):1663–1671. https://doi.org/10.1096/fj.11-193649
Wann AK et al (2012) Primary cilia mediate mechanotransduction through control of ATP-induced Ca^{2+} signaling in compressed chondrocytes. FASEB J 26(4):1663–1671

Wheatley DN, Wang AM, Strugnell GE (1996) Expression of primary cilia in mammalian cells. Cell Biol Int 20(1):73–81

Wilson E, Sudhir K, Ives HE (1995) Mechanical strain of rat vascular smooth muscle cells is sensed by specific extracellular matrix/integrin interactions. J Clin Invest 96(5):2364–2372. https://doi.org/10.1172/jci118293

Wise GE, King GJ (2008) Mechanisms of tooth eruption and orthodontic tooth movement. J Dent Res 87(5):414–434

Wood CR, Rosenbaum JL (2015) Ciliary ectosomes: transmissions from the cell's antenna. Trends Cell Biol 25(5):276–285. https://doi.org/10.1016/j.tcb.2014.12.008./pp46

Wu VM et al (2012) Small molecule inhibitors of smoothened ciliary localization and ciliogenesis. Proc Natl Acad Sci USA 109(34):13644–13649

Wu S et al (2024) LIPUS regulates the progression of knee osteoarthritis in mice through primary cilia-mediated TRPV4 channels. Apoptosis 29(5):785–798

Wysocki A et al (2011) Whole-body vibration therapy for osteoporosis: state of the science. Ann Intern Med 155(10):680–686, w206–13

Xiao Z et al (2023) Genetic interactions between polycystin-1 and Wwtr1 in osteoblasts define a novel mechanosensing mechanism regulating bone formation in mice. Bone Res 11(1):57

Xie Y-F et al (2016) Pulsed electromagnetic fields stimulate osteogenic differentiation and maturation of osteoblasts by upregulating the expression of BMPRII localized at the base of primary cilium. Bone 93:22–32

Xu X et al (2005) PDGFR-α signaling is critical for tooth cusp and palate morphogenesis. Dev Dyn 232(1):75–84

Yamaguchi T et al (2003) Cyclic AMP activates B-Raf and ERK in cyst epithelial cells from autosomal-dominant polycystic kidneys. Kidney Int 63(6):1983–1994

Yamaguchi H et al (2022) Temporospatial regulation of intraflagellar transport is required for the endochondral ossification in mice. Dev Biol 482:91–100

Yamaguchi H et al (2024) The primary cilia: orchestrating cranial neural crest cell development. Differentiation 100818. https://doi.org/10.1016/j.diff.2024.100818./pp32

Yamauchi M, Sricholpech M (2012) Lysine post-translational modifications of collagen. Essays Biochem 52:113–533. https://doi.org/10.1042/bse0520113./pp21

Yan YX et al (2012) Mechanical strain regulates osteoblast proliferation through integrin-mediated ERK activation. PLoS One 7(4):e35709

Yan J-L et al (2015) Pulsed electromagnetic fields promote osteoblast mineralization and maturation needing the existence of primary cilia. Mol Cell Endocrinol 404:132–140

Yan Y et al (2016) Mechanical strain promotes osteoblastic differentiation through integrin-β1-mediated β-catenin signaling. Int J Mol Med 38(2):594–600

Yang L et al (2024) Research progress on the regulatory mechanism of integrin-mediated mechanical stress in cells involved in bone metabolism. J Cell Mol Med 28(7):e18183

Yasuda H et al (1999) A novel molecular mechanism modulating osteoclast differentiation and function. Bone 25(1):109–113

Yong EL, Logan S (2021) Menopausal osteoporosis: screening, prevention and treatment. Singapore Med J 62(4):159–166

Yoo M et al (2020) Integrin-β(1) is required for the renal cystogenesis caused by ciliary defects. Am J Physiol Renal Physiol 318(5):F1306–F1312

You J et al (2001) Osteopontin gene regulation by oscillatory fluid flow via intracellular calcium mobilization and activation of mitogen-activated protein kinase in MC3T3-E1 osteoblasts. J Biol Chem 276(16):13365–13371

Yuan X, Yang S (2016) Primary cilia and intraflagellar transport proteins in bone and cartilage. J Dent Res 95(12):1341–1349. https://doi.org/10.1177/0022034516652383./pp37

Yuan S et al (2012) Target-of-rapamycin complex 1 (Torc1) signaling modulates cilia size and function through protein synthesis regulation. Proc Natl Acad Sci 109(6):2021–2026

Yue L, Berman J (2022) What is osteoarthritis? JAMA 327(13):1300–1300

Zaidel-Bar R et al (2004) Hierarchical assembly of cell-matrix adhesion complexes. Biochem Soc Trans 32(Pt3):416–420

Zeng Q et al (2015) Integrin-β1, not integrin-β5, mediates osteoblastic differentiation and ECM formation promoted by mechanical tensile strain. Biol Res 48(1):25

Zhang H et al (2016) Elevated fibroblast growth factor signaling is critical for the pathogenesis of the dwarfism in Evc2/limbin mutant mice. PLoS Genet 12(12):e1006510

Zhang W et al (2018) Expanding the genetic architecture and phenotypic spectrum in the skeletal ciliopathies. Hum Mutat 39(1):152–166

Zhang X et al (2020) Whole-exome sequencing identified two novel mutations of DYNC2LI1 in fetal skeletal ciliopathy. Mol Genet Genomic Med 8(12):e1524

Zhang Z et al (2022) Piezo1 and IFT88 synergistically regulate mandibular condylar chondrocyte differentiation under cyclic tensile strain. Tissue Cell 76:101781

Zhang Y et al (2023) Primary cilium-mediated mechanotransduction in cartilage chondrocytes. Exp Biol Med (Maywood) 248(15):1279–1287

Zhao S et al (2002) MLO-Y4 osteocyte-like cells support osteoclast formation and activation. J Bone Miner Res 17(11):2068–2079

Zhong C et al (1998) Rho-mediated contractility exposes a cryptic site in fibronectin and induces fibronectin matrix assembly. J Cell Biol 141(2):539–551

Zhou T et al (2020) Piezo1/2 mediate mechanotransduction essential for bone formation through concerted activation of NFAT-YAP1-ß-catenin. elife 9:e52779

Zhou P, Zheng T, Zhao B (2022) Cytokine-mediated immunomodulation of osteoclastogenesis. Bone 164:116540. https://doi.org/10.1016/j.bone.2022.116540./p

Chapter 13
Integrating Mechanical Loading, Mechanotransduction, and Biological Responses in Musculoskeletal Tissues Across the Lifespan: Regulation Influenced by Cells, Extracellular Matrix, and Sex

David A. Hart

Abstract Development of cellular mechanisms to facilitate mechanosensing and mechanotransduction occurred very early in evolution during adaptation to the boundary conditions of Earth, in particular, gravity. Subsequently, with the evolution of increasingly complex organisms leading to *Homo sapiens*, multiple levels of biological regulatory control of mechanical loading were introduced for connective tissues of the musculoskeletal system. Integration of mechanosensing mechanisms with biological regulatory variables exhibits tissue-specific, lifespan-specific, sex specific features. A key integration component is the regulation of response to ground reaction forces. This chapter will discuss various features of this integration and its dynamic nature across the lifespan, and the critical role of the cell-extracellular matrix relationship, and how sex-specific variables appear to influence the functioning of connective tissues, particularly in females, where lifespan transitions (i.e., puberty, pregnancy and lactation, and menopause) appear to contribute to tissue responses to loading.

13.1 Introduction and Background

13.1.1 Evolutionary Context of Mechanosensitivity

The development of self-replicating life on Earth had to contend with the boundary conditions imposed by its location, need for energy, and adaptation to dynamic

D. A. Hart (✉)
Department of Surgery, Faculty of Kinesiology, and McCaig Institute for Bone & Joint Health, University of Calgary, Calgary, Alberta, Canada
e-mail: hartd@ucalgary.ca

M. Kloc, J. Wosik (eds.), *Mechanotransduction and Mechanosensing*, Results and Problems in Cell Differentiation 78, https://doi.org/10.1007/978-3-032-19509-8_13

circumstances. These boundary conditions include light, gravity, geomagnetic field, temperature, and nutrient availability in an aqueous environment (discussed in Hart 2023a, b, 2024a, b). While gravity is fairly constant, light, temperature, and nutrient availability are dynamic.

With development and survival in a marine environment also came the need to develop and use the mechanical environment. Single cells suspended in a marine environment are subjected to hydrodynamic compression with depth, and if they become attached to the ground below the surface of the aqueous environment, they also become subjected to shear forces as tidal and wave action wash over the cells. Underlying these environmental features is gravity, and the need to adapt to this boundary condition was a central tenet of adaptation (Najrana and Sanchez-Esteban 2016). Thus, developing responses to such forces and likely using them in a beneficial manner were central to survival. Interestingly, subjecting current bacteria that are adhered to a surface to an oscillating shear force can lead to the transition of planktonic organisms to a biofilm phenotype (Harrison et al. 2010), and this complex conversion leads to changes in gene expression (Schembri et al. 2002; Vohra et al. 2024; Afrasiabi and Partoazar 2024). Thus, prokaryotic organisms such as bacteria developed mechanosensing abilities and mechanotransduction mechanisms (Kruse and Julicher 2005; Moe and Blount 2005; Hayakawa et al. 2008; Sawada et al. 2012; Harper et al. 2023; Mackey et al. 2025) that have been refined in eukaryotes and eventually, humans. These refinements built on the early influences of gravity and ultimately, ground reaction forces (GRF) that are engaged via movement and mobility.

Early commitment to the development of genetic abilities to pass them on to subsequent generations meant that this genetic commitment regarding mechanical loads was carried forward as multicellular organisms arose. As complex organisms left the marine environment to live on land, adaptations in mechanical response systems and their use were required as gravity forces have now become more prevalent. In addition, mobility and navigation in the environment now became a driving force for survival. However, some of those adaptations may be reversible, as mammals such as whales once lived on land but returned to the marine environment. Others may have adapted further, as whales can function at ocean depths that would crush a human (discussed in Myers et al. 2007a).

Thus, modern humans likely reflect to some degree the commitment of evolutionarily early organisms to mechanisms designed to both adapt to the presence of mechanical loading and use of such loading and forces to regulate their tissues in specific manners. While all tissues are likely subjected to some forms of loading, tissues of the musculoskeletal system (MSK) (i.e., bones, muscles, ligaments, tendons, menisci, and cartilage) are of particular interest as they interface with gravity as an external variable. Humans also exhibit a diurnal rhythm or circadian rhythm that can interface with the MSK system (discussed in Hart et al. 2022). Thus, the various systems associated with responses to boundary conditions of the planet are likely "integrated" to some extent even after extensive evolutionary pressures that led to *Homo sapiens*. Such integration was also likely required in primitive

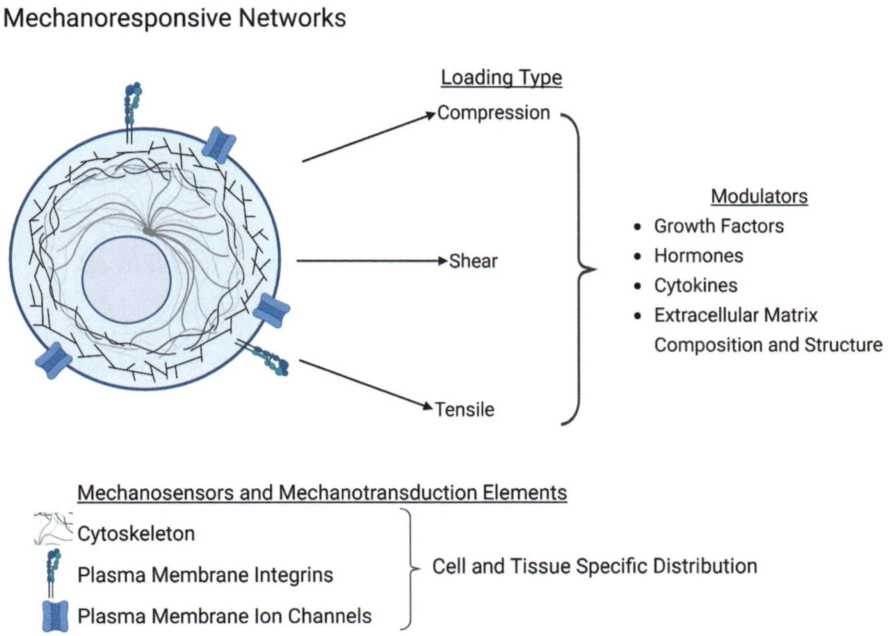

Fig. 13.1 Mechanoresponsive networks consisting of various types of mechanical loading and biological modulators acting on cellular components such as the cytoskeleton and plasma membrane molecules including integrins and mechano-reactive ion channels. (Created with BioRender.com)

single-celled organisms, and bacteria also exhibit circadian rhythms (Fang et al. 2025; Jabbur and Johson 2025).

While humans may retain the basics of mechanoresponsive systems that arose during evolution, clearly the regulation of those early systems is now more complex, with the evolution of a brain that adds neuroregulation to the system and the adaptations required to accommodate the commitment to two-party sex and reproduction. Of course, with the onset of such complexity comes the risk for adverse events or disruptions in the integrity of system functioning. For instance, the average lifespan was considerably shorter 10,000 years ago than it is today, so a system that was designed to function optimally for 40 years may start to exhibit "loss of integrity" when it is common to now live to >80 years of age. The emergence of new information on the differences between biological and chronological age (Zurbuchen et al. 2025; Haugg et al. 2025) adds additional complexity but speaks to human heterogeneity in this regard of MSK system durability.

Thus, as organisms became more complex, they added additional regulatory elements to interface with basic mechanisms of mechanosensing and mechanotransduction. These include biological mediators such as hormones, growth factors, and inflammatory mediators, as well as tissue-specific extracellular matrices (ECM) (Fig. 13.1). In addition, the complexity of the mechanosensing elements likely expanded to be able to functionally address the expansion of the regulatory input, all built on the fundamental need to accommodate the boundary conditions of Earth. As will be discussed below, these response patterns are complex and dynamic across the lifespan. While many insights into some of these complexities have been gained via research, much still remain to be defined and understood.

13.1.2 Different Types of Mechanical Loading Experienced by Cells

13.1.2.1 Hydrostatic Compression

Different types of mechanical loading likely require different systems to both accommodate the loading on cells and the ability to regulate it in a dynamic manner. Thus, cells in articular cartilage are organized mainly as single cells, and these cells predominantly experience compressive loading except in the superficial layer, where the cells also experience shear. However, in other tissues such as menisci of the knee, the tissues are composites of cartilage-like cells in the inner part of the tissue, while the outer portion of the tissue experiences tensile loading (discussed in Sanchez-Adams et al. 2011; Orton et al. 2023; Schwer et al. 2024). In such a construct, the transition zone between the two areas may be critical and vulnerable to unique loading environments.

In addition, static and dynamic hydrostatic compression forces require a cytoskeleton that can allow the cell to function and still undergo alterations to permit cell division with maintenance of integrity and with no compromise in function after the event is completed. This concept led to the development of the tensegrity model (Ingber 2003, 2006, 2008; Chen and Ingber 1999; Ingber et al. 1994, 2014). Thus, a functional cytoskeleton that protects the cell is required, but the regulation of the cytoskeleton must be able to adapt to changes in both the cell and the external environment regarding hydrostatic compression. However, the responses to hydrostatic compression do not function with the cytoskeleton alone, as the plasma membrane is also involved (Ingber 1997; Otoo et al. 2024, 2025). The plasma membrane of eukaryotic cells has folds in it, which allow for changes in volume of the cell, and thus the cytoplasm of the cell can expand and contract as the environment changes, thus likely both protecting the intracellular organelles and providing an adjunct mechanism to the cytoskeleton. Changes in the tension on the plasma membrane can potentially result in the opening or closing of ion transport molecules in the membrane that can facilitate changes in volume (reviewed in Huang et al. 2025a).

In contrast to a static compression environment, exposure of cells to a cyclic and intermittent hydrostatic compression environment requires a system that can react fairly quickly to protect the cell but also allow for the cell to transmit the information to the nucleus, which can permit the cell to modify gene expression (Roelofsen et al. 1995; Tasevski et al. 2005). The molecules involved may be cytoskeletal elements

(Shim et al. 2008; Gould et al. 2021) or others (Feng et al. 2024), depending on the differentiation state of the cell. A loss of mechanical loading also leads to specific expression of catabolism-related molecules that appear to contribute to the "use it or lose it" principle and atrophy (Natsuume et al. 2005). The expression of these catabolism-related molecules can be prevented in vitro via intermittent hydrostatic compression. Thus, through various molecular mechanisms, perturbation of the cytoskeleton, and the plasma membrane of a cell can lead to signal transduction, evoking a cell response to the hydrostatic compression environment.

13.1.2.2 Shear Forces

Shear forces would have been encountered by primitive single-celled organisms via wave action and tidal shifts in an aqueous environment. This also occurs in bone cells due to fluid flow accompanying application of loads to the tissue, such as ground reaction forces (GRF) and by endothelial cells (EC) of the vascular system, as blood flows in a pulsatile manner. There would also be a compressive force exerted on the EC due to blood pressure (Chistiakov et al. 2017), which is not static during a 24-hr day, during exertional activities, and accompanying aging and disease development. As EC are quite heterogeneous based on development of paracrine units with the cells of tissues they infiltrate (Augustin and Koh 2017; Gifre-Renom et al. 2022; Larionov et al. 2024), the impact of shear loading by the movement of blood on the cells may also be heterogeneous directly on the EC (Davies et al. 1995, 2010; Merna et al. 2018) or indirectly due to paracrine interactions in a tissue-specific manner (Bertani et al. 2021). In bone, it is believed that osteocytes are uniquely involved in responsiveness to the shear, a response that subsequently is transmitted to other bone cells (discussed in Pajevic et al. 2013; Murshid 2017). Osteoblast-like cells also respond to fluid flow and shear, and the cytoskeleton is reported to be involved in the cell response (Myers et al. 2007b). Interestingly, disruption of the microfilaments in such cells did not alter the response to shear, but disruption of the microtubule network did lead to loss of responsiveness (Myers et al. 2007b). Thus, different components of the cytoskeleton may be more involved in mechanotransduction than others.

13.1.2.3 Tensile Loading

In multicellular organisms such as humans, cells in tissues such as ligaments and tendons are subjected to tensile loading as a consequence of the tissue's functioning. The structure of the tissue reflects the environment, depending in part on whether it is a high- load or low-load environment. Thus, cells in a ligament such as the anterior cruciate ligament (ACL) of the knee see higher tensile loads than does the medial collateral ligament (MCL), which is a stabilizing ligament. Similarly, cells in the Achilles tendon (AT), which is an energy-storing tissue, experience higher tensile loads than the supraspinatus tendon of the shoulder (discussed in Thornton et al. 2010). The stretching of the tissue, with the initial loss of collagen crimp and subsequent further elongation, can lead to deformation of the cells and activation of signal transduction mechanisms, converting mechanical information to biochemical signals. This deformation in a ligament can lead to elongation of both the whole cell and deformation of the cell nucleus (Matyas et al. 1994). Thus, at high strains, both the plasma membrane and the cytoskeleton are overtly distorted, leading to activation of signal transduction mechanisms.

Of note, it should be mentioned that in tissues such as ligaments and tendons, the cells are linked together in networks (Lo et al. 2002a, b; Chi et al. 2005). During development, the tissues are "cell-rich and extracellular matrix-poor" but then become "cell-poor and matrix-rich" as they grow, mature, and reach skeletal maturity. As the ECM is laid down during growth and maturation, the matrix forms according to the template arising during development, and the cells get further apart. However, many of the cells in ligaments and tendons exist in networks and are not individual and isolated cells (Lo et al. 2002a, b; Chi et al. 2005). The cells are "linked" via connexins, and as such, activation of one cell by loading or initiating a Ca+2 ion flux can also lead to activation of neighboring cells. In such a network of cells, the signal generated by loading a tissue can amplify the mechanically induced signal from one activated cell to other elements of the network via the connexins.

13.2 The Molecules of Mechanotransduction

As multicellular organisms evolved into humans, cells gave up much of their independence to function as a unit. However, they likely retained much of their early genetic commitment to functioning in mechanically active environments, with the cytoskeleton and the plasma membrane as the locations for key molecules that regulate response patterns and incorporate mechanical loading-related commitments with other regulatory systems.

13.2.1 Plasma Membrane-Localized Molecules

13.2.1.1 PIEZO 1 and 2

Two central molecules involved in mechanosensing activities are PIEZO 1 and PIEZO 2 (reviewed in Gottlieb 2017; Borbiro and Rohacs 2017; Saotome et al. 2018; Moroni et al. 2018; Fang et al. 2021; Hendargo et al. 2023). Their discovery led to a Nobel Prize in 2021. They are present on a large variety of cell types, particularly those in tissues of the MSK system (Savadipour et al. 2023a). These mechanically activated ion channels are, therefore, critical components of these cells in very different mechanical environments. These proteins are also found on endothelial cells (Li et al. 2014; Zhang et al. 2017; Beverley et al. 2025), immune cells (Schroder et al. 2025), and cells of the neural system (Mikhailov et al. 2019, 2022). The PIEZO molecules are evolutionarily conserved (discussed in Hendargo et al. 2023) and analogues are found in plants (Zhang et al. 2019) and single-cell eukaryotic cells (discussed in Hendargo et al. 2023). Thus, while central to MSK cell responsiveness to mechanical stresses, the PIEZO molecules are found on a wide variety of cells, including stem cells, where mechanotransduction can influence cell differentiation and fate (Qiu et al. 2023).

The role of the PIEZO function has been well studied for chondrocytes/cartilage (Lee et al. 2017; Du et al. 2020; Zhang et al. 2022; Savadipour et al. 2023b; Nims et al. 2024; Yuan et al. 2025) and bone (Qin et al. 2021; Xu et al. 2021). Adaptations to exercise have also been reported for the PIEZO molecules in chondrocytes and articular cartilage (Jing et al. 2024). PIEZO expression in chondrocytes has also been reported to control endochondral ossification and possibly, osteoarthritis (OA) development in mice (Brylka et al. 2024). Other reports have also characterized PIEZO in OA (Steinecker-Frohnwieser et al. 2023; Zhou et al. 2025) and raised the possibility of targeting ion channels in OA (Zhou et al. 2025). Regarding a role for PIEZO in bone, PIEZO 1 is expressed in osteoblasts and osteocytes, and its inactivation impairs trabecular bone formation (Hendrickx et al. 2021). Therefore, the PIEZO molecules play important roles beyond mere mechanoresponsiveness. As they are tethered to the cytoskeleton (discussed in Lee et al. 2017), they provide a link between mechanosensitivity and the cytoskeleton's role in maintaining shape and function in a mechanically active environment.

Expression of PIEZO genes can be influenced by alterations in environmental conditions. For example, PIEZO can be upregulated in ECs associated with pulmonary hypertension (Wang et al. 2021). In a mouse model of Marfan syndrome, PIEZO1 is reported to attenuate aneurysms (Yang et al. 2025). Mutations in PIEZO 2 result in muscle weakness and scoliosis, as well as proprioception (Haliloglu et al. 2017; Lau et al. 2023). Thus, PIEZO may also exert indirect effect on the MSK system elements via neuro-dependent communication. Finally, the PIEZO function is also reported to be influenced by inflammation (Ren et al. 2024; Zhao et al. 2025a, b, c). While not yet documented, the altered involvement of PIEZO in OA may relate to inflammation as OA appears to be an inflammatory condition (Alad et al. 2025; Batarfi et al. 2025).

13.2.1.2 Transient Receptor Potential Vanilloid Channel 1–4 (TRPV 1–4)

The TRPV ion channels are Ca^{+2} channels expressed in cells of the MSK system (Guilak et al. 2010). In particular, TRPV 4, a complex ion channel (Toft-Bertelsen and MacAulay 2021) is highly expressed on chondrocytes (discussed in Clark et al. 2010; Savadipour et al. 2022; Steinecker-Frohnwieser et al. 2023; Nims et al. 2024; Palmer et al. 2025), as well as fibroblasts of connective tissues (Wang et al. 2024a, b), endothelial cells (Baratchi et al. 2017), and vascular-associated cells (Sharif-Naeini et al. 2008), trabecular network cells of the eye (Lakk and Krizaj 2021), bone (discussed in Liu et al. 2022; Li and Zhang 2025), muscles (Ding et al. 2025), cells of the innate immune system (Michalick and Kuebler 2020; Orsini et al. 2021), and stem cells (Gilchrist et al. 2019; Ferrai and Schulte 2024).

Because of its perceived importance, TRPV 4 has been suggested as a therapeutic target for joint diseases (McNulty et al. 2015) as well as diseases and processes in other tissues (Di et al. 2023). TRPV 4 has been implicated in fibrosis (Zhan and Li 2018), osteoporosis (Liu et al. 2022), endothelial cell dysfunction (Feletou 2016), and metatropic dysplasia via chondrocytes (Hurd et al. 2015).

13.2.1.3 Role of Integrins in Mechanoresponsiveness

Cells in connective tissues of the MSK system interact with the ECM to form linkages with the matrix proteins in specific manners. A major mechanism is via integrins (Jablonski et al. 2014; Jiao et al. 2017; Katoh 2025). These integrins are glycoproteins (Cai et al. 2017) and can interact with various ECM molecules due to the molecules being complexes of heavy and light chains and those are composed of a family of molecules where specificity is dictated by the association of specific heavy and light chains (reviewed in Klein et al. 1990; Tvaroska et al. 2023; Tong et al. 2023; Eble 2025). These integrins are transmembrane proteins linked to the cell cytoskeleton (discussed in Zhang et al. 2025) and can thus interact directly with specific cytoskeletal molecules, such as talin (Zhao et al. 2022; Wang et al. 2024a, b; Mishra et al. 2024), to convey information.

When loads are applied to a connective tissue such as articular cartilage, ligaments and tendons (Chatterjee et al. 2022; Stanczak et al. 2024; Stanczak et al. 2025), or bone (Tollabi et al. 2024; Chen et al. 2025), cells can sense the loads via distortion of the matrix (Salter et al. 2001; Zhao et al. 2025a, b, c), leading to direct mechanotransduction as well as indirect signaling via activation of ion channels (Thodeti et al. 2009; Matthews et al. 2010; Cheng et al. 2023; reviewed in Di et al. 2023; Shi et al. 2025). Integrins also play central roles in mesenchymal stem cell differentiation in response to mechanical signals (Wang et al. 2022).

Because of their importance and potential association with disease and compromised healing responses, some reports have suggested targeting these systems (Yokoyama et al. 2025; Tiskratok et al. 2025). As integrin interactions with the ECM are apparently key to connective tissue regulation, deviations or alterations in the matrix can lead to compromised integrity and function (Zhao et al. 2025a, b, c; Abolhasani et al. 2025). Similarly, mutations in integrins could alter mechanical signal transduction and compromise the regulation of tissue integrity (Dhulipalla and Longmate 2024).

13.2.1.4 Summary

Mechanotransduction involves a variety of molecules at the plasma membrane, including integrins, ion channels, and their interactions, to effectively convey responses to mechanical loading. Effective integration of signals from the ECM and cell-sensing mechanisms is critical for MSK tissue development, growth and maturation, responsiveness to injury, and changing environmental conditions. Loss of integrity of this integration during development, due to mutations or age-related senescence, may lead to altered function and disease development. As discussed below, this system is designed to function dynamically within Earth's boundary conditions, and loss of integrity can result from altering those boundary conditions (e.g., microgravity in spaceflight) or from static conditions such as immobilization or chronic bed rest.

Mechanobiology must also be integrated into the combined responses to other boundary conditions such as gravity, magnetic fields, and circadian rhythms (discussed in Hart 2019, 2021, 2022, 2023a, b, 2024a, b). Finally, cells, tissues, and organs of the MSK system must integrate their mechanotransduction systems with other regulatory systems such as the neuro, endocrine, immune, and anabolic/catabolic mediator molecules, which also contribute to aspects of functional integrity. Such complexity could be quite daunting to understand, but without understanding, one cannot appreciate the role of mechanobiology in the hierarchy of regulation.

Thus, mechanotransduction can involve both the cell's cytoskeleton and the activation of plasma membrane molecules via distortion of the ECM induced by loading, with subsequent effect on nucleus-mediated alterations in gene expression and metabolic responses to the loading environment. Different tissues develop a "set point" for such interactions, dependent on the type and range of loading encountered. Thus, for much of the lifespan, these relationships are dynamic and adaptive. Tissues and MSK organ systems operate within a physiological "window" of cyclic loading (discussed in Hart et al. 2002) that matures from gestation to skeletal maturity and then can decline with aging. Exceeding the mechanical loading limit of a tissue can lead to overuse and overt failure, and as will be discussed, these processes can be influenced by a number of biological and genetic factors, so the regulation is dynamic across the lifespan.

13.3 Mechanical Loading Across the Lifespan: Regulation and Function

While understanding of mechanotransduction and the role of mechanical signals at the level of cells has received considerable research attention, elements of their integration into whole organisms and with more complexity have been somewhat variable. Clearly, the boundary condition of a 1 g environment has been a priority during evolution as it is intrinsically linked to mobility and navigation in the environment (discussed in Hart and Zernicke 2020, 2021, 2025). Commitment of multicellular organisms to mobility required the evolution of both refined mechanical mechanisms at the tissue level and regulatory mechanisms at the whole-organism level with multiple levels of control across the lifespan.

13.3.1 Fetal Development

While the mammalian fetus experiences gravity, early developmental programs appear to be expressed while the fetus is in a protected environment, surrounded by fluid. In this buoyant environment, it is not exposed to ground reaction forces (GRF) whereas the mother is except during bed rest. With the inclusion of joint cavitation and the cardiovascular system, cells across several systems are subjected to various forms of loading (i.e., shear, compression, and tensile). The tissues of the MSK system, such as ligaments, tendons, articular cartilage, menisci, form initially as cell-rich and ECM-poor tissues, but the ECM serves as a template for growth and maturation.

Movement of joint tissues also generates loads that contribute to the further maturation of joints such as the knee as an integrated organ system (Radin et al. 1991; Frank et al. 2004; Loeser et al. 2012; Rollick et al. 2018). During the latter part of human gestation, the fetus continues to grow and integrate anabolic signals, neural input, and mechanical loading in preparation for birth. Thus, the biomechanical environment is fairly protected from GRF during gestation, but biomechanical relationships between tissues continue to mature in its absence.

13.3.2 Post-Birth Growth, Maturation, and Adaptation to GRF

Unlike animals such as zebras, deer, and others who can run and adapt to GRF within minutes to hours after birth, humans cannot walk or run for a protracted period. It appears to take months for the effective integration of muscles and other tissues of the MSK system and neural integration to allow for crawling and then walking and running. Given the time from birth to walking and experiencing GRF in humans can vary between ~7 and 14 months, this maturation sequence does not follow a distinct time-dependent pathway but may be influenced by a variety of factors, mostly unknown. Initially, walking is performed in a directional manner, followed by the ability to navigate the environment and change direction (discussed in Hart 2018). It should be noted that some of this early growth and maturation of the tissues of the MSK system do happen in the presence of a "mini-puberty" due to the onset of increased but transient sex hormone levels soon after birth (discussed in Bidlingmaier 1980; Becker et al. 2015; Schaadt et al. 2015; Borysiak et al. 2015). This early exposure to sex hormones may set the stage for later puberty-related events as all the MSK tissues express sex hormone receptors (discussed in Hart 2022).

Once adapted to GRF, this variable is likely incorporated into continued growth and maturation sequences. This conclusion is supported by studies in very young rabbits in which rigid hindlimb immobilization leads to cessation of growth of the knee ligaments (Walsh et al. 1992, 1993). Thus, in the absence of biomechanical loading, the tissue fails to mature biomechanically and is "arrested' at the point immobilization is initiated. Therefore, even in the presence of systemic anabolic growth mediators, the tissue remains unresponsive. However, the tibia continues to grow, and when immobilization is removed after 6-8 weeks, the joint is no longer functional (Weir 1991). Thus, the bone is not subject to the same constraints as are some of the soft tissues of a joint, and perhaps some aspects of the growth of these soft tissues are in response to the tension provided by the growth of the bone. These studies provide support for the "joint as an organ" concept and point to bone being a unique driver of growth and development in response to increasing mechanical requirements of these organ systems.

The loss of joint function due to continued bone growth after hindlimb immobilization declines with increasing time after birth in the rabbit model (Weir 1991). Therefore, the rate of bone growth rapidly declines with age, and the integration of growth across all tissues of the knee leads to a stronger knee with more complete integration of biology with GRF. Alternatively, bone may be regulated by other factors with higher priority than biomechanical loading (Hart 2023a, b).

An additional factor may also play a role in integrating mechanical loading with the biological response to the loading stimulus. That is, much of the tissue growth during the growth and maturation phase of the life cycle occurs at night, when humans are at rest and GRF is at a minimum (discussed in Hart et al. 2022). This leads to the conclusion that growth in an anabolic environment does not occur effectively when a tissue is actually being subjected to loading but rather, occurs in response to the loading when the impact of the loading is less intense. This may also mean that growth at night is integrated into circadian rhythm-associated molecules and processes (Hart et al. 2022) and may be associated with the need for sleep during periods of rapid growth.

During this period of rapid growth, in some tissues of preclinical models, the unimodal distribution of collagen fibril diameters in tissue arising during development and at birth undergoes a transition to a bimodal distribution with both large and smaller diameter fibrils. This transition is believed to occur when biomechanical demands on tissues require adaptation to increased loads (Lo et al. 2004). Details on how this bimodal distribution arises and is maintained throughout the rest of the lifespan are not well known; however, this likely requires collagen cross-linking. The collagen cross-linking may be mediated by members of the lysyl oxidase family of enzymes (Marturano et al. 2014). Thus, with maturation and prior to puberty, the ECM of MSK tissues undergo a dramatic change to accommodate the changing biomechanical demands on these tissues. However, details regarding the transition from a unimodal distribution of collagen fibrils to a bimodal distribution remain to be elucidated. Whether it relates to the presence of an inhibitor that needs to be removed, an inducer of the transition that is required, or a combination of both options remains to be defined. Furthermore, how the bimodal distribution is maintained within defined limits is also unknown, as is its relationship to the mechanical environment and mechanotransduction.

In the context of growth and maturation, and with increasing mechanical demands, it is also known that tissues such as tendons are heterogeneous with regard to those in different mechanical environments (Berglund et al. 2004). Tendons are also heterogeneous within a specific tendon as well (Huisman et al. 2014), indicating that they are complex structures with unique cell-matrix interactions within a tissue such as the Achilles tendon. Ligaments are also heterogeneous between loading environments and within a ligament (Frank et al. 1999; Frank 2004), again indicating that these tissues grow and adapt to the mechanical loading requirements.

13.3.3 Puberty

The onset of puberty following a period of fairly rapid growth initiates several transitions that affect the biomechanical environment of both males and females (Fig. 13.2). While preparing females for pregnancy and lactation, the onset of puberty in both males and females leads to an additional growth spurt, but now in the context of sex hormones. In males, the onset of puberty leads to increased levels of testosterone and related molecules, which can influence growth in general but particularly at the level of the MSK system.

Thus, with puberty in males, the relatively constant testosterone levels must now be incorporated into the constellation of variables, biological and biomechanical; during adolescence in males, there is an increase in muscle mass and nearly all MSK tissues. This biological growth is accompanied by increased biomechanical loading of these tissues and thus, the hormonal influences have to be incorporated

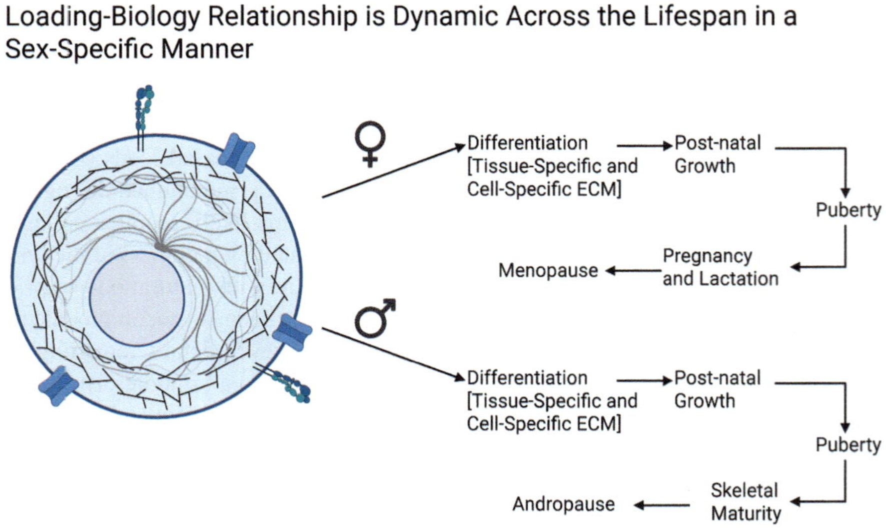

Fig. 13.2 Sex-specific differences in the regulation of mechanical loading and biological factors across the lifespan. (Created with BioRender.com)

with the biomechanical stimuli into the coordinated growth and maturation of joints and MSK functions.

In contrast to males, the onset of puberty in females is accompanied by changes in body shape (i.e., Q-angle), function (as in jumping and landing from a height; (Hewett et al. 2006), growth, and fluctuating sex hormone levels following a menstrual cycle of ~25–30 days. As all of the tissues of the MSK system express hormone receptors for estrogen (cytosolic ER-alpha and ER-beta, plasma membrane-bound ER, PR), they should be responsive to the hormones (reviewed in Hart 2022). Thus, in females, the sex hormone environment is dynamic, and the influence of sex hormones on tissues also varies. Further, as tissue content can vary in the ratio of ER-alpha to ER-beta, the consequences of hormone-receptor interactions can also vary with integration into the biomechanical environment. As hormone receptors alone and hormone-receptor complexes can influence gene expression (reviewed in Hart 2022), the consequences can have prolonged effects on tissue functions, as well as fairly rapid alterations in joint function, such as joint laxity (Park et al. 2009a, b). In the studies by Park et al. (2009a, b), it was noted that ~50% of the young, otherwise healthy females exhibited significant increases in knee joint laxity at the follicular phase of the menstrual cycle, and 50% did not. There were no detectable differences in estrogen levels between those who exhibited increased laxity and those who did not. While the cause for responsiveness versus non-responsiveness remains unknown, the findings indicate that females are heterogeneous with regard to this relationship between laxity and sex hormones. Furthermore, it is not known whether these subsets are defined only at the level of the knee or are also evident at other joints, such as the elbow or shoulder.

In addition to the sex hormones such as estrogen and progesterone on cells of MSK tissues, relaxin has also been implicated in regulating tissues such as ligaments during the menstrual cycle (Gilmer et al. 2020, 2025; Dehghan et al. 2024) and also risk for injury (Berger et al. 2023; Parker et al. 2024). Relaxin is a member of the insulin-like growth factor family of mediators (discussed in De Meyts et al. 2009). In addition, relaxin receptors have been detected on cells of the human ACL (Dragoo et al. 2003; Van Der Westhuizen et al. 2007), as well as other species such as rats (Kim et al. 2016) and dogs (Restucci et al. 2022). In the rat, there appears to be an interplay between sex hormone receptors and relaxin receptors (Dehghan et al. 2015). Thus, there may be direct effects of sex hormones on the cells of MSK tissues as well as indirect effects via the action of mediators such as relaxin on the cell-ECM environment in females compared to males.

While the molecular basis for the alterations in knee laxity is not well defined, as well as which tissues are affected (likely ligaments and the joint capsule), it is likely that the changes must be readily reversible if they occur every month for the 35–40 years a female is experiencing menstrual cycles. As laxity is a low-load behavior of a connective tissue, the changes likely reflect alterations to proteoglycan (PG) rather than collagen integrity, with the latter more responsible for high-load behavior of tissues such as tendons and ligaments. This conclusion is supported by the work of Cunningham (1998), who determined that the extraction of bone-ligament-bone constructs of the rabbit medial collateral ligament with urea led to the removal of some PG, proteins, and glycoproteins, but the structure of the ligament remained intact, with all of the high load behavior retained and only low load behavior was lost. It remains to be determined whether other connective tissues, such as tendons, other ligaments, and menisci, exhibit the same partitioning of mechanical behavior. It also remains to be determined whether there are subsets of cells in these tissues that associate with the major collagen fibrils versus those that associate with PG and soluble proteins and whether these cells respond to alterations in hormone levels and low loads leading to alterations in laxity during the menstrual cycle.

However, one tissue that may be particularly affected during the menstrual cycle is the anterior cruciate ligament (ACL). Interestingly, females are at a higher non-contact risk of tearing the ACL in certain sports that involve cutting maneuvers than their male counterparts (Barlow et al. 2024; Apseloff et al. 2024; John et al. 2025), and it is reported that the incidence is higher during different phases of the menstrual cycle. Thus, changes in the biology of the knee may impair its ability to adapt to an altered biomechanical environment in a subset of females. While there appears to be a higher risk, many female athletes do not tear their ACLs, so other variables may also be at play in these cutting maneuvers.

An additional dysfunctional process that arises post-puberty and predominantly in adolescent females (~90%) is scoliosis of the spine (reviewed in Raggio 2006). In adolescent idiopathic scoliosis (AIS), the intervertebral discs (IVD), predominantly in the thoracic region, grow abnormally, leading to a twisted spinal column. The extent of the abnormality is quite variable, leading to treatments including bracing, rod implantation, or more invasive surgery to restore function (Kuang et al. 2025). The molecular basis for the development of AIS is not known, but genetics has been implicated (Huang et al. 2025b; Tuncay et al. 2025), as has abnormal muscle signaling (Ford et al. 1984; Ohba et al. 2022). In this case, it would appear that biology is overriding any biomechanical considerations, but perhaps the abnormal influence of muscle contractions creates an environment in which growth adapts to these influences, contributing to the development of scoliosis. As a result of scoliosis, the mechanical environment for cells on the concave and convex sides of the curvature would be very different likely altering their relationship with the ECM.

It is of interest that the muscle imbalances in the development of scoliosis and possibly with the ACL occur in a sex-dependent manner (Otsuki et al. 2021) and appear to be potentially due to compromised or altered neuroregulation of muscle. In the case of scoliosis development, the impact of the dysfunction can lead from minor to severe outcomes. Perhaps a deeper understanding of the basis for the development of muscle imbalances could lead to a better understanding of scoliosis onset and future development of interventions implemented prior to puberty or early after. Likely, what is also needed are serum or imaging biomarkers of risk for scoliosis development.

With the onset of puberty, the biological and biomechanical landscapes are altered in both males and females. In females, the biological and biomechanical environments are changed in preparation for reproductive activities, and the growth and maturation associated with these changes also appear to increase the potential for dysfunctional outcomes in the process. Why the risk for such dysfunctional outcomes should be maintained in the populations is not known, but perhaps there are survival advantages to being a heterozygote for such risk, or the outcome in some cases (i.e., muscle imbalance in the femur), the deficiency would not impact normal life activities and only be a problem when engaging in sport activities.

13.3.4 Skeletal Maturity

With the advent of skeletal maturity, the growth plates close and the "set points" for MSK tissues and their interrelationships with organ systems, such as the knee, are established. While such relationships are defined by both biology and biomechanics, the biomechanical environment can be influenced by weight gain/obesity, and for females, pregnancy and lactation can influence both transient weight gain and the biological consequences required for a successful pregnancy and childbirth. For both males and females, exercise can alter muscle mass, thereby altering the ability to apply loads to ligaments, tendons, and bone. Participation in exercise programs to excess can lead to overuse and loss of biological integrity if the biomechanical demands exceed the ability to maintain integrity via repair processes. Thus, even in skeletal maturity, the biomechanical and biological interplay is dynamic, but still the "set points" define the limits to which the mechanical and biological extremes can be applied without leading to loss of integrity (Fig. 13.3). Exceeding the biological limits within a tissue and its ability to repair can lead to loss of tissue integrity with

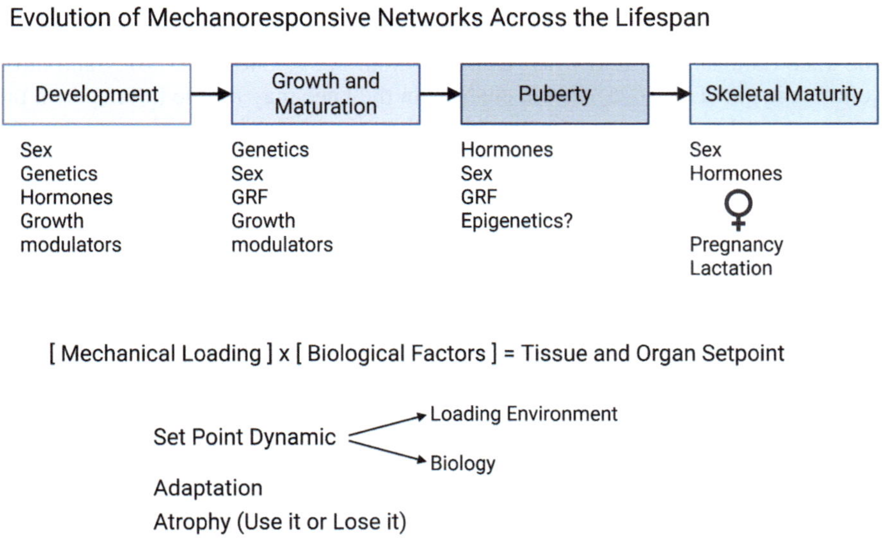

Fig. 13.3 Evolution of mechanoresponsive networks across the lifespan. The interplay between biology and biomechanics. (Created with BioRender.com)

the development of structural alterations and loss of function (discussed in Thornton and Hart 2011; Lin et al. 2025).

13.3.4.1 Pregnancy and Lactation

Pregnancy leads to alterations in both the mechanical and biological environments of a female over the ~9 months of pregnancy. In addition to the weight gain and its effect on tissues of the MSK system, there are substantial alterations to biological regulation that can also impact several of these tissues.

Thus, to meet the demands for Ca^{+2} for skeletal development of the fetus, there is mobilization of this ion from bone (discussed in Kalkwarf and Specker 2002; Wojczakowski and Futyma 2025). While this mobilization results in modest alterations to bone calcium content, in a small subset of females, it can lead to overt osteoporosis (OP) (discussed in Pearson et al. 2004) and thus changes in the biomechanical integrity of bone. This loss of calcium from bone can continue during lactation but appears to be replenished after pregnancy and lactation (Kalkwarf 1999; Pearson et al. 2004). This reversal is of interest as after menopause, bone loss leading to OP is also variable but is unidirectional for progressive loss of integrity with increased risk for fractures (discussed in more detail in later sections). This also implies that the cells replenish the bone to levels comparable with the pre-pregnancy set point thereby re-establishing the mechano-regulatory environment.

The biological changes that occur in pregnancy can impact MSK tissues and their function in different ways during the 9 months. Alterations in hormone level, such as estrogen, progesterone, and others, can affect MSK tissues as they all have receptors for these hormones (discussed in Hart 2022). One noticeable change in many, but not all, pregnant females is an increased joint laxity in joints such as the knee (Dumas and Reid 1997; Charlton et al. 2001; Chu et al. 2019) and other joints (Yoseph et al. 2025). These changes in the knee may be due to effects on the ligaments and perhaps, the joint capsule. While changes to joint laxity and thus the response to mechanical loading in pregnancy would occur over a large part of the pregnancy and be chronic, this would differ from that observed during the menstrual cycle discussed previously. As it is not known whether the subset of females who do not experience change in joint laxity during the menstrual cycle (Park et al. 2009a, b) also do not experience such changes during pregnancy, this issue should be the focus of future research.

For some pregnant females, the increases in joint laxity are most evident in the third trimester. Some evidence indicates the mechanism may involve the mediator relaxin (discussed in Van Der Westhuizen et al. 2007; Yalcinkaya et al. 2025). Therefore, it is not known whether altered sex hormone levels have direct effects on tissues of the MSK system or whether the effects on laxity are indirect via another mediator, such as relaxin (discussed in Sect. 13.3.3). While the involvement of sex hormones, their receptors, relaxin, and its receptors in the menstrual cycle has been noted (discussed in Sect. 13.3.3), the menstrual cycle is ~30 days, and pregnancy is ~9 months, so chronic and prolonged alterations to the ECM and cell behavior in response to mechanical loading may be different in the two circumstances. Therefore, additional investigation is needed to better define the role of relaxin in pregnancy-associated alterations to MSK tissue integrity and cell responses to mechanical stimulation.

Many, but not all, pregnant females develop low back pain (reviewed in Bonnheim et al. 2023; Yoseph et al. 2025). This may result from alterations in the biomechanical environment of the spine as the fetus grows. It may also arise due to the impact of sex hormones, relaxin, or other molecules on the connective tissues of the intervertebral disc. However, this area requires more investigation to generate an improved understanding regarding subsets of females who may be at risk of developing overt disk degeneration during pregnancy.

Thus, pregnancy and lactation can have a significant impact on tissues such as bone as well as soft tissues in an obviously sex-dependent manner, leading to alterations that likely modify the mechanosensing by cells in the presence of endocrine factors and other growth regulating mediators, as well as the increased biomechanical loading due to weight gain from the fetus and related tissues and fluids.

13.3.5 Aging and Menopause: Heterogeneity of Populations

Biological aging leads to alterations in a number of MSK tissues that can impact the functioning of these tissues in their biomechanically active roles. Some tendons are reported to less stiff (Wu et al. 2020; Delabastita et al. 2018), IVD can suffer age-dependent decline in integrity (De Vries et al. 2025), articular cartilage may develop surface defects or fissures (Sacitharan et al. 2012), and skin can become thinner (Zhang et al. 2024a; b; Goldman et al. 2025). Some of the changes in mechanical properties in tissues such as tendons or ligaments may be due to age-dependent changes in the expression of the lubricant PRG4 (Thornton et al. 2015; Reuvers et al. 2011; Hayashi et al. 2013), a molecule that facilitates the sliding of collagen fibrils when loaded. In addition, some reports indicate that the collagen fibril diameter distribution shifts toward smaller fibrils with age (Ishida et al. 2025), a change that may alter the mechanical properties of tissues.

These alterations in function can be asymptomatic and progress without any detectable consequences. It is thus implied that the cells in these tissues are now operating under an abnormal mechanical signaling environment, either through changes in the cell subtypes in the tissues, in the tissue composition, or both. As not all individuals experience these changes as they age, there is considerable heterogeneity in the aging process. This may in part be due to differences between biological and chronological aging (Zurbuchen et al. 2025; Haugg et al. 2025), genetics, and perhaps epigenetics associated with life experiences. Certainly, the experiences of males and females differ regarding reproductive events, and it is clear that some of these clinical outcomes affect the sexes differently. For instance, there is a higher incidence of DDD in the female population than in males (Gautschi et al. 2016a, b), and the incidence of tendon ruptures is higher in males than in females

(Hartman et al. 2024; Gianakos et al. 2024). As females undergo menopause as they age, with an onset at ~45–50 years of age, the hormonal environment also changes in all females, but only a subset develops the clinical conditions (reviewed in Hart 2022, 2024a, b).

The onset of menopause in females is accompanied by a cessation of menstrual cycles and loss of ovarian estrogen and progesterone-secreting cells. However, the testosterone-secreting thecal cells remain functional. After adapting to puberty and integrating new hormonal levels with biomechanical loading of tissues for ~35–40 years, cells and tissues of the MSK system of females are now faced with the loss of these hormonal regulatory influences, and a significant control mechanism is no longer relevant.

While several biological systems are affected in different subsets of females by menopause (discussed in Hart 2024a, b), one tissue that is affected and can have a serious impact on load behavior is bone. Unlike the modifications of bone associated with pregnancy and lactation discussed previously, the impact of menopause on bone in this subset of females is consistent and progressive bone loss leads to OP (discussed in Hart 2023a, b, 2024a, b). Thus, in these affected females, there is a loss of regulation of mechanical loading of bone integrity with the loss of hormonal influences, leading to an imbalance in the relative activity of osteoclasts and osteoblasts. As osteocytes are also considered the mechanosensing cell in bone (reviewed in Yuze et al. 2025), there appears to be a defect in communication between such cells and the effector cells responsible for bone resorption and deposition, respectively. Interestingly, some females with OP appear to lose bone slowly while others lose bone more rapidly (reviewed in Hart 2022), suggesting that genetic factors may also be involved in both the onset of OP and its rate of progression.

The idea that hormones and loading may play roles in OP progression also comes from studies using hormone replacement therapy (HRT) and exercise programs (Hsu et al. 2024; Tang et al. 2025; Kumar et al. 2025). HRT appears to exert a positive effect on bone integrity, and exercise programs also appear to overcome some of the OP-related variables. However, it is not clear whether the exercise programs exert their effect via the mechanical loading on bone or through the release of myokines from muscle (Lau et al. 2022; Kaji 2024) or osteokines from the bone itself (Lau et al. 2022; Zhu et al. 2025) when subjected to exercise. Perhaps some of these myokines and osteokines impact osteocytes and/or osteoblasts to restore pre-menopause levels of regulation. Enhanced understanding of this "muscle-bone" crosstalk may lead to new approaches to integrating it with mechanical signaling to maintain tissue integrity (Dong et al. 2025).

In contrast to all females, some males undergo a gradual loss of sex hormones, termed andropause which can lead to loss of muscle and connective tissue integrity with aging (Ebeling 1998; Huhtaniemi 2014). Therefore, the hormonal environment is also dynamic in some males but differs from that in females. As a subset of males can develop OP, the mechanisms regulating this process in males remain to be elucidated. Thus, the relationship between mechanosensing and mechanotransduction and biological variables is likely "fluid" across the lifespan (Fig. 13.4).

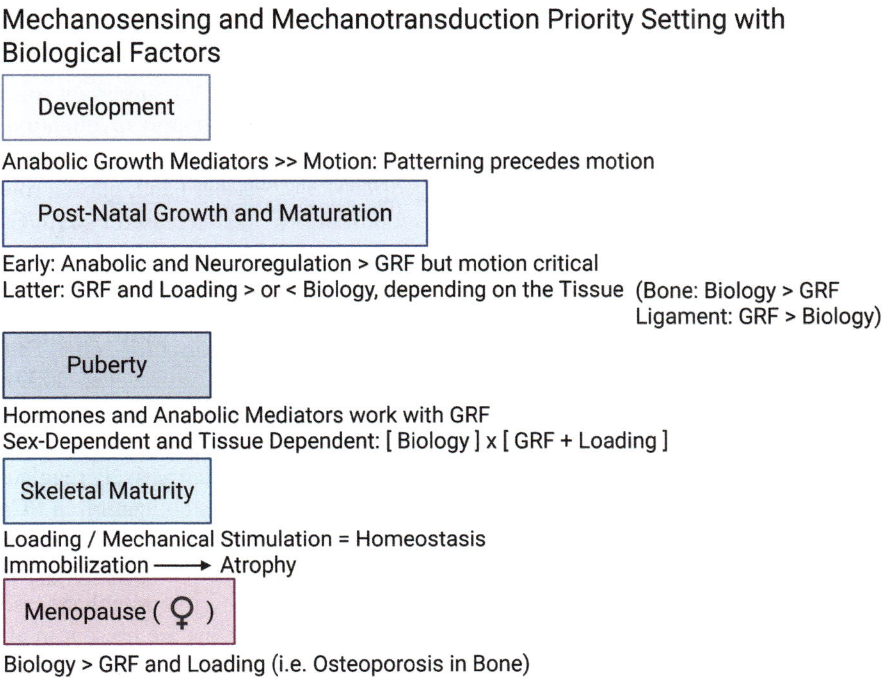

Fig. 13.4 Priority setting across the lifespan regarding the mechanosensing and mechanotransduction relationship with biological variables. The variable dominance of mechanical responses and the dynamic influence of biological factors on MSK tissue regulation. (Created with BioRender.com)

13.3.6 Summary

Tissues of the MSK system are dynamic throughout the lifespan, both at the level of the cells and the ECM, and thus the "Cell-ECM" relationship (Fig. 13.5) and resultant mechanosensing regulation are also dynamic and exhibit sex differences. Therefore, the mechanotransduction environment is dynamic, with cells being influenced by anabolic mediators that regulate the ECM to meet the matrix's mechanical needs and adapt to increasing requirements as the human grows. Subsequently, the cell-ECM relationship is influenced by hormones to adapt to reproductive activities in females, and later, the loss of these hormones with the onset of menopause in females and andropause in males leads to further changes in the cell-ECM relationships. Thus, there are several points where the regulation of cells in mechanically activated tissues exhibits potential sex differences. In some instances, aspects of these sex differences may contribute to clinically significant conditions, such as in menopause and following puberty onset. Thus, these sex-related hormonal influences appear to override the effects of GRFs and mechanotransduction on the maintenance of the integrity of several KSK tissues.

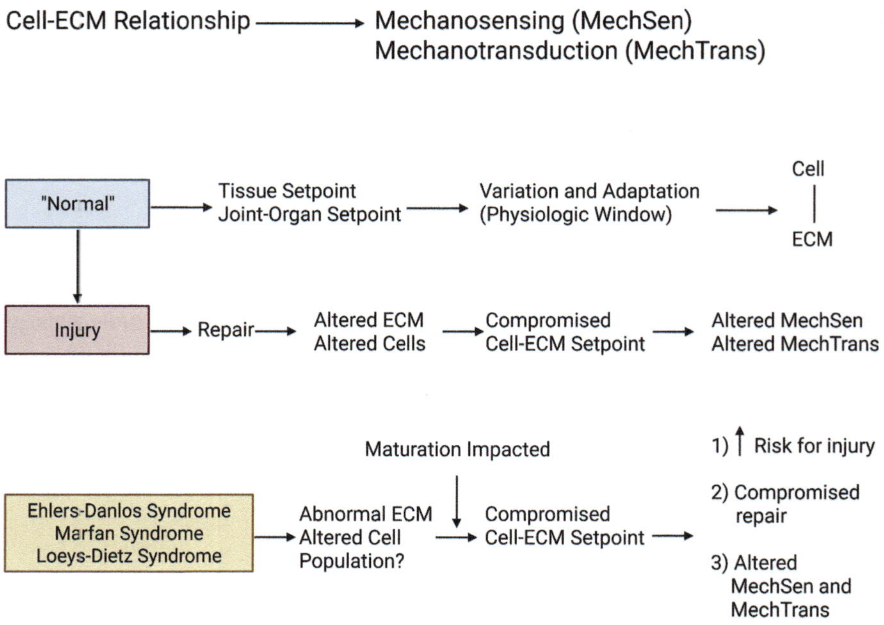

Fig. 13.5 Defining the role of the "Cell-ECM" relationship in mechanosensing and mechanotransduction. The normal sequence for establishing the tissue set point is outlined, as well as resulting from the influence of tissue injury and potential altered ECM due to mutations and/or dysregulated ECM development. (Created with BioRender.com)

13.4 Influence of Alterations to the Mechanical and Biological Environment on MSK Tissues: Insights into Regulation

In addition to the "normal" lifespan being dynamic regarding the MSK tissue cell-ECM-biomechanics relationship and its regulation, many MSK tissues respond to other influences on these relationships, which provide further insights into how this system is regulated and what adaptations are possible.

13.4.1 Modifying the Biomechanical Environment

During growth and maturation, tissues of the MSK system are subjected to tensile, compression, or shear mechanical environments. Some tissues, such as subsets of tendons, have parts that are subjected to tensile loading and others subjected to compressive loading, where they go around a bony prominence (Perez-Castro and Vogel 1999). This leads to differences in the ECM and what the cells are expressing (Vogel 2003, 2004). If one removes the tendon from the bony prominence in a preclinical rabbit model, the fibrocartilage-like area subjected to compressive loading reverts to a tensile-loaded phenotype (Malaviya et al. 2000).

In a somewhat different rabbit model, subjecting a scar developing after a medial collateral ligament injury to a compressive environment, rather than the usual tensile environment, leads to expression of fibrocartilage-like molecules such as aggrecan and collagen II (Majima et al. 2000). Thus, cells in MSK tissues can perceive the type of mechanical loading and respond appropriately to express and secrete the matrix molecules required for the environment. Thus, these cells can distinguish between different types of mechanical loading and respond by adapting or differentiating to meet the situation.

13.4.2 Immobilization

As discussed above, gravity is one of the boundary conditions that defined the evolution of humans and their predecessors (discussed in Hart 2023a, b). Loading of tissues via GRF during mobility is a major mechanical stimulus for mechanotransduction. Thus, movement in a 1 g environment is central to maintaining MSK tissues and the cell-ECM relationship (reviewed in Hart and Zernicke 2021). In this regard, the cell-ECM relationship follows the "Use it or Lose it" paradigm in which loss of loading leads to atrophy and increased loading leads to adaptation within physiological limits. Evidence for this conclusion comes from several areas.

Loss of loading can be achieved by immobilizing a limb with a cast or by pinning it at a fixed angle. Immobilization of the knee of very young rabbits via pinning the limb in flexion leads to a cessation of growth of soft tissues but not the bone (discussed in Sect. 13.3.2). Thus, some tissues respond to immobilization, but the bone in this environment continues to grow, likely due to anabolic factors overriding the loss of mechanical loading in the regulation of the bone. Thus, there is an interplay between mechanotransduction and biological factors that shapes MSK tissue responses under different conditions.

13.4.3 Removal of the GRF Component of Loading

Removal of the GRF component of loading can be accomplished by hindlimb elevation of rodents (Cui et al. 2020; Spatz et al. 2021; Camy et al. 2022), prolonged bed rest (reviewed in Pinto et al. 2023; Zhang et al. 2024a), and space flight/microgravity environments (reviewed in Lau et al. 2022; Winkler 2023; Hart 2023a, b). In all three environments, movement is retained but the influence of gravity and GRF is removed.

Hindlimb unloading leads to disuse atrophy of bone, muscle, cartilage, and ligaments of the limb. However, when returned to a GRF environment with reloading, the atrophy appears to be reversible (Takahashi et al. 2021). The regulation of muscle and bone loss may involve responsiveness to the cytokine irisin (Kawao et al. 2018).

Loss of GRF can also result from prolonged bed rest. This can occur in chronic non-MSK conditions and involve individuals of all ages across the lifespan. Prolonged bed rest in an otherwise healthy population has been compared to the effect of space travel (discussed in Rolvien and Amling 2022). Individuals subjected to prolonged bed rest can move their limbs but are removed from GRF yet remain subjected to other aspects of gravity on other organs. Participants in such studies experience muscle and bone loss within days (Kos et al. 2014). The affected tissues rebound to pre-bed rest levels when re-exposed to GRF.

Space flight also results in loss of GRF leading to atrophy (discussed in Hart 2023a, b). A variety of systems are affected, including muscle, bone, cardiovascular systems, and eyes (reviewed in Rolvien and Amling 2022; Hart 2023a, b). Exercising in space under conditions of an artificial GRF can help retain muscle, but bone integrity is more difficult to retain. Exposure to prolonged microgravity during space flight can also lead to epigenetic changes in peripheral blood cells of astronauts (Ao et al. 2025; Garrett-Bakelman et al. 2019), but it is not known whether other epigenetic alterations to cells of the MSK tissues are also induced by such exposure. Once returned to Earth, muscle, bone, and cardiovascular changes can return to pre-flight conditions in many, but not all, astronauts (discussed in Hart 2023a, b).

Thus, prolonged removal of exposure to GRF such as during space flight, prolonged bed rest, or other means can lead to atrophy of both soft and hard MSK tissues. The cells in these tissues therefore have embedded in their mechanosensing programming, the ability to modulate their metabolic activity to lower the integrity of the Earth-defined set point for a tissue until the cells are again engaged. This process is often reversible once returned to Earth's 1-g environment, so it exhibits elements of plasticity even in the adult human. Whether this apparent plasticity in mechano-regulation declines with aging is not well defined, and it should be noted that atrophy of some tissues can occur with aging, such as muscles, leading to sarcopenia (reviewed in Li et al. 2025, Burton et al. 2025). However, aging and space flight alterations differ but mechanotransduction may be a key commonality between the two processes (Van Rijthoven and van Loon 2025).

Mechanistically, loss of mechanical loading of a tissue leads to derepression of catabolic genes, which can then mediate atrophy-related events (Natsuume et al. 2005). In an intact adult rabbit knee meniscus removed from the animal and maintained in vitro, the induction of genes such as MMP-1, MMP-3, IL-1beta, iNOS, and COX-2 occurred over a 4-hr incubation at pH 7.2 and 37°C. However, expression of collagen I and III, MMP-13, or biglycan was unchanged. Exposure of the menisci in vitro to cyclic hydrostatic compression (1 minute on and 14 minutes off; 1 MPa) completely prevented derepression of the catabolic genes. Additional studies revealed that exposure to 0.5 MPa also prevented derepression, but exposure to 0.25 MPa led to a subset of the catabolic genes not being fully repressed

(Hart and Natsuume; unpublished observations). Thus, mechanical loading with an intermittent signal was sufficient to generate a mechanotransduction signal to maintain repression of this set of genes. Whether similar genes are responsible for atrophy in space flight, hindlimb loading, or bed rest remains to be determined. In addition, it is also not known whether the derepression occurred in cells of the inner, outer, or both parts of the menisci. This may be important as the inner part of the meniscus is subjected to compressive loading in the knee while the outer portion is subjected to tensile loading in vivo.

Based on an extensive literature review, GRFs are an essential factor in the maintenance of mechanically active tissues of the MSK system. Movement of a joint via muscles is not sufficient to maintain tissues, and therefore, the mechanotransduction responses associated with experiencing GRF are essential and developed as a central boundary condition during evolution (the "use it or lose it" principle) (discussed in Hart 2021).

13.5 Injury and Repair: Repair Does Not Recapitulate Development and Maturation

An injury to a tissue of the MSK system in an adult, such as skin, ligament, or tendon, leads to an initial loss of mechanical loading, followed by an inflammatory response and a series of steps leading to tissue repair. This repair tissue is often compromised functionally, structurally, and compositionally over the long term.

The return to functionality of a torn ligament depends in part on the ligament and its location. Thus, a ligament such as the intraarticular anterior cruciate ligament (ACL) of the knee is in a relatively high load environment, while the extraarticular medial collateral ligament (MCL) is in a low load environment.

13.5.1 ACL Healing

The ACL connects the femur to the tibia, with the two insertions very different. Also, the ACL is a two-bundle ligament with the anterior and posterior bundles twisted around each other. The intact adult ACL is under some tension even when not subjected to overt loading. Thus, when a complete mid-substance tear occurs, the torn ends are not in contact, and in the intraarticular environment there are no mechanisms available for the ends to reattach. The torn ends are thus no longer loaded. If the tear occurs at the femoral insertion site, the torn end can attach to a non-normal site in the environment and "heal," but it is not functional (discussed in Lo et al. 1999, 2003).

The loss of the ACL, which is integral to the mechanical loading environment of the knee as an organ system (Radin et al. 1991; Frank et al. 2004; Loeser et al. 2012; Rollick et al. 2018), compromises the knee's integrity. Because of the ACL's central role, attempts to replace a torn ACL with an autologous or allogenic tendon are undertaken. These transplanted tissues are often stronger than the original ACL, stabilize the knee, and, after transplantation, heal into tunnels drilled into the bone and take up load. However, over time, the transplanted tissue appears to be transformed into a tissue more resembling a scar tissue and develops creep, which lessens its ability to stabilize the knee and take up the loads required to maintain integrity (Simonian et al. 2000). Some tendons used for ACL reconstruction in females are more prone to creep than those in males (Miyamoto et al. 2024), potentially indicating that those from females have different set points and loading responses. One feature of this transformation is the conversion of collagen fibril diameters from a bimodal distribution to a unimodal distribution of small fibrils, at least in preclinical models, and this may be associated with infiltration of exogenous cells into the transplanted tissue (Takeuchi et al. 2016). This creep development is also observed in MCL grafts (Boorman et al. 2002; Thornton et al. 2002), so it is not location specific.

This transformation is likely related to both cellular and ECM changes, altering the mechanosensing conditions in the tissue. Details regarding the death of the initially transplanted cells, the influx of perhaps exogenous cells incapable of maintaining the required load-response ECM, and the stimulus for the transformation, remain mostly to be determined. However, some of the stimuli for the transformation may reside in inflammatory processes initiated by the surgery itself or related to mediators in the intraarticular environment due to loading (discussed in Hart and Nakamura 2022; Hart 2025a, b). Some of this conclusion is based on studies in sheep and rabbits where an "idealized" ACL injury leading to induction of subsequent cartilage alterations (O'Brien et al. 2013) can be mitigated by immediate treatment with anti-inflammatory drugs such as corticosteroids (Barton et al. 2018; Heard et al. 2022).

13.5.2 MCL Healing

In contrast to the ACL, a complete rupture of the extraarticular MCL can heal mid-substance tears quite well. Even though the torn or cut ends of the MCL retract slightly as the ligament is under tension, a scar forms rapidly after cutting the MCL in preclinical models such as the rabbit (Frank et al. 1999), and in humans, MCL tears are usually not operated on as they also appear to heal when treated conservatively as the ligament lays on the joint capsule.

Inducing a mid-substance injury to the MCL initiates an inflammatory response that leads to the deposition of a matrix that is somewhat amorphous and contains ~50–50% collagen I and collagen III, compared to the natural ligament, which is 90% collagen I and only 10% collagen III (Boykiw et al. 1998; Frank et al. 1999). This shift to a more collagen III-rich environment is also observed in skin wound healing (Gardeazabal and Izeta 2024), so it is likely a common response to injury in connective tissues. Such a large percentage of collagen III not only affects biomechanical properties but can also alter the engagement of cells that can interact to generate effective mechanotransduction.

After rupture of a ligament, such as the MCL, or a tendon such as the Achilles tendon, the remaining tissue is unloaded for an extended period while the healing process is initiated and progresses. Thus, the initial inflammatory phase of the healing process involves the release of inflammatory mediators that can act on not only cells entering the injury site but also those remaining at the ends of the tissues. This likely means that cells in this environment respond differently from when the tissues are loaded, as mechanical load has been reported to inhibit the actions of the inflammatory mediator IL-1 in some cells (Torzilli et al. 2010). Thus, once the healing process progresses beyond the inflammatory stage to the matrix deposition phase, the scar tissue and the remaining tissue ends will again begin to take up load. While it remains to be demonstrated for the healing of tissues such as ligaments, in some tissues, mechanical loading and responses to anabolic growth factors appear to function beyond additivity (Bonassar et al. 2001). Whether this is due to mechanical loading and growth factor receptors working together, or due to mechanical loading increasing the expression or function of growth factor receptors, remains to be determined. This latter interpretation is somewhat supported by reports that immobilization reduces the expression of neuropeptide receptors in healing rat tendons (Bring et al. 2009) and immobilization of a growing ligament appears to inhibit responses to factors in the anabolic environment (Walsh et al. 1992, 1993). However, direct evidence regarding the reestablishment of loading and the progression of ligament healing is required in future investigations.

As the initial scar tissue does not sense the load, the collagen is not associated with the direction of the ligament, but this changes over time as the scar begins to respond to loading (Liu et al. 1991). The initial matrix deposition exceeds the normal boundary of the ligament but as the ligament begins to sense load, the scar tissue assumes the normal boundary of the ligament by 14 weeks post-injury and the excess matrix not sensing load is resorbed (Frank et al. 1999). Thus, scar tissue is also subjected to the "use it or lose it" paradigm as it matures. During this process, the cellularity of the scar also decreases, but in the presence of an altered matrix with high levels of collagen III, and likely other wound associated molecules (Frank et al. 1999). Whether the cells that remain represent a unique subset of cells that interact with this non-normal matrix to yield a new set point for mechanosensing is not well defined. As all of the above studies were performed with female animals, whether the outcomes would be influenced by sex and hormones is also not known in detail.

The MCL scar tissue increased in strength from 3–14 weeks post-injury but remains mechanically compromised for an extended period of time (reviewed in Hart 2025a, b). Of note, the scar tissue consists of small collagen fibrils early and this persists for years following the injury and never returns to the normal bimodal distribution of large and small fibrils (Frank et al. 1997; reviewed in Frank et al. 1999; Hart 2025a, b).

Why this collagen fibril distribution does not mature into a bimodal distribution is unclear. The basis could reside in the cells that populate the scar being different from those in the normally developing MCL, an ECM developing after injury that does not convey the correct signals to the cells, or the transition from a mainly small collagen fibril environment during growth and maturation may be due to some endocrine-mediated signals to the cells, so it is a time-dependent event. Whatever the reason(s) for the failure to mature in this regard, it compromises the functionality of the scar tissue but does lead to a partially functional tissue, so it likely still provides a survival advantage.

However, one possible explanation for the transformation from a single distribution of small collagen fibrils to a bimodal distribution comes from experiments on MCL scars. The focus was on the expression of a specific small leucine-rich protein (SLRP). SLRPS exist in different classes and have several biological roles, including binding to collagens (reviewed in Kalamajski and Oldberg 2010; Chen and Birk 2013; Halper 2014; Matsushima et al. 2021), and one of them, biglycan, is located on the X chromosome (Geerkens et al. 1999). Scars were treated in vivo with an anti-sense decorin reagent or a nonsense anti-sense reagent and then analyzed using biomechanics to assess functionality and scanning electron microscopy to assess collagen fibril diameters. Treatment of scars with an anti-sense decorin led to increased biomechanical properties and the appearance of large fibril diameter collagen fibril (Nakamura et al. 2000, 2001). In this regard, scar cells do not appear to recapitulate the normal growth and maturation of the ligament, and this may involve decorin expression. Whether this is due to the cells repopulating the scars being very different from those in the normal ligament, the early scar matrix does not resemble the original ECM template, the appropriate matrix-based signals not being recognized by the cells, or other as yet unknown factors, remains to be determined. While the antisense approach does have some limitations regarding delivery to cells in ligaments (Hildebrand et al. 2004), the concept that delivery of appropriate signals that decrease decorin expression could enhance scar properties, more closely reflecting the natural ligament function.

A second approach indicating a role for decorin expression in regulating collagen fibril diameter comes from decorin knockout mice (reviewed in Hwang and Halpert 2021). MSK tissues in such mice have larger collagen fibrils and altered biomechanical properties (Hwang and Halpert 2021). However, levels of biglycan are increased in decorin knockout mice (Zhang et al. 2009), indicating a compensatory increase in biglycan levels. Similarly, levels of another SLRP lumican were reported to be 4x higher in tendons of fibromodulin knockout mice (Svensson et al. 1999). Thus, determining "cause and effect" after in vivo antisense decorin treatment is complex (Hart et al. 2000).

In both tendon and ligament healing, the process leading to scar formation and subsequent scar maturation produces a tissue with partial functionality but one that is compromised compared to normal tissue. Such a compromise may allow some mobility and navigation in the environment, but there is potential for reinjury if the scar is subjected to mechanical environments beyond the compromised limits. The compromised function is likely due in part to impaired signal transduction and to an altered cell-ECM relationship compared to the normal uninjured tissues that arose during development and matured during growth (Fig. 13.5). In addition, injury to some knee ligaments alters the non-injured ligaments of the same limb (Funakoshi et al. 2007), consistent with the joint as an organ concept. Injury to a ligament on one limb can also lead to alterations in ligaments on the contralateral knee, contributing to the "contralateral effect" (Frank et al. 1994).

Based on the above discussion, the cells remaining in a scar after it matures still have an ECM that differs from the natural ECM and also differ from the endogenous cells that populated the tissue originally. Fibroblasts are very heterogeneous (Sempowski et al. 1995; Mascharak et al. 2020; Talbott et al. 2022; Zhao et al. 2025a, b, c), so many fibroblast subsets may repopulate the injury site with the final population representing cells that can function in that environment.

One possibility for the cellular origin in the case of MCL scar is the original epiligament of the tissue. The epiligament is relatively cellular, functions in a low load environment, and has a matrix enriched in collagen III similar to scar tissue (Frank et al. 1999; Frank 2004; Georgiev et al. 2025). Thus, the cells in the mature scar may not be capable of recapitulating the original structural organization of the MCL ECM due to their epiligament origins and be unable to develop bimodal distribution of collagen fibrils. This compromise would likely inhibit a return to the original mechanical response pattern of the MCL (Achari et al. 2011). Whether these changes influence the mechanotransduction environment via expressed integrins and other mechanoresponse elements remains to be determined in detail, but it has been reported that scar maturation appears to require loading to mature (Martinez et al. 2007).

13.6 Influence of ECM-Altering Mutations on ECM Integrity, Cell Function, and Mechano-Regulation

There are several mutations in matrix molecules that appear to increase risk of clinical consequences that range from modest to severe, including death from cardiac events (Elendu et al. 2023; Morisaki 2024). These include conditions such as Ehlers-Danlos syndrome (EDS), Loey-Dietz syndrome (LDS), and Marfan syndrome (MFS) (discussed recently in Hart 2025a, b). The EDS encompasses ~13 different syndromes with most attributed to specific mutations (Syx and Malfait 2024; Malfait et al. 2017; Chiarelli et al. 2019), but those associated with hypermobility EDS (hEDS) and Joint Hypermobility Spectrum disorders have not had the basis clearly identified (Chiarelli et al. 2019; Carroll 2023; Shirvani et al. 2024).

The breadth of molecules affected in these syndromes ranges from subsets of collagens to fibrillins, LOX, TGF-beta receptors, tenascin C, and glycosyltransferases (reviewed in Hart 2025a). Similarly, a variety of tissues can be affected, such as skin, joint tissues, tissues of the cardiovascular system, and other (reviewed in Hart

2025a, b; Syx et al. 2015; Syx and Malfait 2024). Regarding skin, the tissues can be loose, heal poorly, and exhibit abnormal collagen fibril diameters, but the abnormal distribution of collagen fibrils is quite variable (discussed in Syx et al. 2015; Syx and Malfait 2024; Hart 2025a). Those with joint hypermobility can exhibit an increased range of motion compared to those without such mutations, leading to pain, risk for injury, and degenerative diseases such as OA later in life. The increases in range of motion for joints such as the elbow can be observed to be 10–20 degrees and increases for the knee also be quite significant (Golden et al. unpublished observations). Some females with joint hypermobility report increased pain in some joints including the knee, shoulder, and back (Gullapalli and Javed 2023; Garcia et al. 2024). Whether such pain in females is influenced by the menstrual cycle remains unclear.

In contrast, alterations to tissues of the cardiovascular system can lead to aortic dissections, aneurysms, and lethal consequences (Elendu et al. 2023; Morisaki 2024). For some conditions such as MFS, the clinical risk for consequences of the mutations may be observed across the lifespan so it is not restricted to prior to skeletal maturity or after skeletal maturity.

The differences in clinical features among patients with various mutations and within a mutation affecting a single molecule are of interest. It implies that background genes may be compensating for the deficiency induced by the mutated molecule and be dependent in part on whether the mutation leads to expression of an altered intact molecule or a truncated molecule. Thus, in family members it may also be dependent not only on background genes, but also perhaps the sex of the individual. For example, in one family the daughter experienced a cardiac event related to heart mitral valve alterations while two male siblings experienced inguinal aneurysms in the exact same location, but no heart involvement based on echocardiograms (Hart unpublished observations). Thus, in these syndromes and associated with ECM alterations, the tissues as a cell-matrix molecule composite fail to develop, grow, and mature within the same boundaries as does those individuals without these clinical risk factors.

Based on the discussion above, plus points raised recently by Hart (2025a), it would appear that in connective tissues containing the mutated molecules with a dysregulated ECM composition, the template generated is ultimately compromised in its interactions with available cells in the tissues. Whether cells in tissues fail to differentiate properly in this altered environment, or whether different subsets are selected for survival based on biomechanical signaling from the matrix, remains to be determined. However, subsequent fetal development and postnatal growth and maturation appear to generate a tissue environment with a mechanotransduction milieu that differs from the normal state. This may then lead to a tissue, and, in the case of a joint, an organ with a set point for functioning which reflects the joint's extended range of motion or the mechanical limit for cardiovascular tissues.

Also of note, the collagen fibril diameters in the skin of individuals with several of these EDS, MFS, and LDS exhibit an altered distribution with mainly small diameter fibrils but the findings are quite variable (Syx et al. 2015; Syx and Malfait 2024; reviewed in Hart 2025a). Therefore, the maturational sequence leading to a bimodal distribution of collagen fibrils may be variably disrupted. Why this disruption in collagen assembly is so variable is unknown. However, further investigation of the underlying molecular changes may provide clues similar to investigations with MCL scar tissue (Achari et al. 2011).

Relevant to this discussion is the fact that some individuals have secondary derangements in lysyl oxidase (LOX) enzyme regulation (Foehr et al. 2024; discussed in Hart 2025a). LOX enzymes facilitate formation of collagen and elastin cross-links, linkages that provide increased strength to the fibrils (Herchenhan et al. 2015). The LOX enzymes can also be influenced by estrogen (Lee et al. 2015) and therefore could contribute to sex differences in connective tissue structure and function. In addition, secondary dysregulation of decorin metabolism has also been reported in some individuals with mutated molecules and therefore resembles the formation and maturation of scar tissue in this regard (discussed in Sects. 13.5 and 13.5.2).

The ECM of connective tissues in individuals with EDS, MFS, and LDS differs in content and structural organization from that of individuals without such mutations and secondary effects. Because the cells residing in tissues such as skin, ligaments, tendons, as well as tissues of the cardiovascular system are making an ECM that is functionally compromised, either the cells residing in such tissues exhibit a different heterogeneity than those without mutations, or they are similar to normal cell populations but are unable to function properly in the mechanical environment presented by the altered ECM. As the response to a given mutation in these syndromes can be variable, it is likely that the expression of specific background genes can in some individuals compensate for the impact of the mutated molecules. In this regard, cells in these syndromes may interact with the altered ECM environment to exhibit an altered mechanotransduction profile (Fig. 13.5). However, this possibility needs to be further explored in detail via additional research.

13.7 Summary and Conclusions

Mechanosensing mechanisms and mechanotransduction pathways likely developed to address gravity and ground reaction forces during early evolution. The evolution from single cell organisms to complex organisms such as *Homo sapiens* led to the development of multiple levels of regulatory control of mechanosensing and mechanotransduction, several in a sex-specific manner (Fig. 13.2). These regulatory mechanisms involve several biological factors such as hormones, growth factors, cytokines, and mediators integrated into networks with the mechanosensing molecules (i.e., ion channels, integrins, cytoskeleton elements) (Figs. 13.1 and 13.3). These networks sense and respond to various types of mechanical loading (i.e., shear, tensile, compression) so are very adaptable.

Specific points about these mechanosensing and mechanotransduction networks include the following:

1. Mechanosensing and mechanotransduction responses to mechanical loading are central to maintaining mature MSK tissue integrity as there is a commitment to the "use it or lose it" principle.
2. In humans and most other mammals, the integrated mechanosensing and mechanotransduction networks exhibit sex differences and are influenced by sex hormones. As such, the regulatory influences of sex hormones in females are more complex than in males due to transitions in their lifespan: puberty, pregnancy and lactation, and menopause (Figs. 13.2 and 13.3). There is also heterogeneity across tissues in the priority given to biological versus mechanical factors in regulation.
3. The "cell-ECM' relationship is critical to the functioning in tissues of the MSK system that operate in a dynamic mechanical environment. Following injury, the resultant repair tissue does not recapitulate the original structure-function relationship. Mutations in molecules of the ECM or molecules that regulate ECM molecules lead to altered "cell-ECM" relationships that impact normal structure-function and can increase risk for tissue injury under what might be considered normal mechanical loading conditions. These include conditions such as Ehlers-Danlos syndromes, Marfan syndrome, and Loeys-Dietz syndrome (Fig. 13.5).
4. The critical nature of the 3D "cell-ECM" relationship calls into concern findings obtained in vitro regarding mechanosensing and mechanotransduction using 2D cell cultures devoid of normal ECM conditions. Such in vitro findings may not translate well to in vivo environments where additional regulatory elements are in play. Perhaps the use of 3D constructs for in vitro studies may overcome aspects of this 2D limitation.
5. Further understanding of the integration of mechanosensing and mechanotransduction with biological variables is likely required to optimize tissue engineering-based repair of MSK tissues compromised by injury, mutations, or disease. This goal will also need to address sex-specific differences in tissue regulation.

Acknowledgements The author dedicates this chapter to his colleague, collaborator, and friend Dr. Cyril B. Frank (deceased). He also thanks the many other colleagues and trainees who have contributed to the many publications and studies cited in the chapter over the past 35+ years. The author also thanks Emily Gysel for assistance in the preparation of the figures using BioRender. No funding was secured to support the preparation of this chapter.

Conflicts of Interest The author declares he has no conflicts of interest to reveal regarding the publication of this chapter.

References

Abolhasani S, Almadi Y, Rostami Y, Baravar E, Fattahi D (2025) Biomaterials in tissue repair and regeneration: key insights from extracellular matrix biology. Front Med Technol 7:1565810. https://doi.org/10.3389/fmedt.2025.1565810

Achari Y, Chin JWS, Heard BJ, Rattner JB, Shrive NG, Frank CB, Hart DA (2011) Molecular events surrounding collagen fibril assembly in the early healing rabbit medial collateral ligament-failure to recapitulate normal ligament development. Connect Tissue Res 52:301–312. https://doi.org/10.3109/03008207.2010.524719

Afrasiabi S, Partoazar A (2024) Targeting bacterial biofilm-related genes with nanoparticle-based strategies. Front Microbiol 15:1387114. https://doi.org/10.3389/fmich.2024.1387114

Alad M, Yousef F, Epure LM, Lui A, Grant MP, Meerle G, Eliopoulos N, Barralet J, Antoniou J, Mwale F (2025) Unraveling osteoarthritis mechanistic insights and emerging therapies targeting pain and inflammation. Biomolecules. 15:874. https://doi.org/10.3390/biom15060874

Ao X, Parisien M, Cata JP, Montagna F, Vigouroux M, de Araujo LOF, Karaky M, Ingelmo P, Diatchenko L (2025) Longitudinal transcriptomic and epigenetic analysis of the blood in two astronauts. Sci Rep 15:27175. https://doi.org/10.1038/s41598-025-13383-8

Apseloff NA, Hughes JD, Devitt BM, Musahl W (2024) Primary anterior cruciate ligament injuries: extrinsic and intrinasic risk factors. J Am Acad Orthop Surg 33:693–702. https://doi.org/10.5435/JAAAOS-D-24-00341

Augustin HG, Koh GY (2017) Orgaqnotypic vasculature: from descriptive heterogeneity to functional pathophysiology. Science 357:eaal12379. https://doi.org/10.1126/science.aal12379

Baratchi S, Knoerzer M, Khoshmanesh K, Mitchell A, McIntyre P (2017) Shear stress regulates TRPV4 channel clustering and translocation from adherens junctions to the basal membrane. Sci Rep 7:15942. https://doi.org/10.1038/s41598-017-16276-7

Barlow A, Blodgett JM, Williams S, Pedlar CR, Bruinvels G (2024) Injury incidence, severity, and type across the menstrual cycle in female footballers: a prospective three season cohort study. Med Sci Sports Exerc 56:1151–1158. https://doi.org/10.1249/MSS.0000000000003391

Barton KI, Heard BJ, Sevick JL, Martin CR, Shekarforoush SMM, Chung M, Achari Y, Frank CB, Shrive NG, Hart DA (2018) Posttraumatic osteoarthritis development and progression in an ovine model of partial anterior cruciate ligament transection and effect of repeated intra-articular methylprednisolone acetate injections on early disease. Am J Sports Med 46:1596–1605. https://doi.org/10.1177/0363546518765098

Batarfi WA, Yunus MHM, Harmid AA, Maarof M (2025) Breaking down osteoarthritis: exploring inflammatory and mechanical signaling pathways. Life (Basel) 15:1238. https://doi.org/10.3390/life15081238

Becker M, Oehler K, Partsch C-J, Ulmen U, Schmutzler R, Cammann H, Hesse V (2015) Hormonal "Minipuberty" influences the somatic development of boys but not of girls up to the age of 6 years. Clin Endocrinol 83:694–701. https://doi.org/10.1111/cen.12827

Berger GK, Rockov ZA, Bryne C, Trentacosta NE, Stone MA (2023) The role of relaxin in anterior cruciate ligament injuries: a systematic review. Eur J Orthop Surg Traumatol 33:3319–3326. https://doi.org/10.1007/s00590-023-03618-7

Berglund M, Wiig M, Torstensson M, Reno C, Hart DA (2004) Assessment of mRNA levels for matrix molecules and TGF-beta1 in rabbit flexor and peroneus tendons reveals regional differences in steady-state expression. J Hand Surg Br 29:165–169. https://doi.org/10.1016/j.jhsb.2003.09.005

Bertani F, Francesco DD, Corrado MD, Talmon M, Fresu LG, Boccafoschi F (2021) Paracrine shear-stress-dependent signaling from endothelial cells affects downstream endothelial function and inflammation. Int J Mol Sci 22:13300. https://doi.org/10.3390/ijms222413300

Beverley KM, Ahn SJ, Levitan I (2025) Flow-sensitive ion channels in vascular endothelial cells: mechanisms of activation and roles in mechanotransduction. Biophys J S0006-3495(25)00193-6. https://doi.org/10.1016/j.bpj.2025.03.021

Bidlingmaier F (1980) Sex differences in the secretion of gonadotropins and sex hormones in newborns and infants (Article in German). Fortschr Med 98:235–238

Bonassar LJ, Grodzinsky AJ, Frank EH, Davila SG, Bhaktav NR, Trippel SB (2001) The effect of dynamic compression on the response of articular cartilage to insulin-like growth factor −1. J Orthop Res 19:11–17. https://doi.org/10.1016/S0736-0266(00)00004-8

Bonnheim NB, Lazar AA, Kumar A, Akkaya Z, Zhou J, Guo X, O'Neill C, Link TM, Lotz JC, Krug R et al (2023) ISSLS Prize in Bioengineering Science2023: age- and sex-related

differences in lumbar intervertebral disc degeneration between patients with chronic low back pain and asymptomatic controls. Eur Spine J 32:1517–1524. https://doi.org/10.1007/s00586-023-07542-6
Boorman RS, Thornton GM, Shrive NG, Frank CB (2002) Ligament grafts become more susceptible to creep with days after surgery: evidence for early enzymatic degradation of a ligament graft in a rabbit model. Acta Orthop Scand 73:568–574. https://doi.org/10.1080/000164702321022866
Borbiro I, Rohacs T (2017) Regulation of piezo channels by cellular signaling pathways. Curr Top Membr 79:245–261. https://doi.org/10.1016/bs.ctm.2016.10.002
Borysiak A, Hesse V, Wermke P, Hain J, Robb M, Mermke K (2015) Fundamental frequency of crying in two-month old boys and girls: do sex hormones during mini-puberty mediate differences? J Voice 31(128):e21–128.e28
Boykiw R, Sciore P, Reno C, Marchuk L, Frank CB, Hart DA (1998) Altered levels of extracellular matrix molecule mRNA in healing rabbit ligaments. Matrix Biol 17:371–378. https://doi.org/10.1016/s0945-053x(98)90089-0
Bring DK-I, Reno C, Renstrom P, Salo P, Hart DA, Ackermann PW (2009) Joint immobilization reduces the expression of sensory neuropeptide receptors and impairs healing after tendon rupture in a rat model. J Orthop Res 27:274–280. https://doi.org/10.1002/jor.20657
Brylka LJ, Alimy A-R, MEA T-M, Jiang S, Ballhause TM, Baranowsky A, von Kroge S, Delsmann J, Pawlus E, Eghbalian K et al (2024) Piezo1 expression in chondrocytes controls endochondral ossification and osteoarthritis development. Bone Res 12:12. https://doi.org/10.1038/s41413-024-00315-x
Burton AN, Sharibzadeh S, Funnell MP, Khunti K, Watson EL, Wilkinson TJ (2025) The diagnostic accuracy of biological markers of sarcopenia, sarcopenic obesity, and osteosarcopenia: protocol for a systematic review and meta-analysis. Health Sci Rep 8:e71307. https://doi.org/10.1002/hsr2.71307
Cai X, Thinn AMM, Wang Z, Shan H, Zhu J (2017) The importance of N-glycosylation on B3 integrin ligand binding and conformational regulation. Sci Rep 7:4656. https://doi.org/10.1038/s41598-017-04844-w
Camy C, Brioche T, Senni K, Bertaud A, Genovesio C, Lamy E, Fovet T, Chopard A, Pithioux et al (2022) Effects of hindlimb unloading and subsequent reloading on the structure and mechanical properties of Achilles tendon-to-bone attachment. FASEB J 36:e22548. https://doi.org/10.1096/fj.202200713R
Carroll MB (2023) Hypermobility spectrum disorders: a review. Rheumatol Immunol Res 4:60–68. https://doi.org/10.2478/rir-2023-0010
Charlton WP, Coslett-Charlton LM, Cicotti MG (2001) Correlation of estrodiol in pregnancy and anterior cruciate ligament laxity. Clin Orthop Relat Res 387:165–170. https://doi.org/10.1097/00003086-200106000-00022
Chatterjee M, Muljadi PM, Andarawis-Puri N (2022) The role of tendon ECM in mechanotransduction: disruption and repair following overuse. Connect Tissue Res 63:28–42. https://doi.org/10.1080/03008207.2021.1925663
Chen S, Birk DE (2013) The regulatory roles of small leucine-rich proteoglycans in extracellular matrix assembly. FEBS J 280:2120–2137. https://doi.org/10.1111/febs.12136
Chen CS, Ingber DE (1999) Tensegrity and mechanoregulation: from skeleton to cytoskeleton. Osteoarthr Cartil 7:81–94. https://doi.org/10.1053/joca.1998.0164
Chen N, Danalache M, Liang C, Alexander D, Umrath F (2025) Mechanosignaling in osteoporosis: when cells feel the force. Int J Mol Sci 26:4007. https://doi.org/10.3390/ijms26094007
Cheng D, Wang J, Yao M, Cox CD (2023) Joining forces: crosstalk between mechanosensitive PIEZO1 ion channels and integrin-mediated focal adhesions. Biochem Soc Trans 51:1897–1907. https://doi.org/10.1042/BST20230042
Chi SS, Rattner JB, Sciore P, Boorman R, Lo UKY (2005) Gap junctions of the medial collateral ligament: structure, distribution, associations and function. J Anat 207:145–154. https://doi.org/10.1111/j/1469-7580.2005.00440.x

Chiarelli N, Riteli M, Zoppi N, Colombi M (2019) Cellular and molecular mechanisms in the pathogenesis of classical, vascular, and hypermobile Ehlers-Danlos syndromes. Genes (Basel) 10:609. https://doi.org/10.3390/genes10080609

Chistiakov DA, Orekhov AN, Bobryshev YV (2017) Effects of shear stress on endothelial cells: go with the flow. Acta Physiol (Oxf) 219:382–408. https://doi.org/10.1111/apha.12725

Chu SR, Boyer EH, Beynnon B, Segal NA (2019) Pregnancy results in lasting changes in knee joint laxity. PM R 11:1170124. https://doi.org/10.1016/j.pmrj.2018.06.012

Clark AL, Votta BJ, Kumar S, Liedtke W, Guilak F (2010) Chondroprotective role of the osmotically sensitive ion channel transient receptor potential vanilloid 4: age- and sex-dependent progression of osteoarthritis in TRPV4-deficient mice. Arthritis Rheum 62:2973–2983. https://doi.org/10.1002/art.27624

Cui Q, Yang H, Gu Y, Zong C, Chen X, Lin Y, Sun H, Shen Y, Zhu J (2020) RBA sequencing (RNA-seq) analysis of gene expression provides new insights into hindlimb unloading-induced skeletal muscle atrophy. Ann Transl Med 8:1595. https://doi.org/10.21037/atm-20-7400

Cunningham KD (1998) The structure function relationship of the extracellular matrix in the rabbit medial collateral ligament. Master's thesis, University of Calgary, Calgary, AB, Canada

Davies PF, Mundel T, Barbee KA (1995) A mechanism for heterogeneous endothelial responses to flow in vivo and in vitro. J Biomech 28:1553–1560. https://doi.org/10.1016/0021-9290(95)00102-6

Davies PF, Civelek M, Fang Y, Guerraty MA, Passerini AG (2010) Endothelial heterogeneity associated with regional athero-susceptibility and adaptation to disturbed blood flow in vivo. Semin Thromb Hemost 36:265–275. https://doi.org/10.1055/s-0030-1253449

De Meyts P, Gauguin L, Svendsen AM, Sarhan M, Knudsen L, Nohr J, Kiselyov VV (2009) Structural basis of allosteric ligand-receptor interactions in the insulin/relaxin peptide family: implications for other receptor tyrosine kinases and G-protein-coupled receptors. Ann N Y Acad Sci 1160:49–53. https://doi.org/10.1111/j.1749-6632.2009.03837.x

De Vries MHE, Caelers IJMH, van Hemert WLM, Boselie TFM, van Santbrink H (2025) Physiological aging of the lumbar intervertebral disc based on magnetic resonance imaging, a systematic literature review. Medicina (Kaunas) 61:1430. https://doi.org/10.3390/medicina61081430

Dehghan F, Yusof A, Muniandy S, Salleh N (2015) Estrogen receptor (ER)-alpha, beta and progesterone receptor (PR) mediates changes in relaxin receptor (RXFP1 and RXFP2) expression and passive range of motion of rat's knees. Environ Toxicol Pharmacol 40:785–791. https://doi.org/10.1016/j.etap.2015.09.004

Dehghan F, Soon R, Yusof A (2024) Knee laxities changes with sex-steroids throughout the menstrual cycle phases in athlete and non-athlete females. Rev Bras Orthop (Sao Paulo) 59:e29–e37. https://doi.org/10.1055/s-0043-1771007

Delabastita T, Bogaerts S, Vanwanseele B (2018) Age-related changes in Achilles tendon stiffness and impact on functional activities: a systematic review and meta-analysis. J Aging Phys Act 12:1–12. https://doi.org/10.1123/japa.2017-0359

Dhulipalla S, Longmate WM (2024) Integrin mutations in blistering skin diseases and related genetically engineered mouse models. Hum Immunol 85:111175. https://doi.org/10.1016/j.humimm.2024.111175

Di X, Gao X, Peng L, Ai J, Jin X, Qi S, Li H, Wang K, Luo D (2023) Cellular mechanotransduction in health and diseases: from molecular mechanism to therapeutic targets. Signal Transduc Target Ther 8:282. https://doi.org/10.1038/s41392-023-01501-9

Ding X, Zhu C, Lu Q, Zhang Y, Gao B (2025) TRPV1 signaling in skeletal muscle: a mini review of physiological and pathological roles. Cell Calcium 131:103057. https://doi.org/10.1016/j.ceca.2025.101057

Dong Q, Li D, Zhang K, Shi H, Cai M, Li Y, Zhao R, Qin D (2025) Muscle-bone biochemical crosstalk in osteosarcopenia: focusing on mechanisms and potential therapeutic strategies. J Endocrinol 266:e250234. https://doi.org/10.1530/JOE-25-0234

Dragoo JL, Lee RS, Benhaim P, Finerman GAM, Hame SL (2003) Relaxin receptors in the human female anterior cruciate ligament. Am J Sports Med 31:577–584. https://doi.org/10.1177/03635465030310041701
Du G, Li L, Zhang X, Liu J, Hao J, Zhu J, Wu H, Chen W, Zhang Q (2020) Roles of TRPV4 and piezo channels in stretch-evoked Ca2+ response in chondrocytes. Exp Biol Med (Maywood) 245:180–189. https://doi.org/10.1177/1535370219892601
Dumas GA, Reid JG (1997) Laxity of knee cruciate ligaments during pregnancy. J Orthop Sports Phys Ther 26:2–6. https://doi.org/10.2519/jospt.1997.26.1.1
Ebeling PR (1998) Osteoporosis in men. New insights into aetiology, pathogenesis, prevention and management. Drugs Aging 13:421–434. https://doi.org/10.2165/00002512-199813060-00002
Eble JA (2025) Allosteric disulfide bridges in integrins: the molecular switches of redox regulations of integrin-mediated cell functions. Antioxidants (Basel) 14:1005. https://doi.org/10.3390/antiox.14081005
Elendu C, Amaechi DC, Elendu TC, Ibhiedu JO, Torubiri AO, Okaye CK (2023) Comprehensive review of aortic aneurysms, dissections, and cardiovascular complications in connective tissue disorders. Medicine (Baltimore) 102:e36499. https://doi.org/10.1097/MD.0000000000036499
Fang X-Z, Zhou T, Xu J-Q, Wang Y-X, Sun M-M, He Y-J, Pan S-W, Xiong W, Peng Z-K, Gao X-H et al (2021) Structure, kinetic properties and biological function of mechanosensitive piezo channels. Cell Biosci 11:13. https://doi.org/10.1186/s13578-020-00522-z
Fang M, Partch CL, LiWang A, Golden SS (2025) Prokaryotic circadian systems: cyanobacteria and beyond. Ann Rev Microbiol. https://doi.org/10.1146/annurev-micro-041222-022934
Feletou M (2016) Endothelium-dependent hyperpolarization and endothelial dysfunction. J Cardiovasc Pharmacol 67:373–387. https://doi.org/10.1097/FJC.0000000000000346
Feng R, Hu W, Li Y, Yao X, Li J, Li X, Zhang J, Wu Y, Kang F, Dong S (2024) Mechanotransduction in subchondral bone microenvironment and targeted interventions for osteoarthritis. Mechanobiol Med 2:100043. https://doi.org/10.1016/j.mbm.2024.100043
Ferrai C, Schulte C (2024) Mechanotransduction in stem cells. Eur J Cell Biol 103:151417. https://doi.org/10.1016/j.ejcb.2024.151417
Foehr R, Anderson K, Dombrowski O, Foehr A, Foehr ED (2024) Dysregulation of extracellular matrix and lysyl oxidase in Ehlers-Danlos syndrome type IV skin fibroblasts. Orphanet Rare Dis 19:9. https://doi.org/10.1186/s13023-023-03007-7
Ford DM, Bagnall KM, McFadden KD, Greenhill BJ, Raso VJ (1984) Paraspinal muscle imbalance in adolescent idiopathic scoliosis. Spine (Philla Pa) 9:373–376. https://doi.org/10.1097/00007632-198405000-00008
Frank CB (2004) Ligament structure, physiology and function. J Musculoskelet Neuronal Interact 4:199–201
Frank CB, Loitz B, Bray R, Chimich D, King G, Shrive N (1994) Abnormalities of the contralateral ligament after injuries of the medial collateral ligament. A experimental study in rabbits. J Bone Joint Surg Am 76:403–412. https://doi.org/10.2106/00004623-199403000-00011
Frank C, McDonald D, Shrive N (1997) Collagen fibril diameters in the rabbit medial collateral ligament scar: a longer term assessment. Connect Tissue Res 36:261–269. https://doi.org/10.3109/03008209709160226
Frank CB, Hart DA, Shrive NG (1999) Molecular biology and biomechanics of normal and healing ligaments-a review. Osteoarthr Cartil 7:130–140. https://doi.org/10.1053/joca.1998.0168
Frank CB, Shrive NG, Boorman RS, Lo IKY, Hart DA (2004) New perspectives on bioengineering of joint tissues: joint adaptation creates a moving target for engineering replacement tissues. Ann Biomed Eng 32:458–465. https://doi.org/10.1023/b:abme.0000017548.85451.b7
Funakoshi Y, Hariu M, Tapper JE, Marchuk LL, Shrive NG, Kanaya F, Rattner JB, Hart DA, Frank CB (2007) Periarticular ligament changes following ACL/MCL transection in an ovine stifle joint model of osteoarthritis. J Orthop Res 25:997–1006. https://doi.org/10.1002/jor.20370
Garcia N, Lytch A, Cramer E, Black WR, Jones JT (2024) Prevalence of pain and effect of joint strength on perceived physical function in pediatric patients with Ehler-Danlos syndrome. Arch Rehabil Res Clin Transl 7:100416. https://doi.org/10.1016/j.arrct.2014.100416

Gardeazabal L, Izeta A (2024) Elastin and collagen fibres in cutaneous wound healing. Exp Dermatol 33:e15052. https://doi.org/10.1111/exd.15052
Garrett-Bakelman FE, Darshi M, Green SJ, Gur RC, Lin L, Macias BR, McKenna MJ, Meydan C, Mishra T, Nasrini J et al (2019) The NASA twins study: a multidimensional analysis of a year-long human spaceflight. Science 364:eaau8650. https://doi.org/10.1126/science.aau8650
Gautschi OP, Corniola MV, Smoll NR, Joswig H, Schaller K, Hildebrandt G, Stienen MN (2016a) Sex differences in subjective and objective measures of pain, functional impairment, and health-related quality of life in patients with lumbar degenerative disc disease. Pain 157:1065–1071. https://doi.org/10.1097/j.pain.0000000000000480
Gautschi OP, Smoll NR, Corniola MV, Joswig H, Schaller K, Hildebrandt G, Stienen MN (2016b) Sex differences in lumbar degenerative disc disease. Clin Neurol Neurosurg 145:52–57. https://doi.org/10.1016/j.clineuro.2016.04.008
Geerkens C, Vetter U, Just W, Fedarko NS, Fisher LW, Young MF, Termine JD, Robey PG, Wohrle D, Vogel W (1999) The X-chromosomal human biglycan gene BGN subject to X inactivation but is transcribed like an X-Y homologous gene. Hum Genet 96:44–52. https://doi.org/10.1007/BF00214185
Georgiev GP, Gaydarski L, Landzhov B (2025) Should we accept the epiligament theory about the differences in the healing potential of the medial collateral and the anterior cruciate ligament? Biomedicine 13:522. https://doi.org/10.3390/biomedicines13020522
Gianakos AL, Hartman H, Kerkhoffs MMJ, Calder J (2024) Sex differences in biomechanical properties of the Achilles tendon may predispose men to higher risk of injury: a systematic review. J ISAKOS 9:184–191. https://doi.org/10.1016/j.jisako.2023.11.001
Gifre-Renom L, Daems M, Luttun A, Jones EAV (2022) Organ-specific endtothelial cell differentiation and impact of microenvironmental cues on endothelial heterogeneity. Int J Mol Sci 23:1477. https://doi.org/10.3390/ijms23031477
Gilchrist CL, Leddy H, Kaye L, Case ND, Rothenberg KE, Little D, Liedtke W, Hoffman BD, Guilak F (2019) TRPV4-mediated calcium signaling in mesenchymal stem cells regulates aligned collagen matrix formation and vinculin tension. Proc Natl Acad Sci USA 116:1992–1997. https://doi.org/10.1073/pnas.1811095116
Gilmer GG, Washington JK, Roberts MD, Oliver GD (2020) Preliminary evaluation of dynamic knee valgus and serum relaxin concentrations after ACL reconstruction. JB JS Open Access 5:e0060. https://doi.org/10.2106/JBJS.OAQ.1900060
Gilmer G, Crasta N, Tanaka MJ (2025) The effect of sex hormones on joint ligament properties: a systematic review and meta-analysis. Am J Sports Med 53:2738–2748. https://doi.org/10.1177/03635465241310145
Goldman MP, Kilmer SL, Biesman B, McPherson K, Jacobson A (2025) Monopolar radiofrequency-induced fibroblast stimulation for the prevention and improvement of skin laxity. Dermatologic Surg 51(9S):S38–S43. https://doi.org/10.1097/DSS.0000000000004714
Gottlieb PA (2017) A tour de force: the discovery, properties, and function of piezo channels. Curr Top Membr 79:1–36. https://doi.org/10.1016/bs.ctm.2016.11.007
Gould NR, Torre OM, Leseer JM, Stains JP (2021) The cytoskeleton and connected elements in bone cell mechano-transduction. Bone 149:115971. https://doi.org/10.1016/j.bone.2021.115971
Guilak F, Leddy H, Lieddtke W (2010) Transient receptor potential vanilloid 4: the sixth sense of the musculoskeletal system. Ann N Y Acad Sci 1192:404–409. https://doi.org/10.1111/j.1749-6632.2010.05.389.x
Gullapalli PA, Javed S (2023) Multidisciplinary chronic pain management strategies in patients with Ehlers-Danlos syndromes. Pain Manag 13:5–14. https://doi.org/10.2217/pmt-2022-0050
Haliloglu G, Becker K, Temucin C, Talim B, Kucuksahin N, Peregande M, Motameny S, Numberg P, Aydingoz U, Topaloglu H et al (2017) Recessive PIEZO2 stop mutation causes distal arthrogryposis with distal muscle weakness, scoliosis and proprioception defects. J Hum Genetics 62:497–501. https://doi.org/10.1038/jhg.2016.153
Halper J (2014) Proteoglycans and diseases of soft tissues. Adv Exp Med Biol 802:49–58. https://doi.org/10.1007/978-007-7893-1_4

Harper CE, Zhang W, Lee J, Shin J-H, Keller MR, van Wijngaarden E, Chou E, Wang Z, Dorr T, Chen P et al (2023) Mechanical stimuli activate gene expression via a cell envelope stress sensing pathway. Sci Rep 13:13979. https://doi.org/10.1038/s41598-023-40897-w
Harrison JJ, Stremick CA, Turner RJ, Allen ND, Olson ME, Ceri H (2010) Microtiter susceptibility testing of microbes growing on peg lids: a miniaturized biofilm model for high-throughput screening. Nat Protoc 5:1236–1254. https://doi.org/10.1038/nprot.2010.71
Hart DA (2018) Evidence for a potential "knee-eye-brain axis" involved in mobility and navigation control: knee injury and obesity may disrupt axis integrity. J Biomed Sci Eng 11:37–44. https://doi.org/10.4236/jbse.2018.113004
Hart DA (2019) Potential impact of space environments on developmental and maturational programs which evolved to meet the boundary conditions of earth: will maturing humans be able to establish a functional biologic system set point under non-earth conditions? J Biomed Sci Eng 12:500–513. https://doi.org/10.4236/jbise.2019.1212041
Hart DA (2021) Learning from human responses to deconditioning environments: improved understanding of the "use it or lose it" principle. Front Sports Act Living 3:685845. https://doi.org/10.3389/fspor.2021.685845
Hart DA (2022) Sex differences in biological systems and the conundrum of menopause: potential commonalities in post-menopausal disease mechanisms. Int J Mol Sci 23:4119. https://doi.org/10.3390/ijms23084119
Hart DA (2023a) Regulation of bone by mechanical loading, sex hormones, and nerves: integration of such regulatory complexity and implications for bone loss during space flight and post-menopausal osteoporosis. Biomolecules. 13:1136. https://doi.org/10.3390/biom13071136
Hart DA (2023b) *Homo sapiens*- a species not designed for space flight: health risks in low earth orbit and beyond, including potential risks when traveling beyond the geomagnetic field of earth. Life (Basel). 13:757. https://doi.org/10.3390/life13030757
Hart DA (2024a) The heterogeneity of post-menopausal disease risk: could the basis for why only subsets of females are affected be due to a reversible epigenetic modification system associated with puberty, menstrual cycles, pregnancy and lactation, and ultimately, menopause? Int J Mol Sci 25:3866. https://doi.org/10.3390/ijms25073866
Hart DA (2024b) The influence of magnetic fields, including the planetary magnetic field, on complex life forms: how do biological systems function in this field and in electromagnetic fields? Biophysica 4:1–21. https://doi.org/10.3390/biophysica4010001
Hart DA (2025a) Regulation of joint tissues and joint function: is there potential for lessens to be learned regarding regulatory control from joint hypermobility syndromes? Int J Mol Sci 26:1256. https://doi.org/10.3390/ijms26031256
Hart DA (2025b) The use of MSCs, iPSCs, and EVs in the repair of human MSK tissues: is ultimate success dependent on developing excellent implant materials as well as creating an optimal environment for implantation? What is the rationale for these choices? Int J Mol Sci 26:6250. https://doi.org/10.3390/ijms26136250
Hart DA, Nakamura N (2022) Creating an optimal in vivo environment to enhance outcomes using cell therapy to repair/regenerate injured tissues of the musculoskeletal system. Biomedicine 10:1570. https://doi.org/10.3390/biomedicines10071570
Hart DA, Zernicke RF (2020) Optimal human functioning requires exercise across the lifespan: mobility in a 1g environment is intrinsic to the integrity of multiple biological systems. Front Physiol 11:156. https://doi.org/10.3389/fphys.2020.00156
Hart DA, Zernicke RF (2021) Editorial: optimal mobility and function across the lifespan. Front Physiol 12:652640. https://doi.org/10.3389/fphys.2021.652640
Hart DA, Zernicke RF (2025) Editorial: optimal physical activity across the lifespan for people of all abilities. Front Sports Act Living 7:1645447. https://doi.org/10.3389/fspor.2025.1645447
Hart DA, Nakamura N, Marchuk L, Hiroka H, Boorman R, Kaneda Y, Shrive NG, Frank CB (2000) Complexity of determining cause and effect in vivo after antisense gene therapy. Clin Orthop Relat Res 279(suppl):S242–S251. https://doi.org/10.1097/00003086-200010001-00032

Hart DA, Natsuume T, Sciore P, Tasevski V, Frank CB, Shrive NG (2002) Mechanobiology: similarities and differences between in vivo and in vitro analysis at the functional and molecular levels. Recent Res Devel Biophys Biochem 2:153–177

Hart DA, Zernicke RF, Shrive NG (2022) *Homo sapiens* may incorporate daily cycles of "conditioning-deconditioning" to maintain musculoskeletal integrity: need to integrate with biological clocks and circadian rhythm mediators. Int J Mol Sci 23:9949. https://doi.org/10.3390/ijms23179949

Hartman H, Cacace A, Leatherman H, Ashkani-Esfahani S, Guss D, Waryasz G, DiGiovani CW, Gianakos AL (2024) Gender differences in Achilles tendon ruptures-a retrospective study and a review of the literature. J Foot Ankle Surg 63:614–620. https://doi.org/10.1053/j.jfas.2014.04.005

Haugg F, Lee G, He J, Johnson J, Zapaishchykova A, Bitterman DS, Kann BH, Aerts HJWL, Mak RH (2025) Imaging biomarkers of ageing: a review of artificial intelligence-based approaches for age estimation. Lancet Healthy Longev 6:100728. https://doi.org/10.1016/j.lanhl.2025.100728

Hayakawa K, Tatsumi H, Sobabe M (2008) Actin stress fibers transmit and focus force to activate mechanosensitive channels. J Cell Sci 121:496–503. https://doi.org/10.1242/jcs022053

Hayashi M, Zhao C, thorson AR, Chikenji T, Jay GD, An K-N, Amadio PC (2013) The effect of lubricin on the gliding resistance of mouse intrasynovial tendon. PLoS One 8:e83836. https://doi.org/10.1371/journal.pone.0083836

Heard BJ, Barton KI, Abubacker S, Chung M, Martin CR, Schmidt TA, Shrive NG, Hart DA (2022) Synovial and cartilage responsiveness to peri-operative hyaluronic acid +/− dexamethasone administration following a limited injury to the rabbit stifle joint. J Orthop Res 40:838–845. https://doi.org/10.1002/jor.25108

Hendargo KJ, Patel AO, Shukwudozie OS, Moreno-Hagelsieb G, Christen JA, Medrano-Soto A, Saier MH (2023) Sequence similarity among structural repeats in the piezo family of mechanosensitive ion channels. Micro Physiol 33:49–62. https://doi.org/10.1159/000531468

Hendrickx G, Fisher V, Liedert A, von Kroge S, Haffner-Luntzer M, Brylka L, Pawlus E, Schweizer M, Yorgan T, Baranowsky A et al (2021) Piezo1 inactivation in chondrocytes impairs trabecular bone formation. J Bone Miner Res 36:369–384. https://doi.org/10.1002/jbmr.4198

Herchenhan A, Uhlenbrock F, Eliasson P, Weis M, Eyre D, Kadler KE, Magnusson SP, Kjaer M (2015) Lysyl oxidase activity is required for ordered collagen fibrillogenesis by tendon cells. J Biol Chem 290:16440–16450. https://doi.org/10.1074/jbc.M115.641670

Hewett TE, Myer GD, Ford KR, Slauterbeck JR (2006) Preparticipation physical examination using a box drop vertical jump test in young athletes: the effects of puberty and sex. Clin J Sports Med 16:298–304. https://doi.org/10.1097/00042752-200607000-00003

Hildebrand KA, Frank CB, Hart DA (2004) Gene intervention in ligament and tendon: current status, challenges, future directions. Gene Ther 11:368–378. https://doi.org/10.1038/sj.gt.3302198

Hsu H-H, Chiu C-Y, Chen W-C, Yang Y-R, Wang R-Y (2024) Effects of exercise on bone density and physical performance in postmenopausal women: a systematic review and meta-analysis. PM R 16:1358–1383. https://doi.org/10.1002/pmrj.13206

Huang J, Shi H, Shang J, Fu G, Zhou P (2025a) Ion channels of the plasma membrane and endolyosomes: roles in osmoregulation and volume regulation in immune cells. Pharmacol Res 221:107951. https://doi.org/10.1016/j.phrs.2025.107951

Huang J, Xie R, Li J, Yao X, Gu Y, Yan W, Dou C, He Y, Luo P, Gao Y et al (2025b) The bioinformatics analysis of chromatin regulators and immune-related genes in adolescent idiopathic scoliosis based on transcriptome data. Medicine (Baltimore) 104:e45038. https://doi.org/10.1097/MD.0000000000045038

Huhtaniemi IT (2014) Andropause-lessons from the European male ageing study. Ann Endocrinol (Paris) 75:128–131. https://doi.org/10.1016/j.ando.2014.03.005

Huisman ES, Andersson G, Scott A, Reno CR, Hart DA, Thornton GM (2014) Regional molecular and cellular differences in the female rabbit Achilles tendon complex: potential implications for understanding responses to loading. J Anat 224:538–547. https://doi.org/10.1111/joa.12168

Hurd L, Kirwin SM, Boggs M, Mackenzie WG, Bober MB, Funanage VL, Duncan RL (2015) A mutation in TRPV4 results in altered chondrocyte calcium signaling in severe metatropic dysplasia. Am J Med Genet A 167A:2286–2293. https://doi.org/10.1002/ajmg.a.37182

Hwang CT, Halpert J (2021) Proteoglycans and diseases of soft tissues. Adv Exp Med Biol 1348:127–138. https://doi.org/10.1007/978-3-30-80614-9_5

Ingber DE (1997) Integrins, tensegrity, and mechanotransduction. Gravit Space Biol Bull 10:49–55

Ingber DE (2003) Tensegrity II. How structural networks influence cellular information processing networks. J Cell Sci 116:1397–1408. https://doi.org/10.1242/jcs.00360

Ingber DE (2006) Mechanical control of tissue morphogenesis during embryological development. Int J Dev Biol 50:255–266. https://doi.org/10.1387/ijdb.052044di

Ingber DE (2008) Tensegrity-based mechanosensing from macro to micro. Prog Biophys Mol Biol 97:163–179. https://doi.org/10.1016/j.pbiomolbio.2008.02.005

Ingber DE, Dike L, Hansen L, Karp S, Liley H, Maniotis A, McNamee, Mooney D, Plopper G, Sims J et al (1994) Cellular tensegrity: exploring how mechanical changes in the cytoskeleton regulate cell growth, migration, and tissue pattern during morphogenesis. Int Rev Cytol 150:173–224. https://doi.org/10.1016/s0074-7696(08)61542-9

Ingber DE, Wang N, Stamenovic D (2014) Tensegrity, cellular biophysics, and the mechanics of living systems. Rep Prog Phys 77:046603. https://doi.org/10.1088/0034-4885/77/4/046603

Ishida Y, Takata Y, Ishikawa T, Kimura M, Takemoto N, Nishimura M, Ozaki N, Demura S, Nakase J (2025) Age-related changes in the quadriceps tendon: collagen fibril diameter decreases with aging. Asia Pac J Sports Med Arthrosc Rehabil Technol 41:20–25. https://doi.org/10.1016/j.asmart.2025.06.002

Jabbur ML, Johson CH (2025) Could photoperiodic responses have evolved before the emergence of circadian clocks? New Phytol. https://doi.org/10.1111/nph.70598

Jablonski CL, Ferguson S, Pozzi A, Clark AL (2014) Integrin Alpha1Beta1 participates in chondrocyte transduction of osmotic stress. Biochem Biophys Res Commun 445:184–190. https://doi.org/10.1016/j.bbrc.2014.01.157

Jiao R, Cui D, Wang SC, Li D, Wang Y-F (2017) Interactions of the mechanosesitive channels with extracellular matrix, integrins, and cytoskeletal network in osmosensation. Front Mol Neurosci 10:96. https://doi.org/10.3389/fnmol.2017.00096

Jing X, Kotelsky A, Zhang Y, Dirksen R, Mannava S, Buckley M, Lee W (2024) Medhcano-adaptation: exercise-driven Piezo1 & Piezo2 augmentation and chondroprotection in articular cartilage. bioRxiv (Preprint) 2024.08.02.606183. https://doi.org/10.1101/2024.08.02.606183

John G, AlNadwi A, Antoun TGA, Ahmetov II (2025) Injury prevention strategies in female football players: addressing sex-specific risks. Sports (Basel) 13:39. https://doi.org/10.3390/sports13020039

Kaji H (2024) Crosstalk between muscle and bone. J Bone Miner Metab 42:391–398. https://doi.org/10.1007/s00774-023-01472-5

Kalamajski S, Oldberg A (2010) The role of small leucine-rich proteoglycans in collagen fibrillogenesis. Matrix Biol 29:248–253. https://doi.org/10.1016/j.matbio.2010.01.001

Kalkwarf HJ (1999) Hormonal and dietary regulation of changes in bone density during lactation and after weaning in women. J Mammary Gland Biol Neoplasia 4:319–329. https://doi.org/10.1023/a:1018780425600

Kalkwarf HJ, Specker BL (2002) Bone mineral changes during pregnancy and lactation. Endocrine 17:49–53. https://doi.org/10.1385/ENDO.17:1:49

Katoh K (2025) Integrin and its associated proteins as a mediator for mechano-signal transduction. Biomolecules 15:166. https://doi.org/10.3390/biom15020166

Kawao N, Moritake A, Tatsumi K, Kaji H (2018) Roles of irisin in the linkage from muscle to bone during mechanical unloading in mice. Calcif Tissue Int 103:24–34. https://doi.org/10.1007/s00223-018-0387-3

Kim JH, Lee SK, Lee SK, Kim JH, Fredericson M (2016) Relaxin receptor RXFP1 and RXFP2 expression in ligament, tendon, and shoulder joint capsule of rats. J Korean Med Sci 31:983–988. https://doi.org/10.3346/jkms.2016.31.6.983

Klein CE, Steinmayer T, Mattes JM, Kaufmann R, Weber L (1990) Integrins of normal human epidermis: differential expression, synthesis and molecular structure. Br J Dermatol 123:171–178. https://doi.org/10.1111/j.1365-2133.1990.tb01844.x

Kos O, Hughson RL, Hart DA, Clement G, Frings-Meuthen P, Linnarsson D, Paloski WH, Rittweger J, Wuyts F, Zange J et al (2014) Elevated serum soluble CD200 and CD200R as surrogate markers of bone loss under bed rest conditions. Bone 60:33–40. https://doi.org/10.1016/j.bone.2013.12.004

Kruse K, Julicher F (2005) Oscillations in cell biology. Curr Opin Cell Biol 17:20–26. https://doi.org/10.1016/j.ceb.2004.12.007

Kuang H, Chen L, Huang M, Chen J (2025) Management of adolescent scoliosis: a comprehensive review of etiology and rehabilitation. Front Pediatr 13:1596400. https://doi.org/10.3389/fped.2025.1596400

Kumar S, Smith C, Clifton-Bligh R, Beck BR, Girgis CM (2025) Exercise for postmenopausal bone health-can we raise the bar? Curr Osteoporos Rep 23:20. https://doi.org/10.1007/s11914-025-00912-7

Lakk M, Krizaj D (2021) TRPV4-rho signaling drives cytoskeletal and focal adhesion remodeling in trabecular meshwork cells. Am J Physiol Cell Physiol 320:C1013–C1030. https://doi.org/10.1152/ajpcell.00599.2020

Larionov A, Hammer CM, Fiedler K, Filgueira L (2024) Dynamics of endothelial cell diversity and plasticity in health and disease. Cells 13:1276. https://doi.org/10.3390/cells13151276

Lau P, Vico L, Rittweger J (2022) Dissociation of bone resorption and formation in spaceflight and simulated microgravity: potential role of myokines and osteokines? Biomedicine 10:342. https://doi.org/10.3390/biomedicines10020342

Lau KK, Law KKP, Kwan KYH, Cheung JPY, Cheung KMC (2023) Proprioception-related gene mutations in relation to the aetiopathogenesis of idiopathic scoliosis: a scoping review. J Orthop Res 41:2694–2702. https://doi.org/10.1002/jor.25626

Lee CA, Lee-Barthel A, Marquino L, Sandoval N, Marcotte GR (2015) Estrogen inhibits lysyl oxidase and decreases mechanical function in engineered ligaments. J Appl Physiol (1985) 118:1250–1257. https://doi.org/10.1152/japplphysiol.00823.2014

Lee W, Guilak F, Liedtke W (2017) Role of piezo channels in joint health and injury. Curr Top Membr 79:263–273. https://doi.org/10.1016/bs.ctm.2016.10.003

Li H, Zhang R (2025) The role of calcium ions and the transient receptor potential vanilloid (TRPV) channel in bone remodelling and orthodontic tooth movement. Mol Biol Rep 52:297. https://doi.org/10.1007/s11033-025-10399-1

Li J, Hou B, Tumova S, Muraki K, Bruns A, Ludlow MJ, Sedo A, Hyman AJ, McKwown L, Young RS et al (2014) Piezo1 integration of vascular architecture with physiological force. Nature 515:279–282. https://doi.org/10.1038/nature13701

Li N, Chen Y, Wang Q, Liu X, Han C, Qu C, Guan X, Zou W, Wang X, Li A et al (2025) Microenvironment-driven satellite cell regeneration and repair in aging-related sarcopenia: mechanisms and therapeutic frontiers. Stem Cell Res Ther 16:545. https://doi.org/10.1186/s13287-025-04481-5

Lin LM, de Cassia Marqueti R, Bonelli HM, Parreno J, Silbernagel KG, Elliott DM (2025) Early adaptive and late degenerative tendon response to overload in rat model of synergist ablation. J Physiol. https://doi.org/10.1113/JP289173

Liu ZQ, Rangayyan RM, Frank CB (1991) Statistical analysis of collagen alignment in ligaments by scale-space analysis. IEEE Trans Biomed Eng 38:580–588. https://doi.org/10.1109/0.81583

Liu N, Lu W, Dai X, Qu X, Zhu C (2022) The role of TRPV channels in osteoporosis. Mol Biol Rep 49:577–585. https://doi.org/10.1007/s11033-021-06794-z

Lo IK, de Maat GH, Valk JW, Frank CB (1999) The gross morphology of torn human anterior cruciate ligaments in unstable knees. Arthroscopy 15:301–306. https://doi.org/10.1016/s0749-8063(00)70039-3

Lo IKY, Ou Y, Rattner J-P, Hart DA, Marchuk LL, Frank CB, Rattner JB (2002a) The cellular networks of normal ovine medial collateral and anterior cruciate ligaments are not

accurately recapitulated in scar tissue. J Anat 200:283–296. https://doi.org/10.1046/j.1469-7580.2002.00024.x
Lo IK, Chi S, Ivie T, Frank CB, Rattner JB (2002b) The cellular matrix: a feature of tensile bearing dense soft connective tissues. Histol Histopathol 17:523–537. https://doi.org/10.14670/HH-17.523
Lo IKY, Marchuk L, Hart DA, Frank CB (2003) Messenger ribonucleic acid levels in disrupted human anterior cruciate ligaments. Clin Orthop Relat Res 407:249–258. https://doi.org/10.1097/00003086-200302000-00034
Lo IKY, Marchuk LL, Leatherbottom KE, Frank CB, Hart DA (2004) Collagen fibrillogenesis and mRNA levels in the maturing rabbit medial collateral ligament and patellar tendon. Connect Tissue Res 45:11–22. https://doi.org/10.1080/03008200490278070
Loeser RF, Goldring SR, Scanzello CR, Goldring MB (2012) Osteoarthritis: a disease of the joint as an organ. Arthritis Rheum 64:1697–1707. https://doi.org/10.1002/art.34453
Mackey C, Feng Y, Liang C, Liang A, Tian H, Narayan OP, Dong J, Tai Y, Hu J, Mu Y et al (2025) Mechanical modulation, physiological roles, and imaging innovations of intracellular calcium waves in living systems. Cancers (Basel) 17:1851. https://doi.org/10.3390/cancers17111851
Majima T, Marchuk LL, Sciore P, Shrive NG, Frank CB, Hart DA (2000) Compressive compared with tensile loading of medial collateral ligament scar in vitro uniquely influences mRNA levels for aggrecan, collagen type II, and collagenase. J Orthop Res 18:524–531. https://doi.org/10.1002/jor.1100180403
Malaviya P, Butler DL, Boivin GP, Smith FN, Brry FP, Murphy JM, Vogel KG (2000) An in vivo model for load-modulated remodeling in the rabbit flexor tendon. J Orthop Res 18:116–125. https://doi.org/10.1002/jor.1100180117
Malfait F, Francomano C, Byers P, Belmont J, Berglund B, Black J, Bloom L, Bowen JM, Brady AF, Burrows NP et al (2017) The 2017 international classification of the Ehlers-Danlos syndromes. Am J Med Genet C Semin Med Genet 175:8–26. https://doi.org/10.1002/ajmg.c.31552
Martinez DA, Vailas AC, Vanderby R, Grindeland RE (2007) Temporal extracellular matrix adaptations in ligament during wound healing and hindlimb unloading. Am J Physiol Regul Integr Comp Physiol 293:R1552–R1560. https://doi.org/10.1152/ajpregu.00423.2007
Marturano JE, Xylas JF, Sridharan GV, Georgakoudi I, Kuo CK (2014) Lysyl oxidase-mediated collagen crosslinks may be assessed as markers of functional properties of tendon tissue formation. Acta Biomater 10:1379–1379. https://doi.org/10.1016/j.actbio.2013.11.024
Mascharak S, des Jardins-Park HE, Longaker MT (2020) Fibroblast heterogeneity in wound healing: hurdles to clinical translation. Trends Mol Med 26:1101–1106. https://doi.org/10.1016/j.molmed.2020.07.008
Matsushima N, Miyashita H, Kretsinger RH (2021) Sequence features, structure, ligand interaction and diseases in small leucine rich repeat proteoglycans. J Cell Commun Signal 15:519–531. https://doi.org/10.1007/s12079-021-00616-4
Matthews BD, Thodeti CK, Tydell JD, Mammoto A, Overby DR, Ingber DE (2010) Ultra-rapid activation of TRPV4 ion channels by mechanical forces applied to cell surface beta1 integrins. Integr Biol Chem 2:435–442. https://doi.org/10.1039/c0ib00034e
Matyas JR, Edwards P, Miniaci A, Shrive N, Wilson J, Bray R, Frank C (1994) Ligament tension affects nuclear shape in situ: an in vitro study. Connect Tissue Res 31:45–53. https://doi.org/10.3109/03008209409005634
McNulty AL, Leddy HA, Liedtke W, Giulak F (2015) TRPV4 as a therapeutic target for joint diseases. Nauyn Schmiedebers Arch Pharmacol 388:437–450. https://doi.org/10.1007/soo210-014-1078-x
Merna N, Wong AK, Barahona V, Llanos P, Kunar B, Palikuqi B, Ginsberg M, Rafii S, Rabbany SY (2018) Laminar shear stress modulates endothelial luminal surface stiffness in a tissue-specific manner. Microcirculation 25:e12455. https://doi.org/10.1111/micc.12455
Michalick L, Kuebler WM (2020) TRPV4- A missing link between mechanosensation and immunity. Front Immuol 11:413. https://doi.org/10.3389/fimmu.2020.00413

Mikhailov N, Leskinen J, Fagerlund I, Poguzhelskaya E, Giniatullina R, Gafurov O, Malm T, Karjalainen T, Grohn O, Giniatullin R (2019) Mechanosensitive meningeal nociception via piezo chanels: implications for pulsatile pain in migraine? Neuropharmacology 149:113–123. https://doi.org/10.1016/j.neuropharm.2019.02.015

Mikhailov N, Plotnikova L, Singh P, Giniatullin R, Hamalainen RH (2022) Functional characterization of mechanosensitive piezo1 channels in trigeminal and somatic nerves in a neuron-on-chip model. Int J Mol Sci 23:1370. https://doi.org/10.3390/ijms23031370

Mishra J, Chakaborty S, Niharika, Roy A, Manna S, Baral T, Nandi P, Patra SK (2024) Mechanotransduction and epigenetic modulations of chromatin: role of mechanical; signals in gene regulation. J Cell Biochem 125:e30531. https://doi.org/10.1002/jcb.30531

Miyamoto N, Miyamoto-Mikami E, Kinoshita M, Akazawa T, Kimura N, Nishio H, Kobayashi Y, Nagao M, Takazawa N (2024) Age-dependence of semitendinosus tendon properties used for anterior cruciate ligament reconstruction differs in males and females. Clin Biomech (Bristol) 119:106329. https://doi.org/10.1016/j.clinbiomech.2024.106329

Moe P, Blount P (2005) Assessment of potential stimuli for mechano-dependent gating of Mscl: effect of pressure, tension, and lipid headgroups. Biochemistry 44:12239–12244. https://doi.org/10.1021/bi0509649

Morisaki H (2024) Hereditary aortic aneurysms and dissections: clinical diagnosis and genetic testing. Ann Vasc Dis 17:128–134. https://doi.org/10.3400/avd.ra.24-00013

Moroni M, Servin-Vences MR, Fleischer R, Sanchez-Carranza O, Lewin GR (2018) Voltage gating of mechanosensitive PIEZO channels. Nat Commun 9:1096. https://doi.org/10.1038/s41467-18-03502-7

Murshid SA (2017) The role of osteocytes during experimental orthodontic tooth movement: a review. Arch Oral Biol 73:25–33. https://doi.org/10.1016/j.archoralbio.2016.09.001

Myers KA, Rattner JB, Shrive NG, Hart DA (2007a) Hydrostatic pressure sensation in cells: integration into the tensegrity model. Biochem Cell Biol 85:543–551. https://doi.org/10.1139/o07-108

Myers KA, Rattner JB, Shrive NG, Hart DA (2007b) Osteoblast-like cells and fluid flow: cytoskeleton-dependent shear sensitivity. Biochem Biophys Res Commun 364:214–219. https://doi.org/10.1016/j.bbrc.2007.09.109

Najrana T, Sanchez-Esteban J (2016) Mechanotransduction as an adaptation to gravity. Front Pediatr 4:140. https://doi.org/10.3389/fped.2016.00140

Nakamura N, Hart DA, Boorman RS, Kaneda Y, Shrive NG, Marchik L, Shino K, Ochi T, Frank CB (2000) Decorin antisense gene therapy improves functional healing of early rabbit ligament scar with enhanced collagen fibrillogenesis in vivo. J Orthop Res 18:517–523. https://doi.org/10.1002/jor.1100180402

Nakamura N, Hart DA, Frank CB, Marchuk LL, Shrive NG, Ota N, Taira K, Yoshikawa H, Kaneda Y (2001) Efficient transfer of intact oligonucleotides into the nucleus of ligament scar fibroblasts by HVJ-cationic liposomes is correlated with effective antisense gene inhibition. J Biochem 129:755–759. https://doi.org/10.1093/oxfordjournals.jbiochem.a002916

Natsuume T, Majima T, Reno C, Shrive NG, Frank CB, Hart DA (2005) Menisci of the rabbit knee require mechanical loading to maintain homeostasis: cyclic compression in vitro prevents derepression of catabolic genes. J Orthop Res 10:396–405. https://doi.org/10.1007/s00776-005-0912-x

Nims R, Palmer DR, Kassab J, Zhang B, Guilak F (2024) The chondrocyte "mechanome": activation of the mechanosensitive ion channels TRPV4 and PIEZO1 drives unique transcriptional signatures. FASEB J 38:e23778. https://doi.org/10.1096/fj.202400883R

O'Brien EJO, Beveridge JE, Huebner KD, Heard BJ, Tapper JE, Shrive NG, Frank CB (2013) Osteoarthritis develops in the operated joint of an ovine model following ACL reconstruction with immediate anatomic reattachment of the native ACL. J Orthop Res 31:35–43. https://doi.org/10.1002/jor.22187

Ohba T, Goto G, Tanaka N, Oda K, Katsu M, Takei H, Koyama K, Oba H, Garo H (2022) Upper extremity skeletal muscle mass asymmetry exacerbated by shoulder imbalance in Lenke1A adolescent idiopathic scoliosis. J Clin Med 11:7117. https://doi.org/10.3390/jcm11237117

Orsini EM, Perelas A, Southern DB, Grove LM, Oliman MA, Scheraga RG (2021) Stretching the function of innate immune cells. Front Immunol 12:767319. https://doi.org/10.3389/fimmu.2021767319
Orton K, Batchelor W, Ziebarth NM, Best TM, Travascio F, Jackson AR (2023) Biomechanical properties of porcine meniscus as determined via AFM: effect of region, compartment and anisotropy. PLoS One 18:e0280616. https://doi.org/10.1371/journal.pone.0280616
Otoo BS, Moo EK, Komeili A, Hart DA, Herzog W (2024) Chondrocyte deformation during the unloading phase of cyclic compression loading. J Biomech 171:112179. https://doi.org/10.1016/j.jbiomech.2024
Otoo BS, Moo EK, Komeili A, Hart DA, Herzog W (2025) From fluctuations to stability: in-situ chondrocyte response to cyclic compressive loading. J Biomech 186:112734. https://doi.org/10.1016/j.jbiomech.2025.112734
Otsuki R, Del Bel MJ, Benoit DI (2021) Sex differences in muscle activation patterns associated with anterior cruciate ligament injury during landing and cutting tasks: a systematic review. J Electromyogr Kinesiol 60:102583. https://doi.org/10.1016/j.jelekin.2021.102583
Pajevic PD, Spatz JM, Garr J, Adamson C, Misener L (2013) Osteocyte biology and space flight. Curr Biotechnol 2:179–183. https://doi.org/10.2174/22115501113029990017
Palmer DR, Nims R, Zhang B, Guilak F (2025) Activation of the mechanosensitive ion channels TRPV4 and PIEZO1 downregulates key regulatory systems in the chondrocyte mechanome. Connect Tissue Res 66:239–262. https://doi.org/10.1080/03008207.2025.2498512
Park S-K, Stefanyshyn DJ, Ramage B, Hart DA, Ronsky JL (2009a) Relationship between knee joint laxity and knee joint mechanics during the menstrual cycle. Br J Sports Med 43:174–179. https://doi.org/10.1136/bjsm.2008.049270
Park S-K, Stefanyshyn DJ, Lotiz-Ramage B, Hart DA, Ronsky JL (2009b) Changing hormone levels during the menstrual cycle affect knee laxity and stiffness in healthy female subjects. Am J Sports Med 37:588–598. https://doi.org/10.1177/0363546508326713
Parker EA, Duchma KR, Meyer AM, Wolf BR, Westermann RW (2024) Menstrual cycle hormone relaxin and ACL injuries in female athletes: a systematic review. Iowa Orthop J 44:113–123
Pearson D, Kaur M, San P, Lawson N, Baker P, Hosking D (2004) Recovery of pregnancy mediated bone loss during lactation. Bone 34:570–578. https://doi.org/10.1016/j.bone.2003.11.005
Perez-Castro AV, Vogel KG (1999) In situ expression of collagen and proteoglycan genes during development of fibrocartilage in bovine deep flexor tendon. J Orthop Res 17:1390148. https://doi.org/10.1002/jor.11001701200
Pinto AJ, Bergouignan A, Dempsey PC, Roschel H, Owen N, Gualano B, Dunstan DW (2023) Physiology of sedentary behavior. Physiol Rev 103:2561–2622. https://doi.org/10.1152/physrev.00022.2022
Qin L, He T, Chen S, Yang D, Yi W, Cao H, Xiao G (2021) Roles of mechanosensitive channel Piezo1/2 proteins in skeleton and other tissues. Bone Res 9:44. https://doi.org/10.1038/s1413-021-00168-8
Qiu X, Deng Z, Wang M, Feng Y, Bi L, Li L (2023) Piezo protein determines stem cell fate by transmitting mechanical signals. Hum Cell 36:540–553. https://doi.org/10.1007/s13577-022-00853-8
Radin EL, Burr DB, Caterson B, Fyhrie D, Brown TD, Boyd RD (1991) Mechanical determinants of osteoarthrosis. Semin Arthritis Rheum 21(3 Suppl 2):12–21. https://doi.org/10.1016/0049-0172(91)90036-y
Raggio CL (2006) Sexual dimorphism in adolescent idiopathic scoliosis. Orthop Clin North Am 37:555–558. https://doi.org/10.1016/j.ocl.2006.09.010
Ren J, Zhan C, Lin Y (2024) Inflammation alters the expression and activity of the mechanosensitive ion channels in periodontal ligament cells. Eur J Orthod 47:cjae079. https://doi.org/10.1093/ejo079
Restucci B, Sgadari M, Fatone G, Valle GD, Aragosa F, Caterino C, Ferrara G, Niebauer GW (2022) Immunoexpression of relaxin and its receptors in stifle joints of dogs with cranial cruciate ligament disease. Animals (Basel) 12:819. https://doi.org/10.3390/ani12070819

Reuvers J, Thoreson AR, Zhao C, Zhang L, Jay GD, An K-N, Warman ML, Amadio RC (2011) J Struct Biol 176:41–45. https://doi.org/10.1016/j.jsb.2011.07.013

Roelofsen J, Klein-Nulend J, Burger EH (1995) Mechanical stimulation by intermittent hydrostatic compression promotes bone-specific gene expression in vitro. J Biomech 28:1493–1503. https://doi.org/10.1016/0021-9290(95)00097-6

Rollick NC, Lemmex DR, Ono Y, Reno CR, Hart DA, Lo IK, Thornton GM (2018) Gene-expression changes in the knee-joint tissues with aging and menopause: implications for the joint as an organ. Clin Interv Aging 13:365–375. https://doi.org/10.2147/CIA.S151453

Rolvien T, Amling M (2022) Disuse osteoporosis: clinical and mechanistic insights. Calcif Tissue Int 110:592–604. https://doi.org/10.1007/s00223-021-00836-1

Sacitharan PK, Snelling SJB, Edwards JR (2012) Aging mechanisms in arthritic disease. Discov Med 14:345–352

Salter DM, Millward-Sadler SJ, Wright MO (2001) Integrin-interleukin-4 mechanotransduction pathways in human chondrocytes. Clin Orthop Relat Res 391(suppl):S49–S60. https://doi.org/10.1097/00003086-200010001-00006

Sanchez-Adams J, Willard VB, Athanasiou KA (2011) Regional variation in the mechanical role of knee meniscus glycoaminoglycans. J Appl Physiol (1985) 111:1590–1596. https://doi.org/10.1152/japplphysiol.00848.2011

Saotome K, Murthy SE, Kefauver JM, Whitwam T, Patapoutian A, Ward AB (2018) Structure of the mechanically activated ion channel piezo1. Nature 554:481–486. https://doi.org/10.1038/nature25453

Savadipour A, Nims R, Katz DB, Giulak F (2022) Regulation of chondrocyte biosynthetic activity by dynamic hydrostatic pressure: the role of TRP channels. Connect Tissue Res 63:69–81. https://doi.org/10.1080/033008207.2020.1871475

Savadipour A, Nims RJ, Rashidi N, Garcia-Castorena JM, Tang R, Marushack G, Oswald SJ, Liedtke WB, Guilak F (2023a) Membrane stretch as the mechanism of activation of PIEZO1 ion channels in chondrocytes. Proc Natl Acad Sci USA 120:e2221958120. https://doi.org/10.1073/pnas.2221958120

Savadipour A, Palmer D, Ely EV, Collins KH, Garcia-Castorena JM, Harissa Z, Kim YS, Oestrich A, Qu F, Rashidi N et al (2023b) The role of PIEZO ion channels in the musculoskeletal system. Am J Physiol Cell Physiol 324:C728–C740. https://doi.org/10.1152/ajpcell.00544.2022

Sawada Y, Murase M, Sokabe M (2012) The gating mechanism of the bacterial mechanosensitive channel *Mscl* revealed by molecular dynamics simulations: from tension sensing to channel opening. Channels (Austin) 6:317–331. https://doi.org/10.4161/chan.21895

Schaadt G, Hesse V, Friederici AD (2015) Sex hormones in early infancy seem to predict aspects of later language development. Brain Lang 141:70–76. https://doi.org/10.1016/j.bandl.2014.11.015

Schroder A, Engelhardt H, Nogueira A, Clausen B, Kirschneck C, Jantsch J, Proff P, Rennere K, Paddenberg-Schubert E (2025) The mechanosensitive PIEZO1 channel contributes to the reaction of RAW264.7 macrophages to mechanical strain. Mediat Inflamm 2025:9998838. https://doi.org/10.1155/mi/9998838

Schembri MA, Givskov M, Klemm P (2002) An attractive surface: gram-negative bacterial biofilms. Sci STKE 2002(132):re6. https://doi.org/10.1126/stke.2002.132.e6

Schwer J, Ignatius A, Seitz AM (2024) The biomechanical properties of human menisci: a systematic review. Acta Biomater 175:1–26. https://doi.org/10.1016/j.actbio.2023.12.010

Sempowski GD, Borrello MA, Bliedon TM, Barth RK, Phipps RP (1995) Fibroblast heterogeneity in the healing wound. Wound Repair Regen 3:120–131. https://doi.org/10.1046/j.1524-475X.1995

Sharif-Naeini R, Dedman A, Folgering JHA, Duprat F, Patel A, Nilius B, Honore E (2008) TRP channels and mechanosensory transduction: insights into the arterial myogenic response. Pflugers Arch 456:529–540. https://doi.org/10.1007/s00424-007-0432-y

Shi Y, Yue J, Wang Z, He S, Sa G (2025) Current status and future direction of mechanotransduction mechanisms: perspectives from cell membrane to nucleus. Biochem Biophys Res Commun 781:152524. https://doi.org/10.1016/j.bbrc.2025.152524

Shim JW, Wise DA, Elder SH (2008) Effect of cytoskeletal disruption on mechanotransduction of hydrostatic pressure by C3H10T1/2 murine fibroblasts. Open Orthop J 2:155–162. https://doi.org/10.2174/1874325000802010155

Shirvani P, Shirvani A, Holick MF (2024) Decoding the genetic basis of mast cell hypersensitivity and infection risk in hypermobile Ehlers-Danlos syndrome. Curr Issues Mol Biol 46:11613–11629. https://doi.org/10.3390/cimb46100689

Simonian PT, Levine RE, Wright TM, Wickiewicz TL, Warren RF (2000) Response of hamstring and patellar tendon grafts for anterior cruciate ligament reconstruction during cyclic tensile loading. Am J Knee Surg 13:8–12

Spatz JM, Ko FC, Ayturk UM, Warman ML (2021) RNAseq and RNA molecular barcoding reveal differential gene expression in cortical bone following hindlimb unloading in female mice. PLoS One 16:e0250715. https://doi.org/10.1371/journal.pone.0250715

Stanczak M, Kacprzak B, Gawda P (2024) Tendon cell biology: effect of mechanical loading. Cell Physiol Biochem 58:677–701. https://doi.org/10.33594/000000743

Stanczak M, Bialy M, Hagner-Derengowska M (2025) Ligament cell biology: effect of mechanical loading. Cell Physiol Biochem 59:252–295. https://doi.org/10.33594/000000773

Steinecker-Frohnwieser B, Lohberger B, Toegel S, Windhager R, Glantz V, Kratchmann C, Leithner A, Weigl L (2023) Activation of the mechanosensitive ion channels Piezo1 and TRPV4 in primary human healthy and osteoarthritis chondrocytes exhibits ion channel crosstalk and modulates gene expression. Int J Mol Sci 24:7868. https://doi.org/10.3390/ijms2409868

Svensson L, Aszodi A, Reinholt FP, Fassler R, Heinegard D, Oldberg A (1999) Fibromodulin-null mice have abnormal collagen fibrils, tissue organization, and altered lumican deposition in tendon. J Biol Chem 274:9636–9647. https://doi.org/10.1074/jbc.274.14.9636

Syx D, Malfait F (2024) Pathogenic mechanisms in genetically defined Ehlers-Danlos syndromes. Trends Mol Med 30:824–843. https://doi.org/10.1016/j.molmed.2024.06.001

Syx D, Van Damme T, Symoens S, Mailburg MC, vande Laar I, Morton J, Suri M, Del Campo M, Hausser I, Hermann-Le T et al (2015) Genetic heterogeneity and clinical variability in musculocontractural Ehlers-Danlos syndrome caused by impaired dermatan sulfate biosynthesis. Hum Mutat 36:535–547. https://doi.org/10.1002/humu.22774

Takahashi I, Matsuzaki T, Kuroki H, Hoso M (2021) Physiological reloading recovers histologically disuse atrophy of the articular cartilage and bone by hindlimb suspension in rat knee joint. Cartilage 13(2_suppl):1530S–1535S. https://doi.org/10.1177/19476035211063857

Takeuchi H, Niki Y, Matsunari H, Umeyama K, Nagashima H, Enomoto H, Toyama Y, Matsumoto M, Nakamura M (2016) Temporal changes in cellular repopulation and collagen fibril remodeling and regeneration after allograft anterior cruciate ligament reconstruction: an experimental study using Kusabira-Oragnge transgenic pigs. Am J Sports Med 44:2375–2383. https://doi.org/10.1177/0363546516650881

Talbott HE, Mascharak S, Griffin M, Wan DC, Longraker MT (2022) Wound healing, fibroblast heterogeneity, and fibrosis. Cell Stem Cell 29:1161–1180. https://doi.org/10.1016/j.stem.2022.07.006

Tang Y, Ma R, Zhang L, Sun X, Wang Y (2025) Effectiveness and safety of hormone replacement therapy in the treatment of menopausal syndrome: a meta-analysis. Am J Transl Res 17:1–15. https://doi.org/10.62347/UGLT3830

Tasevski V, Sorbetti JM, Chiu SS, Shrive NG, Hart DA (2005) Influence of mechanical and biological signals on gene expression in human MG-63 cells: evidence for a complex interplay between hydrostatic compression and vitamin D3 or TGF-beta1 on MMP-1 and MMP-3 mRNA levels. Biochem Cell Biol 83:96–107. https://doi.org/10.1139/o04-124

Thodeti CK, Matthews B, Ravi A, Mammoto A, Ghosh K, Bracha AL, Ingber DE (2009) TRPV4 channels mediate cyclic strain-induced endothelial cell reorientation through integrin-to integrin signaling. Circ Res 104:123–130. https://doi.org/10.1161/CIRVRESADA.108.192930

Thornton GM, Hart DA (2011) The interface of mechanical loading and biological variables as they pertain to the development of tendinosis. J Musculoskelet Neuronal Interact 11:94–105

Thornton GM, Boorman RS, Shrive NG, Frank CB (2002) Medial collateral ligament autografts have increased creep response for at least two years, and early immobilization makes this worse. J Orthop Res 20:346–352. https://doi.org/10.1016/S0736-0266(01)00100-0
Thornton GM, Shao X, Chung M, Sciore P, Boorman RS, Hart DA, Lo IKY (2010) Changes in mechanical loading lead to tend specific alterations in MMP and TIMP expression: influence of stress deprivation and intermittent cyclic hydrostatic compression on rat supraspinatus and Achilles tendons. Br J Sports Med 44:698–703. https://doi.org/10.1136/bjsm.2008.050575
Thornton GM, Lemmex DB, Ono Y, Beach CJ, Reno CR, Hart DA, Lo IKY (2015) Aging affects mechanical properties and lubricin/PRG4 gene expression in normal ligaments. J Biomech 48:3306–3311. https://doi.org/10.1016/j.jbiomech.2015.06.005
Tiskratok W, Chuinsiri N, Limraksasin P, Kyawsoewin M, Jitprasertwong P (2025) Extracellular matrix stiffness: mechanotransduction and mechanobiological response-driven strategies for biomedical applications targeting fibroblast inflammation. Polymers (Basel) 17:822. https://doi.org/10.3390/polym17060822
Toft-Bertelsen TL, MacAulay N (2021) TRPing to the point of clarity: understanding the function of the complex TRPV4 channel. Cells 10:165. https://doi.org/10.3390/cells10010165
Tollabi M, Poursalehi Z, Mehrafshar P, Bakhtian R, Sarmadi VH, Tayebi L, Haramshahi SMA (2024) Insight into the role of integrins and integrins-targeting biomaterials in bone regeneration. Connect Tissue Res 65:343–363. https://doi.org/10.1080/03008207.2024.2306002
Tong D, Soley N, Kolasangiani R, Schwartz MA, Bidone TC (2023) Integrin alpha IIbB3 intermediates: from molecular dynamics to adhesion assembly. Biophys J 122:533–543. https://doi.org/10.1016/j.bpj.2022.12.2022
Torzilli PA, Bhargave M, Park S, Chen CTC (2010) Machanical load inhibits IL-1 induced matrix degradation in articular cartilage. Osteoarthr Cartil 18:97–105. https://doi.org/10.1016/j.joca.2009.07.012
Tuncay IO, Lee EK, Gustafson A, Lee Y, Jung D, Koh J-Y, Lee W, Lee S, Shazand K (2025) Whole genome sequencing in adolescent idiopathic scoliosis cohort implicates multiple biological pathways. NPJ Senon Med 10:67. https://doi.org/10.1038/s41525-025-00520-5
Tvaroska I, Kozmon S, Kona J (2023) Molecular models insights into the structure and behavior of integrins: A review. Cells 12:324. https://doi.org/10.3390/cells12020324
Van Der Westhuizen ET, Summers RJ, Halls ML, Bathgate RAD, Sexton PM (2007) Relaxin receptors—new drug targets for multiple disease states. Curr Drug Targets 8:91–104. https://doi.org/10.2174/138945007779315650
Van Rijthoven S, van Loon JJWA (2025) Aging and altered gravity: a cellular response. FASEB J 39:e70777. https://doi.org/10.1096/fj.202402989R
Vogel KG (2003) Tendon structure and response to changing mechanical load. J Musculoskelet Neuronal Interact 3:323–325
Vogel KG (2004) What happens when tendons bend and twist? Proteoglycans. J Musculoskel Neuronal Interact 4:202–203
Vohra M, Kour A, Kalia NP, Kumar M, Sharma S, Jaglan S, Kamath N, Sharma S (2024) A comprehensive review of genomics, transcriptomics, proteomics, and metabolomic insights into the differentiation of Pseudomonas aeruginosa from the planktonic to biofilm state: a multi-omics approach. Int J Biol Macromol 257:128563. https://doi.org/10.1016/j.ijbiomac.2023.128563
Walsh S, Frank C, Hart D (1992) Immobilization alters cell metabolism in an immature ligament. Clin Orthop Relat Res 277:277–288
Walsh S, Frank C, Hart D (1993) Knee immobilization inhibits biomechanical maturation of the rabbit medial collateral ligament. Clin Orthop Relat Res 297:253–261
Wang Z, Chen J, Babicheva A, Jain P, Rodriguez M, Ayon R, Ravellette KS, Wu L, Balistrien F, Tang H et al (2021) Endothelial upregulation of mechanosensitive channel Piezo1 in pulmonary hypertension. Am J Physiol Cell Physiol 32:C1010–C1027. https://doi.org/10.1152/ajpcell.00147.2021
Wang L, Zheng F, Song R, Zhuang L, Yang M, Suo J, Li L (2022) Integrins in the regulation of mesenchymal stem cell differentiation by mechanical signals. Stem Cells Rev. Rep. 18:126–141. https://doi.org/10.1007/s12015-021-1-260-5

Wang Q, Ji C, Smith P, McCulloch CA (2024a) Impact of TRP channels on extracellular matrix remodeling: focus on TRPV4 and collagen. Int J Mol Sci 25:3566. https://doi.org/10.3390/ijms25073566
Wang Y, Huang H, Wang H, Jia C, Liao B, Long Y, Yu F, Nie Y (2024b) Talin mechanotransduction in disease. Int J Biochem Cell Biol 166:106490. https://doi.org/10.1016/j.biocel.2023.106490
Weir TMR (1991) Recovery of a growing ligament following immobilization in a rabbit model. Master's thesis, University of Calgary, Calgary, AB, Canada
Winkler LH (2023) Human physiological limitations to long-term spaceflight and living in space. Aerosp Med Hum Perform 94:444–456. https://doi.org/10.3357/AMHP.6190.2023
Wojczakowski W, Futyma K (2025) Vitamin D and calcium levels related to bone mineral density during pregnancy and postpartum. Ginekol Pol 96:613–617. https://doi.org/10.5603/gpl.102995
Wu R, De Vito G, Delahunt E, Ditroilo M (2020) Age-related changes in motor function (I). Mechanical and neuromuscular factors. Int J Sports Med 41:709–719. https://doi.org/10.1055/a-1144-3408
Xu X, Liu S, Liu H, Ru K, Jia Y, Wu Z, Liang S, Khan Z, Chen Z, Qian A et al (2021) Piezo channels: awesome mechanosensitive structures in cellular mechanotransduction and their role in bone. Int J Mol Sci 22:6429. https://doi.org/10.3390/ijms22126429
Yalcinkaya B, Sezgin EA, Sacinti KG, Ozcakar L (2025) Neuromusculoskeletal disorders in pregnancy revisited: insights and clinical implications. Jt Dis Relat Surg 36:741–750. https://doi.org/10.52312/jdrs.2025.2242
Yang P, Liu H, Wang S, Xiao X, Jiang L, Le S, Chen S, Ye P, Xia J (2025) PIEZO1 attenuates Marfan syndrome aneurysm development through TGF-B signaling pathway inhibition via TGFBR2. Eur Heart J 46:958–974. https://doi.org/10.1093/eurheartj/ehae786
Yokoyama Y, Domkam N, Kabir H, Mansour A, Tsukamoto S, Yerima G, Adachi T, Mofrad MRK (2025) Emerging mechanomedicines informed by mechanotransduction along the integrin-cytoskeleton-nucleus axis. APL Bioeng 9:021503. https://doi.org/10.1063/5.0255473
Yoseph ET, Taiwo R, Kiapurour A, Touponse G, Massad E, Theologitas M, Wu JY, Williamson T, Zygourakis CC (2025) Pregnancy-related spinal biomechanics: a review of low back pain and degenerative spine disease. Bioengineering (Basel) 12:858. https://doi.org/10.3390/bioengineering12080858
Yuan G, Xiong Z, Ke X, Wang G, Liu X, Li Z (2025) Exploring the multifactorial regulation of PIEZO1 in chondrocytes: mechanisms and implications. Int J Med Sci 22:3393–3411. https://doi.org/10.7150/ijms.111082
Yuze M, Hu J, Jun L, Cheng X, Tianwen X, Junquing Z (2025) Osteocytes function as biomechnical signaling hubs bridging mechanical stress sensing and systemic adaptation. Front Physiol 16:1629273. https://doi.org/10.3389/fphys.2025.1629273
Zhan L, Li J (2018) The role of TRPV4 in fibrosis. Gene 642:1–8. https://doi.org/10.1016/j.gene.2017.10.067
Zhang G, Chen S, Goldoni S, Calder BW, Simpson HC, Owens RT, McQuillan DJ, Young MF, Iozzo RV, Birk DE (2009) Genetic evidence for the coordinated regulation of collagen fibrillogenesis in the cornea by decorin and biglycan. J Biol Chem 284:8888–8897. https://doi.org/10.1074/jbc.M806590200
Zhang T, Chi S, Jiang F, Zhao Q, Xiao B (2017) A protein interaction mechanism for suppressing the mechanosensitive piezo channel. Nat Commun 8:1797. https://doi.org/10.1038/s41467-017-01712-z
Zhang Z, Tong X, Liu S-Y, Chai L-X, Zhu F-F, Zhang X-P, Zou J-Z, Wang X-B (2019) Genetic analysis of piezo-like protein suppressing systemic movement of plant viruses in Arabodopsis thaliana. Sci Rep 9:3187. https://doi.org/10.1038/s41598-019-39436-3
Zhang M, Meng N, Wang X, Chen W, Zhang Q (2022) TRPV4 and PIEZO channels mediate the mechanosensing of chondrocytes to the mechanical microenvironment. Membranes (Basel) 12:237. https://doi.org/10.3390/membranes12020237

Zhang J, Gao Y, Yan J (2024a) Roles of myokines and muscle-derived extracellular vesicles in musculoskeletal deterioration under disuse conditions. Meta 14:88. https://doi.org/10.3390/metabo14020088
Zhang J, Yu H, Man M-Q, Hu L (2024b) Aging in the dermis: fibroblast senescence and its significance. Aging Cell 23:14054. https://doi.org/10.1111/acel.14054
Zhang H, Yang M, Kim SH, Li ITS (2025) Integrin force loading rate in mechanobiology: from model to molecular measurement. QRB Discov 6:e9. https://doi.org/10.1017/qed.2024.28
Zhao Y, Lykov N, Tzeng C (2022) Talin-1 interaction network in cellular mechanotransduction (review). Int J Mol Med 49:60. https://doi.org/10.3892/ijmm.2022.5116
Zhao H, Yao Y, Jiang F, Zhu X, Yu T, Xiao X (2025a) Heterogeneity of fibroblasts from different anatomical sites in healthy human knee ligaments revealed by single-cell RNA sequencing. Genomics 117:111040. https://doi.org/10.1016/j.ugeno.2025.1111040
Zhao J, Xiong Y, Liu Y, Ling J, Liu S, Wang W (2025b) Endothelium piezo1 deletion alleviates experimental varicose veins by attenuating perivenus inflammation. Mol Cell Biochem 480:2423–2435. https://doi.org/10.1007/s11010-024-05115-9
Zhao T, Huang Y, Zhu J, Qin Y, Wu H, Zhai Q, Li S, Qin X, Wang D et al (2025c) Extracellular matrix signaling cues: biological functions, diseases, and therapeutic targets. MedComm (2020) 6:e70281. https://doi.org/10.1002/mco2.70281
Zhou R, Hu W, Waxman SG, Liu C-J (2025) Ion channels as therapeutic targets in osteoarthritis. Trends Pharmacol Sci 46:792–813. https://doi.org/10.1016/j.tips.2025.06.003
Zhu S, Storlino G, Kumar S (2025) Editorial: the endocrine role of the musculoskeletal system. Front Endocrinol (Lausanne) 16:1552950. https://doi.org/10.3389/fendo.2025.1552950
Zurbuchen R, von Daniken A, Janka H, van Wolff M, Stute P (2025) Methods for the assessment of biological age-a systematic review. Maturitas 195:108215. https://doi.org/10.1016/j.maturitas.2025.108215

Chapter 14
Epigenetic Regulation of RUNX2 in Bone Mechanobiology

Maria A. Katsianou, Antonios N. Gargalionis, Athanasios G. Papavassiliou, and Efthimia K. Basdra

Abstract Bone is a highly mechanosensitive tissue that adapts its function in response to mechanical cues. These cues are converted into biomechanical signals that regulate osteogenic gene expression through mechanotransduction. Bone homeostasis and the long-term bone adaptation to its mechanical environment depend on the proper integration of mechanical signals with transcriptional programs that regulate osteogenesis. Runt-related transcription factor 2 (RUNX2) is a pioneer transcription factor, a mechanosensitive molecule, which coordinates osteoblast differentiation and bone remodeling. RUNX2 activity in mechanically induced environments is subject to epigenetic regulation. Mechanical cues modulate RUNX2 through epigenetic mechanisms, including DNA methylation, histone modifications, chromatin remodeling, and non-coding RNAs. Proper epigenetic coordination ensures the stability of osteogenic gene expression. Any disruption of this mechanotransduction-epigenetic route leads to bone diseases such as osteoporosis, osteoarthritis, and dysfunctional fracture repair. Moreover, epigenetic reprogramming of RUNX2 is increasingly evident in bone tumors, particularly osteosarcoma, where aberrant RUNX2 activity is associated with cell proliferation, tumor progression, survival, metastasis, and loss of differentiation. This chapter examines current evidence on RUNX2 epigenetic regulation in response to mechanical forces, discusses how its orchestration relates to bone disease and bone tumor development, and underscores potential therapeutic opportunities that arise from targeting specific epigenetic mechanisms.

Authors Maria A. Katsianou and Antonios N. Gargalionis have equally contributed to this chapter.

M. A. Katsianou · A. G. Papavassiliou (✉) · E. K. Basdra
Department of Biological Chemistry, Medical School, National and Kapodistrian University of Athens, Athens, Greece
e-mail: papavas@med.uoa.gr

A. N. Gargalionis
Laboratory of Clinical Biochemistry, 'Attikon' University General Hospital, Medical School, National and Kapodistrian University of Athens, Athens, Greece

M. Kloc, J. Wosik (eds.), *Mechanotransduction and Mechanosensing*, Results and Problems in Cell Differentiation 78, https://doi.org/10.1007/978-3-032-19509-8_14

14.1 Introduction

Bone is a mechanosensitive, highly remodeling tissue that is subjected to various mechanical cues, including fluid shear stress, cyclic strain, matrix stiffness, and mechanical loading. Mechanical forces regulate skeletal architecture and control cellular processes involved in bone formation and resorption (Wang et al. 2022). Osteocytes act as the primary mechanosensors in bone, while mesenchymal stem cells (MSCs) and primary osteoblasts function as secondary mechanosensors (Wang et al. 2022). MSCs and osteoblasts rely on mechanotransduction to define lineage destination, initiate matrix deposition, and adapt bone to matrix environment through signaling pathways mediated by forces, such as extracellular signal-regulated kinase (ERK)-mitogen-activated protein kinase (MAPK) and downstream effectors, including Runt-related transcription factor 2 (RUNX2) and Yes-associated protein/transcriptional co-activator with PDZ-binding motif (YAP/TAZ) (Liu et al. 2023).

A pivotal transcription factor required for MSCs' commitment is RUNX2. Its presence is essential for osteoblast gene expression, and even small variability can lead to skeletal abnormalities (Fig. 14.1) (Komori 2020). RUNX2 activity must be tightly controlled, as a prolonged expression can lead to an impairment of osteoblast maturation (Komori 2020).

RUNX2 is a coordinator of osteogenic transcription and chromatin accessibility during cell fate (Fig. 14.1) (Komori 2024; Hojo et al. 2022). Therefore, epigenetic regulation has emerged as a crucial player in osteogenic gene expression, particularly RUNX2. Epigenetic processes such as DNA methylation, histone modifications, chromatin remodeling, and non-coding RNA-mediated regulation affect transcriptional potential, without altering the DNA sequence (Gargalionis et al. 2024). Recent research supports that mechanical forces are interrelated with epigenetics and can directly affect epigenetic states, altering RUNX2 expression in a force-dependent manner (Fig. 14.1). Epigenetic modifications, triggered mechanically, connect external signals to bone-forming osteogenic pathways. These mechanically induced nuclear changes link epigenetic regulation and biomechanics, enabling long-term bone adaptation, through persistent osteogenic gene expression. Skeletal malformations such as osteoporosis, osteopenia, poor fracture healing, and osteoarthritis may arise when this connection is ruptured (Zhu et al. 2024).

We summarize the latest findings on how mechanical signals influence epigenetic responses, thereby modulating *RUNX2* expression and its implications for bone pathology and bone disorders. The understanding of this mechanobiology-epigenetic mechanism axis may assist therapies that aim at restoring osteogenesis in mechanically induced environments.

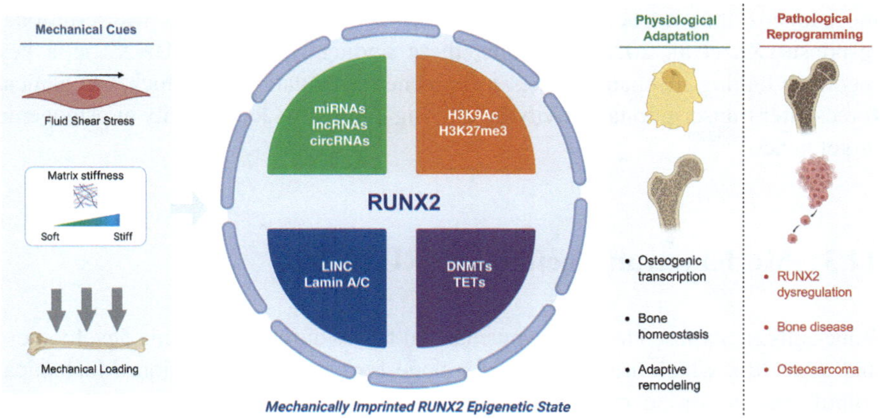

Fig. 14.1 Mechanical cues imprint RUNX2 epigenetic memory to control bone fate and disease. Mechanical stimuli, such as shear stress, extracellular matrix stiffness, and mechanical loading, are integrated at the nuclear level via RUNX2. Mechanical forces establish epigenetic modulators, including non-coding RNAs, histone modifications, DNA methylation, and chromatin remodeling, to facilitate RUNX2-dependent transcription. Balanced epigenetic regulation promotes physiological osteogenesis and bone homeostasis, whereas abnormal regulation drives bone disease and tumorigenesis. circRNAs, circular RNAs; DNMTs, DNA methyltransferases; H3K9ac, histone H3 lysine 9 acetylation; H3K27me3, histone H3 lysine 27 trimethylation; lncRNAs, long non-coding RNAs; LINC, linker of nucleoskeleton and cytoskeleton; lncRNAs, long non-coding RNAs; miRNAs, microRNAs; RUNX2, Runt-related transcription factor 2; TET, ten-eleven translocation. Created in BioRender. Gargalionis, A. (2025) https://BioRender.com/n9so6wq

14.2 RUNX2 and Regulation of Osteogenesis

During early osteogenesis, the fate of MSCs is driven by RUNX2, which helps secure MSCs' differentiation away from alternative fates such as adipocytes or chondrocytes, in part by enabling osteogenic chromatin accessibility (Hojo et al. 2022). *RUNX2* expression is higher in pre-osteoblasts during their maturation into osteoblasts and decreases in osteocytes (Park et al. 2001). RUNX2, through transcriptional control of osteogenic genes such as *osteocalcin* (*bone gamma-carboxyglutamate protein*; *BGLAP*), *type I collagen type I alpha 1 chain* (*COL1A1*), and *alkaline phosphatase* (*ALPL*), tightly regulates matrix deposition and mineralization. RUNX2 dosage is tightly regulated; reduced expression characterizes defective ossification, whereas overexpression disrupts osteoblast maturation and bone homeostasis (Yamashita et al. 2025).

RUNX2 is continually active in adult bone remodeling and responds to mechanical loading that enhances RUNX2 nuclear localization, inducing its transcriptional activity (Fig. 14.1) (Gargalionis et al. 2024; Somemura et al. 2021). Mechanotransduction pathways, including the integrin-focal adhesion kinase (FAK)-MAPK and YAP/TAZ, influence RUNX2 responsiveness to mechanical cues (Han et al. 2024; Wang et al. 2020). These mechanical effects are mediated through epigenetic mechanisms that impact *RUNX2* promoter accessibility and its transcriptional response (Xu et al. 2023). Together, these findings highlight RUNX2 as a key mediator linking mechanical cues to epigenetic regulation, in which mechanical forces alter transcriptional activity by altering chromatin accessibility at osteogenic target genes.

14.3 Mechanotransduction and RUNX2

Bone cells are exposed to various mechanical forces during embryonic development and adult life, which are necessary for bone formation and function. Mechanical stimuli can be varied ranging from biophysical cues such as stress, strain, and fluid flow, forces attributed to cell movement and cell-cell or cell-extracellular matrix (ECM) interactions, as well as compressive forces, fluid shear stress (FSS), hydrostatic pressure, or loading/unloading (Wang et al. 2022).

Piezo 1 and Piezo 2 are core force sensors essential for osteogenesis and osteoblast differentiation (Jiang et al. 2024). The absence of Piezo1 or Piezo1/2 in MSCs or pre-osteoblasts impairs osteoblast differentiation which is associated with fractures. Notably, deleting *Piezo1/2* rendered bone unresponsive to mechanical force, such as unloading, thereby preventing further bone loss that is normally triggered by mechanical load loss, leading to increased bone resorption (Wang et al. 2020; Song et al. 2020). Mechanistically, Piezo1/2 transmit FSS and ECM stiffness signals activating Ca^{2+} influx to induce calcineurin, which facilitates triggering of nuclear factor of activated T-cells, cytoplasmic 1 (NFATc1), YAP1, and β-catenin transcription factors, as well as NFAT/YAP1/β-catenin complex formation. YAP1 and β-catenin activities were suppressed in Piezo1- and Piezo1/2-defective bones, which were restored by increased β-catenin activities. NFAT/YAP1/β-catenin function upstream of RUNX2 to enhances its transcriptional potential, fostering osteogenesis under mechanical strain. Thus, RUNX2 is a key downstream effector that translates Piezo-mediated mechanosensing into osteoblast differentiation (Zhou et al. 2020). Integrin-based adhesions mediate the transmission of mechanical forces from the extracellular matrix to intracellular signaling pathways. More specifically, mechanical loading activates integrin-FAK signaling, which regulates MAPK pathways including ERK and p38. These can alter RUNX2 phosphorylation, nuclear architecture, and transcriptional activity (Wang et al. 2022; Zhou et al. 2020; Zhang et al. 2021). Piezo-mediated calcium signaling, along with integrin mechanotransduction, allows bone cells to integrate mechanical cues and orchestrate osteoblast differentiation and bone formation (Zhou et al. 2020). Overall, mechanosensitive ion channels transmit extracellular mechanical stimuli to alter intracellular calcium influx and potentiate transcription factors that enhance RUNX2 activity and regulate osteogenesis. Integrin-based adhesion complexes sense mechanical inputs to further regulate osteogenic transcription via kinase pathways which modulate RUNX2 phosphorylation. Both mechanisms provide a mechanistic basis for mechanoregulation of osteogenic transcription.

14.4 Epigenetic Regulation of RUNX2 in Response to Mechanical Cues

Epigenetic modifications during osteoblast development play a crucial role in gene regulation, by altering chromatin accessibility to transcriptional machinery, thereby controlling gene expression during osteogenesis. Epigenetic regulation, therefore, enables mechanical cues to be imprinted through non-coding RNAs, histone modifications, DNA methylation, and chromatin remodeling to sustain osteogenic gene expression (Gargalionis et al. 2024).

14.4.1 Non-coding RNAs

14.4.1.1 MicroRNAs (miRNAs)

miRNAs are non-coding small RNAs (21–25 nucleotides in length) that negatively regulate gene expression at the post-transcriptional (epigenetic) level in a variety of biological processes such as apoptosis, cell proliferation, and differentiation, by targeting the 3′-untranslated region (UTR) of mRNA molecules (Lytle et al. 2007; Lim et al. 2005). miRNAs are key post-transcriptional epigenetic regulators of RUNX2 and are implicated in osteogenic differentiation and bone formation (Fig. 14.1).

miR-20a, miR-96, and miR-21 promote osteogenic differentiation, accompanied by elevated RUNX2 levels (Zhang et al. 2011; Sun et al. 2015; Yang et al. 2014). Osteoblast differentiation is inhibited by miR-138 in bone marrow mesenchymal stromal cells (BMSCs) by suppressing the phosphorylation of downstream targets, including FAK, ERK1/2, and RUNX2. Notably, miR-138 downregulates RUNX2, which is essential for the platelet-derived growth factor (PDGF)-mediated inhibition of osteogenic differentiation of BMSCs. miRNAs act as mechanosensitive molecules in fluid shear stress (FSS) and microgravity models (Qu et al. 2014). Specifically, miR-33-5p promotes osteogenic differentiation targeting Hmga2, and is associated with elevated *RUNX2*, *Osterix* (*OSX*), and *alkaline phosphatase* (*ALP*) expression (Wang et al. 2016).

On the other hand, osteogenic differentiation is inhibited by miR-103a, miR-133a-5p, miR-224–5p, and miR-503-5p in mechanical environments (Zuo et al. 2015; Zhang et al. 2018; Ding et al. 2023). Importantly, variations in mechanical forces elicit distinct epigenetic regulation of RUNX2 mediated by miRNAs. miR-103a is downregulated in a cyclical mechanical stretching model by targeting its host gene *pank3,* followed by an upregulation of RUNX2 protein levels (Zuo et al. 2015). miR-224-5p, however, provides an osteoporosis relevant example of RUNX2 epigenetic repression, as RUNX2 and SP7 are direct targets of miR-224-5p. miR-224-5p inhibits osteogenic regulation in osteoporotic patients, highlighting its therapeutic and diagnostic potential for osteoporosis (Ding et al. 2023). In another study using a mechanical stretch model in which miR-503-5p was overexpressed using an agomir, BMSCs' osteogenic differentiation was suppressed, resulting in decreased *RUNX2* expression and osteoblastic features (Liu et al. 2017).

14.4.1.2 Long Non-coding RNAs (lncRNAs) and Circular RNAs (circRNAS)

lncRNAs are a family of transcripts longer than 200 nucleotides, which are either non-coding or code for short micropeptides. Their role is important in stabilizing the expression of genes involved in osteogenesis, such as *RUNX2* (Fig. 14.1) (Ouyang et al. 2022). lncRNA HOTAIRM1 has recently been identified as a key player in the regulation of osteoblast differentiation. Inhibition of HOTAIRM1 is associated with reduced calcium deposition and ALP activity. Molecularly, it was concluded that HOTAIRM1 promotes c-Jun activity by recruiting the acetyltransferase p300, leading to upregulated RUNX2 (Fu et al. 2019). Another lncRNA, AK045490, had an inhibitory effect on osteoblast progression and bone formation. Mechanistically, AK045490 reduced osteogenic capacity by inhibiting β-catenin/T-cell factor 1 (TCF1)-dependent RUNX2 signaling (Li et al. 2019). A further study in MSC cells found that the overexpression of the lncRNA brain-derived neurotrophic factor-antisense (BDNF-AS) suppressed RUNX2, thereby inhibiting osteogenic differentiation (Feng et al. 2018).

CircRNAs participate in epigenetic regulation by sponging miRNAs, interacting with RNA-binding proteins, or encoding peptides. For instance, circRNA CDR1 is a critical regulator of osteogenic differentiation by modulating the miR-7-5p/WNT5B axis (Chen et al. 2020). Liu et al. discovered that circRNA AFF4 modulates osteogenic differentiation in BMSCs by regulating miR-135a-5p/FNDC5/Irisin axis (Liu et al. 2021). A recent study revealed that circ-3626 promotes osteogenesis by enhancing BMSCs' osteoblastic capacity in in vitro and in vivo, as it triggers RUNX2 activity by inhibiting miR-338-3p to promote bone formation (Chen et al. 2024). lncRNAs and circRNAs under mechanical stimulation act as epigenetic regulators that stabilize RUNX2-dependent osteogenesis.

14.4.2 Chromatin Remodeling

RUNX2 is a pioneer transcription factor that facilitates chromatin accessibility and induces osteoblast differentiation by regulating cell-type-dependent gene expression. Through interaction with co-factors such as Sp7 (also called Osterix) and histone-regulatory enzymes, RUNX2 promotes the formation of enhancer "hotspots," which activate osteoblast-specific transcriptional pathways (Komori 2024). Mechanical forces can be transmitted to the nucleus via the cytoskeleton and the linker of nucleoskeleton and cytoskeleton (LINC) complex, inducing changes in chromatin accessibility at specific gene targets. Disruption of the LINC complex in in vivo and in vitro osteogenic models results in impaired osteogenic differentiation and disrupted bone formation (Fig. 14.1) (Birks and Uzer 2021; Birks et al. 2023).

RUNX2 activation of enhancer regions depends on chromatin remodeling maintained by nuclear envelope proteins, including A/C and Nesprin, both vital for nuclear architecture and chromatin organization. RUNX2 deficiency disrupts Lamin A/C binding in a spatial manner to chromatin transcription start sites (TSS), leading to impaired gene transcription and osteogenic differentiation (Hojo et al. 2022). The nuclear environment and membrane, as well as the nuclear morphology, are affected by a defective RUNX2. For example, nuclear morphology is characterized by less stiff nuclei and lower intracellular mechanical tension altering their ability to respond to mechanical cues. The recovery of Nesprin in RUNX2-deficient cells restored the defective nuclear rigidity and intracellular tension, proving that the cytoskeleton and the nucleus are related (Hojo et al. 2022). It is therefore established that LINC is a critical mechanosensitive linker that links mechanical forces to RUNX2-dependent regulation of osteogenesis (Fig. 14.1) (Gargalionis et al. 2024). As seen, mechanical cues are sensed at the cell membrane and transmitted through the cytoskeleton and the nucleus, ultimately influencing chromatin remodeling.

14.5 RUNX2-Related Epigenetic Mechanisms and Bone Tumorigenesis

RUNX2 is implicated in bone disease and bone tumor pathogenesis through epigenetic dysregulation. Abnormal DNA methylation, histone modifications, chromatin remodeling, and non-coding RNAs activity modulate osteogenic gene expression, impairing bone remodeling and promoting pathogenesis such as osteoporosis, defective fracture healing, and osteoarthritis (Fig. 14.1) (Green et al. 2023).

Aberrant *RUNX2* expression has been observed in osteosarcoma and other tumors where high RUNX2 levels correlate with poor prognosis, metastasis, and aggressive tumorigenesis (Katsianou et al. 2025; Vitale et al. 2025; Zhou et al. 2025). Epigenetic mechanisms, histone methylation, and acetylation at RUNX2 target regions influence tumor relevant transcription. Genome-wide studies in osteosarcoma (OS) highlight dysfunctional histone modifications, including altered activity of histone methyltransferases such as EZH2 (Yang et al. 2025). For example, H3K9 trimethylation is catalyzed by the histone methyltransferase SET domain bifurcated histone lysine methyltransferase 1 (SETDB1), which inhibits gene transcription by forming heterochromatin. Dysfunctional SETDB1 activity is associated with OS, highlighting the importance of a tightly regulated H3K9 methylation in bone-associated genes (Strepkos et al. 2021).

Furthermore, studies in thyroid and breast cancer suggest a chromatin-based mechanism of RUNX2 activation in bone tumors, where enhancers looping to the *RUNX2* promoter are regulated by bromodomain-containing protein 4 (BRD4) and c-Jun (43). RUNX2 acts as an epigenetic survival regulator in p53-deficient cancer cells, by epigenetically potentiating MYC transcription (Shin et al. 2016). Enhancer-associated long non-coding RNAs (elncRNA) can also epigenetically regulate RUNX2, as shown by the RUNX2-associated lncRNA RAIN, which cooperates functionally with RUNX2 to orchestrate the transcription of genes linked to migration, matrix remodeling, and loss of differentiation in cancer (Vitale et al. 2024).

OS implicates RUNX2 epigenetic regulation in therapeutic schemes. miR-150 holds a tumor suppressor and chemoprotective role by targeting RUNX2 in OS, reducing growth and sensitizing tumors to chemotherapy, implying its therapeutic potential (Ling et al. 2020). In pre-clinical studies of primary bone cancer, a RUNX2 inhibitor, CADD522, was developed and proved to reduce proliferation in bone cancer cells, including osteosarcoma models. Even though it has yet to enter clinical trials, its future role is promising, as it blocks RUNX2-related RNA interactions, thereby slowing tumor progression (Green et al. 2023). Lastly, bromodomain and extra-terminal motif (BET) bromodomain inhibitor I, like those targeting BRD4, can also affect epigenetic regulators of RUNX2, as demonstrated in thyroid and breast cancer (Sancisi et al. 2017). Overall, these findings indicate that RUNX2 dysregulation is involved in bone tumorigenesis, and its inhibition could pose a promising therapeutic target.

14.6 Conclusion

Epigenetic regulation of RUNX2 maintains bone mechanobiology by integrating mechanical cues, controlling osteogenic differentiation, and bone homeostasis. Mechanical cues are transferred from the extracellular environment to the nucleus, where epigenetic modifications, including histone acetylation and methylation, DNA methylation, chromatin remodeling, and non-coding RNAs, orchestrate RUNX2 expression and activity. RUNX2 epigenetic regulation stabilizes osteogenic gene expression. Any dysregulation of RUNX2 epigenetic control contributes to bone disorders including osteoporosis, impaired fracture healing, and osteoarthritis, as well as bone tumorigenesis. A deeper understanding of RUNX2 epigenetic regulation in response to mechanical cues may support the development of targeted epigenetic and mechano-therapeutic strategies for bone diseases and bone-associated malignancies.

References

Birks S, Uzer G (2021) At the nuclear envelope of bone mechanobiology. Bone. 151:116023. Epub 20210526. https://doi.org/10.1016/j.bone.2021.116023

Birks S, Howard S, O'Rourke C, Thompson WR, Lau A, Uzer G (2023) Osterix-driven LINC complex disruption in vivo diminishes bone microarchitecture in 8-week male mice but not after 6-week voluntary wheel running. bioRxiv. https://doi.org/10.1101/2023.08.24.554623

Chen G, Wang Q, Li Z, Yang Q, Liu Y, Du Z et al (2020) Circular RNA CDR1as promotes adipogenic and suppresses osteogenic differentiation of BMSCs in steroid-induced osteonecrosis of the femoral head. Bone. 133:115258. Epub 20200201. https://doi.org/10.1016/j.bone.2020.115258

Chen G, Wang S, Wei R, Liu Y, Xu T, Liu Z et al (2024) Circular RNA circ-3626 promotes bone formation by modulating the miR-338-3p/Runx2 axis. Joint Bone Spine 91(2):105669. Epub 20231130. https://doi.org/10.1016/j.jbspin.2023.105669

Ding S, Ma Y, Yang J, Tang Y, Jin Y, Li L et al (2023) MiR-224-5p inhibits osteoblast differentiation and impairs bone formation by targeting Runx2 and Sp7. Cytotechnology. 75(6):505–516. Epub 20230905. https://doi.org/10.1007/s10616-023-00593-z

Feng X, Lin T, Liu X, Yang C, Yang S, Fu D (2018) Long non-coding RNA BDNF-AS modulates osteogenic differentiation of bone marrow-derived mesenchymal stem cells. Mol Cell Biochem. 445(1–2):59–65. Epub 20171215. https://doi.org/10.1007/s11010-017-3251-2

Fu L, Peng S, Wu W, Ouyang Y, Tan D, Fu X (2019) LncRNA HOTAIRM1 promotes osteogenesis by controlling JNK/AP-1 signalling-mediated RUNX2 expression. J Cell Mol Med. 23(11):7517–7524. Epub 20190911. https://doi.org/10.1111/jcmm.14620

Gargalionis AN, Adamopoulos C, Vottis CT, Papavassiliou AG, Basdra EK (2024) Runx2 and polycystins in bone mechanotransduction: challenges for therapeutic opportunities. Int J Mol Sci 25(10) Epub 20240513. https://doi.org/10.3390/ijms25105291

Green D, Singh A, Tippett VL, Tattersall L, Shah KM, Siachisumo C et al (2023) YBX1-interacting small RNAs and RUNX2 can be blocked in primary bone cancer using CADD522. J Bone Oncol 39:100474. Epub 20230305. https://doi.org/10.1016/j.jbo.2023.100474

Han J, Zhang J, Zhang X, Luo W, Liu L, Zhu Y et al (2024) Emerging role and function of Hippo-YAP/TAZ signaling pathway in musculoskeletal disorders. Stem Cell Res Ther 15(1):386. Epub 20241029. https://doi.org/10.1186/s13287-024-04011-9

Hojo H, Saito T, He X, Guo Q, Onodera S, Azuma T et al (2022) Runx2 regulates chromatin accessibility to direct the osteoblast program at neonatal stages. Cell Rep 40(10):111315. https://doi.org/10.1016/j.celrep.2022.111315

Jiang Y, Lin H, Chen Y, Lan Y, Wang H, Li T et al (2024) Piezo1 contributes to alveolar bone remodeling by activating beta-catenin under compressive stress. Am J Orthod Dentofacial Orthop 165(4):458–470. Epub 20240106. https://doi.org/10.1016/j.ajodo.2023.10.020

Katsianou MA, Andreou D, Korkolopoulou P, Vetsika EK, Piperi C (2025) Epigenetic modifications in osteosarcoma: mechanisms and therapeutic strategies. Life (Basel) 15(8) Epub 20250728. https://doi.org/10.3390/life15081202

Komori T (2020) Molecular mechanism of Runx2-dependent bone development. Mol Cells 43(2):168–175. https://doi.org/10.14348/molcells.2019.0244

Komori T (2024) Regulation of skeletal development and maintenance by Runx2 and Sp7. Int J Mol Sci 25(18) Epub 20240920. https://doi.org/10.3390/ijms251810102

Li D, Tian Y, Yin C, Huai Y, Zhao Y, Su P et al (2019) Silencing of lncRNA AK045490 promotes osteoblast differentiation and bone formation via beta-catenin/TCF1/Runx2 signaling Axis. Int J Mol Sci 20(24) Epub 20191210. https://doi.org/10.3390/ijms20246229

Lim LP, Lau NC, Garrett-Engele P, Grimson A, Schelter JM, Castle J et al (2005) Microarray analysis shows that some microRNAs downregulate large numbers of target mRNAs. Nature. 433(7027):769–773. Epub 20050130. https://doi.org/10.1038/nature03315

Ling Z, Fan G, Yao D, Zhao J, Zhou Y, Feng J et al (2020) MicroRNA-150 functions as a tumor suppressor and sensitizes osteosarcoma to doxorubicin-induced apoptosis by targeting RUNX2. Exp Ther Med 19(1):481–488. Epub 20191122. https://doi.org/10.3892/etm.2019.8231

Liu L, Liu M, Li R, Liu H, Du L, Chen H et al (2017) MicroRNA-503-5p inhibits stretch-induced osteogenic differentiation and bone formation. Cell Biol Int 41(2):112–123. Epub 20161219. https://doi.org/10.1002/cbin.10704

Liu C, Liu AS, Zhong D, Wang CG, Yu M, Zhang HW et al (2021) Circular RNA AFF4 modulates osteogenic differentiation in BM-MSCs by activating SMAD1/5 pathway through miR-135a-5p/FNDC5/Irisin axis. Cell Death Dis 12(7):631. Epub 20210618. https://doi.org/10.1038/s41419-021-03877-4

Liu Z, Wang Q, Zhang J, Qi S, Duan Y, Li C (2023) The Mechanotransduction signaling pathways in the regulation of osteogenesis. Int J Mol Sci 24(18) Epub 20230920. https://doi.org/10.3390/ijms241814326

Lytle JR, Yario TA, Steitz JA (2007) Target mRNAs are repressed as efficiently by microRNA-binding sites in the 5' UTR as in the 3' UTR. Proc Natl Acad Sci USA 104(23):9667–9672. Epub 20070529. https://doi.org/10.1073/pnas.0703820104

Ouyang X, Ding Y, Yu L, Xin F, Yang X (2022) LncRNA TUG regulates osteogenic differentiation of bone marrow mesenchymal stem cells via miRNA-204/SIRT 1. J Musculoskelet Neuronal Interact 22(3):401–410

Park MH, Shin HI, Choi JY, Nam SH, Kim YJ, Kim HJ et al (2001) Differential expression patterns of Runx2 isoforms in cranial suture morphogenesis. J Bone Miner Res 16(5):885–892. https://doi.org/10.1359/jbmr.2001.16.5.885

Qu B, Xia X, Wu HH, Tu CQ, Pan XM (2014) PDGF-regulated miRNA-138 inhibits the osteogenic differentiation of mesenchymal stem cells. Biochem Biophys Res Commun. 448(3):241–247. Epub 20140502. https://doi.org/10.1016/j.bbrc.2014.04.091

Sancisi V, Manzotti G, Gugnoni M, Rossi T, Gandolfi G, Gobbi G et al (2017) RUNX2 expression in thyroid and breast cancer requires the cooperation of three non-redundant enhancers under the control of BRD4 and c-JUN. Nucleic Acids Res 45(19):11249–11267. https://doi.org/10.1093/nar/gkx802

Shin MH, He Y, Marrogi E, Piperdi S, Ren L, Khanna C et al (2016) A RUNX2-mediated epigenetic regulation of the survival of p53 defective cancer cells. PLoS Genet 12(2):e1005884. Epub 20160229. https://doi.org/10.1371/journal.pgen.1005884

Somemura S, Kumai T, Yatabe K, Sasaki C, Fujiya H, Niki H et al (2021) Physiologic mechanical stress directly induces bone formation by activating glucose transporter 1 (glut 1) in osteoblasts, inducing signaling via NAD+-dependent deacetylase (Sirtuin 1) and runt-related transcription factor 2 (Runx2). Int J Mol Sci 22(16) Epub 20210823. https://doi.org/10.3390/ijms22169070

Song J, Liu L, Lv L, Hu S, Tariq A, Wang W et al (2020) Fluid shear stress induces Runx-2 expression via upregulation of PIEZO1 in MC3T3-E1 cells. Cell Biol Int 44(7):1491–1502. Epub 20200328. https://doi.org/10.1002/cbin.11344

Strepkos D, Markouli M, Klonou A, Papavassiliou AG, Piperi C (2021) Histone methyltransferase SETDB1: a common denominator of tumorigenesis with therapeutic potential. Cancer Res. 81(3):525–534. Epub 20201028. https://doi.org/10.1158/0008-5472.CAN-20-2906

Sun Y, Xu L, Huang S, Hou Y, Liu Y, Chan KM et al (2015) Mir-21 overexpressing mesenchymal stem cells accelerate fracture healing in a rat closed femur fracture model. BioMed Res Int 2015:412327. Epub 20150323. https://doi.org/10.1155/2015/412327

Vitale E, Manicardi V, Gugnoni M, Torricelli F, Donati B, Muccioli S et al (2024) Exploring the transcriptional cooperation between RUNX2 and its associated elncRNA RAIN. Cell Death Dis 15(9):673. Epub 20240914. https://doi.org/10.1038/s41419-024-07058-x

Vitale E, Gugnoni M, Manicardi V, Muccioli S, Torricelli F, Donati B et al (2025) RUNX2 cooperates with SREBP1 to rewire cancer metabolism and promote aggressiveness. J Exp Clin Cancer Res 44(1):298. Epub 20251031. https://doi.org/10.1186/s13046-025-03549-7

Wang H, Sun Z, Wang Y, Hu Z, Zhou H, Zhang L et al (2016) miR-33-5p, a novel mechano-sensitive microRNA promotes osteoblast differentiation by targeting Hmga2. Sci Rep. 6:23170. Epub 20160316. https://doi.org/10.1038/srep23170

Wang L, You X, Lotinun S, Zhang L, Wu N, Zou W (2020) Mechanical sensing protein PIEZO1 regulates bone homeostasis via osteoblast-osteoclast crosstalk. Nat Commun 11(1):282. Epub 20200115. https://doi.org/10.1038/s41467-019-14146-6

Wang L, You X, Zhang L, Zhang C, Zou W (2022) Mechanical regulation of bone remodeling. Bone Res 10(1):16. Epub 20220218. https://doi.org/10.1038/s41413-022-00190-4

Xu X, Wang W, Zou J, Kratz K, Deng Z, Lendlein A et al (2023) Histone modification of osteogenesis related genes triggered by substrate topography promotes human mesenchymal stem cell differentiation. ACS Appl Mater Interfaces. 15(25):29752–29766. Epub 20230613. https://doi.org/10.1021/acsami.3c01481

Yamashita Y, Hayashi M, Liu A, Sasaki F, Tsuchiya Y, Takayanagi H et al (2025) Fam102a translocates Runx2 and Rbpjl to facilitate Osterix expression and bone formation. Nat Commun 16(1):9. Epub 20250102. https://doi.org/10.1038/s41467-024-55451-z

Yang M, Pan Y, Zhou Y. miR-96 promotes osteogenic differentiation by suppressing HBEGF-EGFR signaling in osteoblastic cells. FEBS Lett 2014;588(24):4761–4768. Epub 20141115. doi:https://doi.org/10.1016/j.febslet.2014.11.008

Yang Y, Tang X, Liu Z (2025) Multi-omics analysis of histone-related genes in osteosarcoma: a multidimensional integrated study revealing drug sensitivity and immune microenvironment characteristics. Technol Cancer Res Treat 24:15330338251336275. Epub 20250417. https://doi.org/10.1177/15330338251336275

Zhang JF, Fu WM, He ML, Xie WD, Lv Q, Wan G et al (2011) MiRNA-20a promotes osteogenic differentiation of human mesenchymal stem cells by co-regulating BMP signaling. RNA Biol. 8(5):829–838. Epub 20110728. https://doi.org/10.4161/rna.8.5.16043

Zhang W, Wu Y, Shiozaki Y, Sugimoto Y, Takigawa T, Tanaka M et al (2018) miRNA-133a-5p inhibits the expression of osteoblast differentiation-associated markers by targeting the 3' UTR of RUNX2. DNA Cell Biol 37(3):199–209. Epub 20180123. https://doi.org/10.1089/dna.2017.3936

Zhang G, Li X, Wu L, Qin YX (2021) Piezo1 channel activation in response to mechanobiological acoustic radiation force in osteoblastic cells. Bone Res 9(1):16. Epub 20210310. https://doi.org/10.1038/s41413-020-00124-y

Zhou T, Gao B, Fan Y, Liu Y, Feng S, Cong Q et al (2020) Piezo1/2 mediate mechanotransduction essential for bone formation through concerted activation of NFAT-YAP1-ss-catenin. Elife. 9. Epub 20200318. https://doi.org/10.7554/eLife.52779

Zhou P, Wei Q, Wang H, Yu T, Chen L, Huang Q et al (2025) RUNX2 mediated epithelial mesenchymal transition induced by TGF beta and Smad signaling promotes malignant progression in glioma. Sci Rep 15(1):26781. Epub 20250723. https://doi.org/10.1038/s41598-025-09167-9

Zhu S, Chen W, Masson A, Li YP (2024) Cell signaling and transcriptional regulation of osteoblast lineage commitment, differentiation, bone formation, and homeostasis. Cell Discov 10(1):71. Epub 20240702. https://doi.org/10.1038/s41421-024-00689-6

Zuo B, Zhu J, Li J, Wang C, Zhao X, Cai G et al (2015) microRNA-103a functions as a mechanosensitive microRNA to inhibit bone formation through targeting Runx2. J Bone Miner Res 30(2):330–345. https://doi.org/10.1002/jbmr.2352

Chapter 15
Mechanobiology of Myelin Generation/Regeneration in Health and Disease

Zehra G. Morcimen-Gulicli, Yannis F. Missirlis, and Aylin Sendemir

Abstract The myelin sheath is a multilayered glial membrane that surrounds axons providing insulation and rapid propagation of action potentials and contributing to structural support in both the central nervous system (CNS) and the peripheral nervous system (PNS). Therefore, myelin formation plays a crucial role in maintaining the integrity, connectivity, and functionality of the nervous system, as well as neuronal regeneration after injury or damage. Myelination in early developmental stages requires the molecular and mechanical exogenous cues provided mainly by the axons of the neurons, coupling the myelin-generating oligodendrocytes in the CNS and Schwann cells in the PNS. Complementary to early development, adaptive myelination dynamically regulates myelin remodeling in response to neuronal activity, ensuring optimal synaptic function and plasticity. Adaptive myelination is shown to be extremely mechanosensitive, requiring strict spatial constraints, stiffness, and topography for cell plasticity and myelination. Consequently, myelin regeneration is limited after traumatic injury and in demyelinating neurodegenerative diseases such as multiple sclerosis (MS). In addition to the inflammatory niche created around damaged myelin sheath, it is often overlooked that altered mechanical properties of the surrounding tissue are also pivotal in the inhibition of remyelination.

This review focuses on the effects of mechanical cues on the polarization and plasticity of myelin generating cells, as well as on the induction of the myelin sheath

Z. G. Morcimen-Gulicli
Department of Bioengineering, Faculty of Engineering, Ege University, Izmir, Türkiye

Y. F. Missirlis
Department of Mechanical Engineering & Aeronautics, University of Patras, Patras, Greece

Sabanci University Nanotechnology Research and Application Center (SUNUM), Istanbul, Türkiye

A. Sendemir (✉)
Department of Bioengineering, Faculty of Engineering, Ege University, Izmir, Türkiye

Department of Biomedical Technologies, Graduate School of Natural and Applied Sciences, Ege University, Izmir, Türkiye
e-mail: aylin.sendemir@ege.edu.tr

M. Kloc, J. Wosik (eds.), *Mechanotransduction and Mechanosensing*, Results and Problems in Cell Differentiation 78, https://doi.org/10.1007/978-3-032-19509-8_15

during normal development and in pathological conditions. It also aims to give insights into potential mechanoepigenetic therapeutic approaches for demyelinating conditions.

15.1 Introduction

Myelin is a multi-lamellar, lipid-rich structure developed by the plasma membrane of specialized glial cells, namely, oligodendrocytes (OLs) in the central nervous system (CNS) and Schwann cells (SCs) in the peripheral nervous system (PNS). The main function of the myelin sheath is insulation of axons for faster and more efficient conduction of action potential, while it also provides metabolic and homeodynamic support to axons. While the high lipid content of myelin gives it a unique insulating capacity, its protein composition provides stability of the sheath.

The unmyelinated nervous system begins myelination after birth, and myelination continues throughout adolescence and adulthood (Scantlebury et al. 2014). In 1854, Rudolf Virchow named the structures in the brain's nucleus "myelin," derived from the Greek word myelos, meaning marrow (Boullerne 2016). It was hypothesized then that myelin is an insulating material secreted by neurons. A century later, silver carbonate histological staining by Hortega and Penfield demonstrated that myelin was not of neuronal origin but was formed by OLs (Groh and Leblanc 2025). This staining showed that OLs were more densely present in white matter than in gray matter.

Myelin generation by OLs is estimated to reach up to 5000 μm^2 of myelin per day per OL (Pfeiffer et al. 1993); therefore, OLs are among the most productive cellular membrane producers. Myelination of axons in the CNS exhibits a wide variation in the number of myelin sheaths, sheath length, sheath thickness, and distribution along their length. Changes in any of these sheath parameters, including myelin regeneration (remyelination) after injury, have significant effects on altering the underlying axon's conduction velocity. The internodes, or sheaths, that form around axons facilitate cell-cell interactions that support saltatory conduction. Nerve conduction is facilitated by the concentration of voltage-gated sodium channels in the spaces between these sheaths, called "nodes of Ranvier." When this mechanism was discovered, it was thought that internode length and thickness were optimized for maximum conduction velocity and reduced energy consumption and that they remained constant (Huxley and Stämpfli 1949; Rushton 1951). However, subsequent observations suggest that OLs and myelination are altered throughout our lifetimes and serve other roles.

Despite their different origins, SCs and OLs share similarities in the overall organization of their myelin structures and their conserved roles in saltatory conduction (Brown and Macklin 2020).

New OLs develop throughout life from actively dividing precursor cells (OPCs) in the CNS (Hughes et al. 2013; Young et al. 2013). OLs and OPCs are known to be sensitive to environmental conditions and mechanical cues including ECM stiffness, mechanical strain, physical alterations, and macromolecular crowding. The sensitivity of OPCs to biophysical signals is crucial to myelin pathology and demyelinating diseases (Yang et al. 2025). Axon diameter and spacing are also among the most fundamental biophysical cues affecting OLs. Under healthy conditions, axon diameter in the CNS ranges from 0.1 to 5.0 μm (Liewald et al. 2014), while axon swelling and increased axon diameter have been observed in some diseases such as amyotrophic lateral sclerosis (ALS) (Gatto et al. 2018; Coleman 2022) and multiple sclerosis (MS) (Fisher et al. 2007; Singh et al. 2017). Axon damage and loss of axon density are hallmarks of progressive MS, particularly for chronically demyelinated axons (Correale et al. 2019). Changes in brain stiffness also directly affect the physiology of the OL and OPCs. Imaging studies have shown correlations between progressive MS (Wuerfel et al. 2010; Streitberger et al. 2012) and changes in the structural integrity and mechanical properties of brain tissue, with the Young's modulus (E) of the brain parenchyma decreasing from 240 Pa to 120 Pa in response to acute demyelination. The decrease in stiffness with disease progression has been supported by in vivo demyelination models (Schregel et al. 2012; Millward et al. 2015).

SCs originate from the trunk neural crest and migrate ventrally to the PNS, where they develop into glia associated with axons. Migration and microenvironmental signals determine the type of SC they will differentiate into (myelinating vs. non-myelinating). SCs provide the sole protection against physiological mechanical stresses to which the PNS is exposed. SC mechanosensitivity is important for PNS development and myelination and also plays a role in PNS mechanoprotection (Rosso et al. 2017b). The basal lamina surrounding SCs protects neurons from compression and deformation by acting as a mechanical shield with components such as collagen and laminin. Furthermore, this mechanosensitivity directly affects SC migration, proliferation, and differentiation potential (Rosso et al. 2017a).

In 2015, Bechler and Byrne developed a model called "smart wiring," which postulates that myelination has two stages: innate and adaptive. In the latter, active axons become more myelinated. This model explains why a single signal necessary and sufficient for CNS myelination has not been identified and suggests that myelination, like synapses, may contribute to brain activity through activity-dependent changes (Bechler and Byrne 2015; Bechler et al. 2018). While it is known that the innate wiring of myelin varies and changes in response to neuronal signals, how myelin is formed through adaptive myelination is less well-known. The concept of adaptive myelination has emerged from various observations, such as the enhancement of motor learning by newly differentiated cells (McKenzie et al. 2014; Xiao et al. 2016), the reduction of myelination in the cortex by environmental factors such as social isolation (Liu et al. 2012; Makinodan et al. 2012), the enhancement of myelination by optogenetic stimulation (Gibson et al. 2014), or the changes in axon tracts resulting from increased myelination (Sampaio-Baptista et al. 2013).

Adaptive myelination plays a pivotal role in dynamically regulating myelin remodeling in response to neuronal activity and environmental cues, thereby maintaining optimal synaptic function and neural plasticity. This process exhibits pronounced mechanosensitivity, necessitating stringent spatial organization, substrate stiffness, and topographical cues to facilitate cellular plasticity and effective myelination. Consequently, myelin regeneration remains markedly constrained following traumatic injury or within demyelinating neurodegenerative disorders such as MS. In addition to the inflammatory niche that develops around damaged myelin sheaths, alterations in the mechanical properties of the surrounding tissue are a critical yet often underappreciated factor in inhibiting remyelination.

The differentiation of OLs from OPCs to mature myelinated cells involves several stages during which the cells undergo biochemical and morphological changes. OL differentiation is regulated by soluble factors such as growth factors (e.g., neurotrophin-3, brain-derived neurotrophic factor, transforming growth factor-beta1, ciliary neurotrophic factor) (Stankoff et al. 2002; Woolley et al. 2008; Xiao 2023) and hormones (e.g., triiodothyronine (T3), insulin, estrogen, progesterone, glucocorticoids) (Emery 2010; Schumacher et al. 2012; Barateiro and Fernandes 2014; Zendedel et al. 2016; Miguel-Hidalgo et al. 2019). Additionally, insoluble factors such as ECM proteins and microenvironmental cues directly influence OL differentiation.

SC maturation is also mediated by axon-derived signals (NRG1) and paracrine/endocrine factors (BDNF, IGF-1, T3/T4, glucocorticoids). NRG1 type III–ErbB2/3 signaling is the primary determinant of myelinating fate, while the ECM and Notch/cAMP pathways regulate the process (Rosso et al. 2017b).

Numerous studies have been published on the influence of biophysical properties of the extracellular environment on myelin-producing cells' differentiation and maturation, demonstrating the importance of ECM stiffness and topography, mechanical forces, and spatial constraints. For example, it has been shown that in vitro OL differentiation and maturation are supported by substrates within the brain's stiffness range (0.1 to 10 kPa) (Zhang et al. 2011; Lourenço and Grãos 2016; Zhang and Weickenmeier 2023) and that the myelination process is influenced by the diameter of axon-like fibers (Lee et al. 2013, Bechler and Byrne 2015; Redmond et al. 2016). The initiation of SC myelination was also linked to mechanical cues from basal lamina through Laminin-211 polymerization (Petersen et al. 2015). These results have created a field of research linking mechanotransduction to the mechanobiology of myelin-producing cells and demyelinating disorders.

15.2 Myelin Structure and Composition

15.2.1 Myelin Structure

The structure of myelin was revealed by electron microscopy, which showed that myelin consists of a dense lipid layer. This layered dense band consists of tightly packed cytoplasmic surfaces and outer membranes, the intraperiod stripe (Stadelmann et al. 2019). Distinct lamellae are present at the terminal edges of each myelin segment, and these cytoplasm-containing myelin lamellae connect to the

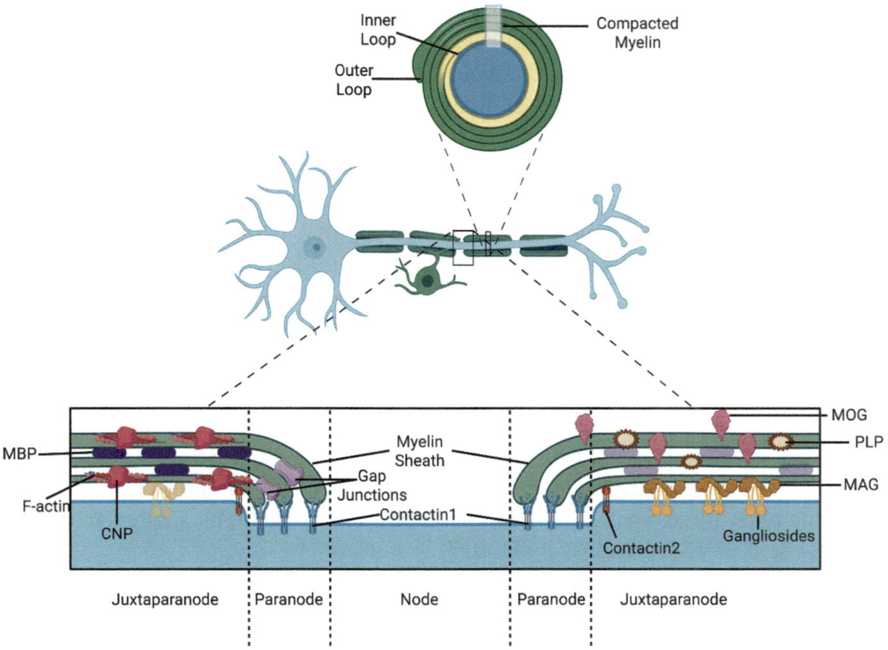

Fig. 15.1 Illustrative representation of the myelin compact structure in CNS showing the localization of MBP, CNP, PLP, MAG, and MOG. Created with BioRender

axon in terminal loops at a point called the "paranodal area." Between the paranodes of the two separate myelin sheaths lies a nodal region where axons constrict (Inouye et al. 2014) (Fig. 15.1).

Myelin biogenesis is associated with substantial membrane synthesis, and the membrane consists of 40% water, 40–50% lipids, and 10–20% protein (Weickenmeier et al. 2017).

In the CNS, OLs myelinate small-diameter axons, forming numerous short internodes and thin myelin layers, while OLs myelinating large-diameter axons have fewer but longer internodes and thick myelin sheaths (Stadelmann et al. 2019). This compact structure of the myelin membrane, formed by adjacent layers, is crucial for its insulating properties. It facilitates efficient and rapid nerve conduction but impairs molecular exchange within the myelin sheath and the axon it envelops. However, cytoplasmic vacuoles and connexin (Cx)-containing gap junctions in the myelin sheath enable the transport of organelles and small molecules across the internodes, thus greatly shortening the transit time of molecules across the myelin sheath (Stadelmann et al. 2019).

During myelination of the PNS, unlike OLs, SCs perform a "radial sorting" of axons, where a group of immature SCs deposits a basal lamina rich in laminin around them and produce lamellopodia-like extrusions to detect axon diameters. Unlike OLs that can myelinate multiple axons, SCs only wrap around single axons based on the selection through radial sorting (> 1 μm in rodents) (Feltri et al. 2016; Herbert and Monk 2017). There remains a non-myelinating population of SCs after radial sorting that form Remak bundles. Therefore, there is a 1:1 ratio between the SCs and axon numbers in the PNS. Remyelination after injury in the PNS is faster and more plausible due to the high plasticity of SCs compared to OLs (Pereira et al. 2012).

Myelin basic protein (MBP), located between the two lipid bilayers, is a multifunctional protein and essential for the myelin membranes' tight adhesion and compaction. MBP is an unstructured polypeptide chain that forms both α-helical and β-sheet structures when associated with membranes (Nawaz et al. 2013; Raasakka et al. 2017). It interacts with polyanionic proteins such as actin, tubulin, calmodulin, and negatively charged lipids (Boggs and Rangaraj 2000). The interaction of MBP with the membrane is based on electrostatic forces between the negatively charged head groups (phosphatidylserine and phosphatidylinositol 4,5-bisphosphate) in lipids and the basic structures of MBP (Boggs et al. 2012). By binding to the cytosolic membrane surface, opposing charges are neutralized, allowing other forces such as hydrogen bonding and hydrophobic factors, to emerge (Nawaz et al. 2013; Raasakka et al. 2017).

The mechanisms that ensure close interactions at the outer surfaces of myelin lamellae are less well understood. One view is that the loss of repulsive charges from the outer myelin surface creates the weak interactions that contribute to membrane tightness. Proteolipid protein (PLP) and its isoform DM-20 are the most abundant transmembrane proteins in CNS myelin and are thought to facilitate tight association of membrane sheaths through their hydrophilic extracellular domains (Ruskamo et al. 2022).

The radial component is a network of interlamellar tight junctions composed of claudin-11 that connects the outer leaflets of myelin lamellae (Boggs et al. 2008). Claudin-11 is the most abundant myelin protein after PLP1 and MBP, accounting for approximately 7% of the total myelin protein. Claudin-11 increases the conduction speed in small-diameter axons by preserving the integrity of the myelin barrier and helps ensure normal walking and motor functions (Gjervan et al. 2024). Interestingly, these Claudin-11-rich interlamellar strands are not present in PNS myelins (Martini et al. 2010).

MBP facilitates membrane zippering at the cytoplasmic surfaces of the myelin bilayer, resulting in the formation of a narrow, dense protein layer that restricts the entry of most proteins into the myelin sheath (Bakhti et al. 2014; Stadelmann et al. 2019). Mechanisms against this MBP-mediated membrane compression exist to protect cytoplasmic compartments within myelin and maintain molecular exchange. One of these is the OL-specific protein 2′,3'-Cyclic nucleotide 3′-phosphodiesterase (CNP), which facilitates the formation of intact cytoplasmic regions in adult myelin and forms a structure that can antagonize the adhesion forces exerted by polymerizing MBP molecules by directly associating with the actin cytoskeleton (Snaidero et al. 2017). CNP, in conjunction with the actin cytoskeleton, forms the support that keeps the cytoplasmic leaflets at a sufficient distance to prevent membrane compression (Snaidero and Simons 2017).

The paranodal axoglial junction provides electrical insulation by restricting current within the myelin sheath and connects the myelin sheath to the axon. To do this, the junction anchors the paranodal membrane in close proximity to the axon, creating a barrier that separates the nodes from the internodal axon. Small gaps between the paranodal rings allow diffusion between the axon and myelin (Bhat 2003). This structure consists of a cell adhesion complex containing contactin (Cnt)-related protein and Cnt-1 on the axonal side and neurofascin on the glial side (Bhat 2003; Zou et al. 2017; Stadelmann et al. 2019). These complexes are anchored to a submembranous layer composed of ankyrinG on the glial side and the scaffolding protein spectrin on the axonal side (Teliska and Rasband 2021). The subcortical axonal cytoskeleton consists of repeating structures of short actin filaments, capped at their barbed ends with α-adducin, and connected by spectrin tetramers (Hofmann et al. 2022).

In the internodal region of the myelinated axon, Cnt is found as a thin band along the border, while myelin-associated glycoprotein (MAG) is localized along the adaxonal myelin membrane and interacts with neuronal gangliosides. Several key kinases are involved in MAG-triggered signaling. MAG promotes axon growth cone directionality by increasing the tyrosine phosphorylation of focal adhesion kinase (FAK) via β1-integrin (Dashiell et al. 2002). Furthermore, MAG increases the expression of neurofilament and microtubule-associated proteins by activating Cyclin-dependent kinase 5 (Cdk5) and extracellular signal-regulated kinases (ERK) 1/2 (Dashiell et al. 2002). Cdk5 and ERK1/2 induce a mechanism that contributes to the maturation of the axonal cytoskeleton and thickening of the axon diameter (Dashiell et al. 2002; Cortés et al. 2019).

Myelin oligodendrocyte glycoprotein (MOG) is expressed later than other myelin proteins and is therefore a marker of maturation and myelin compaction. Primates, unlike mice and rats, express multiple MOG isoforms with the same extracellular Ig domain but differently spliced C-terminal internal domains (Reindl and Waters 2019). MOG and its isoforms do not interact directly with the cytoskeleton; some studies suggest MOG is a cellular receptor, adhesion molecule, or regulator of microtubule stability (Bauer et al. 2009; Reindl and Waters 2019).

One of the most fundamental structural differences between SCs and OLs is the absence of PLP, MOG, and CNP proteins in myelin (Fig. 15.2). These OL proteins are replaced by myelin protein zero (P0/MPZ) and peripheral myelin protein 22 (PMP22), which form the compact structure of peripheral myelin and ensure its stability in SCs. MBP and MAG are common to both systems (Salzer 2015). P0 is a transmembrane adhesion molecule of the immunoglobulin gene superfamily. It compacts myelin lamellae by zipping them together. PMP22 is among the compact myelin proteins and is an integral membrane protein that plays a critical role in the stability and compactness of peripheral myelin (Salzer 2015)

Another supporting component of the axon-myelin structure is the filamentous scaffolds found in the innermost layer of the myelin. These filaments are composed of various septin monomers. Septins are filament-forming proteins and are associated with the plasma membrane. A study examining their role in micron-scale curvature recognition concluded that septins provide a mechanism for cell surface

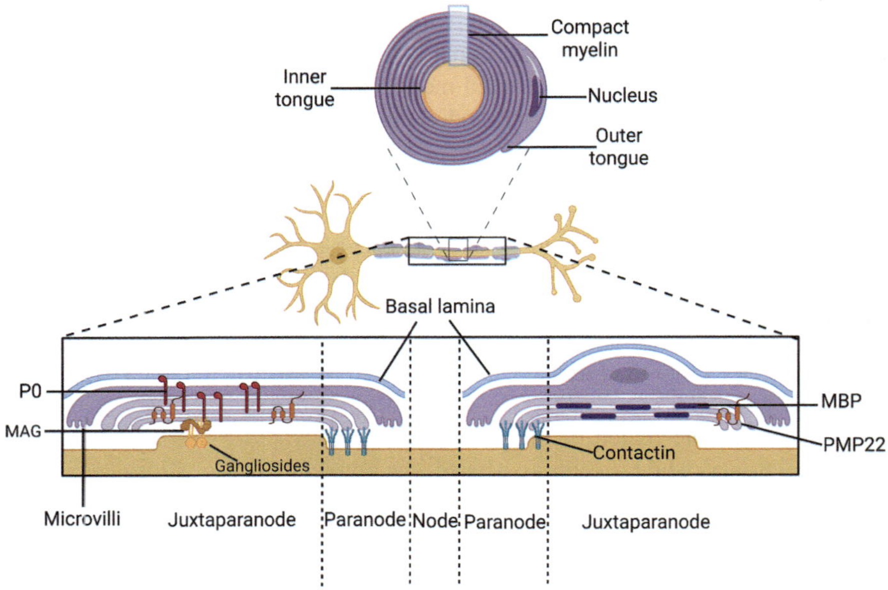

Fig. 15.2 Illustrative representation of the myelin compact structure in PNS showing the localization of MBP, PMP2, P0, and MAG. Created with BioRender

shape recognition (Bridges et al. 2016). There are 13 septin genes in humans, and alternative septin complexes composed of distinct septin proteins may also play a role in curvature sensing (Valadares et al. 2017). Septin complexes act as hinges that interact with curved membranes and thus are thought to play a role in curvature sensing through septin-lipid interactions that are geometrically enabled (Bridges et al. 2016). The formation of myelin septin filaments is a late stage of myelin maturation. These submembranous septin/aniline scaffolds are essential for normal axon/myelin structure in vivo, as their deficiency has been shown to cause specific structural deterioration in the myelin sheath, myelin outer whorls, and nerve conduction velocity (Patzig et al. 2016). Septins are of interest in this context because they can directly interact with the tubulin and actin cytoskeletons. The interactions of septins with actin and tubulin related to surface curvature changes and the possible cytoskeletal cascade that develops with it are unknown. Septin deficiency causes myelin to detach from the axon and form outward folds (Alkhanjari et al. 2025).

Lipids also play an important role in myelin formation. The intermolecular cohesive forces between lipid molecules stabilize the myelin structure. Myelin is largely composed of saturated, long-chain fatty acids. Glycosphingolipids constitute 20% of the total lipids, and cholesterol constitutes 40% (Schmitt et al. 2015). This high lipid content directly affects the membrane's physical structure, including thickness and density (Giussani et al. 2021). In a study using ceramides, another compound containing long-chain fatty acids, low levels of galactosylceramide, and sulfatide were found in mice lacking the ceramide synthase gene; however, relatively normal myelin was observed, with mild structural defects such as focal separation of lamellae (Coetzee et al. 1996). Another class of lipids found in myelin structures is ethanolamine plasmalogens, which contain a vinyl ether bond and an ester bond. These two bonds allow for both the closer alignment of the two acyl chains in plasmalogens and, by eliminating the carbonyl oxygen in the bond structure, increase hydrophilicity and lead to stronger intermolecular hydrogen bonding between the head groups (Rog and Koivuniemi 2016). Therefore, plasmalogens can increase myelin density and stability (Nunes 2024).

Due to its unique structural properties, cholesterol is one of the few lipid classes that cannot be replaced by other lipids (Ho et al. 2022). Myelination can occur despite cholesterol deficiency, but it is severely delayed (Barnes-Vélez et al. 2023). OLs can obtain the cholesterol necessary for myelin production from environmental sources, particularly astrocytes (Camargo et al. 2017). Cholesterol supply from these external sources allows OLs to initiate myelination and plays a critical role in maintaining myelin sheath integrity. Squalene synthase is a key component of cholesterol biosynthesis, and its deficiency in OLs slows myelination; however, some myelin sheaths can develop with nearly normal morphology (Mathews and Appel 2016). These findings suggest that OLs can compensate for the deficiency by taking up significant amounts of cholesterol from astrocytes under physiological conditions (Mathews and Appel 2016; Camargo et al. 2017; Berghoff et al. 2022).

Myelin structure is resistant to lipid degradation, possibly because lipids act and function collectively. Thus, after the synthesis of one lipid class is eliminated, synthesis of structurally related lipids increases and maintains their function. How myelin-producing cells detect the deficiency of one component and regulate the synthesis of another structurally related molecule remains unknown.

Myelination is regulated by both intrinsic and extrinsic cues. The idea that genetic and epigenetic processes might influence myelination has been controversial for many years in part because observations indicate that OLs possess an innate capacity for myelination. Findings that regional constraints affect OPC differentiation (Rosenberg et al. 2008) and that OLs envelop even fixed axons or bare polymer fibers of appropriate diameters (Lee et al. 2013, Bechler and Byrne 2015, Redmond et al. 2016) have contributed to the idea that myelination is independent of active neuronal signaling. In vitro studies have also highlighted the cell-specific nature of the myelin segments produced by OLs. For example, OLs derived from mouse spinal cord OPCs were observed to form longer myelin sheaths on polymer fibers compared to OLs derived from cortical OPCs (Bechler and Byrne 2015). While cortical OPC-derived OLs increased myelination on laminin-coated surfaces, spinal cord OPC-derived cells were unaffected by this ECM protein. These findings highlight the region-specific intrinsic properties of myelinating cells and contribute to the literature examining intrinsic heterogeneity within the OL lineage (Hill et al. 2014, Tomassy & Fossati 2014, Marques et al. 2016). Although OLs appear to possess inherent knowledge to initiate myelination, intrinsic cues alone do not regulate myelin microstructure. Bechler and Byrne (2015) proposed an innate myelination mechanism independent of neuronal activity and an adaptive

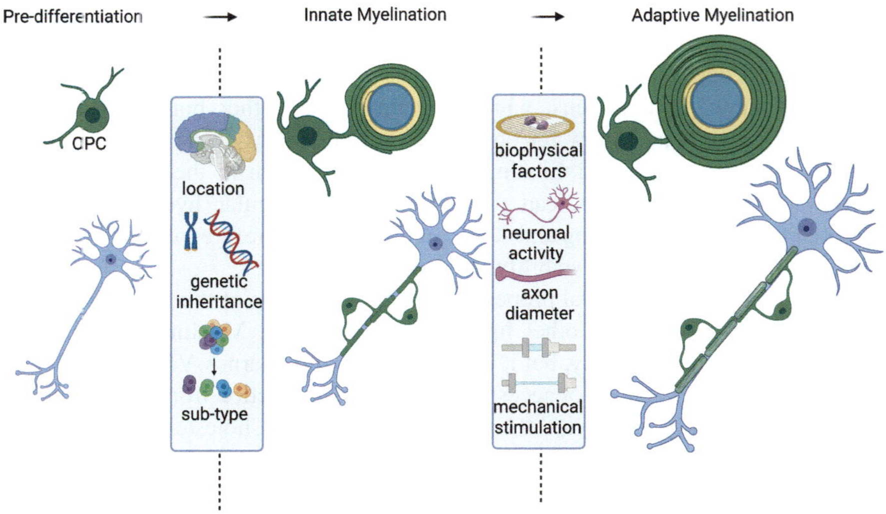

Fig. 15.3 Innate myelination mechanisms are independent of neuronal activity and adaptive myelination mechanisms are regulated by neuronal activity, biophysical factors, and mechanical stimulation. Created with BioRender

myelination mechanism regulated by neuronal activity and epigenetic factors, as discussed below (Bechler and Byrne 2015) (Fig. 15.3).

15.2.2 Innate Myelination

The intrinsic myelination properties of OLs were revealed when spatial constraints were observed to influence OPC differentiation (Rosenberg et al. 2008; Bechler and Byrne 2015; Lee et al. 2013; Redmond et al. 2016). OLs possess intrinsic signals to initiate at least the first steps of myelination (Mount and Monje 2017). Studies have also shown that the capacity for myelin formation depends on the cell's origin.

The hypothesis that intrinsic differences between OL subpopulations could lead to different myelin profiles has been tested in two ways. First, OLs from the rodent optic nerve, which normally myelinate relatively smooth, small-caliber axons, were transplanted into the spinal cord, which has much larger and more diverse axon calibers, and myelination was demonstrated (Fanarraga et al. 1998). Second, OLs from the ventral spinal cord of the zebrafish, grafted onto large-caliber axons, also demonstrated myelin formation (Almeida et al. 2011). The effect of OL phenotype was most clearly observed when OLs from mouse spinal cords developed longer myelin sheaths on polymer microfilaments compared to those isolated from the cortex (Bechler and Byrne 2015). These experiments revealed that OLs derived from rodent cortex and spinal cord maintain distinctly programmed behaviors that determine their myelin sheath lengths. Spinal cord OLs synthesize significantly longer sheaths than cortical OLs in the same physical and chemical environment (Bechler and Byrne 2015). It is unknown when OPCs become intrinsically differentiated or what the molecular differences are between them.

Cross-transplantation is one method used to study heterogeneity between white matter and gray matter OLs and has been shown that OPCs have a differential myelin-generating capacity. Heterotopic and orthotopic transplantation of OPCs from cortical gray or white matter to the same or opposite sites has shown that their proliferation and differentiation rates vary with location (Viganò et al. 2013). Heterogeneous OL populations influence CNS myelin sheath diversity. Lineage-tracing studies of OL heterogeneity have demonstrated the existence of dorsal and ventral precursor cell populations originating from distinct subventricular zones at different time points during mouse development (Tripathi et al. 2011, Rowitch and Kriegstein, 2010). Ablation studies have reported that in the deficiency of each subtype, OPCs redifferentiate into OL in different CNS regions, maintaining myelination integrity (Kessaris et al. 2006). This finding suggests that the loss of one cell population may be compensated to some extent. It is unknown whether these populations have an independent phenotype, an epigenetic program, or possess functional differences.

At the transcriptional level, single-cell RNA sequencing of OL lineage cells in CNS gray matter regions has revealed six transcriptionally distinct OL clusters during the developmental myelination period (Marques et al. 2016). The extent to which these six populations have different myelination potential is unknown.

15.2.3 Adaptive Myelination

Myelin-producing cells exhibit sensitivity to various environmental cues, including ECM stiffness, mechanical strain, physical perturbations, and macromolecular crowding. The sensitivity of myelin-producing cells to biophysical signals plays a crucial role in myelin pathology and demyelinating diseases (Yang et al. 2025). The fact that OL myelination is influenced by neuronal activity and requires physical dimensions to form myelin sheaths (Lee et al. 2012) supports a fundamental myelinating behavior in which the myelin sheath can adapt to change its length or properties (Bechler and Byrne 2015). One prediction of the adaptive myelination hypothesis is that a predetermined myelin pattern is established by a cluster of independent pathways and signals. The adaptive model is defined by the persistence of myelin changes into adulthood, with an animal's experience and neuronal activity shaping the cellular and ultrastructural properties of myelin.

The differentiation of precursor cells to mature myelinated cells involves several stages during which the cells undergo biochemical and morphological changes. In recent years, a growing number of studies have revealed signaling pathways underlying cellular responses to forces and other biophysical stimuli and their importance in cell fate decisions, particularly for OL differentiation and myelination. These interactions can be classified under the following titles:

15.2.3.1 Integrins and ECM Interactions

Integrins are heterodimeric transmembrane receptors that facilitate cellular adhesion by binding to ECM proteins such as laminin, fibronectin, or collagen. They are central to mechanotransduction because they physically link the ECM with the actin cytoskeleton. Other elements, such as glycocalyxes, tight junctions, desmosomes, and adherent junctions, also play a crucial role in mechanotransduction, particularly in sensing fluid shear stress or tissue tension (Seetharaman and Etienne-Manneville 2018). A fundamental view of mechanobiology is the development of focal adhesions, which are multiprotein clusters composed of integrins and various intracellular adaptor proteins that function as cellular anchors and sensors (Li et al. 2016; Sun et al. 2016; Seetharaman and Etienne-Manneville 2018). In the intracellular space, adaptor proteins such as talin and vinculin bind to integrins and, in turn, mobilize the actin cytoskeleton. Other proteins associated with focal adhesions, such as FAKs and Src family kinases, control cytoskeleton dynamics, cellular spreading, and contraction by regulating Rho family GTPases (Seetharaman and Etienne-Manneville 2018). This leads to stronger binding of actin filaments to FAs and enhanced cellular contraction, mediated by actin-associated motor proteins such as nonmuscle myosin-II (Vicente-Manzanares et al. 2009). This mechanism supports adherent cells in recognizing ECM-induced forces and sensing substrate stiffness.

Although interactions at the intercellular interface or between cells and the extracellular environment are at the root of mechanotransduction signaling, these stimuli are known to propagate from the cytoskeleton and the nucleoskeleton-cytoskeleton-linker (LINC) complex to the nucleus, where nuclear lamins influence nuclear and chromosomal architecture, chromatin modulation, and gene expression. This cascade of events, starting from the integrin force sensing and continuing with the collaborative work of mechanical and biochemical signaling in chromatin remodeling, is termed "mechanoepigenetics" (Missirlis 2016). Histone deacetylases, microRNAs, and other epigenetic modulators of chromatin structure are shown to be critical for myelination (Jacob et al. 2011; Chen et al. 2023; Berry and Wang 2020), but the specific function of the LINC complex in myelinating cell development remains undefined. The discovery of transcription factors, such as YAP/TAZ, that are sensitive to substrate stiffness and cell shape has provided further insight into how gene regulation occurs as feedback to biophysical cues (Swift et al. 2013).

OLs express defined integrin subtypes ($\alpha v\beta 1$, $\alpha v\beta 3$, $\alpha v\beta 5$, $\alpha v\beta 8$, and $\alpha 6\beta 1$) depending on maturation and ECM composition, acting as mechanosensitive receptors that perceive the mechanical properties of the ECM (O′ Meara et al. 2011). The expression of integrins varies with the differentiation stage. While $\alpha v\beta 8$-integrin and α6β1-integrin are highly expressed in OPCs, αvβ3-integrin is expressed at intermediate levels as maturation progresses, and αvβ5 and αvβ8-integrin are expressed at the OL maturity stage (O′ Meara et al. 2011). Different signaling pathways and cellular responses are activated depending on combination of integrins and different ECM proteins. For example, fibronectin activates αvβ3, which stimulates the Src family kinase Lyn. This cascade triggers proliferation and survival pathways in the cell and supports the maintenance of the progenitor state of OPCs (Lourenço et al. 2016). OL differentiation is promoted by the laminin-2/merosin interaction, which activates α6β1 and Syn-Fyn, initiating a cascade that induces MBP expression and cell maturation (Relucio et al. 2009).

Although both OLs and SCs were shown to require laminin-binding receptors and dystroglycans associated with integrins for myelin formation (Colognato et al. 2007), there may be differences between the integrin types that interact with these laminin-binding receptors in OLs and SCs, particularly because SCs require a laminin-rich basal lamina as a prerequisite for myelin formation. β4 integrin subunit has been identified to function in the production and stabilization of myelin in SCs (Feltri et al. 1994), and the absence of α6β4 integrin was shown to cause abnormal myelin folding (Nodari et al. 2008). Integrin β1 subunit was also associated with laminin binding and was shown to be responsible for the lamellipodia extension of SCs for radial sorting (Berti et al. 2011). Defective laminin signal transduction due to integrin β1 deficiency was associated with several demyelinating pathologies in PNS (Feltri et al. 2002; Feltri and Wrabetz 2005).

15.2.3.2 Cytoskeletal Regulation

Myelination depends heavily on the organization of the actin cytoskeleton. Rho GTPases control cytoskeletal extensions and cellular morphology by regulating actin polymerization/depolymerization dynamics. Within the Rho GTPase family, activation of RhoA induces contraction of actin filaments and actomyosin, while activation of Rac and Cdc42 induces the formation of filopodia and lamellipodia in various cell types. Fyn activation in OLs and SCs is also associated with Rho GTPases. Fyn inhibits RhoA via p190RhoGAP and promotes Cdc42 and Rac, inducing morphological differentiation. During maturation, both OLs and SCs undergo cytoskeletal rearrangements and morphological changes toward branched cells and ultimately myelin-rich membranous structures upon full maturation. Furthermore, activation of the Fyn pathway has been shown to increase the myelin sheaths in laminin-2-coated fibers (Bechler and Byrne 2015).

Interestingly, mature OLs are composed of two main cytoskeletal components: filamentous (F)-actin and microtubules (Wilson and Brophy 1989). F-actin promotes migration by playing a role in the formation of filopodia and lamellipodia in immature OLs. OL maturation occurs with morphological changes following increased microtubule stabilization. Increased morphological complexity is associated with the inactivation of RhoA and a resulting decrease in actinomyosin contractility by the RhoA GTPase-activated ROCK. Simultaneously, inhibition of myosin-II induced OL maturation, increased MBP expression, and cellular branching, and even enhances remyelination after lysolecithin-induced demyelination (Rusielewicz et al. 2014; Domingues et al. 2020). While actin dynamics is known to play a crucial role in myelin formation and axon myelin wrapping, the process remains poorly understood. Actin depolymerizing drugs have been shown to promote the extension of the developing myelin sheath in vitro and in vivo. These actin depolymerizing agents have also been linked to increasing actin dynamics by promoting repeated polymerization/depolymerization cycles at the leading edge (inner tongue) that guide cell extension (Brown and Macklin 2020).

It has been shown that F-actin-rich lamellipodia-like extensions form during the initial phase of myelination but later cease during axon wrapping and are completely disintegrated during active myelination. F-actin levels are inversely proportional to MBP expression and membrane tension; therefore, actin cytoskeletal disruption caused by the change of F-actin to G-actin appears to be crucial for myelin wrapping (Nawaz et al. 2015; Brown and Macklin 2020).

Actin is dynamically remodeled by members of the ADF/cofilin family of polymerizing factors, which regulate the Actin-related protein (Arp) 2/3 complex, and depolymerizing factors, which disassemble actin and release actin monomers for reassembly (Bauer et al. 2009). Arp2/3 and ADF/cofilin1 also play important roles during myelination. A high F-actin/G-actin ratio is known to support myelination. Arp2/3 modulates this ratio and is a key factor supporting lamellipodia, which are essential for initiation of myelination. ADF/cofilin1, on the other hand, plays a role in F-actin turnover and, by acting inversely, contributes to the decreased F-actin/G-actin ratio observed in differentiated OLs (Nawaz et al. 2015). Cofilin activity can be regulated by plasma membrane binding. In mature OLs, MBP competes with cofilin for binding to phosphatidylinositol 4,5-bisphosphate (PI (4,5) P2) on the plasma membrane, releasing cofilin and inducing the disassembly of F-actin. This contributes to membrane compaction during axon myelination (Nawaz et al. 2009).

The role of actin assembly/disassembly was also shown for SCs. Inhibition of actin polymerization was shown to delay SC myelination (Fernandez-Valle et al. 1997; Jin et al. 2011). Rho family GTPases Rac1 and Cdc42, which actively regulate actin dynamics, were also implicated in myelin formation and stabilization in SCs (Benninger et al. 2007; Krause et al. 2008).

Recently, a cytoskeletal protein group (Talin1 and Talin2) that connects integrins with actin at focal adhesions has attracted attention with its possible role in disease mechanotransduction (Bate et al. 2012; Wang et al. 2024). Talins are present in all cell types and are actively involved in cell attachment to the ECM. Talin has been shown to trigger integrin activation, focal adhesion assembly, and the assembly of other adhesome components through various interactions, including Ca^{2+} signaling and the Wnt/β-catenin pathway (Klapholz and Brown 2017). In particular, the association of Talin1 with the integrin subunit β1 (Nieves et al. 2010; Zhan et al. 2018; Lu et al. 2022) suggests a possible role in myelination.

15.2.3.3 Piezo-Type Mechanosensitive Ion Channel Regulation

Piezo channels are mechanosensitive ion channels embedded in the cell membrane that serve as principal cellular mechanotransducers, converting mechanical stimuli into electrochemical signals. When the lipid bilayer is stretched or deformed, membrane tension increases inducing conformational changes in the Piezo channels. These changes open the central pore, allowing cation flux (primarily Ca^{2+} and Na^{+}) into the cell. Expression of Piezo1 was identified in rodent OPCs in a 2019 study (Segel et al. 2019). Besides its central role in neural-glial communication, Piezo1 is a key mediator through which glial cells sense matrix stiffness. While OPCs arrest differentiation and proliferation on stiffer substrates, inhibition of Piezo1 enables OPCs to differentiate and proliferate in stiff microenvironments (Segel et al. 2019).

Piezo1 and Piezo2 are also highly expressed mechanosensitive ion channels in the PNS. While Piezo1 is associated with inhibition of myelination, Piezo2 was related to myelin formation in SCs (Acheta et al. 2022).

In addition, pharmaceutical activation of Piezo1 channels has been shown to induce demyelination whereas their inhibition has been shown to prevent demyelination in rodents. This suggests that Piezo1 inhibition may exert a protective effect by inhibiting Ca^{2+} signaling, which in turn prevent myelin from activating lipid degradation (Velasco-Estevez et al. 2020).

Similarly, it was shown that graphene-coated nerve guidance conduits trigger SCs myelination through Piezo1 activation concurrent with NFAT and Krox-20 (Zhan et al. 20252025).

15.2.3.4 Neuronal Activity and Soluble Factors

OL and SC differentiation is regulated by many growth factors and hormones. Proteomic studies of myelin have contributed to unraveling the mechanisms associated with it. In addition to cytoskeletal and structural proteins, synaptic proteins (synaptophysin, synaptosomal-associated protein, syntaxin, etc.) and electrogenic channels (voltage-dependent channels, ion-dependent channels (sodium, potassium, calcium) have been identified (Gopalakrishnan et al. 2013). Studies suggest the existence of myelin signaling proteins that may respond to changes in axon activity required for adaptive myelination. OL N-methyl-D-aspartate (NMDA) receptors have been suggested to detect vesicular glutamate release into active axons (Micu et al. 2016). Inhibition of this vesicular release in vitro in OLs or zebrafish led to decreased MBP mRNA translation or sheath formation, respectively (Hines et al. 2015; Mensch 2015). NMDA has also been suggested to play a role in myelin compaction and in calibrating the metabolic support that OL provides to axons (Frühbeis et al. 2013; Saab et al. 2016).

In addition to studies focusing on signals previously thought to be essential for myelination, some studies have identified several other factors, such as brain-derived neurotrophic factor (BDNF) and gamma-aminobutyric acid (GABA), that may promote myelin adaptation. BDNF secreted from cortical neurons following optogenetic stimulation induced myelination of OLs through adaptive myelin-associated NMDA receptors (Lundgaard et al. 2013). Global BDNF knockout mice developed both axonal diameter regression and hypomyelination (Djalali et al. 2005), while loss of the OL-specific BDNF receptor TrkB resulted in thinner myelin production consistent with its role in adaptive myelination (Wong 2013).

15.2.3.5 ECM Mechanical Properties

Numerous studies have been published on the importance of mechanical properties of the extracellular environment in myelination, demonstrating the importance of ECM stiffness and topography, tensile forces, and spatial constraints. For example, it has been shown that in vitro OL differentiation and maturation are supported by substrates within the brain's stiffness range (Jagielska et al. 2012) and that the myelination process is altered by the diameter of axon-like fibers (Jagielska et al. 2012). These results have created a field of research linking mechanotransduction to OL biology and demyelinating disorders. Further discussion on the effects of ECM mechanical properties is provided in the section Mechano-Effect on Myelinating Cell Development and Differentiation.

15.2.3.6 Geometric and Topographic Properties

If the initial development of myelin sheaths is an intrinsic process dependent solely on the availability of a suitable substrate, then myelination of all axons and dendrites in the CNS would be expected, while there exist unmyelinated axons, and the majority of dendrites are unmyelinated. Experiments in which the PI3 kinase signaling inhibitor PTEN was blocked in neurons provided further evidence of a threshold mechanism regulating in vivo. This mechanism, dependent on size, regulates axon diameter, causing previously unmyelinated axons in the cerebellum to increase in size and also become myelinated (Goebbels et al. 2017). One of the first studies demonstrating the influence of mechanical cues during OPC differentiation used topographic features that mimic axon topography and regulate OPC alignment and migration (Webb et al. 1995). It was reported that OLs form parallel extensions on narrow, regular topographic surfaces, similar to the axon arrangement in the intact optic nerve and that these cells do not exhibit high levels of F-actin filaments or prominent FAK; thus, alignment occurs through topographic guidance and not mechanical deformation (Webb et al. 1995).

The thickness of myelin sheaths (number of lamellae) can be calculated by the ratio of the axon diameter to the outer diameter of the myelin sheath, defined as the "g ratio" for axon diameters, and is averaged with a value of 0,7 (Gow 2025). Early studies also showed that internode length increases with axon diameter (Friede and Bischhausen 1982; Butt et al. 1998). Experiments performed on polymer fibers suggest that the physical diameter of the fiber itself may serve as a signal determining myelin sheath length without any requirement for the axon-related molecules. Rodent OPCs varied their myelin sheath length according to microfibers of different diameters in the absence of axonal molecular signals (Bechler and Byrne 2015). Although it remains to be determined whether the number of myelin lamellae is regulated solely by fiber size, this study suggests that OLs possess size-sensitive mechanisms as demonstrated in microfiber culture studies. However, while many studies have shown a linear correlation between axon diameter and internode length or thickness (Villalón et al. 2018), as would be predicted by a simple curvature-sensing model, exceptions to this rule exist. The smallest and largest axons (diameters <1 μm and >8 μm) show nonlinear changes in sheath length or thickness. Samples that vary in both length and thickness suggest that factors beyond physical cues further guide or fine-tune sheath dimensions (Bechler et al. 2018).

How OLs precisely detect radial axon size is not fully understood, but some suggest that they use features such as curvature (Osso and Chan 2017). The effect of surface curvature on OL cell membrane curvature may directly modulate the cell's ability to wrap axons with high curvature or subthreshold diameters. One hypothesis for OLs' perception of curvature is that they sense the axon's size by sensing the length of one wrap of their membrane around an axon. After completing a full wrap, the cell membrane will contact itself. If the membrane length falls below the threshold, it will retract, preventing wrapping. Once this threshold is exceeded, the length-sensing mechanism initiates rapid suppression of the self-propulsion mechanism during the wrapping process, allowing the suprathreshold peripheral axons to myelinate (Osso and Chan 2017; Domingues et al. 2018). Retraction of an extended OL membrane upon contact with another extension of the same OL has been observed in the adult mouse cortex (Osso and Chan 2017). Similar mechanisms could also be used to relate internode length to radial axon size.

In a recent study on SCs, it was shown that anisotropic topography created by incorporating single-layer graphene (SLG) and nanodiamond (ND) into nanogrooved polycaprolactone (PCL) fibers enhances SC myelination capacity, while quenching fibroblast collagen production (Zhan et al. 2025).

15.3 Mechano-effect on Myelinating Cell Development and Differentiation

Cells have developed sensors for biophysical cues from the extracellular microenvironment, such as from shear stress, strain, and other mechanical signals. Extracellular mechanical stimuli, along with substrate stiffness, geometric structures that determine cell shape, and topographical elements of the extracellular environment, can be converted into biochemical signals (Table 15.1). The process of transducing these biophysical factors into biochemical signals is defined as "mechanotransduction" (Martino et al. 2018; Jaalouk and Lammerding 2009). Cells change shape in response to biophysical and biochemical conditions, creating topologies

Table 15.1 Effects of mechanical environment, cytoskeletal status, and neurotransmitter/neurotrophic factors on OPC/OL and SC differentiation

Factor/stimulus	Mechanism	Effect on OLs	References
ECM stiffness and substrate properties	Soft substrates (1–10 kPa)	Promote OL and SC differentiation	Moore et al. (2010), Chatelin et al. (2010), Eyckmans et al. (2011), Urbanski et al. (2016), and Montgomery et al. (2025)
	Stiff substrates (10 kPa and above)	Reduce OPC and SC differentiation	
Integrins and ECM interactions	Fibronectin/αvβ3 interaction	Maintenance of OPC progenitor status	Relucio et al. (2009), O′ Meara et al. (2011), and Lourenço et al. (2016)
	Laminin-2/α6β1 cascade triggers Syn-Fyn	Induces MBP expression, promoting OL differentiation	
Cytoskeletal regulation	Stiff substrate > RhoA activation > inhibition of actin polymerization	Decreased OL and SC differentiation	Benninger et al. (2007), Krause et al. (2008), Dupont et al. (2011), Rusielewicz et al. (2014), Bechler and Byrne (2015), Jin et al. (2011), Domingues et al. (2020), and Brown and Macklin (2020)
	Soft substrate>RhoA inhibition> Fyn activation	OL and SC differentiation induced	
	Soft substrate >NMM-II inhibition > reduced actomyosin contractility	Increased MBP expression and cellular branching	
	Stiff substrate > increased F-actin/G-actin ratio	Decreased OL and SC differentiation	
YAP/TAZ and nuclear mechanotransduction	Stiff substrate>increased F-actin polymerization> decreased YAP/TAZ phosphorylation> YAP/TAZ translocation in nucleus	Inhibition of OL myelin synthesis and regeneration	Swift et al. (2013) and Shimizu et al. (2019)
	Stiff substrate> increased YAP expression> increased c-Jun	Increased SC remyelination	Poitelon et al. (2016) and Lopez-Anido et al. (2016)
	YAP/TAZ inhibition in mouse OLs	Normal OL maturation and myelination, decreased re-myelination after cuprizone-induced demyelination	Hong et al. (2024)

Piezo1 regulation	Stiff substrate>Piezo1 activation	Inhibits OPC and SC differentiation and proliferation	Segel et al. (2019), Velasco-Estevez et al. (2020), and Velasco-Estevez et al. (2022)
	Stiff substrate>Piezo1 inhibition (GsMTx4)	OPCs differentiation and proliferation on stiff substrate and induces myelin regeneration	
	Soft substrate > decreased Piezo1, increased Piezo2 > YAP/TAZ translocation in cytoplasm	Increased SC myelination	Acheta et al. (2022)
Neuronal activity and soluble factors	Soft substrate >NMDA-mediated controlled Ca^{2+} signaling	Induced OL differentiation	Frühbeis et al. (2013) and Saab et al. (2016)
	Soft substrate>inactivation of YAP/TAZ > BDNF signaling> controlled Ca^{2+} signaling	Induced OL and SC differentiation	Wong (2013), Lundgaard et al. (2013), and Rosso et al. (2017b)
Geometric and topographic properties	>0.4 μm allow	Myelin sheath formation	Lee et al. (2012) and Bechler and Byrne (2015)
	>1 μm	SC myelin sheath formation	Taveggia et al. (2005) and Stassart et al. (2018)
	Radial axon size or curvature	Regulation of myelin formation	Osso and Chan (2017) and Domingues et al. (2018)
Mechanical tension	Static, uniaxial stretching>nuclear/epigenetic regulation	OPC cell proliferation and differentiation	Kim et al. (2020)

with microscale variation as described in the adaptive myelination section. While individual proteins can sense membrane curvature at the nanometer scale (Bridges et al. 2016), it is unknown whether a cell can use nanometer-scale components to sense microscale curvatures.

Cells themselves generate force in their cytoplasm and exert it on their surroundings, e.g., ECM via integrins or with other cells, similar or different ones, immune cells, for example, through tight junctions and gap junctions, or via transient cell adhesion molecules. Collective cell behavior is distinctly different from individual cell responses to mechanical stimuli (Bayir et al. 2019), which is particularly important in PNS myelin formation through collaborative efforts of SCs in the production of basal lamina, which heavily contributes to SC plasticity, active axonal growth, and remyelination after injury. Basal lamina is composed of laminin and collagen cross-linked with proteoglycans and nidogens, which gives extra resilience against external mechanical stresses (Bunge et al. 1980).

Studies to date have shown that changes in brain tissue mechanical properties directly impact developmental processes. Reported shear modulus and Young's modulus in the healthy CNS range from 0.1 to 10 kPa (Zhang et al. 2011). It is important to note that the hardness of brain tissue compared to other organs and tissues is low, which means that the experimental assessments are affected by all measurement conditions, including the applied stress, shear intensity, duration, and sample form (in vivo, in vivo, in vitro), limiting the reproducibility of measurements.

Weickenmeier et al. demonstrated that brain stiffness changes during development, with both gray and white matter tissues stiffening significantly after maturation. Gray matter stiffness (measured by indentation testing on brain slices) doubled from 0.31 ± 0.20 kPa prenatally to 0.68 ± 0.20 kPa postnatally, while white matter stiffness tripled from 0.45 ± 0.18 kPa prenatally to 1.33 ± 0.64 kPa postnatally. White matter myelin content also increased from $58 \pm 2\%$ to $74 \pm 9\%$. This study demonstrated that myelin is important for protecting neurons against physical forces and for providing a strong microstructural network that stiffens the white matter tissue as a whole (Weickenmeier et al. 2017).

The stiffness of CNS tissues and the mechanical stresses and forces to which it is exposed change temporally throughout development. The spinal cord is exposed to as little as 10% mechanical strain, while the forces involved in development can reach up to 1000 nN. Similarly, CNS tissues are exposed to mechanical stresses such as tension, compression, and shear, both during development and following external mechanical trauma. In cases such as tissue swelling, strains can reach up to 100% (Makhija et al. 2018).

Numerous studies have investigated the effects of substrate stiffness on OL differentiation and OPC maturation and examined the molecular pathways involved in regulatory mechanisms (Lourenço et al. 2016). OPCs have been reported to be mechanosensitive, and their viability, proliferation, migration, and differentiation are affected by substrate stiffness (Jagielska et al. 2012). Coating compatible substrates with ECM molecules for OL differentiation will directly influence the process, as it also affects surface stiffness (Lourenço et al. 2016). OPCs cultured on soft substrates have been shown to have reduced expression of one of the mechanosensitive transcriptional regulators (YAP) compared to those on stiff substrates (Lourenço et al. 2016).

It is consistent with the observation that low RhoA activity and low actinomyosin contractility play a positive role in this process due to the influence of substrate stiffness during OL differentiation (Wang et al. 2008). Apparently, soft substrates enable OPC differentiation by reducing actinomyosin contractility, supporting the cell's low contractility, as observed in several cell types (Moore et al. 2010; Eyckmans et al. 2011). OPCs have been shown to exhibit increased cell surface area on substrates with a Young's modulus (~6 kPa) close to mammalian brain stiffness (Chatelin et al. 2010) compared to softer substrates. This effect has been associated with the most favorable actinomyosin contraction (Kippert et al. 2009).

Progenitor status of rat OPCs has been shown to be preserved on substrates with a stiffness similar to rat brain tissue (~6,5 kPa) (Juge et al. 2016) functionalized with fibronectin (Colognato and Tzvetanova 2011). OPCs showed enhanced differentiation when cultured on substrates of the same stiffness but coated with laminin-2/merosin compared to substrates coated with poly-D-lysine. Subsequently, OPC differentiation was reduced in cells cultured on softer (2,5 kPa) or stiffer (10 kPa–10 GPa) substrates (Lourenço and Grãos 2016). In this study, increased MBP and PLP expression and mature morphology of the cells demonstrated the importance of incorporating compatible substrates with ECM proteins for the full maturation of OLs. A similar approach demonstrated a significant increase in morphological complexity in OPCs undergoing differentiation on soft matrices (1,5 kPa) compared to cells cultured on stiff substrates (30 kPa) in an NMM-II-dependent manner (Urbanski et al. 2016). Differentiation of OPCs on soft substrates resulted in a higher percentage of mature OLs and lower Olig1 expression compared with those on hard substrates (RIP is an immunocytochemical marker for myelinating OL) (Dupont et al. 2011).

While studies have shown that substrate stiffness influences OL differentiation, the effects of mechanical stiffness on the myelination phase of the OL were not demonstrated before Ong et al. (2020). It was shown that a stiff mechanical microenvironment inhibits OL myelination in a manner that is independent of and distinct from OL differentiation and YAP has been identified as influencing OL myelination through mechanotransduction (Ong et al. 2020).

Similarly, SC myelination was linked to translocation of YAP and TAZ molecules to the nucleus via Hippo pathway in response to mechanical stimulation during development (Poitelon et al. 2016; Lopez-Anido et al. 2016) and the epithelial polarity factor Crb3 that is highly regulated by mechano-sensitive mechanisms during nerve stretching (Tricaud 2018). A recent work has shown the effects of dynamic changes of matrix stiffness on SC myelination using a UV-tunable polydimethylsiloxane (PDMS) substrate. By dynamically changing stiffness, they simulated the mechanical microenvironmental changes in the matrix during PNS injury. They compared a soft (4 kPa) and a hard (22 kPa) substrate, to a substrate whose stiffness gradually increased from 4 kPa to 22 kPa through a 7-day period.

SCs showed increased YAP expression and fluctuations in c-Jun production, as the stiffness increased in the dynamic environment, suggesting that prolonged exposure to high stiffness—as experienced in injury—hinders remyelination after injury (Montgomery et al. 2025). Earlier it was shown that stiffer matrices induced higher expression of transcription factors such as Sox2, Oct6, and Krox20 and subsequently reduced c-Jun expression (Rosso et al. 2022). c-jun is considered an important regulator of SC remyelination, although the literature reports conflicting results regarding its effects on gene expression in various pathological conditions (Figlia 2018).

In vitro tension-compression and shear experiments are generally conducted with laboratory-developed reactors or microfluidic chips. The strain is typically applied to the surfaces to which the OLs adhere. Researchers have examined changes in cell proliferation and differentiation by applying physiological amounts of uniaxial tensile strain to OPCs on PDMS surfaces for varying durations. Sustained tensile strain of 10–15% suppressed OPC proliferation while promoting their differentiation into OLs. This strain-induced response depended on specific interactions between OPCs and extracellular matrix ligands. The applied strain also altered nuclear morphology and chromatin organization, leading to mechanoepigenetic changes, including increased histone deacetylation—changes consistent with enhanced OL differentiation (Jagielska et al. 2017). Similarly, 10% strain significantly accelerated differentiation, accompanied by increased production of the intracellular microtubule network (Makhija et al. 2018). Another study using acute compression on OPCs reported that the cells exhibited changes at the nuclear and epigenetic levels leading to OPC differentiation and myelin formation and identified KASH domain- containing protein SYNE1 as a key signaling molecule for mechanotransduction in OLs (Hernandez et al. 2016). Researchers investigating the role of YAP showed that under both tension and shear stress, OPCs induced the phosphorylation of p130Cas and talin-associated assembly of vinculin, suggesting a key role for the mechanoreceptor p130Cas in OPC morphogenesis and maturation (Shimizu et al. 2019).

In a study that applied non-physiological stretching to OLs on a PDMS membrane to model traumatic brain injury, this mechanical effect led to intracellular Ca^{2+} accumulation in OLs, activating the Erk1/2 pathway, resulting in myelin loss in mature OLs. While stretch-induced MBP loss was reported to be transient and remyelination was observed, targeted inhibition of the Ca^{2+}-Erk1/2 signaling axis has been proposed as a potential therapeutic strategy to preserve myelin integrity or increase myelin stability after injury (Kim et al. 2020).

The geometric properties and 3D structure of the CNS also modulate cell differentiation and cellular responses. Studies have even demonstrated the necessity of spatial constraints. It has been reported that OPCs require the presence of neighboring cells to differentiate (Rosenberg et al. 2008). A similar approach has shown that increased OPC differentiation with high cell density is paralleled by an increase in the number of chromocenters in the cell nuclei. This result demonstrates that spatial density is transmitted to the nucleus through a mechanoepigenetic effect via chromatin changes (Hernandez et al. 2016).

One of the most critical topographic features for OLs is the structure of axons they are engulfing. Studies investigating the need for axonal signals for the initiation of OL myelination have shown that the myelination is not dependent on axonal signals and that OLs can, in fact, form myelin on surfaces such as paraformaldehyde-fixed axons (Rosenberg et al. 2008). Nanofibers have also been used to guide cells and provide surface cues that mimic axons. Various materials, such as polymers, carbon, and glass, that mimic the fundamental physical properties of neuronal axons have been used to separate molecular cues from biophysical properties and to screen for their effects on myelin formation. Bullock and Rome (1990) and Howe (2006) were pioneers in these studies and were first to use randomly oriented glass microfibers as axon models. Howe has shown that the degree of wrapping by OLs varies with fiber coating (Howe 2006). Because axonal diameter regulates myelination (thicker axons indicate increased myelination), nanofibers with diameters within the axonal range have been created, and nanofibers with diameters of 2–4 μm have been shown to enhance the differentiation of OPCs and the myelin sheath compared with fibers with smaller diameters (<0.5 μm) (Lee et al. 2012; Bechler and Byrne 2015). Using a microfiber assay, it was shown that only fibers with diameters greater than 0.4 μm allow sheath formation (Lee et al. 2012). Graphene-nanofiber scaffolds have also been shown to significantly enhance the differentiation of neural stem cells (NSCs) into OLs (Shah et al. 2014).

The axon diameter threshold required for SCs to initiate the myelin sheath formation process has been reported to be >1 μm. This indicates that fibers with axon diameters below 1 μm are generally located within Remak bundles and are unsuitable for myelination (Taveggia et al. 2005; Stassart et al. 2018).

Electrospun biocompatible polymers poly (ε-caprolactone) (PCL), poly-L-lactic acid (PLLA), and poly (lactic-co-glycolic) acid (PLGA) are widely used in regenerative studies. Researchers have developed a water-soluble core-shell fiber scaffold using coaxial electrospinning, providing a potential platform for OL myelination. The high-molecular-weight polymer poly-L-lactic acid (PLLA) was used as the core, while a natural polymer material such as HA, sodium alginate (SA), or chitosan (CS) was used as the shell. Myelin formation in the modified fibers with softer surfaces was found to be more successful than on bare PLLA (Liu et al. 2021).

15.4 Demyelination and Pathologic Mechano-environment

Demyelinating diseases are idiopathic disorders characterized by myelin loss (Cañellas et al. 2007; Hu and Lucchinetti 2009). Myelin loss is a pathological condition with multiple potential causes and results from the deterioration of myelin structures, leading to severe impairment of nerve impulse transmission in the CNS. Multiple sclerosis (MS), Marburg disease, neuromyelitis optica (NMO), acute disseminated encephalomyelitis (ADEM), Charcot-Marie-Tooth disease, and acute hemorrhagic leukoencephalitis (AHL) are examples of CNS demyelinating diseases (Coutinho Costa et al. 2023; Ruskamo et al. 2017).

Cells respond nonlinearly to mechanical signals like environmental stiffness, tension, or pressure, such as stiffening or loosening of the cytoskeleton and altering adhesion mechanisms. This creates larger-than-expected changes in the viscoelastic properties and the mechanical microenvironment of CNS or PNS tissues, directly contributing to the pathogenesis of diseases. Pathological mechanical changes are followed by biochemical changes such as myelin loss, accumulation of ECM components, and gliosis.

Besides inflammatory foci with demyelination and reactive gliosis, acute demyelination may occur as myelin breaks down from its lamellar structure and transforms into a structure composed of small vesicles. The immune system produces antibodies against the myelin sheath, such as anti-MBP and anti-MOG, which combine with the complement system and are phagocytosed by macrophages (Frohman et al. 2006; Bando 2020).

All typical histopathological findings in MS contribute to tissue integrity loss. Studies using magnetic resonance elastography (MRE) have shown that brain viscoelasticity values are reduced in MS patients. Using the MRE technique, Wuerfel et al. demonstrated viscoelastic changes in 45 MS patients and linked changes in astroelasticity to reactive astrogliosis (Wuerfel et al. 2010). Reactive astroglial proliferation is a fundamental feature of neuroinflammatory diseases. Previous studies have demonstrated that glial cells are biomechanically less rigid than neurons, a finding consistent with in vivo observations (Lu et al. 2006).

A study examining the elastic properties of the corpus callosum in a cuprizone-induced mouse demyelination model demonstrated an initial increase followed by a decrease in stiffness and a steady increase in ECM deposition (Zhang and Weickenmeier 2023). Human MRE studies comparing the brains of healthy volunteers and MS patients also demonstrated a decrease in parenchymal stiffness (Wuerfel et al. 2010). These findings of decreased mechanical stiffness are contradictory, given the increased deposition of ECM components observed in MS lesions and animal models of demyelination. Increased ECM deposition is generally associated with increased tissue stiffness, as seen in injury-induced tissue fibrosis and tumor progression (Handorf et al. 2015; Deng et al. 2022).

Acute demyelinating lesions exhibited lower stiffness than healthy tissue in in vivo models, with stiffness values returning to nearly normal levels throughout the remyelination process. Urbanski's 2019 study comparing acute and chronic demyelinating lesions demonstrated that tissue stiffness could be relatively restored in both cases (Urbanski et al. 2019). Because increased matrix stiffness has been shown to inhibit OL differentiation, while softer matrices promote it, these changes may influence the potential for remyelination between acute soft ECM and chronic hard ECM demyelinating lesions (Lourenço et al. 2016). An ex vivo study showed that mouse brain tissue stiffness is affected by age and the degree of myelination (Eberle et al. 2020). Acute cuprizone-induced demyelination has been reported to cause a decrease in tissue stiffness, particularly in the white matter, where myelin loss is most pronounced. While there was no difference in tissue stiffness in the shiverer mouse model, researchers who encountered such a difference in the acute model have proposed several hypotheses regarding the possible causes of this difference (Eberle et al. 2020). The first of these is the change in stiffness due to the loss of axon connections and tissue structure disruption that develops with OL death in the acute phase. The second is the change in stiffness due to an active glial response and glial scar formation with increased astrocytes and microglia in the cuprizone model. Finally, the change in stiffness is affected by changes in ECM components. For example, hyaluronan or fibronectin accumulated in the demyelinated area may have contributed to softening due to its viscoelastic properties (Eberle et al. 2020). Another group conducting similar research linked changes in stiffness to decreases in BDNF and phosphorylated cAMP response element-binding protein (pCREB) levels after cuprizone treatment (Hahn et al. 2022). These changes were reported to compromise neuronal survival, synaptic stability, and axonal structural integrity, thereby decreasing tissue mechanical stiffness.

One study examined the expression of the Piezo1 mechanoreceptor in human OLs and in brain tissues of MS patients and healthy individuals and investigated its potential effects on OL proliferation, maturation, and migration. The study determined that Piezo1 is expressed in human OLs and OPCs and that inhibiting Piezo1 with GsMTx4 increased OL proliferation and migration. In contrast, activation of Piezo1 with Yoda-1 produced opposing effects, and Piezo1 expression decreased with in vitro maturation of OLs (Velasco-Estevez et al. 2022). It was also determined that Piezo1 expression differed between healthy and MS brains, lower in the white matter of MS brains compared to the white matter of healthy controls OLs (Velasco-Estevez et al. 2022). Xie et al.'s (2024) study also reported that Piezo1 plays a role in microglia activation and demyelination of OLs in animal models of sepsis-associated encephalopathy, leading to cognitive impairment. Consequently, this suggests that the Piezo1-mediated signaling pathway may be used to regulate OL functions and may play an important role in MS pathophysiology (Yang et al. 2022).

Piezo1 not only plays a direct role in myelin formation and destruction but also appears to regulate the immune response associated with MS. In mice with an EAE model, Piezo1-deficient T cells have been shown to have a milder course of the disease. This results in increased TGFβ signaling and an increase in the overall number of regulatory T cells (Tregs) (Jairaman et al. 2021).

These studies demonstrate that Piezo1 is a critical regulator of OL and SC biology and can have both positive and negative effects in different contexts. While decreased Piezo1 expression in MS brains suggests that this channel may support OL functions under certain conditions, its activation has also been reported to increase inflammation, leading to demyelination and cognitive impairment. Therefore, targeting Piezo1 should be carefully considered contextually; while its inhibition may be beneficial for promoting OL proliferation and migration, further research is needed to better understand its role in demyelination pathophysiology.

It is known that after traumatic brain injury (TBI), axons and myelin sheaths swell due to the pressure within the resulting scar (Maxwell 2013). In severe TBI models, mechanical strain causes damage to the mitochondria of cortical neurons, accumulation of reactive oxygen species, release of inflammatory cytokines and pro-apoptosis factors, and cell death (Ji et al. 2012).

15.5 Potential Mechanoepigenetic Therapeutic Approaches for Remyelination

When myelin structures fail to repair after injury, axons and ultimately neurons progressively degenerate, leading to the deterioration of CNS and PNS functions. Contrary to the accepted view that the CNS has little regenerative capacity, a partial regenerative response occurs following injury to the myelin-forming cells of the CNS, namely, OLs (Unal et al. 2019). This process is called "remyelination." Although remyelination is more commonly achieved in PNS, for many patients, CNS remyelination ultimately fails, and the metabolic support normally provided by myelin sheaths to axons is lost, leading to progressive axonal and neural degeneration. The study of remyelination is a vital area of research in its own right. Remyelination is a neuroprotective process and an important treatment strategy for diseases characterized by myelin loss, which impose a significant financial burden on society.

In an interesting study, induced pluripotent stem cells (IPSCs) derived from skins of MS patients were differentiated into OPCs. These OPCs were able to mature into healthy OLs, produce myelin and perform normal interactions with axons and other glia in an immune-deficient mouse model (Mozafari et al. 2020). This study suggests that the remyelination failure in MS is not due to an intrinsic OL pathology but rather related to more complex mechanisms attributable to extrinsic factors, such as pathological microenvironment. The hallmarks of the pathological microenvironment around demyelinating axons include modified ECM by reactive astrocytes oversecreting ECM molecules like MMPs, laminin, fibronectin, and tenascin-C, a process that is usually called "glial scar formation" (Herthum et al. 2022). The inhibitory microenvironment created by this altered ECM leads to progressive neuronal degeneration. Besides molecular alterations, mechanical properties of this glial scar possibly play a crucial role against remyelination. These observations lead to an optimistic point of view, suggesting that epigenetic approaches targeting regulation of the mechanical properties of the pathological microenvironment, or adjusting the reactions of the myelin-producing cells against these mechanical cues, hold great promise as therapeutic options in demyelinating diseases.

Traditionally, in vitro cell culture relies on the use of polystyrene, which is a polymer that cannot simulate the in vivo environment, being six times stiffer than brain tissue. Hydrogels developed to induce remyelination of OLs in in vitro demyelinating disease models hold great potential for regenerative studies. To accommodate the very low stiffness of CNS tissue, biocompatible materials based on hydrogels of synthetic and/or natural polymers have begun to be developed. These include biomaterials that mimic CNS ECM proteins, such as hyaluronic acid (HA), laminin, fibronectin, and collagen, as well as biopolymers based on polyacrylamide (PAA), polyethylene glycol (PEG), alginate, and gelatin (Jagielska et al. 2012; Kubinova 2017; Espinosa-Hoyos et al. 2018; Unal et al. 2020).

Designing mechanosensitive surfaces that induce the differentiation or remyelination of myelin-producing cells may offer solutions to many demyelinating diseases. Such biomaterials when transplanted into the pathological site can provide a favorable environment for cell survival, growth, and differentiation, as well as trigger remyelination. Results from implanting soft hydrogels that prevent glial scar formation and induce neurite formation at the injury site in CNS tissue have been promising (Urbanski et al. 2016; Segel et al. 2019). However, the correct choice of biomaterials that induce cell survival and myelin formation is still to be determined. Material stiffness can also trigger gliosis and inflammation, and understanding cell-substrate interactions is crucial.

Researchers who developed PAA gels with stiffnesses ranging from 0.1 to 70 kPa, higher than the Young's modulus of brain tissue (0.1 to 10 kPa), have shown that OL differentiation potential is modulated by substrate stiffness in this range (Jagielska et al. 2012). Similar studies using PAA gels have also shown that mechanical changes alter OL differentiation (Urbanski et al. 2016; Segel et al. 2019). Working with a mixture of PEG diacrylate with gelatin and HA gels, Li et al. (2013) developed an injectable hydrogel spanning an elastic modulus range of 0.0048 to 1.60 kPa and demonstrated that it induced remyelination of OPCs in a demyelinating lesion environment (Li et al. 2013). 3D-printed artificial axons designed to mimic the cylindrical geometry and mechanical stiffness of axons were developed with HDDA-PEG (Espinosa-Hoyos et al. 2018). By adjusting the artificial axon stiffness between healthy (0.4 kPa) and pathological values (140 kPa), the researchers examined the adhesion and myelination potential of OLs (Espinosa-Hoyos et al. 2018). 3D noramine and hyaluronic acid composite hydrogels that exhibit mechanical properties similar to those of CNS tissues have been shown to be suitable for OPC growth and proliferation, while softer, larger-network hydrogels have been shown to support OPC metabolic health. Future exogenous OPC transplantation and the use of OPCs to direct remyelination in vivo have been proposed in preclinical and clinical studies (Unal et al. 2020).

Yang et al. also investigated the relationship between OL myelin formation and axon mechanical stiffness. They tested the effects of several promyelinating compounds on artificial axons of different stiffnesses (13 kPa and 0.8 kPa), diameters (3 to 15 μm), and inter axonal gaps have been developed. Rat OLs were cultured for 7 days, and the myelin wrapping index (WI) was measured by MBP immunostaining. Their findings indicated that some compounds induced higher WI in stiffer axons, while others were more effective in softer axons. For example, ketoconazole, T3, quetiapine, clemastine, benztropine, and miconazole produced stronger responses in stiffer axons, while amorolfine and bazedoxifene were more effective in softer axons. Tasin-1 was identified as the compound with the highest WI at both stiffnesses, but the effectiveness of other compounds varied depending on axon stiffness. A key conclusion of the study is that ignoring axon mechanical properties in drug screening may lead to the oversight or misjudgment of some potentially effective compounds (Yang et al. 2025).

Functionalized biomaterials are also being used to modulate the microenvironment that inhibits regeneration. One of these studies examined OL regeneration in a mouse SCI model using miR-219/miR-338-loaded fiber-hydrogel scaffolds. miR-219 and miR-338, which suppress the expression of OL differentiation inhibitors such as Foxj3, Hes5, Sox6, and ZFP238, were found to be effective in promoting OL differentiation and myelination (Wei et al. 2025). The use of composite fiber-hydrogel scaffolds not only enabled efficient delivery of miR-219/miR-338 but also served as a structural matrix mimicking the native ECM (Wei et al. 2025).

Piezo1 plays a crucial role in OL-mediated myelin formation and regeneration. When OLs begin to wrap axons, Piezo1 regulates myelin formation by sensing mechanical forces. However, Piezo1 has also been reported to be a negative regulator of myelin formation in OLs. High Piezo1 activity after axonal injury may exacerbate nerve damage and demyelination. The Piezo1 inhibitor GsMTx4 prevents chemically induced demyelination and neuronal damage, whereas Yoda1, in contrast, has been observed to directly cause axonal damage and demyelination (Velasco-Estevez et al. 2022). Furthermore, it has been suggested that a stiff mechanical microenvironment may inhibit myelin regeneration by directing YAP to the nuclear region through Piezo1 activation (Zhang et al. 2025). Therefore, inhibition of Piezo1 may be considered a potential therapeutic strategy to promote myelin regeneration.

An interesting study on adult mouse brains showed that inhibition of YAP/TAZ gene expression did not have a significant effect on animal viability and normal myelination, but it decreased remyelination after cuprizone-induced demyelination (Hong et al. 2024). This work points out the fact that YAP/TAZ translocation in and out of the nucleus may successfully be used as a possible epigenetic therapy target to induce remyelination. Mechanoepigenetic regulation of tendon functionality by altering chromatin accessibility targeting YAP/TAZ has been previously shown (Jones et al. 2023).

15.6 Conclusions

There is extensive data showing that both OL and SC myelination depend on extrinsic factors, particularly mechanical cues from the microenvironment during healthy development, as well as in demyelinating pathologies. Increasing evidence supports that myelin producing cells are highly mechanosensitive cells. The concept of "adaptive myelination" in OLs that defines myelin as a dynamic structure ever-modeling in response to environmental cues also contributes to our understanding of the mechanobiology of myelination. This dependence on mechano-signaling highlights a new concept of epigenetic targets as novel therapeutic approaches for remyelination.

It is not yet well-studied whether PNS has a similar adaptive myelination mechanism to that observed in the CNS. Particularly in PNS, being more accessible to surgical intervention, biomaterial-based approaches regulating the SC mechanical environment, such as bioengineered scaffolds and nerve conduits with optimal mechanical properties for remyelination, hold great promise for triggering and improving regeneration.

More attention must be given to novel, nanoparticle, and/or extracellular vesicle-based delivery approaches for epigenetic modifiers such as miRNAs for demyelinating CNS pathologies. CRISR-based approaches can be explored for chromatin modulation, particularly reversing the effects of pathological mechanical environment through mechanosensitive factors, including, but not limited to, Piezo1, HDAC, Talin, Hippo, YAP/Taz, SYNE1, etc.

Further research is crucial for understanding the cytoskeletal reorganization during remyelination after injury or disease and decipher the correct targets for epigenetic modifications. Besides remyelinating diseases, novel findings in myelin mechanobiology might shed light on some other developmental disorders related to nerve conduction, such as cognitive disabilities, autism, and schizophrenia.

Acknowledgments This article is partially based upon work from COST Action CA22153 (EuroCurvoBioNet) supported by COST (European Cooperation in Science and Technology).

Author Contributions ZGM, YFM, and AS designed the manuscript, searched, and analyzed the literature. ZGM and AS wrote the manuscript and designed the figures and the table. AS and YFM corrected and revised the manuscript. All authors have read and agreed to the published version of the manuscript.

Funding Statement The authors received no specific funding for this study. Zehra G. Morçimen-Güliçli is supported by the TUBITAK-BIDEB 2211-A National Scholarship Program for Ph.D. Students and the YOK 100/2000 Scholarship in the field of smart and innovative materials.

Data Availability Statement Not applicable.

Conflicts of Interest The authors declare no conflict of interest.

References

Acheta J, Bhatia U, Haley J, Hong J, Rich K, Close R et al (2022) Piezo channels contribute to the regulation of myelination in Schwann cells. Glia 70(12):2276–2289

Alkhanjari RR, Alhajeri MM, Bhamidimarri PM, Alhosani K, Kashir J, Torii T, Hamdan H (2025) Septins in the nervous system: from cytoskeletal dynamics to neurological disorders. Cell Commun Signal 23(1):425

Almeida RG, Czopka T, Ffrench-Constant, C, Lyons DA (2011) Individual axons regulate the myelinating potential of single oligodendrocytes in vivo. Development 138(20):4443–4450

Bakhti M, Aggarwal S, Simons M (2014) Myelin architecture: zippering membranes tightly together. Cell Mol Life Sci 71(7):1265–1277

Bando Y (2020) Mechanism of demyelination and remyelination in multiple sclerosis. Clin Experim Neuroimmunol 11:14–21

Barateiro A, Fernandes A (2014) Temporal oligodendrocyte lineage progression: in vitro models of proliferation, differentiation and myelination. Biochimica et Biophysica Acta (BBA) Molecular Cell Res 1843(9):1917–1929

Barnes-Vélez JA, Yasar FBA, Hu J (2023) Myelin lipid metabolism and its role in myelination and myelin maintenance. Innovation (Camb) Int J Hydrogen Energ 4(1):100360

Bate N, Gingras AR, Bachir A, Horwitz R, Ye F, Patel B et al (2012) Talin contains a C-terminal calpain2 cleavage site important in focal adhesion dynamics. PLoS One 7(4):e34461

Bauer NG, Richter-Landsberg C, Ffrench-Constant C (2009) Role of the oligodendroglial cytoskeleton in differentiation and myelination. Glia 57(16):1691–1705

Bayir E, Sendemir A, Missirlis YF (2019) Mechanobiology of cells and cell systems, such as organoids. Biophys Rev 11(5):721–728

Bechler ME, Byrne L (2015) CNS myelin sheath lengths are an intrinsic property of oligodendrocytes. Curr Biol 25(18):2411–2416

Bechler ME, Swire M, ffrench-Constant C (2018) Intrinsic and adaptive myelination—a sequential mechanism for smart wiring in the brain. Dev Neurobiol 78(2):68–79

Berghoff SA, Spieth L, Saher G (2022) Local cholesterol metabolism orchestrates remyelination. Trends Neurosci 45(4):272–283

Berry K, Wang J, Lu QR (2020) Epigenetic regulation of oligodendrocyte myelination in developmental disorders and neurodegenerative diseases, vol 9. F1000Research. F1000-Faculty

Berti C, Bartesaghi L, Ghidinelli M, Zambroni D, Figlia G, Chen ZL et al (2011) Non-redundant function of dystroglycan and β1 integrins in radial sorting of axons. Development 138(18):4025–4037

Bhat MA (2003) Molecular organization of axo-glial junctions. Curr Opin Neurobiol 13(5):552–559

Boggs JM, Rangaraj G (2000) Interaction of lipid-bound myelin basic protein with actin filaments and calmodulin. Biochemistry 39(26):7799–7806

Boggs JM, Gao W, Hirahara Y (2008) Myelin glycosphingolipids, galactosylceramide and sulfatide, participate in carbohydrate–carbohydrate interactions between apposed membranes and may form glycosynapses between oligodendrocyte and/or myelin membranes. Biochimica et Biophysica Acta (BBA)-General Subjects 1780(3):445–455

Boggs JM, Rangaraj G, Dicko A (2012) Effect of phosphorylation of phosphatidylinositol on myelin basic protein-mediated binding of actin filaments to lipid bilayers in vitro. Biochimica et Biophysica Acta (BBA)-Biomembr 1818(9):2217–2227

Boullerne AI (2016) The history of myelin. Exp Neurol 283:431–445

Bridges AA, Jentzsch MS, Oakes PW, Occhipinti P, Gladfelter AS (2016) Micron-scale plasma membrane curvature is recognized by the septin cytoskeleton. J Cell Biol 213(1):23–32

Brown TL, Macklin WB (2020) The actin cytoskeleton in myelinating cells. Neurochem Res 45(3):684–693

Bullock PN, Rome LH (1990) Glass micro-fibers: a model system for study of early events in myelination. J Neurosci Res 27(3):383–393

Bunge MB, Williams AK, Wood PM, Uitto J, Jeffrey JJ (1980) Comparison of nerve cell and nerve cell plus Schwann cell cultures, with particular emphasis on basal lamina and collagen formation. J Cell Biol 84(1):184–202

Butt AM, Ibrahim M, Berry M (1998) Axon-myelin sheath relations of oligodendrocyte unit phenotypes in the adult rat anterior medullary velum. J Neurocytol 27(4):205–217

Camargo N, Goudriaan A, van Deijk ALF, Otte WM, Brouwers JF, Lodder H et al (2017) Oligodendroglial myelination requires astrocyte-derived lipids. PLoS Biol 15(5):e1002605

Cañellas AR, Gols AR, Izquierdo JR, Subirana MT, Gairin XM (2007) Idiopathic inflammatory-demyelinating diseases of the central nervous system. Neuroradiology 49(5):393–409

Chatelin S, Constantinesco A, Willinger R (2010) Fifty years of brain tissue mechanical testing: from in vitro to in vivo investigations. Biorheology 47(5–6):255–276

Chen L, Lu W, Yang Z, Yang S, Li C, Shi X, Tang Y (2011) Age-related changes of the oligodendrocytes in rat subcortical white matter. Anat Rec Adv Integr Anat Evol Biol 294(3):487–493

Chen Z, Sun Z, Fan Y, Yin M, Jin C, Guo B et al (2023) Mimicked spinal cord fibers trigger axonal regeneration and remyelination after injury. ACS Nano 17(24):25591–25613
Coetzee T, Fujita N, Dupree J, Shi R, Blight A, Suzuki K et al (1996) Myelination in the absence of galactocerebroside and sulfatide: normal structure with abnormal function and regional instability. Cell 86(2):209–219
Coleman MP (2022) Axon biology in ALS: mechanisms of axon degeneration and prospects for therapy. Neurotherapeutics 19(4):1133–1144
Colognato H, Tzvetanova ID (2011) Glia unglued: how signals from the extracellular matrix regulate the development of myelinating glia. Dev Neurobiol 71(11):924–955
Colognato H, Galvin J Wang Z, Relucio J, Nguyen T, Harrison D, ..., French-Constant C (2007) Identification of dystroglycan as a second laminin receptor in oligodendrocytes, with a role in myelination
Correale J, Marrodan M, Ysrraelit MC (2019) Mechanisms of neurodegeneration and axonal dysfunction in progressive multiple sclerosis. Biomedicine 7(1):14
Cortés N, Guzmán-Martínez L, Andrade V, González A, Maccioni RB (2019) CDK5: a unique CDK and its multiple roles in the nervous system. J Alzheimer's Dis 68(3):843–855
Coutinho Costa VG, Araújo SES, Alves-Leon SV, Gomes FCA (2023) Central nervous system demyelinating diseases: glial cells at the hub of pathology. Front Immunol 14:1135540
Dashiell SM, Tanner SL, Pant HC, Quarles RH (2002) Myelin-associated glycoprotein modulates expression and phosphorylation of neuronal cytoskeletal elements and their associated kinases. J Neurochem 81(6):1263–1272
Deng B, Zhao Z, Kong W, Han C, Shen X, Zhou C (2022) Biological role of matrix stiffness in tumor growth and treatment. J Transl Med 20(1):540
Djalali S, Höltje M, Grosse G, Rothe T, Stroh T, Grosse J et al (2005) Effects of brain-derived neurotrophic factor (BDNF) on glial cells and serotonergic neurones during development. J Neurochem 92(3):616–627
Domingues HS, Cruz A, Chan JR, Relvas JB, Rubinstein B, Pinto IM (2018) Mechanical plasticity during oligodendrocyte differentiation and myelination. Glia 66(1):5–14
Domingues HS, Urbanski MM, Macedo-Ribeiro S, Almaktari A, Irfan A, Hernandez Y et al (2020) Pushing myelination–developmental regulation of myosin expression drives oligodendrocyte morphological differentiation. J Cell Sci 133(15):jcs232264
Dupont S, Morsut L, Aragona M, Enzo E, Giulitti S, Cordenonsi M et al (2011) Role of YAP/TAZ in mechanotransduction. Nature 474(7350):179–183
Eberle D, Fodelianaki G, Kurth T, Jagielska A, Möllmert S, Ulbricht E et al (2020) Acquired demyelination but not genetic developmental defects in myelination leads to brain tissue stiffness changes. Brain Multiphysics 1:100019
Emery B (2010) Regulation of oligodendrocyte differentiation and myelination. Science 330(6005):779–782
Espinosa-Hoyos D, Jagielska A, Homan KA, Du H, Busbee T, Anderson DG et al (2018) Engineered 3D-printed artificial axons. Sci Rep 8(1):478
Eyckmans J, Boudou T, Yu X, Chen CS (2011) A hitchhiker's guide to mechanobiology. Dev Cell 21(1):35–47
Fanarraga ML, Griffiths IR, Zhao M, Duncan ID (1998) Oligodendrocytes are not inherently programmed to myelinate a specific size of axon. J Comp Neurol 399(1):94–100
Feltri ML, Wrabetz L (2005) Laminins and their receptors in Schwann cells and hereditary neuropathies. J Peripher Nerv Syst 10(2):128–143
Feltri ML, Scherer SS, Nemni R, Kamholz J, Vogelbacker H, Scott MO et al (1994) β4 integrin expression in myelinating Schwann cells is polarized, developmentally regulated and axonally dependent. Development 120(5):1287–1301
Feltri ML, Porta DG, Previtali SC, Nodari A, Migliavacca B, Cassetti A et al (2002) Conditional disruption of β1 integrin in Schwann cells impedes interactions with axons. J Cell Biol 156(1):199–210
Feltri ML, Poitelon Y, Previtali SC (2016) How Schwann cells sort axons: new concepts. Neuroscientist 22(3):252–265

Fernandez-Valle C, Gorman D, Gomez AM, Bunge MB (1997) Actin plays a role in both changes in cell shape and gene-expression associated with Schwann cell myelination. J Neurosci 17(1):241–250

Figlia G (2018) C-Jun in Schwann cells: stay away from extremes. J Neurosci 38(14):3388

Fisher E, Chang A, Fox RJ, Tkach JA, Svarovsky T, Nakamura K et al (2007) Imaging correlates of axonal swelling in chronic multiple sclerosis brains. Ann Neurol: Off J Am Neurol Assoc Child Neurol Soc 62(3):219–228

Friede RL, Bischhausen R (1982) How are sheath dimensions affected by axon caliber and internode length? Brain Res 235(2):335–350

Frohman EM, Havrdova E, Lublin F, Barkhof F, Achiron A, Sharief MK et al (2006) Most patients with multiple sclerosis or a clinically isolated demyelinating syndrome should be treated at the time of diagnosis. Arch Neurol 63(4):614–619

Frühbeis C, Fröhlich D, Kuo WP, Amphornrat J, Thilemann S, Saab AS et al (2013) Neurotransmitter-triggered transfer of exosomes mediates oligodendrocyte–neuron communication. PLoS Biol 11(7):e1001604

Gatto RG, Li W, Gao J, Magin RL (2018) In vivo diffusion MRI detects early spinal cord axonal pathology in a mouse model of amyotrophic lateral sclerosis. NMR Biomed 31(8):e3954

Gibson EM, Purger D, Mount CW, Goldstein AK, Lin GL, Wood LS et al (2014) Neuronal activity promotes oligodendrogenesis and adaptive myelination in the mammalian brain. Science 344(6183):1252304

Giussani P, Prinetti A, Tringali C (2021) The role of sphingolipids in myelination and myelin stability and their involvement in childhood and adult demyelinating disorders. J Neurochem 156(4):403–414

Gjervan SC, Ozgoren OK, Gow A, Stockler-Ipsiroglu S, Pouladi MA (2024) Claudin-11 in health and disease: implications for myelin disorders, hearing, and fertility. Front Cell Neurosci 17:1344090

Goebbels S, Wieser GL, Pieper A, Spitzer S, Weege B, Yan K et al (2017) A neuronal PI (3, 4, 5) P3-dependent program of oligodendrocyte precursor recruitment and myelination. Nat Neurosci 20(1):10–15

Gopalakrishnan G, Awasthi A, Belkaid W, De Faria Jr O, Liazoghli D, Colman DR, Dhaunchak AS (2013) Lipidome and proteome map of myelin membranes. J Neurosci Res 91(3):321–334

Gow A (2025) Understanding the myelin g ratio from first principles, its derivation, uses and artifacts. ASN Neuro 17(1):2445624

Groh AM, Leblanc R (2025) Ghost cells: Wilder Penfield and the characterization of glia and glial pathology, 1924–1932. J Hist Neurosci 34(1):1–28

Hahn KR, Kim W, Jung HY, Kwon HJ, Kim DW, Hwang IK, Yoon YS (2022) Comparison of the effects of cuprizone on demyelination in the corpus callosum and hippocampal progenitors in young adult and aged mice. Neurochem Res 47(4):1073–1082

Handorf AM, Zhou Y, Halanski MA, Li WJ (2015) Tissue stiffness dictates development, homeostasis, and disease progression. Organogenesis 11(1):1–15

Herbert AL, Monk KR (2017) Advances in myelinating glial cell development. Curr Opin Neurobiol 42:53–60

Hernandez M, Patzig J, Mayoral SR, Costa KD, Chan JR, Casaccia P (2016) Mechanostimulation promotes nuclear and epigenetic changes in oligodendrocytes. J Neurosci 36(3):806–813

Herthum H, Hetzer S, Scheel M, Shahryari M, Braun J, Paul F, Sack I (2022) In vivo stiffness of multiple sclerosis lesions is similar to that of normal-appearing white matter. Acta Biomater 138:410–421

Hill RA, Patel KD, Goncalves CM, Grutzendler J, Nishiyama A (2014) Modulation of oligodendrocyte generation during a critical temporal window after NG2 cell division. Nat Neurosci 17(11):1518–1527

Hines JH, Ravanelli AM, Schwindt R, Scott EK, Appel B (2015) Neuronal activity biases axon selection for myelination in vivo. Nat Neurosci 18(5):683–689

Ho WY, Hartmann H, Ling SC (2022) Central nervous system cholesterol metabolism in health and disease. IUBMB Life 74(8):826–841

Hofmann M, Biller L, Michel U, Bähr M, Koch JC (2022) Cytoskeletal assembly in axonal outgrowth and regeneration analyzed on the nanoscale. Sci Rep 12(1):14387

Hong J, Kirkland JM, Acheta J, Marziali LN, Beck B, Jeanette H et al (2024) YAP and TAZ regulate remyelination in the central nervous system. Glia 72(1):156–166

Howe CL (2006) Coated glass and vicryl microfibers as artificial axons. Cells Tissues Organs 183(4):180–194

Hu, W., & Lucchinetti, C. F. (2009). The pathological spectrum of CNS inflammatory demyelinating diseases. In Semin Immunopathol. 31, 4, pp. 439–453). Berlin/Heidelberg: Springer-Verlag

Hughes EG, Kang SH, Fukaya M, Bergles DE (2013) Oligodendrocyte progenitors balance growth with self-repulsion to achieve homeostasis in the adult brain. Nat Neurosci 16(6):668–676

Huxley AF, Stämpfli R (1949) Evidence for saltatory conduction in peripheral myelinated nerve fibres. J Physiol 108(3):315

Inouye H, Liu J, Makowski L, Palmisano M, Burghammer M, Riekel C, Kirschner DA (2014) Myelin organization in the nodal, paranodal, and juxtaparanodal regions revealed by scanning x-ray microdiffraction. PLoS One 9(7):e100592

Jaalouk DE, Lammerding J (2009) Mechanotransduction gone awry. Nat Rev Mol Cell Biol 10(1):63–73

Jacob C, Lebrun-Julien F, Suter U (2011) How histone deacetylases control myelination. Mol Neurobiol 44(3):303–312

Jagielska A, Norman AL, Whyte G, Vliet KJV, Guck J, Franklin RJ (2012) Mechanical environment modulates biological properties of oligodendrocyte progenitor cells. Stem Cells Dev 21(16):2905–2914

Jagielska A, Lowe AL, Makhija E, Wroblewska L, Guck J, Franklin RJ et al (2017) Mechanical strain promotes oligodendrocyte differentiation by global changes of gene expression. Front Cell Neurosci 11:93

Jairaman A, Othy S, Dynes JL, Yeromin AV, Zavala A, Greenberg ML et al (2021) Piezo1 channels restrain regulatory T cells but are dispensable for effector CD4+ T cell responses. Sci Adv 7(28):eabg5859

Ji X, Tian Y, Xie K, Liu W, Qu Y, Fei Z (2012) Protective effects of hydrogen-rich saline in a rat model of traumatic brain injury via reducing oxidative stress. J Surg Res 178(1):e9–e16

Jin F, Dong B, Georgiou J, Jiang Q, Zhang J, Bharioke A et al (2011) N-WASp is required for Schwann cell cytoskeletal dynamics, normal myelin gene expression and peripheral nerve myelination. Development 138(7):1329–1337

Jones DL, Hallström GF, Jiang X, Locke RC, Evans MK, Bonnevie ED et al (2023) Mechanoepigenetic regulation of extracellular matrix homeostasis via yap and Taz. Proc Natl Acad Sci 120(22):e2211947120

Juge L, Pong AC, Bongers A, Sinkus R, Bilston LE, Cheng S (2016) Changes in rat brain tissue microstructure and stiffness during the development of experimental obstructive hydrocephalus. PLoS One 11(2):e0148652

Kessaris N, Fogarty M, Iannarelli P, Grist M, Wegner M, Richardson WD (2006) Competing waves of oligodendrocytes in the forebrain and postnatal elimination of an embryonic lineage. Nat Neurosci 9(2):173–179

Kim J, Adams AA, Gokina P, Zambrano B, Jayakumaran J, Dobrowolski R et al (2020) Mechanical stretch induces myelin protein loss in oligodendrocytes by activating Erk1/2 in a calcium-dependent manner. Glia 68(10):2070–2085

Kippert A, Fitzner D, Helenius J, Simons M (2009) Actomyosin contractility controls cell surface area of oligodendrocytes. BMC Cell Biol 10(1):71

Klapholz B, Brown NH (2017) Talin–the master of integrin adhesions. J Cell Sci 130(15):2435–2446

Kubinova S (2017) Extracellular matrix based biomaterials for central nervous system tissue repair: the benefits and drawbacks. Neural Regen Res 12(9):1430–1432

Lee S, Leach MK, Redmond SA, Chong SC, Mellon SH, Tuck SJ et al (2012) A culture system to study oligodendrocyte myelination processes using engineered nanofibers. Nat Methods 9(9):917–922

Lee S, Chong SC, Tuck SJ, Corey JM, Chan JR (2013) A rapid and reproducible assay for modeling myelination by oligodendrocytes using engineered nanofibers. Nat Protoc 8(4):771–782
Li X, Liu X, Cui L, Brunson C, Zhao W, Bhat NR et al (2013) Engineering an in situ crosslinkable hydrogel for enhanced remyelination. FASEB J 27(3):1127
Li Z, Lee H, Zhu C (2016) Molecular mechanisms of mechanotransduction in integrin-mediated cell-matrix adhesion. Exp Cell Res 349(1):85–94
Liewald D, Miller R, Logothetis N, Wagner HJ, Schüz A (2014) Distribution of axon diameters in cortical white matter: an electron-microscopic study on three human brains and a macaque. Biol Cybern 108(5):541–557
Liu J, Dietz K, DeLoyht JM, Pedre X, Kelkar D, Kaur J et al (2012) Impaired adult myelination in the prefrontal cortex of socially isolated mice. Nat Neurosci 15(12):1621–1623
Liu Z, Wang J, Chen H, Zhang G, Lv Z, Li Y et al (2021) Coaxial electrospun PLLA fibers modified with water-soluble materials for oligodendrocyte myelination. Polymers 13(20):3595
Lopez-Anido C, Poitelon Y, Gopinath C, Moran JJ, Ma KH, Law WD et al (2016) Tead1 regulates the expression of peripheral myelin protein 22 during Schwann cell development. Hum Mol Genet 25(14):3055–3069
Lourenço T, Grãos M (2016) Modulation of oligodendrocyte differentiation by mechanotransduction. Front Cell Neurosci 10:277
Lourenço T, Paes de Faria J, Bippes CA, Maia J, Lopes-da-Silva JA, Relvas JB, Grãos M (2016) Modulation of oligodendrocyte differentiation and maturation by combined biochemical and mechanical cues. Sci Rep 6(1):21563
Lu YB, Franze K, Seifert G, Steinhäuser C, Kirchhoff F, Wolburg H et al (2006) Viscoelastic properties of individual glial cells and neurons in the CNS. Proc Natl Acad Sci 103(47):17759–17764
Lu F, Zhu L, Bromberger T, Yang J, Yang Q, Liu J et al (2022) Mechanism of integrin activation by Talin and its cooperation with kindlin. Nat Commun 13(1):2362
Lundgaard I, Luzhynskaya A, Stockley JH, Wang Z, Evans KA, Swire M et al (2013) Neuregulin and BDNF induce a switch to NMDA receptor-dependent myelination by oligodendrocytes. PLoS Biol 11(12):e1001743
Makhija E, Jagielska A, Zhu L, Bost AC, Ong W, Chew SY et al (2018) Mechanical strain alters cellular and nuclear dynamics at early stages of oligodendrocyte differentiation. Front Cell Neurosci 12:59
Makinodan M, Rosen KM, Ito S, Corfas G (2012) A critical period for social experience–dependent oligodendrocyte maturation and myelination. Science 337(6100):1357–1360
Martini R, Groh J, Bartsch U, Armati P, Mathey E (2010) Comparative biology of Schwann cells and oligodendrocytes. In: The biology of oligodendrocytes, pp 19–48
Martino F, Perestrelo AR, Vinarský V, Pagliari S, Forte G (2018) Cellular mechanotransduction: from tension to function. Front Physiol 9:824
Marques S, Zeisel A, Codeluppi S, Van Bruggen D, Mendanha Falcão A, Xiao L et al (2016) Oligodendrocyte heterogeneity in the mouse juvenile and adult central nervous system. Science 352(6291):1326–1329
Mathews ES, Appel B (2016) Cholesterol biosynthesis supports myelin gene expression and axon ensheathment through modulation of P13K/Akt/mTor signaling. J Neurosci 36(29):7628–7639
Maxwell WL (2013) Damage to myelin and oligodendrocytes: a role in chronic outcomes following traumatic brain injury? Brain Sci 3(3):1374–1394
McKenzie IA, Ohayon D, Li H, Paes J, de Faria B, Emery KT, Richardson WD (2014) Motor skill learning requires active central myelination. Science 346(6207):318–322
Mensch S (2015) Investigating axon-oligodendrocyte interactions during myelinated axon formation in vivo. Doctoral dissertation, University of Edinburgh, Scotland
Micu I, Plemel JR, Lachance C, Proft J, Jansen AJ, Cummins K et al (2016) The molecular physiology of the axo-myelinic synapse. Exp Neurol 276:41–50
Miguel-Hidalgo JJ, Carter K, Deloach PH, Sanders L, Pang Y (2019) Glucocorticoid-induced reductions of myelination and connexin 43 in mixed central nervous system cell cultures are prevented by mifepristone. Neuroscience 411:255–269

Millward JM, Guo J, Berndt D, Braun J, Sack I, Infante-Duarte C (2015) Tissue structure and inflammatory processes shape viscoelastic properties of the mouse brain. NMR Biomed 28(7):831–839
Missirlis YF (2016) Mechanoepigenetics. Front Cell Develop Biol 4(113):1–4
Moore SW, Roca-Cusachs P, Sheetz MP (2010) Stretchy proteins on stretchy substrates: the important elements of integrin-mediated rigidity sensing. Dev Cell 19(2):194–206
Montgomery A, Westphal J, Bryan AE, Harris GM (2025) Dynamically changing extracellular matrix stiffness drives Schwann cell phenotype. Matrix Biol Plus 25:100167
Mount CW, Monje M (2017) Wrapped to adapt: experience-dependent myelination. Neuron 95(4):743–756
Mozafari S, Starost L, Manot-Saillet B, Garcia-Diaz B, Xu YKT, Roussel D et al (2020) Multiple sclerosis iPS-derived oligodendroglia conserve their properties to functionally interact with axons and glia in vivo. Sci Adv 6(49):eabc6983
Nawaz S, Kippert A, Saab AS, Werner HB, Lang T, Nave KA, Simons M (2009) Phosphatidylinositol 4, 5-bisphosphate-dependent interaction of myelin basic protein with the plasma membrane in oligodendroglial cells and its rapid perturbation by elevated calcium. J Neurosci 29(15):4794–4807
Nawaz S, Schweitzer J, Jahn O, Werner HB (2013) Molecular evolution of myelin basic protein, an abundant structural myelin component. Glia 61(8):1364–1377
Nawaz S, Sánchez P, Schmitt S, Snaidero N, Mitkovski M, Velte C et al (2015) Actin filament turnover drives leading edge growth during myelin sheath formation in the central nervous system. Dev Cell 34(2):139–151
Nieves B, Jones CW, Ward R, Ohta Y, Reverte CG, LaFlamme SE (2010) The NPIY motif in the integrin β1 tail dictates the requirement for Talin-1 in outside-in signaling. J Cell Sci 123(8):1216–1226
Nodari A, Previtali SC, Dati G, Occhi S, Court FA, Colombelli C et al (2008) α6β4 integrin and dystroglycan cooperate to stabilize the myelin sheath. J Neurosci 28(26):6714–6719
Nunes LGA (2024) Selenoprotein I is essential for proper central nervous system development: a study on ethanolamine phospholipid deficiency in oligodendrocyte development and function. Doctoral dissertation, University of Hawai'i, Manoa
O′ Meara RW, Michalski JP, Kothary R (2011) Integrin signaling in oligodendrocytes and its importance in CNS myelination. J Signal Transduct 2011(1):354091
Ong W, Marinval N, Lin J, Nai MH, Chong YS, Pinese C et al (2020) Biomimicking fiber platform with tunable stiffness to study mechanotransduction reveals stiffness enhances oligodendrocyte differentiation but impedes myelination through YAP-dependent regulation. Small 16(37):2003656
Osso LA, Chan JR (2017) Architecting the myelin landscape. Curr Opin Neurobiol 47:1–7
Patzig J, Erwig MS, Tenzer S, Kusch K, Dibaj P, Möbius W et al (2016) Septin/anillin filaments scaffold central nervous system myelin to accelerate nerve conduction. Elife 5:e17119
Pereira JA, Lebrun-Julien F, Suter U (2012) Molecular mechanisms regulating myelination in the peripheral nervous system. Trends Neurosci 35(2):123–134
Petersen SC, Luo R, Liebscher I, Giera S, Jeong SJ, Mogha A et al (2015) The adhesion GPCR GPR126 has distinct, domain-dependent functions in Schwann cell development mediated by interaction with laminin-211. Neuron 85(4):755–769
Pfeiffer SE, Warrington AE, Bansal R (1993) The oligodendrocyte and its many cellular processes. Trends Cell Biol 3(6):191–197
Poitelon Y, Lopez-Anido C, Catignas K, Berti C, Palmisano M, Williamson C et al (2016) YAP and TAZ control peripheral myelination and the expression of laminin receptors in Schwann cells. Nat Neurosci 19(7):879–887
Raasakka A, Ruskamo S, Kowal J, Barker R, Baumann A, Martel A et al (2017) Membrane association landscape of myelin basic protein portrays formation of the myelin major dense line. Sci Rep 7(1):4974

Redmond SA, Mei F, Eshed-Eisenbach Y, Osso LA, Leshkowitz D, Shen YAA et al (2016) Somatodendritic expression of JAM2 inhibits oligodendrocyte myelination. Neuron 91(4):824–836

Reindl M, Waters P (2019) Myelin oligodendrocyte glycoprotein antibodies in neurological disease. Nat Rev Neurol 15(2):89–102

Relucio J, Tzvetanova ID, Ao W, Lindquist S, Colognato H (2009) Laminin alters fyn regulatory mechanisms and promotes oligodendrocyte development. J Neurosci 29(38):11794–11806

Rog T, Koivuniemi A (2016) The biophysical properties of ethanolamine plasmalogens revealed by atomistic molecular dynamics simulations. Biochimica et Biophysica Acta (BBA)-Biomembranes 1858(1):97–103

Rosenberg SS, Kelland EE, Tokar E, De La Torre AR, Chan JR (2008) The geometric and spatial constraints of the microenvironment induce oligodendrocyte differentiation. Proc Natl Acad Sci 105(38):14662–14667

Rosso G, Liashkovich I, Young P, Rohr D, Shahin V (2017a) Schwann cells and neurite outgrowth from embryonic dorsal root ganglions are highly mechanosensitive. Nanomedicine 13:493–501. https://doi.org/10.1016/j.nano.2016.06.011

Rosso G, Young P, Shahin V (2017b) Implications of schwann cells biomechanics and mechanosensitivity for peripheral nervous system physiology and pathophysiology. Front Mol Neurosci 10:345

Rosso G, Wehner D, Schweitzer C, Möllmert S, Sock E, Guck J, Shahin V (2022) Matrix stiffness mechanosensing modulates the expression and distribution of transcription factors in Schwann cells. Bioeng Transl Med 7(1):e10257

Rowitch DH, Kriegstein AR (2010) Developmental genetics of vertebrate glial–cell specification. Nature 468(7321):214–222

Rushton WAH (1951) A theory of the effects of fibre size in medullated nerve. J Physiol 115(1):101

Rusielewicz T, Nam J, Damanakis E, John GR, Raine CS, Melendez-Vasquez CV (2014) Accelerated repair of demyelinated CNS lesions in the absence of non-muscle myosin IIB. Glia 62(4):580–591

Ruskamo S, Nieminen T, Kristiansen CK, Vatne GH, Baumann A, Hallin EI et al (2017) Molecular mechanisms of Charcot-Marie-tooth neuropathy linked to mutations in human myelin protein P2. Sci Rep 7(1):6510

Ruskamo S, Raasakka A, Pedersen JS, Martel A, Škubník K, Darwish T et al (2022) Human myelin proteolipid protein structure and lipid bilayer stacking. Cell Mol Life Sci 79(8):419

Saab AS, Tzvetavona ID, Trevisiol A, Baltan S, Dibaj P, Kusch K et al (2016) Oligodendroglial NMDA receptors regulate glucose import and axonal energy metabolism. Neuron 91(1):119–132

Salzer JL (2015) Schwann cell myelination. Cold Spring Harb Perspect Biol 7(8):a020529

Sampaio-Baptista C, Khrapitchev AA, Foxley S, Schlagheck T, Scholz J, Jbabdi S et al (2013) Motor skill learning induces changes in white matter microstructure and myelination. J Neurosci 33(50):19499–19503

Scantlebury N, Cunningham T, Dockstader C, Laughlin S, Gaetz W, Rockel C et al (2014) Relations between white matter maturation and reaction time in childhood. J Int Neuropsychol Soc 20(1):99–112

Schmitt S, Castelvetri LC, Simons M (2015) Metabolism and functions of lipids in myelin. Biochimica et Biophysica Acta (BBA) Mol Cell Biol Lipids 1851(8):999–1005

Schregel K, Wuerfel née Tysiak E, Garteiser P, Gemeinhardt I, Prozorovski T, Aktas O et al (2012) Demyelination reduces brain parenchymal stiffness quantified in vivo by magnetic resonance elastography. Proc Natl Acad Sci 109(17):6650–6655

Schumacher M, Hussain R, Gago N, Oudinet JP, Mattern C, Ghoumari AM (2012) Progesterone synthesis in the nervous system: implications for myelination and myelin repair. Front Neurosci 6:10

Seetharaman S, Etienne-Manneville S (2018) Integrin diversity brings specificity in mechanotransduction. Biol Cell 110(3):49–64

Segel M, Neumann B, Hill MF, Weber IP, Viscomi C, Zhao C et al (2019) Niche stiffness underlies the ageing of central nervous system progenitor cells. Nature 573(7772):130–134

Shah S, Yin PT, Uehara TM, Chueng STD, Yang L, Lee KB (2014) Guiding stem cell differentiation into oligodendrocytes using graphene-nanofiber scaffolds. Adv Mater (Deerfield Beach, Fla) 26(22):3673

Shimizu T, Osanai Y, Tanaka KF, Thai TQ, Abe M, Natsume R et al (2019) Mechanical regulation of oligodendrocyte morphology and maturation by the mechanosensor p130Cas. J Neurochem 150(2):158–172

Singh S, Dallenga T, Winkler A, Roemer S, Maruschak B, Siebert H et al (2017) Relationship of acute axonal damage, Wallerian degeneration, and clinical disability in multiple sclerosis. J Neuroinflammation 14(1):57

Snaidero N, Simons M (2017) The logistics of myelin biogenesis in the central nervous system. Glia 65(7):1021–1031

Snaidero N, Velte C, Myllykoski M, Raasakka A, Ignatev A, Werner HB et al (2017) Antagonistic functions of MBP and CNP establish cytosolic channels in CNS myelin. Cell Rep 18(2):314–323

Stadelmann C, Timmler S, Barrantes-Freer A, Simons M (2019) Myelin in the central nervous system: structure, function, and pathology. Physiol Rev 99:1381

Stankoff B, Aigrot MS, Noël F, Wattilliaux A, Zalc B, Lubetzki C (2002) Ciliary neurotrophic factor (CNTF) enhances myelin formation: a novel role for CNTF and CNTF-related molecules. J Neurosci 22(21):9221–9227

Stassart RM, Möbius W, Nave KA, Edgar JM (2018) The axon-myelin unit in development and degenerative disease. Front Neurosci 12:467

Streitberger KJ, Sack I, Krefting D, Pfüller C, Braun J, Paul F, Wuerfel J (2012) Brain viscoelasticity alteration in chronic-progressive multiple sclerosis. PLoS One 7(1):e29888

Sun Z, Guo SS, Fässler R (2016) Integrin-mediated mechanotransduction. J Cell Biol 215(4):445–456

Swift J, Ivanovska IL, Buxboim A, Harada T, Dingal PDP, Pinter J et al (2013) Nuclear lamina scales with tissue stiffness and enhances matrix-directed differentiation. Science 341(6149):1240104

Taveggia C, Zanazzi G, Petrylak A, Yano H, Rosenbluth J, Einheber S et al (2005) Neuregulin-1 type III determines the ensheathment fate of axons. Neuron 47(5):681–694

Teliska LH, Rasband MN (2021) Spectrins. Curr Biol 31(10):R504–R506

Tomassy GS, Fossati V (2014) How big is the myelinating orchestra? Cellular diversity within the oligodendrocyte lineage: facts and hypotheses. Front Cell Neurosci 8:201

Tricaud N (2018) Myelinating Schwann cell polarity and mechanically-driven myelin sheath elongation. Front Cell Neurosci 11:414

Tripathi RB, Clarke LE, Burzomato V, Kessaris N, Anderson PN, Attwell D, Richardson WD (2011) Dorsally and ventrally derived oligodendrocytes have similar electrical properties but myelinate preferred tracts. J Neurosci 31(18):6809–6819

Unal DB, Caliari SR, Lampe KJ (2019) Engineering biomaterial microenvironments to promote myelination in the central nervous system. Brain Res Bull 152:159–174

Unal DB, Caliari SR, Lampe KJ (2020) 3D hyaluronic acid hydrogels for modeling oligodendrocyte progenitor cell behavior as a function of matrix stiffness. Biomacromolecules 21(12):4962–4971

Urbanski MM, Kingsbury L, Moussouros D, Kassim I, Mehjabeen S, Paknejad N, Melendez-Vasquez CV (2016) Myelinating glia differentiation is regulated by extracellular matrix elasticity. Sci Rep 6(1):33751

Urbanski MM, Brendel MB, Melendez-Vasquez CV (2019) Acute and chronic demyelinated CNS lesions exhibit opposite elastic properties. Sci Rep 9(1):999

Valadares NF, d'Muniz Pereira H, Ulian Araujo AP, Garratt RC (2017) Septin structure and filament assembly. Biophys Rev 9(5):481–500

Velasco-Estevez M, Gadalla KK, Liñan-Barba N, Cobb S, Dev KK, Sheridan GK (2020) Inhibition of Piezo1 attenuates demyelination in the central nervous system. Glia 68(2):356–375

Velasco-Estevez M, Koch N, Klejbor I, Caratis F, Rutkowska A (2022) Mechanoreceptor Piezo1 is downregulated in multiple sclerosis brain and is involved in the maturation and migration of oligodendrocytes in vitro. Front Cell Neurosci 16:914985
Vicente-Manzanares M, Choi CK, Horwitz AR (2009) Integrins in cell migration–the actin connection. J Cell Sci 122(2):199–206
Viganò F, Möbius W, Götz M, Dimou L (2013) Transplantation reveals regional differences in oligodendrocyte differentiation in the adult brain. Nat Neurosci 16(10):1370–1372
Villalón E, Barry DM, Byers N, Frizzi K, Jones MR, Landayan DS et al (2018) Internode length is reduced during myelination and remyelination by neurofilament medium phosphorylation in motor axons. Exp Neurol 306:158–168
Wang H, Tewari A, Einheber S, Salzer JL, Melendez-Vasquez CV (2008) Myosin II has distinct functions in PNS and CNS myelin sheath formation. J Cell Biol 182(6):1171–1184
Wang Y, Huang H, Weng H, Jia C, Liao B, Long Y et al (2024) Talin mechanotransduction in disease. Int J Biochem Cell Biol 166:106490
Webb A, Clark P, Skepper J, Compston A, Wood A (1995) Guidance of oligodendrocytes and their progenitors by substratum topography. J Cell Sci 108(8):2747–2760
Wei JU, Huang C, Singh R, Chueakula N, Le PHA, Chew SY (2025) Sustained MicroRNA delivery enhanced remyelination and functional recovery after spinal cord injury. Biomaterials 123565
Weickenmeier J, de Rooij R, Budday S, Ovaert TC, Kuhl E (2017) The mechanical importance of myelination in the central nervous system. J Mech Behav Biomed Mater 76:119–124
Wilson R, Brophy PJ (1989) Role for the oligodendrocyte cytoskeleton in myelination. J Neurosci Res 22(4):439–448
Wong J (2013) Neurotrophin signaling and Alzheimer's disease neurodegeneration– focus on BDNF/TrkB signaling. In: Trends in cell signal pathways in neuronal fate decision, pp 181–194
Woolley AG, Tait KJ, Hurren BJ, Fisher L, Sheard PW, Duxson MJ (2008) Developmental loss of NT-3 in vivo results in reduced levels of myelin-specific proteins, a reduced extent of myelination and increased apoptosis of Schwann cells. Glia 56(3):306–317
Wuerfel J, Paul F, Beierbach B, Hamhaber U, Klatt D, Papazoglou S et al (2010) MR-elastography reveals degradation of tissue integrity in multiple sclerosis. NeuroImage 49(3):2520–2525
Xiao J (2023) Thirty years of BDNF study in central myelination: from biology to therapy. J Neurochem 167(3):321–336
Xiao L, Ohayon D, McKenzie IA, Sinclair-Wilson A, Wright JL, Fudge AD et al (2016) Rapid production of new oligodendrocytes is required in the earliest stages of motor-skill learning. Nat Neurosci 19(9):1210–1217
Xie D, Ma Y, Gao C, Pan S (2024) Piezo1 activation on microglial cells exacerbates demyelination in sepsis by influencing the CCL25/GRP78 pathway. Int Immunopharmacol 142:113045
Yang K, He X, Wu Z, Yin Y, Pan H, Zhao X, Sun T (2022) The emerging roles of piezo1 channels in animal models of multiple sclerosis. Front Immunol 13:976522
Yang M, Martin CJ, Kowsari K, Jagielska A, Van Vliet KJ (2025) Myelin ensheathment and drug responses of oligodendrocytes are modulated by stiffness of artificial axons. PLoS One 20(1):e0290521
Young KM, Psachoulia K, Tripathi RB, Dunn SJ, Cossell L, Attwell D et al (2013) Oligodendrocyte dynamics in the healthy adult CNS: evidence for myelin remodeling. Neuron 77(5):873–885
Zendedel A, Kashani IR, Azimzadeh M, Pasbakhsh P, Omidi N, Golestani A et al (2016) Regulatory effect of triiodothyronine on brain myelination and astrogliosis after cuprizone-induced demyelination in mice. Metab Brain Dis 31(2):425–433
Zhan R, Wang F, Wu Y, Wang Y, Qian W, Liu M et al (2018) Nitric oxide induces epidermal stem cell de-adhesion by targeting integrin β1 and Talin via the cGMP signalling pathway. Nitric Oxide 78:1–10
Zhan L, Wang X, Lv Y, Deng J, Nan L, Ke Q et al (2025) Anisotropic single-layer graphene/nanodiamond loaded PCL conduits provide biophysical cues to manipulate nerve biomechanics and bioelectric function in the restoration of nerve microenvironment. Adv Funct Mater 35(26):2419411

Zhang X, Weickenmeier J (2023) Brain stiffness follows cuprizone-induced variations in local myelin content. Acta Biomater 170:507–518

Zhang J, Green MA, Sinkus R, Bilston LE (2011) Viscoelastic properties of human cerebellum using magnetic resonance elastography. J Biomech 44(10):1909–1913

Zhang Y, Yang X, Wang C, Hu J, Deng S, Lan Q (2025) Mechanosensitive Piezo1 channel: an emerging target in demyelination disease. Front Cell Neurosci 19:1556892

Zou Y, Zhang WF, Liu HY, Li X, Zhang X, Ma XF et al (2017) Structure and function of the contactin-associated protein family in myelinated axons and their relationship with nerve diseases. Neural Regen Res 12(9):1551–1558

Chapter 16
Application of Bioactive Natural Compounds to Promote Cell Proliferation and Migration in the Wound Healing Process

Kamila Kozłowska-Tomczyk, Zuzanna Flis, Justyna Gogola-Mruk, Piotr Szatkowski, Edyta Molik, and Anna Ptak

Abstract Wound healing is a complex biological process that restores the integrity and strength of the skin. Keratinocytes and fibroblasts play pivotal roles in tissue repair through their interactions and signaling pathways. Keratinocytes proliferate to re-epithelialize the wound, while fibroblasts contribute to the formation of new extracellular matrix. Although cell proliferation is essential for wound healing, excessive proliferation can lead to scar formation. Therefore, strategies that balance proliferation and migration, along with proper mechanotransduction, are crucial for efficient wound repair. Sheep and camel milk contain numerous bioactive substances, such as insulin, lactoferrin, proline, and conjugated linoleic acid, that have demonstrated promising effects on wound healing. While direct evidence on skin cell proliferation and migration is limited, the properties of these bioactive substances suggest beneficial effects in these areas. Therefore, incorporating these milks into hydrogel dressings can promote wound healing. Moreover, improving the cell-material mechano-interaction in such hydrogels could lead to the development of novel regenerative medical therapies.

K. Kozłowska-Tomczyk · J. Gogola-Mruk · A. Ptak (✉)
Laboratory of Physiology and Toxicology of Reproduction, Institute of Zoology and Biomedical Research, Faculty of Biology, Jagiellonian University, Krakow, Poland
e-mail: anna.ptak@uj.edu.pl

Z. Flis · E. Molik
Department of Animal Biotechnology, Faculty of Animal Science, University of Agriculture in Krakow, Krakow, Poland

P. Szatkowski
Department of Glass Technology and Amorphous Coatings, Faculty of Materials Science and Ceramics, AGH University of Krakow, Krakow, Poland

M. Kloc, J. Wosik (eds.), *Mechanotransduction and Mechanosensing*, Results and Problems in Cell Differentiation 78, https://doi.org/10.1007/978-3-032-19509-8_16

16.1 Introduction

The skin serves as a protective barrier against the external environment. To prevent the risk of infection, any damaged tissue must be repaired promptly. Wound healing is a dynamic, multi-step biological process coordinated through complex interactions among diverse cell types and signaling networks (Takayama and Aoki 2012; Mayya et al. 2021). Skin regeneration begins with hemostasis, followed by inflammation, proliferation, and matrix remodeling (Wilkinson and Hardman 2020). The crucial role in the wound healing process is hemostasis, which controls bleeding and prevents infection, thereby creating a controlled environment that supports subsequent phases. This step involves complex crosstalk between platelets, coagulation mechanisms, and immune responses (Häkkinen et al. 2015; Hassan and Ismail 2019). During the inflammatory phase, immune cells such as neutrophils and macrophages are recruited to the damaged tissue and play a role in eliminating pathogens and debris (Wang et al. 2022; Cioce et al. 2024; Raziyeva et al. 2021; Boniakowski et al. 2017). However, sustained inflammatory responses can contribute to the development of chronic wounds, which are characterized by slow healing and an increased risk of pathological scarring or fibrosis (Shukla et al. 2019; Krejner et al. 2016). The proliferation phase involves multiple cellular activities. Fibroblasts migrate and proliferate to produce collagen and the extracellular matrix, while keratinocytes restore the epidermal barrier during re-epithelialization (Wang et al. 2023; Kim et al. 2020). Blood vessels formed during angiogenesis supply nutrients and oxygen to the tissue (Raina et al. 2021). Myofibroblast-driven contraction helps to reduce the wound size and facilitate closure (Lacina et al. 2024). Matrix remodeling encompasses the dynamic interplay between the extracellular matrix (ECM) and multiple cell types. This process is important for maintaining both structural integrity and normal tissue function (Abdelwahab et al. 2025). Understanding the biological cascade of wound healing is fundamental to the developing of efficient therapeutic approaches.

16.2 Mechanisms and Factors Influencing Wound Healing

16.2.1 Cell Migration and Proliferation

16.2.1.1 Fibroblasts

Cell migration and proliferation play a crucial role in the restoration of wounded tissue, governed by the extracellular matrix (ECM), cell-to-cell signaling, and mechanical stimuli (Takayama and Aoki 2012; Valero et al. 2014). Fibroblasts and epithelial cells migrate to the site of injury to initiate tissue repair. Along with keratinocytes, they proliferate to fill the wound gap and replace the damaged cells (Kananivand et al. 2025; Kiwanuka et al. 2012; Choi et al. 2014). Fibroblasts are primarily responsible for producing and organizing collagen, the principal structural component of the ECM, thereby ensuring the mechanical stability and integrity of tissues (Yasmin et al. 2025; Plikus et al. 2021). Beyond ECM synthesis, fibroblasts actively remodel the matrix through mechanisms such as phagocytosis and intracellular degradation. This continuous remodeling is fundamental to effective tissue regeneration and the maintenance of homeostasis (Marconi et al. 2021). Moreover, integrins, which are a family of ECM-binding receptors, mediate crucial interactions between fibroblasts and their surrounding matrix, thereby regulating fibroblast behavior, signaling pathways, and ECM remodeling (Veerman and Prakash 2023). During the process of tissue repair, fibroblasts undergo a phenotypic transition into myofibroblasts. These cells exhibit greater proliferative and synthetic potential (Darby and Hewitson 2007). The formation of granulation tissue and subsequent wound contraction occurs largely due to the activity of myofibroblasts, which display both morphological and biochemical characteristics of fibroblasts and smooth muscle cells. Myofibroblasts can originate from different sources: resident connective tissue fibroblasts near the wound, mesenchymal stem cells, pericytes, bone marrow-derived circulating fibrocytes, and epithelial cells (Ko et al. 2019; Marconi et al. 2021).

16.2.1.2 Fibroblasts-Related Signaling Pathways

Numerous studies have demonstrated that various signaling pathways regulate fibroblast migration and proliferation. A crucial role in fibroblast activity during wound healing is played by the Wnt/β-catenin pathway. Experiments using a Wnt pathway inhibitor and β-catenin knockdown demonstrated inhibited migration of these cells (Wang et al. 2017). Studies investigating mesenchymal stem cell-derived exosomes revealed that they induce Wnt pathway activation in recipient cells, promoting wound closure; this effect was abolished upon application of a Wnt signaling inhibitor (Li et al. 2022). Another essential pathway for migration and proliferation, as well as for cell survival is the PI3K/Akt pathway (Kasowanjete et al. 2024). Studies using a Rac1 inhibitor, a downstream effector of PI3K, demonstrated that this kinase is essential for fibroblast migration (Kanazawa et al. 2010). A study by Zhang and colleagues demonstrated that Notoginsenoside Ft1 enhanced proliferation and collagen synthesis in human dermal fibroblast cells (HDF-a) through the PI3K/AKT/mTOR signaling pathway (Zhang et al. 2016). It has also been demonstrated that mTORC1 activation in fibroblasts stimulates both cellular proliferation and collagen synthesis, facilitating wound closure, while simultaneously posing a risk of excessive scarring and fibrotic tissue formation (Hu et al. 2020). The JNK pathway, a downstream effector of PI3K, is also involved in fibroblast migration, and its inhibition impairs this process (Kanazawa et al. 2010). The ERK signaling pathway promotes the expression of c-Fos and FAK, proteins that are fundamental to the regulation of cellular migration and proliferative responses, which was confirmed in rat studies (Cheng et al. 2004). Conversely,

Table 16.1 Signaling pathways and molecules that are involved in the fibroblasts and keratinocytes proliferation and migration

Cell type	Effect	Signaling pathways	Signaling molecules
Fibroblasts 	Proliferation migration	Wnt/β-catenin (Wang et al. 2017; Li et al. 2022)	bFGF, marine polysaccharides (Wang et al. 2017; Kalavathy et al. 2025)
		PI3K/Akt (Kasowanjete et al. 2024; Kanazawa et al. 2010; Zhang et al. 2016; Hu et al. 2020)	Rac 1, Notoginsenoside Ft1, bFGF, PDGF (Kanazawaet al. 2010; Zhang et al. 2016; Kim et al. 2008)
		ERK (Cheng et al. 2004)	bFGF (Kanazawa et al. 2010)
		JAK2/STAT3 (Vij et al. 2008)	hBD-3 and PDGF (Vij et al. 2008; Takahashi et al. 2021)
	Migration	JNK (Kanazawa et al. 2010)	PI3K and Rac1 (Kanazawa et al. 2010)
		NF-κB (Kondaveeti et al. 2025)	Marine polysaccharides (Chang et al. 2025)
Keratinocytes 	Proliferation migration	EGFR (Talati et al. 2018; Wu et al. 2023)	Thrombin, PAR-1, Gintonin, Pt5-1c,HB-EGF (Talati et al. 2018; Wu et al. 2023; Won et al. 2023)
		Notch1/Twist1 (Hsu et al. 2017)	FIR (Hsu et al. 2017)
		BMP (Lewis et al. 2015)	Noggin (Lewis et al. 2015)
		PI3K/Akt (Zhang et al. 2015)	Marine polysaccharides, KGF, miR-126 (Kondaveeti et al. 2025; Zhang et al. 2015; Chang et al. 2019)
		RhoA-JNK (Zhang et al. 2011)	Activin B (Zhang et al. 2011)
	Migration	Calcineurin/NFAT (Brun et al. 2014)	T-plastin (Brun et al. 2014)

Abbreviations: *Bfgf* basic fibroblast growth factor, *PDGF* platelet-derived growth factor, *hBD-3* human β-defensin-3, *PAR-1* protease-activated receptor 1, *KGF* keratinocyte growth factor, *EGFR* epidermal growth factor receptor, *HB-EGF* heparin-binding EGF-like growth factor *FIR* far infrared, *BMP* bone morphogenetic proteins

NF-κB activation induces the expression of pro-inflammatory cytokines and other mediators that promote fibroblast migration and contribute to wound healing. It has been shown that fucoidan inhibited NF-κB pathway activation, consequently decreasing the expression of pro-inflammatory cytokines and mitigating oxidative stress at the wound site (Kondaveeti et al. 2025) (Table 16.1).

16.2.1.3 Mechanotransduction in Fibroblasts Migration and Proliferation

Mechanosensing and mechanotransduction play essential roles in enabling fibroblasts to respond to mechanical signals from their surroundings, thereby regulating their migration. The key sites of mechanosensing are focal adhesions (FAs), which connect ECM with the actomyosin network. This is a main mechanotransduction pathway that generates signals driving cell migration. Recent studies suggest that microtubules take part in mechanosensing by controlling α-tubulin acetylation, thereby tuning the mechanosensitivity of focal adhesions and the activity of RhoA, a key regulator of actomyosin contractility (Seetharaman et al. 2022). RhoA, a member of the Rho GTPase family, is also a main mediator in mechanotransduction from the cell surface directly to the nucleus. Interfering with the linkage between the nucleus and the cytoskeleton, or completely removing the nucleus, decreases the levels of active RhoA, which in turn modifies cell mechanics and weakens the cellular response to mechanical cues (Burridge et al. 2019).

The FAK-ERK-MCP1 (focal adhesion kinase–extracellular signal-regulated kinases–monocyte chemoattractant protein-1) axis is a key mechanobiology mediator in skin, activated after injury. Mechanical forces intensify FAK activation via phosphorylation. FAK signaling couples ECM stress to the cytoskeleton, activating inflammatory cascades. Wong et al. showed that mechanical forces regulate scar formation through this pathway and that local FAK inhibition can effectively decouple mechanical tension from fibrosis. FAK inhibitor (FAKI) hydrogels have been successfully applied in wound therapy, resulting in reduced scarring and accelerated closure (Yao et al. 2024; Zhou et al. 2024). The Wnt/β-catenin pathway is another major contributor to mechanotransduction in wound healing. In response to physical force and Wnt signaling activation, β-catenin accumulates in the nucleus. Notably, fibroblasts with high β-catenin are observed in hypertrophic scars and keloids, implicating it in fibrotic skin diseases. Mechanotransduction activates canonical Wnt/β-catenin signaling, promoting lymphatic vascular patterning and valve development (Cha et al. 2016; Fan et al. 2024; Jia et al. 2023; Li et al. 2023).

Substrate stiffness also influences the dynamics of ERK signaling, primarily through receptor-ligand interactions. In soft microenvironments, EGFR expression is reduced, and the spatial pattern of EGF binding to the cell membrane is altered. As a result, ERK signaling is attenuated both with and without EGF stimulation. These findings indicate that the mechanical properties of the microenvironment may regulate ERK activity by influencing receptor availability and distribution (Farahani et al. 2021). Furthermore, studies have shown that fibroblasts cultured on stiff substrates exhibit higher ERK activation, which correlates with increased cell proliferation and survival, whereas on soft substrates, ERK activation is reduced, leading to decreased cell proliferation (Klein et al. 2009).

NF-κB signaling is another mechanotransduction pathway influencing fibroblast proliferation. NF-κB is activated in response to mechanical stress and various other stimuli, triggering the transcription of genes involved in cell proliferation and cell cycle regulation, such as CD1 (Kumar and Boriek 2003; Sée et al. 2004; Zuo et al. 2025).

16.2.1.4 Keratinocytes

Keratinocytes, the primary cells of the epidermis, migrate from the wound margins to repopulate the wound bed. Their coordinated migration and proliferation are essential for the effective re-epithelialization process, which involves the restoration of the wound surface by newly formed epithelial cells (Hu et al. 2012; Rousselle et al. 2019). Keratinocytes become activated in response to mechanical stress, electrical gradient, and exposure to pathogens, growth factors, cytokines, or hydrogen peroxide. Upon activation, keratinocytes at the wound margin undergo a partial epithelial–mesenchymal transition (EMT), acquiring enhanced migratory and invasive properties that support effective wound closure (Wilkinson and Hardman 2020; Shaw and Martin 2016; Li et al. 2007). Trailing keratinocytes adjust cell–cell adhesion to coordinate movement with the migrating epithelial sheet, while simultaneously secreting matrix metalloproteinases and depositing new ECM proteins to restore the basement membrane (Rousselle et al. 2019; Shaw and Martin 2016). Conversely, the extracellular matrix serves not only as a structural scaffold but also as a dynamic source of biochemical cues that promote keratinocyte migration and proliferation (Rousselle et al. 2019). Key components such as dermatopontin (DPT) and matrix metalloproteinases actively participate in this process (Wager and Leavesley 2015). It has been reported that DPT enhances keratinocyte migratory activity but does not promote cell proliferation. Its effect is mediated through paracrine signaling originating from dermal cells (Krishnaswamy and Korrapati 2014). Through integrin receptors, keratinocytes establish interactions with the structural proteins of the provisional matrix, thereby coordinating cell adhesion and movement (Di Persio et al. 2016; Koivisto et al. 2014). Hair follicle stem cells are activated to proliferate, with their progeny migrating either from the follicle in partial-thickness wounds or from the epidermal margin in full-thickness wounds to contribute to wound re-epithelialization. Specific stem cell niches are selectively activated or recruited to contribute to wound re-epithelialization (Wilkinson and Hardman 2020; Ito et al. 2005, 2007).

16.2.1.5 Keratinocyte-Related Signaling Pathways

A crucial pathway in keratinocytes associated with migration and proliferation is EGFR signaling. It has been reported that EGFR stimulation activates ERK1/2, Akt, MAPK, and STAT3 pathways, thereby promoting keratinocyte migration and proliferation (Talati et al. 2018; Wu et al. 2023). Activation of the Notch1 signaling pathway enhances keratinocyte migration and proliferation via the Notch1/Twist1 axis. It has been shown that Notch1 in keratinocytes plays an essential role in far-infrared-induced wound healing (Hsu et al. 2017). As in fibroblasts, activation of the PI3K/Akt pathway in keratinocytes facilitates both migration and proliferation, underscoring its importance in epidermal regeneration (Zhang et al. 2015). A keratinocyte-specific pathway is the Calcineurin/NFAT signaling cascade, which is associated with their growth, differentiation, and migration. Studies have confirmed that T-plastin, regulated downstream of the Calcineurin/NFAT pathway, is involved in this process (Brun et al. 2014). Studies on mice have shown that the RhoA-JNK pathway is also involved in keratinocyte proliferation and migration. RhoA has been reported to accelerate wound healing by increasing p-JNK and p-cJun expression, whereas its inhibition retards the healing process (Zhang et al. 2011).

16.2.1.6 Mechanotransduction in Keratinocytes Migration and Proliferation

Mechanosensing and mechanotransduction play very important roles also in keratinocyte migration and proliferation involved in the wound healing process. Similar to fibroblasts, FAs in keratinocytes are also involved in mechanotransduction signaling required for cell migration. It has been reported that the loss of intraflagellar transport protein (IFT20) results in a lower level and impaired reformation of FAs, as well as disrupted endocytosis and recycling of β1 integrins. As a result, hair follicle stem cells (HF-SCs) lose their ability to migrate during wound healing in in vivo studies. These data demonstrate the importance of this mechano-signaling in epidermal regeneration (Su et al. 2020).

The PI3K/Akt pathway is essential in epidermal keratinocyte differentiation, activated early in this process through EGFR, E-cadherin adhesion, and Src kinases. PI3K blockade causes keratinocyte death, underscoring the importance of PI3K/Akt signaling for proper cell function. Experiments show that stretching keratinocytes activates Akt, generating proliferative and anti-apoptotic signals, which inhibits cell loss and promotes tissue preservation (Zhang et al. 2024a). In another study, it has been also reported that accelerated stress relaxation facilitates cell cycle progression through a growth-dependent PI3K/AKT–p27Kip1 signaling cascade regulated by stretch-activated channels, which provides complementary information to the previous results (Di-Luoffo et al. 2021; Nam et al. 2019).

The RhoA/ROCK pathway is also central to mechanotransduction in repair. Animal data show that the RhoA-JNK axis regulates keratinocyte proliferation and migration; RhoA speeds healing by elevating p-JNK and p-cJun, while its inhibition slows the process. Mechanical strain activates RhoA via phosphorylation, stimulating downstream signaling that controls actin reorganization and cellular contractility (Park et al. 2023; Sivaraj et al. 2021).

16.2.1.7 Mesenchymal Stem Cells

Mesenchymal stem cells (MSCs) are involved in multiple mechanisms during the wound healing process. MSCs release a broad spectrum of growth factors, cytokines, and extracellular vesicles that influence the wound microenvironment. Through these paracrine mechanisms, MSC-derived factors facilitate angiogenesis, suppress inflammatory responses, and stimulate the proliferation and migration of cells critical to wound regeneration (Hocking and Gibran 2010; Dittmer and Leyh 2014; Bluguermann et al. 2013). They are engaged in tissue formation and re-epithelialization by promoting fibroblast and keratinocyte proliferation (Sriramulu et al. 2020; Kananivand et al. 2025; Wang et al. 2022a). Besides affecting other cells, MSCs can transform into multiple cell types, such as fibroblasts, keratinocytes, and endothelial cells, which play crucial roles in tissue repair, in wound contraction, re-epithelialization, and angiogenesis. They also move toward the wound area, where they actively participate in the regeneration of damaged tissue (Kananivand et al. 2025; Srinivasan et al. 2018; Wu et al. 2007). Studies in mice showed that MSCs, like other cell types, participate in remodeling the ECM by stimulating collagen synthesis and reinforcing the organization of repaired tissue. These actions help limit fibrosis and enhance the mechanical strength of the wound (Kananivand et al. 2025; Casado et al. 2014).

16.2.1.8 Mechanotransduction in MSCs

Mechanotransduction plays a critical role in regulating the function of MSCs during wound healing. These cells are sensitive to mechanical stimuli such as fluid flow and matrix stiffness, which can modulate their differentiation and functional properties (Raman et al. 2022). Similar to keratinocytes and fibroblasts, focal adhesions (FAs) play an important role in this process by influencing MSC differentiation. The formation of focal adhesions involves numerous structural proteins, such as integrin β-subunits, vinculin, talin, and elements of the actin cytoskeleton (Steward and Kelly 2015).

The composition of ECM significantly affects MSCs behavior due to mechanical forces. Several studies show that ECM stiffness determines differentiation via transcriptional co-activators YAP and TAZ (Dupont et al. 2011; Tschumperlin et al. 2013). Park et al. reported that stiff substrates influence the expression of smooth muscle cell marker ASMA in MSCs, whereas soft substrates result in the expression of chondrogenic and adipogenic markers. The same group showed that TGF-β has an impact on the higher expression of the smooth muscle cell marker in cells cultured on stiff substrates. Meanwhile, soft substrates contribute to suppressed adipogenic marker expression and increased chondrogenic marker expression. These data show that mechanotransduction, together with biochemical stimuli, influences MSCs cellular responses (Park et al. 2011; Tschumperlin et al. 2013).

16.2.2 Growth Factors, Cytokines, and Other Factors

Growth factors and cytokines are key mediators of tissue repair. Basic fibroblast growth factor (bFGF) stimulates fibroblast proliferation and migration by activating multiple signaling pathways, such as Wnt/β-catenin, PI3K/Akt, and ERK (Wang et al. 2017; Kanazawa et al. 2010). Recent studies have also shown that the Wnt/β-catenin pathway can be stimulated by marine polysaccharides derived from *Codium tomentosum* (Kalavathy et al. 2025). Moreover, marine polysaccharides from *Kappaphycus alvarezii* modulate the NF-κB pathway (Chang et al. 2025). Platelet-derived growth factor (PDGF), a chemoattractant for fibroblasts during wound healing, also plays a crucial role in fibroblast proliferation and migration (Wilkinson and Hardman 2020). It has been reported that PDGF stimulation enhances PI3-kinase signaling and promotes migration (Kim et al. 2008). Moreover, PDGF stimulation activates STAT3 and induces SOCS3 expression promoting IL-8 chemokine release in human corneal fibroblasts. PDGF-driven proliferation, migratory activity, and IL-8 production in these cells were mediated through the JAK2–STAT3 signaling pathway (Vij et al. 2008). Recent studies indicate that human β-defensin-3 (hBD-3) also acts as an activator of JAK2 and STAT3 by phosphorylating fibroblast growth factor receptor 1 (FGFR1). As a result, scientists have proposed hBD-3 as a potential therapeutic agent for wound healing due to its ability to promote tissue repair, angiogenesis, and fibroblast activation (Takahashi et al. 2021). The EGFR signaling pathway in keratinocytes can be stimulated by multiple molecular factors, thereby enhancing their migration and proliferation. One of these factors is thrombin, activating the protease-activated receptor 1 (PAR-1), which subsequently transactivates the epidermal growth factor receptor (EGFR), leading to Smad2 phosphorylation and ERK1/2 activation, thereby facilitating keratinocyte migration and proliferation (Talati et al. 2018). Another EGFR activator, gintonin, a ginseng-derived compound, triggers the release of heparin-binding EGF-like growth factor (HB-EGF), which in turn promotes keratinocyte migration and proliferation (Won et al. 2023). Additionally, EGFR can be activated by the antimicrobial peptide Pt5-1c, which stimulates the Akt, MAPK, and STAT3 signaling pathways (Wu et al. 2023). Similar to fibroblasts, the PI3K/Akt pathway in keratinocytes can be modulated by marine polysaccharides. Studies have shown that polysaccharides such as ulvan and carrageenan promote keratinocyte migration and proliferation (Kondaveeti et al. 2025). Another factor that stimulates this pathway in keratinocytes is keratinocyte growth factor (KGF). It has been reported that KGF enhances keratinocyte proliferation, migration, and adhesion by interacting with the keratinocyte growth factor receptor (KGFR) and activating downstream signaling cascades, including the PI3K-AKT-mTORC1 pathway (Zhang et al. 2015). Furthermore, studies using the human HaCaT keratinocytes cell line have shown that overexpression of miR-126 activates the PI3K/Akt pathway, thereby enhancing cell viability, colony formation, and migration (Chang et al. 2019). An example of a factor that modulates keratinocyte migration and proliferation through the RhoA-JNK-cJun signaling pathway is Activin B. Activin B promotes keratinocyte and hair follicle cell proliferation at wound sites and accelerates wound closure through the RhoA-JNK-cJun signaling pathway. Inhibition of RhoA or JNK disrupts this process, highlighting the critical role of activin B in epithelial wound repair and its potential therapeutic application (Zhang et al. 2011). In contrast, bone morphogenetic proteins (BMPs) inhibit keratinocyte proliferation and migration, and elevated levels of BMP ligands have been observed in chronic wounds. BMP antagonists, such as Noggin, promote keratinocyte migration and proliferation. Accordingly, modulation of BMP signaling with inhibitors such as Noggin may offer a novel therapeutic strategy to enhance cutaneous wound repair (Lewis et al. 2015). Collectively, these findings highlight the complex network of molecular regulators that control fibroblast and keratinocyte behavior during the wound healing process (Table 16.2).

16.2.2.1 Role of Growth Factors and Cytokines in Mechanotransduction

Mechanical signals originating from damaged tissue, such as stiffness, stretching, shearing, tension, and load, are detected by cells and converted into biochemical signals through a process known as mechanotransduction. These signals regulate the activation, production, and response of growth factors, cytokines, and other mediators, including extracellular matrix proteins and enzymes. These substances then regulate cell proliferation and migration, as well as altering the extracellular matrix structure and cell contractility. This creates a feedback loop that controls the wound healing or fibrosis process (Yin et al. 2022).

TGF-β1 and mechanical tension are both essential for the differentiation of fibroblasts into myofibroblasts, which contract in order to close the wound. However, excessive tension and high levels of TGF-β1 contribute to scarring and fibrosis (Van de Water et al. 2013).

Integrins/focal adhesion complexes (FAs) bind to the ECM. Under force, they aggregate, activating FAK/Src. This then recruits talin and vinculin, promoting cytoskeletal remodeling via the MAPK/PI3K pathway (Katoh 2025). The mechanical forces exerted by integrins and matrix stiffening promote the activation of latent TGF-β (Yin et al. 2022; Pang et al. 2023). Blocking αv integrins may reduce TGF-β activation and fibrosis (Pang et al. 2023).

The mechanical signaling pathway involving YAP/TAZ also works with TGF-β to sustain the process of fibrosis (Yin et al. 2022). YAP/TAZ (Hippo pathway effectors) activity links the mechanical state to the production of signals that promote repair. In the case of cellular tension or high stiffness, YAP/TAZ undergo nuclear translocation. This contributes to the upregulation of transcription of proliferative and pro-matrix genes and the modulation of growth factor expression (e.g., TGF-β) (Yu Lin et al. 2025; Dupont et al. 2011). Growth factors, particularly TGF-β1, promote ECM deposition and cross-linking, which increases tissue stiffness. In addition, they increase the activation of mechanosensors and YAP/TAZ signaling. Positive feedback can cause the repair process to shift toward fibrosis, and breaking this loop can be used for therapeutic purposes (Ragazzini et al. 2022). Reducing the

Table 16.2 Key growth factors and cytokines and their fundamental roles in mechanotransduction

Growth factors and cytokines	Mechanotransduction role	Outcome in wound healing
TGF-β	Present in ECM in a latent form; physically activated by mechanical tension induced by av. integrins, induction of fibrosis (Yin et al. 2022; Pang et al. 2023; Van De Water et al. 2013)	Myofibroblast formation, ECM deposition, scarring (Yin et al. 2022; Pang et al. 2023; Van De Water et al. 2013; Ragazzini et al. 2022)
PDGF	Acts on fibroblasts sensitive to mechanical stimuli; increased matrix stiffness intensifies its effect (Chandra et al. 2025)	Fibroblast proliferation and migration into wound (Chandra et al. 2025)
VEGF	VEGF signaling enhanced by tension; endothelial cells respond to mechanical stretching (Miller and Sewell-Loftin 2022; Husain et al. 2022)	Angiogenesis and vascular remodeling (Miller and Sewell-Loftin 2022; Husain et al. 2022)
FGF/EGF	Cytoskeletal tensions necessary to effectively transmit signaling regulating keratinocyte proliferation and migration (Abdelhakim and Ogawa 2025; Pang et al. 2023)	Re-epithelialization (Abdelhakim and Ogawa 2025; Pang et al. 2023)
IL-1β/TNF-α	Increase level under mechanical injury/strain; sensitize cells to mechanical stress (Patel et al. 2025; Gracey et al. 2020)	Early inflammation and cell recruitment (Patel et al. 2025; Gracey et al. 2020)
IL-6	Upregulated by stretching; stimulates fibroblast activation (Skutek et al. 2001; Johnson et al. 2020)	Transition from inflammation to repair (Skutek et al. 2001; Johnson et al. 2020)
IL-10	Anti-inflammatory cytokine; reduction of fibrosis caused by mechanical stress (Patilas et al. 2024)	Tissue homeostasis and regeneration (Patilas et al. 2024)

Abbreviations: *TGF-β* transforming growth factor beta, *PDGF* platelet-derived growth factor, *VEGF* vascular endothelial growth factor, *FGF* fibroblast growth factor, *EGF* epidermal growth factor, *IL-1β* Interleukin 1 Beta, *IL-6* Interleukin 6, *IL-10* interleukin 10, *TNF-α* tumor necrosis factor-alpha

load along with targeted modulation of mechanotransduction (integrin/YAP/TGF-β axis) and delivery of growth factors tailored to the specific stage of wound healing is the most evidence-based conceptual approach to wound treatment while minimizing scarring (Yin et al. 2022). Furthermore, YAP inhibitors (e.g., verteporfin) show promising effects in preventing scarring and promoting regenerative healing in preclinical models (Mascharak et al. 2025).

Endothelial cells respond to stretching and stiffness. Mechanical stresses regulate endothelial cell alignment and stabilize VEGF receptors to regulate angiogenesis in the granulation tissue. The PI3K/Akt/mTOR pathway, amplified by VEGF-A stimulation, is activated on stiffer substrates and correlates with increased endothelial cell proliferation and capillary formation (Miller and Sewell-Loftin 2022; Husain et al. 2022).

PDGF promotes the recruitment and proliferation of fibroblasts and mesenchymal cells in response to mechanical stimuli, and its levels are often elevated in areas of stress (Chandra et al. 2025). Moreover, FGF promotes cytoskeletal reorganization necessary for wound closure during keratinocyte and fibroblast proliferation. Mechanical stresses imposed on tissue enhance FGF and EGF signaling, which are important in the process of re-epithelialization (Abdelhakim and Ogawa 2025). During re-epithelialization, EGF enhances cell migration by activating ERK and Pi3K pathways when cells detect mechanical stretching. Mechanical tension in the extracellular matrix informs the cell where to move and how strongly to attach (Pang et al. 2023).

However, the level of growth factors and mechanics must be balanced. Too low mechanical tension and growth factor levels lead to poor regeneration or chronic wounds. On the other hand, excessive mechanical tension and too high TGF-β levels result in hypertrophic scars or keloids (Barnes et al. 2018; Zhou et al. 2023).

The utilization of biomaterials and dressings that are capable of adjusting stiffness or topography, or that can deliver growth factors in a stage-specific manner (e.g., hydrogels with sustained release of growth factors), has been demonstrated to improve healing outcomes in wound models (Wang et al. 2025).

Mechanical signals have been demonstrated to modify immune signals and the resolution of inflammation by regulating cytokines, which have a role in the coordination of inflammation. Mechanical stiffness has been demonstrated to prolong the inflammatory response. In cases of soft, susceptible wounds, the process of inflammation is known to subside more rapidly. This phenomenon has been observed to be concomitant with enhanced regeneration (Wong et al. 2011). Mechanical stress raises pro-inflammatory cytokines IL-1β and TNF-α in fibroblasts. This reaction is a key element in the early inflammatory process, in which stress sensitizes cells to release these molecules, which in turn attract other immune cells to the site of injury. This creates a cycle that can perpetuate inflammation and potentially lead to further tissue damage (Patel et al. 2025; Gracey et al. 2020). Stretching also increases IL-6 levels, which in turn activate fibroblasts and promote the transition from the inflammatory phase to the reparative phase of tissue healing (Skutek et al. 2001). This process involves IL-6 signaling triggering fibroblast activation and the production and remodeling of the extracellular matrix (ECM). While this is critical for repairing damaged tissue, excessive ECM production can contribute to fibrosis (Johnson et al. 2020). However, IL-10 is an anti-inflammatory cytokine which reduces fibrosis caused by mechanical stress. This leads to the maintenance of tissue homeostasis and regeneration by balancing the pro- and anti-inflammatory responses (Patilas et al. 2024).

16.2.3 *Inflammation*

Inflammation is an essential part of the wound healing process, triggering the immune response required to eliminate pathogens and start tissue regeneration. The effective resolution of inflammation is vital for the transition to the proliferative phase. If inflammation is not properly resolved, it may result in the development of chronic wounds (Wang et al. 2022b). Imbalance in the immune response during wound healing, characterized by increased necrotic tissue, insufficient blood supply, elevated levels of pro-inflammatory cytokines, proteases, reactive oxygen species (ROS), and other signaling molecules, as well as infections with diverse pathogens, interferes with proper immune cell recruitment, disturbs proteolytic homeostasis, and hinders new blood vessel formation. As a result, the wound becomes arrested in the inflammatory stage, which contributes to delayed repair or chronic wound formation (Schilrreff and Alexiev 2022; Raziyeva et al. 2021). Naturally derived bioactive compounds with antioxidant, anti-inflammatory, and antimicrobial effects have potential to control chronic inflammation and support wound healing (Schilrreff and Alexiev 2022). It has been reported that wounds in diabetic patients often exhibit sustained and chronic inflammation, which disrupts the normal healing process. This is driven by hyperglycemia-related oxidative stress, immune system abnormalities, and compromised formation of new blood vessels (Porel et al. 2025; Yue et al. 2022; D'souza and Addepalli 2018). Moreover, studies have shown that diabetes can disrupt the normal progression of wound healing by impairing cellular proliferation (Campbell and Newgard 2021). To sum up, maintaining a balanced inflammatory response is crucial for progressing through the subsequent phases of healing and achieving proper wound repair.

16.3 Cell Models in Wound Healing Research

16.3.1 *In Vitro Cell Models*

In vitro cell models are commonly used in standard experiments in wound healing research. Due to their simplicity and low cost, 2D cell monolayer assays are among the most popular methods. Wound healing assays are the most affordable, well-established techniques for examining cell proliferation, migration, polarity, extra-cellular matrix remodeling, and actin cytoskeleton organization. In this method, a scratch is made in a confluent cell layer, creating an open gap that cells progressively fill as they migrate. The duration of gap closure depends on the cell type, culture conditions, and the initial wound width (Tong and Wang 2017). Some quantitative software tools have been developed to enhance the analysis of cell migration into artificially created gaps in confluent monolayers. For example, the ATCA tool was designed for image analysis, providing high accuracy in identifying cell-free areas and offering screening capabilities (Yin et al. 2023). Because migrating cells are surrounded by other cells and the ECM in vivo, 3D culture models much better reflect the natural environment. These models also enable more precise simulation of the mechanical forces involved. They serve as powerful tools for investigating mechanotransduction processes during wound healing, in which physical stimuli regulate cellular migration, proliferation, and differentiation through biochemical signaling pathways (Kural and Billiar 2013; Lam and Calvo 2019). Mechanotransduction signaling that facilitates fibroblast migration can accelerate wound closure. For example, softer substrates combined with optimal mechanical cues can promote faster and more directed fibroblast movement (Mkrtschjan et al. 2018). Furthermore, modifying the wound microenvironment, through strategies such as incorporating fibrillar scaffolds or applying controlled mechanical stimulation, can enhance fibroblast motility and support more efficient tissue regeneration (Qin et al. 2015). Building on these approaches, 3D wound healing assays commonly employ constructs that mimic human skin. These are often called skin substitutes or bioengineered skin (Stamm et al. 2016). Numerous products have been developed as grafts to treat acute and chronic wounds, including Hyalograft 3D, Apligraf, and the TissueTech Autograft System (Groeber et al. 2011). Most constructs used in wound healing assays are designed as bilayer systems that integrate both dermal and epidermal elements. The bottom layer replicates the dermis and typically consists of a scaffold populated with fibroblasts. Commonly used scaffolds include collagen-based hydrogels, but alternatives such as decellularized dermis, synthetic polymers, and glycosaminoglycans can also be used. On top of this dermal-like layer, an epidermal layer of keratinocytes is cultured, which may be derived from either primary cells or immortalized cell lines (Stamm et al. 2016). Co-culture systems incorporate different cell types to more closely replicate the in vivo environment. For example, a 3D model combining keratinocytes and fibroblasts can promote keratinocyte migration and provide a more faithful representation of the wound healing process (Iyer et al. 2018). Recent studies have shown that human gingival mesenchymal stem cells (hGMSCs) incorporated into 3D bioengineered Matrigel scaffolds also exhibit strong potential for supporting local tissue regeneration and stimulating collagen fiber formation in wound models (Xu et al. 2024). Another promising approach in wound healing research is the use of MSC-conditioned medium, which contains a wide range of bioactive molecules secreted by MSCs. Studies have shown that this medium enhances the proliferation and migration of dermal fibroblasts, and wounds treated with it heal faster and exhibit less scarring (Li et al. 2017). Thus, the field of wound healing is advancing dynamically, driven by the integration of diverse stem cell-based models, innovative biomaterials, and cutting-edge biotechnological approaches. Ongoing scientific efforts are essential to address existing limitations, such as ensuring cell viability, optimizing scaffold design, and improving functional tissue integration. With continued research and technological progress, these strategies hold great promise for translating experimental findings into effective clinical therapies for tissue repair and wound healing.

16.4 Properties and Therapeutic Potential of Natural Bioactive Substances of Sheep and Camel Milk in the Treatment of Wounds

In recent years, there has been a growing interest in using sheep and camel milk in regenerative medicine and biomedicine. The milk of these animals contains valuable natural bioactive substances that support tissue regeneration. It is particularly important in the treatment of difficult-to-heal wounds, diabetic wounds, and cancer-related wounds (Szatkowski et al. 2024a). The milk of these animals is distinguished by its rich composition of bioactive substances, including lactoferrin, proline, conjugated linoleic acid (CLA), and orotic acid, as well as vitamins and trace elements that exhibit strong anti-inflammatory and antioxidant properties. The insulin-like proteins present in sheep and camel milk may also play a key role in supporting wound healing, including that of diabetic wounds. These bioactive peptides may enhance immune and anti-inflammatory responses in wounds, thereby accelerating wound healing (Ebaid et al. 2015). Sheep and camel milk contain significantly higher concentrations of bioactive compounds than cow's milk, making these two raw materials particularly promising for use in designing hydrogel dressings for treating wounds of various origins. Thanks to their wound healing properties, natural milk components used in hydrogels can improve patients' quality of life and increase the effectiveness of treatment for chronic and difficult-to-heal wounds (Szatkowski et al. 2024a). Due to its natural origin and non-toxicity, milk has a high biological potential and numerous applications in cosmetics and dermatology.

16.4.1 Proline

The amino acid proline is found in high concentrations in milk, particularly sheep's milk. It plays a key role as a precursor of arginine and an initiator of protein synthesis, mainly collagen (Flis et al. 2023). Sheep milk contains 537 mg of proline per 100 g (Nong and Hsu 2021), while camel milk contains 129 mg of proline per g of casein (Rafiq et al. 2016). Due to its properties, proline may play a significant role in skin regeneration, particularly in wound healing and extracellular matrix reconstruction. Collagen is the main extracellular protein in the granulation tissue of a healing wound, and its synthesis increases immediately after injury, contributing to the strength and integrity of the tissue matrix (Badr 2013). Proline is essential for active collagen synthesis, and its local availability in the wound can accelerate this process (Thangavel et al. 2017). Proline is also a precursor of hydroxyproline, a key compound for the structure of the skin's extracellular matrix (Kumar Srivastava et al. 2016). In wounds where collagen synthesis is impaired, such as diabetic wounds or cancer-affected tissues, the supply of proline may support matrix reconstruction. Studies on a mouse model of diabetes have shown that hydroxyproline from camel milk supports tissue regeneration (Badr 2013). Local administration of proline improves healing of postoperative wounds in rats. A study by Aydin et al. (2019) observed faster wound healing, manifested by accelerated re-epithelialization and increased wound contraction. Proline also improves the quality of newly formed tissue and increases skin elasticity and stability (Rithika et al. 2025). Proline also influences the immune response and exhibits antibacterial and antioxidant effects in the wound environment (Flis et al. 2023). Compared to cow milk, camel milk exhibits greater antioxidant activity, which is likely due to its casein structure and higher proline content (Izadi et al. 2019). Incorporating proline into a hydrogel matrix may be a promising strategy for designing effective wound dressing materials for diabetic wounds. Studies have shown that proline-enriched hydrogels exhibit enhanced antioxidant activity, improved cytocompatibility, and enhanced cell proliferation (Thangavel et al. 2017). Furthermore, they demonstrate greater thermal stability than conventional hydrogels, which is essential for their technological properties (Thangavel et al. 2017). Proline is characterized by high biocompatibility and high water solubility, which further facilitates the development of new hydrogels (Zhu et al. 2011).

16.4.2 Lactoferrin

Available literature highlights the importance of lactoferrin, present in both sheep and camel milk, as a key component in the treatment of difficult-to-heal wounds of various origins, skin diseases, and cancers, including melanoma (Flis et al. 2025). Lactoferrin is a glycoprotein present in milk that performs many beneficial functions for the body. It exhibits antibacterial, antiviral, antifungal, immunoregulatory, and anti-inflammatory properties, making it a promising bioactive compound for wound healing (Flis et al. 2023; Brouwer et al. 2025). Lactoferrin content is 0.7–0.9 g/kg in sheep milk, 0.2–0.9 g/kg in camel milk, and 0.02–0.5 g/kg in cow milk (Alichanidis et al. 2016). Sheep and camel milk often exhibit stronger antimicrobial activity than cow milk, which is attributed to higher concentrations of lactoferrin and other whey proteins, among other things (El-Baky et al. 2021; Wiesner and Vilcinskas 2010). Lactoferrin exhibits antimicrobial activity through two main mechanisms (Cheng et al. 2025). The first involves binding iron, which limits the availability of this element to bacteria and fungi, hindering their growth and development (Brouwer et al. 2025; Moatsou and Sakkas 2019). The second mechanism involves direct interaction of lactoferrin with microorganisms, leading to their destabilization and lysis (Kell et al. 2020). These mechanisms work synergistically to provide effective defense against pathogens. Additionally, lactoferrin exhibits antiviral activity by blocking viral attachment to host cells, interfering with viral entry, inhibiting viral replication, and modulating the host immune response (Brouwer et al. 2025). Thanks to its numerous health-promoting properties, lactoferrin may play a significant role in supporting the healing of chronic wounds. Research by León Malaga et al. (2023) demonstrated that local administration of this bioactive protein supports wound healing. Lactoferrin derived from colostrum or milk promotes the regeneration, proliferation, and migration of keratinocytes and fibroblasts, among other things, by inhibiting inflammatory responses (Cheng et al. 2025; Takayama and Aoki 2012). Furthermore, it supports the formation of granulation tissue and stimulates the synthesis of collagen and hyaluronan. Chronic inflammation persists in the wound environment, resulting in a weakened immune response and a high risk of infection. Thanks to its anti-inflammatory properties, lactoferrin supports repair and protective processes, restoring the balance between defense mechanisms and tissue regeneration. Numerous studies on the use of lactoferrin in wound therapy are currently underway. Proposed solutions include gels with silver nanoparticles containing lactoferrin (Geneidy et al. 2025), exosomal lactoferrin derived from colostrum (Aydin et al. 2019), and lactoferrin-based biosensors (Pardoux et al. 2020). Particular attention has been paid to hydrogel dressings containing lactoferrin from sheep and camel milk, which represent a promising strategy for treating difficult-to-heal wounds, including diabetic wounds. These dressings provide continuous release of bioactive substances, supporting healing processes, eliminating bacterial infections, and reducing inflammatory reactions in the wound environment (Lin et al. 2025). Lactoferrin incorporated into hydrogels enhances the stability and functional properties of the dressing. Studies by Wang et al. (2022b) have shown that hydrogels containing lactoferrin and lithium magnesium silicate exhibit hemostatic and antibacterial properties and support healing and tissue regeneration in infected wounds.

16.4.3 Insulin

Insulin and insulin-like peptides play an important role in regenerative strategies, particularly in the treatment of chronic wounds. Sheep and camel milk provide numerous bioactive substances, such as insulin-like peptides, lactoferrin, proline, and CLA, that can work together to promote tissue repair. Incorporating these bioactive ingredients into dressings can promote healing where conventional therapies are ineffective. Modern approaches, such as cell therapies, can be costly, and their safety is still under investigation (Emanuelli et al. 2016). Insulin is a peptide hormone and growth factor that primarily regulates blood glucose levels. It also has anabolic effects and promotes regenerative processes. Insulin's role in wound healing includes stimulating protein synthesis, cell migration, and tissue growth. Insulin acts through several mechanisms: it regulates oxidative and inflammatory responses and stimulates cell migration and re-epithelialization. Insulin also modulates signaling pathways associated with IGF-1, VEGF, and TGF-β1 receptors, which facilitates angiogenesis and granulation tissue formation (Azevedo et al. 2016; Ching et al. 2011). In diabetic wounds, local insulin administration of insulin can reverse an impaired inflammatory response and accelerate healing by improving granulation tissue growth (Zhang and Lv 2016). Insulin goes far beyond its classical glucose-metabolizing role. Insulin promotes the migration and differentiation of keratinocytes and fibroblasts, increases collagen synthesis, inhibits apoptosis, reduces inflammatory cytokine production, and supports extracellular matrix formation (Ching et al. 2011). Sheep and camel milk contain insulin and natural insulin-like proteins, which have been shown to improve glucose metabolism (Behrouz et al. 2022; Bakri et al. 2023). Numerous studies support the efficacy of camel milk in the treatment of diabetes (Abdalla et al. 2015; Agrawal et al. 2020). Camel milk is characterized by a high concentration of insulin (approximately 52–59 units/liter) and other insulin-like substances (Agrawal et al. 2011; Ashraf et al. 2021). Sheep milk is also a promising source of insulin (approximately 0.44 μg/l) and peptides with antidiabetic and antihypertensive effects (Flis et al. 2023; Zhang et al. 2024b). This is also important for supporting the wound healing process. Numerous studies have demonstrated that local insulin administration has a beneficial effect on the healing of diabetic, burn, acute, and chronic wounds (Ramirez-GarciaLuna et al. 2023). Experimental studies have shown that local insulin administration accelerates wound healing in diabetic and non-diabetic rats (Apikoglu-Rabus et al. 2010). In second-degree burn models, insulin has been shown to shorten the inflammatory phase and improve re-epithelialization, granulation, wound contraction, and collagen deposition (Azevedo et al. 2019). Clinical reports and systematic reviews highlight that local insulin can improve postoperative wound healing in patients with diabetes, including by regulating E-cadherin and Ki-67 expression and promoting angiogenesis, while simultaneously having no effect on systemic glucose levels (Khattab et al. 2025). Insulin and insulin-like peptides from sheep and camel milk represent a promising component of supportive wound therapy. Their metabolic, anti-inflammatory, and anabolic properties can support healing processes. Combined with modern wound dressings, insulin can help restore metabolic balance and repair impaired cell signaling, particularly in burn, postoperative, and diabetic wounds (Hrynyk and Neufeld 2014). Integrating insulin with dressings, particularly hydrogel dressings, opens the door to precise and long-term wound therapy.

16.4.4 Orotic Acid

Orotic acid is an important component of milk and dairy products, with the highest concentration in ruminant milk being found in sheep milk (Bhattacharya et al. 2006; Flis and Molik 2021; Flis et al. 2022). The orotic acid content in the milk of different ruminant species varies. Sheep milk contains up to 20–400 mg/L of orotic acid, whereas its content in cow milk is lower, 20–100 mg/L (EFSA 2009). Camel milk is also an extremely important source of this nutrient. Orotic acid has several biological properties, including anticancer potential, support for circulatory system function, and participation in the body's detoxification processes (Marynowicz et al. 2023; Czauderna et al. 2021; Löffler et al. 2015). Furthermore, it can modulate inflammatory responses and support skin regeneration, making it a promising compound with a wide range of therapeutic applications (Flis et al. 2025). Orotic acid is involved in the regulation of genes crucial for cell, tissue, and organ development and is of great interest in medical research and drug development (Löffler et al. 2015; Nath et al. 2013). Due to its beneficial effects on the skin, it has also been used in the cosmetic and pharmaceutical industries (Parfenyuk et al. 2024). Orotic acid participates in the regeneration of epithelial cells, supports the function of the epidermal barrier, and supports skin DNA repair mechanisms (West et al. 2017). It also protects against oxidative stress, which helps maintain skin elasticity and integrity and delays the aging process (Parfenyuk et al. 2024; Jastrzębska et al. 2017). Orotic acid and its salts or derivatives have potential applications in the prevention and treatment of wounds, scars, atopic dermatitis, eczema, and psoriasis (Park et al. 2010). Research by Kim et al. (2023) indicates that orotic acid plays a significant role in promoting the balanced distribution of beneficial microbial species in the skin environment. Studies on dairy cows have shown that orotic acid concentrations in serum and milk are higher in animals with subclinical mastitis compared to healthy cows (Karatas et al. 2008). Elevated levels of orotic acid suggest that it may participate in tissue repair and the body's response to damage, indicating its potential as a biomarker of metabolic status and as a mechanism supporting tissue regeneration, including wound healing (Karatas et al. 2008). Its unique combination of anticancer, skin-protective, and microbiome-regulating properties makes orotic acid a promising candidate for innovative dermatological applications, particularly in skin cancer prevention and diabetic wound healing. Sheep milk, as the richest natural source of orotic acid, may provide a basis for the development of new bioactive dressing formulations with improved pharmacological properties.

16.4.5 Linoleic Acid

Camel and sheep milk fat contains the highest concentration of conjugated linoleic acid (CLA) among the milk of other ruminants. Studies on CLA content have shown that sheep and camel milk have the highest concentrations (0.8%), compared to goat milk (0.7%) and cow milk (0.7%), which is the most widely available (Alichanidis et al. 2016; Shibani et al. 2011). Milk fat from these animals also contains significant amounts of vaccenic acid, which is the physiological precursor of CLA (Alichanidis et al. 2016; Molik et al. 2021). The most important and functional CLA isomers are cis-9, trans-11, and trans-10, cis-12, which constitute approximately 85% and 10% of all natural CLA isomers, respectively (JL 2019). These isomers have anti-inflammatory, ant-obesity, antidiabetic, and anticancer effects (Basak and Duttaroy 2020). The cis-9, trans-11, and trans-10, cis-12 isomers also affect T lymphocyte populations and immunoglobulin subclasses (Bassaganya-Riera et al. 2012). The most biologically active isomer of CLA is cis-9, trans-11, which has been shown to inhibit the occurrence and progression of skin cancer (Wang and Jones 2004). In the study by Park et al. (2010), CLA supplementation accelerated skin wound healing by regulating antioxidant and anti-inflammatory functions. These studies encourage analysis of the superficial effects of CLA in milk on wound healing. In recent years, there has been growing interest in the therapeutic potential of CLA, for the treatment of inflammatory dermatoses, skin cancers, and difficult-to-heal wounds, including diabetic ulcers and postoperative wounds. In vitro and in vivo studies indicate that CLA inhibits melanoma cell proliferation (MacDonald 2000). The study by Hwang et al. (2007) demonstrated that local administration of the cis-9, trans-11 isomer in the HR-1 mouse model inhibits NF-κB activation and COX-2 expression by blocking the IKK and Akt pathways. This study provides evidence for the molecular mechanisms underlying CLA's anti-inflammatory and antitumor effects in the skin of hairless mice (Hwang et al. 2007). In a study by Tang et al. (2021), local application of CLA in a mouse model of atopic dermatitis improved skin barrier function and modulated the inflammatory response. CLA reduced the production of cytokines and pro-inflammatory enzymes (including IL-6, IL-1β, TNF-α, iNOS, and COX-2) and inhibited NF-κB pathway activation, indicating its potent anti-inflammatory properties (Tang et al. 2021). Additionally, CLA has been observed to have a beneficial effect on the epidermal barrier, including maintaining physiological pH and increasing skin hydration (Tang et al. 2021). Chronic inflammation in the wound area typically leads to insufficient and delayed healing (Hallan et al. 2025). Due to its anti-inflammatory properties, CLA present in milk may modulate the inflammatory response within the wound, limiting cytokine and pro-inflammatory enzyme expression, thereby promoting accelerated healing, improved tissue regeneration, and restoration of the skin barrier. Bergamo et al. (2013) suggested that combining the trans-10, cis-12 isomer with other bioactive substances may be an effective therapeutic strategy in the prevention and treatment of melanoma. The lipid sphere of camel milk exhibits anti-adhesive and antibacterial properties. Mansour et al. (2015) described a cream containing 25% sheep milk fat globules that accelerates burn wound healing, reduces inflammation, and accelerates wound contraction. Incorporating natural bioactive substances from sheep or camel milk into hydrogel dressings is increasingly promising strategy for treating difficult-to-heal wounds of various origins.

16.4.6 Role of Milk Compounds in Mechanosensing and Mechanotransduction

In the area of wound healing, mechanotransduction, i.e., the processing of mechanical signals into cellular signals, plays a key role in regulating the migration, proliferation, differentiation of fibroblasts and keratinocytes, and the remodeling of the extracellular matrix (ECM) (Kuehlmann et al. 2020; Mascharak et al. 2022; Mascharak et al. 2025; Yin et al. 2022). Milk bioactive compounds can modulate these processes in various ways: by influencing the mechanical properties of the microenvironment, by interacting with surface receptors and messenger proteins, by modifying the composition of the ECM and proteoglycans, and by affecting mechanosensory channels (e.g., Piezo, TRP) and their modulation by lipids (Romero et al. 2019; Yin et al. 2022). Dressings containing bioactive milk substances may alter matrix stress, relaxation, and stiffness gradients, thereby directly affecting cell mechanical sensations (Carton et al. 2024). Therefore, hydrogel design must enable the tuning of elasticity, relaxation, and dynamic deformation while simultaneously controlling the release of bioactive components (Carton et al. 2024). Therefore, the design of the dressing should strive for synergy by matching the mechanical properties of the hydrogel with the controlled release of bioactive milk components. This approach aims to promote cell migration and proliferation while inhibiting chronic inflammation and excessive fibrogenesis (Mascharak et al. 2025).

Milk lipid composition, including unsaturated fatty acids, can influence the mechanical sensitivity of cell membranes and mechanosensory channels by modifying their mechanical properties. Studies have shown that changes in membrane lipid profile (including dietary fat) can modulate Piezo1's response to mechanical stimuli (Romero et al. 2019). In the context of milk, unsaturated omega-3/omega-6 fatty acids and CLA can alter the distribution of membrane lipids and thus influence membrane stiffness and tension. CLA, as an isomer of linoleic acid present in milk fat, potentially lowers the threshold for Piezo channel activation by increasing membrane elasticity, which can modify Ca^{2+} ion flux and the intensity of mechanical signals in wound cells (Romero et al. 2019). This may ultimately lead to modulation of fibroblast migration and angiogenesis by altering the mechanosensory activity of the endothelium.

Milk proteins and peptides, such as lactoferrin, β-lactoglobulin, and caseins, can interact directly or indirectly with integrin receptors, growth factor receptors, and focal adhesion components. By modulating the activity of kinases such as focal adhesion kinase (FAK), Src, and ERK, these compounds can influence the mechanical coupling between the ECM and the cytoskeleton. In this context, insulin, present in milk, may play a role as an indirect regulator of mechanotransduction (Farach-Carson et al. 2024). By activating the PI3K/Akt and MAPK pathways, insulin increases integrin expression and cytoskeletal reorganization, which enhances force transmission between the matrix and the cell interior (Farach-Carson et al. 2024). This action promotes keratinocyte proliferation and migration and modulates cellular tension, thus influencing the mechanosensitivity of skin cells in the healing wound.

Milk bioactive compounds can influence the composition of the ECM and proteoglycans, which alters the stiffness and structure of the mechanical microenvironment. Proteoglycans modulate water retention, stress distribution, and force distribution within the ECM (Farach-Carson et al. 2024). If milk bioactive compounds influence the expression or degradation of proteoglycans, for example, by binding to ECM enzymes or modulating their transcription, they may indirectly alter the mechanical context in which cells receive mechanical signals, which may influence tissue migration, contraction, and remodeling in the wound. In this regard, orotic acid, naturally occurring in milk, may indirectly influence mechanical processes by promoting nucleotide synthesis and increasing fibroblast proliferation, which in turn enhances the generation of cytoskeletal tension and promotes ECM remodeling (Vogel et al. 2015).

Within mechanosensory ion channels, milk bioactive substances may act as modulators. TRPC3 expression increases under stretching conditions, promoting fibronectin production and enhancing profibrotic signals (Yin et al. 2022). Modulation of the lipid membrane by CLA, along with insulin-induced activation of the PI3K/Akt pathway, may jointly enhance cellular mechano-ionic responses (Zhong et al. 2013; Romero et al. 2019).

In summary, milk bioactive components such as CLA, orotic acid, and insulin, among others, may interact with mechanotransduction signals in healing. Integrating these mechanical and metabolic perspectives deepens our understanding of how milk components support tissue regeneration and remodeling.

16.5 Mechanotransduction and Hydrogel Design

Mechanotransduction occurs through integrins, focal adhesion complexes, and the regulation of YAP/TAZ, depending on extracellular matrix (ECM) stiffness and cytoskeletal tension (Zebda et al. 2012). In the context of hydrogel dressings, it is crucial that the hydrogel matrix itself is designed to transmit optimal mechanical stimuli that promote cell migration and proliferation. Simultaneously, it must not provoke mechanosensory over activation, leading to impairment of the wound healing process (Yin et al. 2022). When designing hydrogel dressings, it is crucial to consider appropriate material stiffness. Excessive stiffness can lead to over activation of mechanosensory YAP/TAZ proteins, which in turn can induce excessive extracellular matrix production and scarring (Liu et al. 2015). Studies have shown that modifying the material of hydrogel dressings can significantly increase the elastic modulus and improve mechanical stability under strain and temperature changes (Szatkowski et al. 2024b).

16.5.1 Influence of Matrix Stiffness on Healing

Mechanotransduction occurs through integrins, focal adhesion complexes, and the regulation of YAP/TAZ, depending on extracellular matrix (ECM) stiffness and cytoskeletal tension (Zebda et al. 2012). In the context of hydrogel dressings, it is crucial that the hydrogel matrix itself is designed to transmit optimal mechanical stimuli that promote cell migration and proliferation. Simultaneously, it must not provoke mechanosensory over activation, leading to impairment of the wound healing process (Yin et al. 2022). When designing hydrogel dressings, it is crucial to consider appropriate material stiffness. Studies have shown that modifying the material of hydrogel dressings can significantly increase the elastic modulus and improve mechanical stability under strain and temperature changes (Szatkowski et al. 2024b). Excessive stiffness can lead to over activation of mechanosensory YAP/TAZ proteins, which in turn can induce excessive extracellular matrix production and scarring (Liu et al. 2015). Mechanosensitive proteins YAP and TAZ are the main regulators of mechanotransduction in response to cytoskeleton organization and matrix stiffness. In murine models of hypertrophic scarring, persistent mechanical loading of wounds led to a fibrotic response and hypertrophic scar formation, confirming that mechanomodulation is linked with scarring through the stimulation of inflammation. In pathological conditions such as human hypertrophic scars, the elastic modulus may reach up to 3 MPa, versus 0.2–0.5 MPa in normal skin (Yuan et al. 2024; Lin et al. 2024). The mechanical properties of hydrogels impact fibroblast-to-myofibroblast differentiation, which is central to wound contraction but also excessive scarring. Rigid matrices promote α-SMA (alpha-smooth muscle actin) bundle formation in fibroblasts, increasing contractility and fibrosis. Hydrogels of appropriately tuned stiffness can inhibit excessive fibroblast activation, as confirmed by studies where tailored hydrogel dressings reduced scar area in full thickness wound models (Ishfaq et al. 2023; Chen et al. 2024).

16.5.2 Mechanical Properties in Dressings for Movable Body Regions

For wounds on movable body parts (e.g., joints), the adhesive and elastic properties of hydrogels are particularly important. Highly stretchable hydrogels (up to 4100% elongation, 550 kPa tensile strength) can accommodate body movement while maintaining integrity. Such hydrogels, based on hydrophobic interactions of cellulose nanofibers and polyacrylamide, provide both mechanical strength and self-healing (Liang et al. 2024). The mechanical characteristics of hydrogels also affect angiogenesis and new vessel formation. Studies indicate that optimal stiffness promotes angiogenesis by regulating VEGF and endothelial cell activity; excessive rigidity may inhibit angiogenesis and delay repair (Zhang et al. 2023).

16.5.3 Mechanotransduction in Regenerative Healing

Inhibition of mechanotransduction can promote regenerative repair. Chemical inhibition of YAP using verteporfin blocks postnatal Engrailed-1 (En-1) expression and enhances fibroblast-mediated healing. Treated wounds regain normal hair follicles, glands, and matrix microstructure, along with mechanical strength equivalent to unwounded skin within 30 days. Similar phenotypes have been reported in mice with selective fibroblast YAP knockout or ablation of En-1+ fibroblasts, suggesting En-1 may be a mechanosensitive scar regulator (Li et al. 2024; Zhang et al. 2024c).

16.6 Conclusions

Sheep and camel milk are rich sources of natural bioactive substances that are non-toxic, biocompatible with human tissue, and exhibit broad therapeutic potential. The lyophilization process minimizes nutrient loss and preserves their biological properties. Studies have shown that lyophilized milk preserves the chemical composition of raw milk, allowing its use in the design of functional hydrogels enriched with bioactive milk components (Molik et al. 2024). Hydrogel dressings provide a moist environment conducive to wound healing, accelerate re-epithelialization, and improve cell proliferation, outperforming conventional dressings in supporting regenerative processes. The mechanical properties of a hydrogel, including elastic modulus, extensibility, and strength, play a key role in mechanotransduction. Excessive stiffness can lead to the activation of YAP/TAZ proteins, fibroblast differentiation into myofibroblasts, and scarring (Mascharak et al. 2022). In turn, a well-designed, highly elastic hydrogel can support cell migration and proliferation while minimizing the risk of mechanosensory overactivation. Moreover, appropriately selected stiffness of the hydrogel matrix can inhibit the mechanotransduction pathway FAK-ERK-MCP-1, thus limiting inflammation and the formation of excessive extracellular matrix (Kuehlmann et al. 2020). Hydrogels are characterized by good ductility, a high water content, and favorable oxygen transport, making them among the most promising materials for wound dressings (Zhang et al. 2022). Additionally, they demonstrate improved biodegradability, biocompatibility, and colloidal stability and can support healing processes both through their own properties and as carriers for bioactive substances (Zhang et al. 2022). Hydrogels are safe for health and can positively impact the healing of all types of wounds (Zhang et al. 2022). Furthermore, milk processing can further enhance its bioactive properties, increasing both its technological functionality and biological activity (Flis et al. 2023; Lordan et al. 2019). The introduction of hydrogels enriched with bioactive substances derived from sheep and camel milk enables local, targeted support of wound healing processes. These dressings are natural, biocompatible, biodegradable, and superior to traditional dry dressings, which do not support tissue regeneration and can cause trauma during wound healing (Szatkowski et al. 2024a, 2024b). Due to their safety, biological activity, and ease of manufacture, such dressings have significant therapeutic and market potential, responding to the growing consumer demand for natural products for the treatment of difficult-to-heal wounds.

Competing Interests This study was funded by National Centre for Research and Development, NCBiR, Poland, grant number INNOGLOBO/3/73/HONTGR/2024, entitled "Hydrogel Dressing for Hard-to-Heal Wounds Containing Natural Ingredients from Sheep's and Camel's Milk."

The authors have no conflicts of interest to declare that are relevant to the content of this chapter.

References

Abdalla EB, Anis Ashmawy AEH, Farouk MH, Abd El-Rahman Salama O, Khalil FA, Seioudy AF (2015) Milk production potential in Maghrebi she-camels. Small Rumin Res 123(1):129–135. https://doi.org/10.1016/j.smallrumres.2014.11.004

Abdelhakim M, Ogawa R (2025) Emerging therapies in chronic wound healing: advances in stem cell therapy, growth factor modulation, mechanical strategies and adjuvant interventions. Dermatol Ther (Heidelb):1–13. https://doi.org/10.1007/s13555-025-01564-2

Abdelwahab MM, Abohashema DM, Abdelwahab MS, Fayed MT, Ghanem A, Mohamed OAA (2025) The role of the extracellular matrix (ECM) in wound healing (i.e., matrix metalloproteinases (MMPs) and growth factors). In: Ahmed MG, Ashique S, Farid A, Zengin G (eds) Nanotechnology in wound healing: innovations, applications, and future perspectives, 1st edn. CRC Press, Boca Raton, pp 25–45

Agrawal RP, Jain S, Shah S, Chopra A, Agarwal V (2011) Effect of camel milk on glycemic control and insulin requirement in patients with type 1 diabetes: 2-years randomized controlled trial. Eur J Clin Nutr 65(9):1048–1052. https://doi.org/10.1038/ejcn.2011.98

Agrawal RP, Agrawal R, Faris MAIE, Ghazzawi HA (2020) Potential anti-diabetic effect of camel milk. In: Handbook of research on health and environmental benefits of camel products. IGI Global Scientific Publishing, pp 185–196

Alichanidis E, Moatsou G, Polychroniadou A (2016) Composition and properties of non-cow Milk and products. In: Tsakalidou E, Papadimitriou K (eds) Non-bovine milk and milk products. Academic Press, San Diego, pp 81–116

Apikoglu-Rabus S, Izzettin FV, Turan P, Ercan F (2010) Effect of topical insulin on cutaneous wound healing in rats with or without acute diabetes. Clin Exp Dermatol 5(2):180–185. https://doi.org/10.1111/j.1365-2230.2009.03419.x

Ashraf A, Mudgil P, Palakkott A, Iratni R, Gan CY, Maqsood S, Ayoub MA (2021) Molecular basis of the anti-diabetic properties of camel milk through profiling of its bioactive peptides on dipeptidyl peptidase IV (DPP-IV) and insulin receptor activity. J Dairy Sci 104(1):61–77. https://doi.org/10.3168/jds.2020-18627

Authority (EFSA) EFS (2009) Orotic acid salts as sources of orotic acid and various minerals added for nutritional purposes to food supplements. EFSA J 7(7):1187

Aydin H, Tatar C, Savas OA, Karsidag T, Ozer B, Dursun N, Bekem A, Unal A, Tuzun IS (2019) The effects of local and systemic administration of proline on wound healing in rats. J Investig Surg 32(6):523–529. https://doi.org/10.1080/08941939.2018.1441342

Azevedo F, Pessoa A, Moreira G, Santos MD, Liberti E, Araujo E, Carvalho C, Saad M, Lima MH (2016) Effect of topical insulin on second-degree burns in diabetic rats. Biol Res Nurs 18(2):181–192. https://doi.org/10.1177/1099800415592175

Azevedo FF, Moreira GV, Teixeira CJ, Pessoa AFM, Alves MJ, Liberti EA, Carvalho CRO, Araújo EP, Saad MJA, Lima MHM (2019) Topical insulin modulates inflammatory and proliferative phases of burn-wound healing in diabetes-induced rats. Biol Res Nurs 21(5):473–484. https://doi.org/10.1177/1099800419864443

Badr G (2013) Camel whey protein enhances diabetic wound healing in a streptozotocin-induced diabetic mouse model: the critical role of β-Defensin-1, −2 and −3. Lipids Health Dis 12(1):46. https://doi.org/10.1186/1476-511X-12-46

Bakri HMH, Hakami S, Dira MA, Dira HA, Naseeb W, Maadi M, Mobarki K, Madkhali R, Hakami TM, Sultana F (2023) A systematic review about the consequences of diabetes on peri-implant therapy. Saudi J Oral Dent Res 8(11):345–351. https://doi.org/10.36348/sjodr.2023.v08i11.003

Barnes LA, Marshall CD, Leavitt T, Hu MS, Moore AL, Gonzalez JG, Longaker MT, Gurtner G (2018) Mechanical forces in cutaneous wound healing: emerging therapies to minimize scar formation. Adv Wound Care (New Rochelle) 7:47–56. https://doi.org/10.1089/wound.2016.0709

Basak S, Duttaroy AK (2020) Conjugated linoleic acid and its beneficial effects in obesity, cardiovascular disease, and cancer. Nutrients 12(7):1913. https://doi.org/10.3390/nu12071913

Bassaganya-Riera J, Hontecillas R, Horne WT, Sandridge M, Herfarth HH, Bloomfeld R, Isaacs KL (2012) Conjugated linoleic acid modulates immune responses in patients with mild to moderately active Crohn's disease. Clin Nutr 31(5):721–727. https://doi.org/10.1016/j.clnu.2012.03.002

Behrouz S, Saadat S, Memarzia A, Sarir H, Folkerts G, Boskabady MH (2022) The antioxidant, anti-inflammatory and immunomodulatory effects of camel Milk. Front Immunol 13:855342. https://doi.org/10.3389/fimmu.2022.855342

Bergamo P, Cocca E, Palumbo R, Gogliettino M, Rossi M, Palmieri G (2013) RedOx status, proteasome and APEH: insights into anticancer mechanisms of t10,c12-conjugated linoleic acid isomer on A375 melanoma cells. PLoS One 8(11):e80900. https://doi.org/10.1371/journal.pone.0080900

Bhattacharya A, Banu J, Rahman M, Causey J, Fernandes G (2006) Biological effects of conjugated linoleic acids in health and disease. J Nutr Biochem 17(12):789–810. https://doi.org/10.1016/j.jnutbio.2006.02.009

Bluguermann C, Wu L, Petrigliano F, McAllister D, Miriuka S, Evseenko DA (2013) Novel aspects of parenchymal-mesenchymal interactions: from cell types to molecules and beyond. Cell Biochem Funct 31(4):271–280. https://doi.org/10.1002/cbf.2950

Boniakowski AE, Kimball AS, Jacobs BN, Kunkel SL, Gallagher KA (2017) Macrophage-mediated inflammation in normal and diabetic wound healing. J Immunol 199:17–24. https://doi.org/10.4049/jimmunol.1700415

Brouwer C, Welling MM, Alwasel S, Boekhout T (2025) Potential health benefits of lactoferrin and derived peptides—how to qualify as a medical device? Crit Rev Microbiol 18:1–25. https://doi.org/10.1080/1040841X.2025.2466465

Brun C, Demeaux A, Guaddachi F, Jean-Louis F, Oddos T, Bagot M, Bensussan A, Jauliac S, Michel L (2014) T-plastin expression downstream to the calcineurin/NFAT pathway is involved in keratinocyte migration. PLoS One 9(9):e104700. https://doi.org/10.1371/journal.pone.0104700

Burridge K, Monaghan-Benson E, Graham DM (2019) Mechanotransduction: from the cell surface to the nucleus via RhoA. Philos Trans R Soc Lond Ser B Biol Sci 374(1779):20180229. https://doi.org/10.1098/rstb.2018.0229

Campbell JE, Newgard CB (2021) Mechanisms controlling pancreatic islet cell function in insulin secretion. Nat Rev Mol Cell Biol 22(2):142–158. https://doi.org/10.1038/s41580-020-00317-7

Carton F, Rizzi M, Canciani E, Sieve G, Di Francesco D, Casarella S, Di Nunno L, Boccafoschi F (2024) Use of hydrogels in regenerative medicine: focus on mechanical properties. Int J Mol Sci. 25(21):11426. https://doi.org/10.3390/ijms252111426

Casado JG, Blazquez R, Jorge I, Alvarez V, Gomez-Mauricio G, Ortega-Muñoz M, Vazquez J, Sanchez-Margallo FM (2014) Mesenchymal stem cell-coated sutures enhance collagen depositions in sutured tissues. Wound Repair Regen 22(2):256–264. https://doi.org/10.1111/wrr.12153

Cha B, Geng X, Mahamud MR, Fu J, Mukherjee A, Kim Y, Jho EH, Kim TH, Kahn ML, Xia L, Dixon JB, Chen H, Srinivasan RS (2016) Mechanotransduction activates canonical Wnt/β-catenin signaling to promote lymphatic vascular patterning and the development of lymphatic and lymphovenous valves. Genes Dev 30(12):1454–1469. https://doi.org/10.1101/gad.282400.116

Chandra P, Faizan M, Porwal M, Sharma H, Sachan N (2025) An overview and review of growth factors in wound healing: emerging trends and innovations. Curr Diabetes Rev. https://doi.org/10.2174/0115733998332692241202072249
Chang L, Liang J, Xia X, Chen X (2019) miRNA-126 enhances viability, colony formation, and migration of keratinocytes HaCaT cells by regulating PI3 K/AKT signaling pathway. Cell Biol Int 43(2):182–191. https://doi.org/10.1002/cbin.11088
Chang VS, Lim LH, Su SF, HuangFu J, Teo SS (2025) The wound-healing effect of Kappaphycus alvarezii: in vitro and in vivo. Next Res 2(2):100246. https://doi.org/10.1016/j.nexres.2025.100246
Chen S, Luo Y, He Y, Li M, Liu Y, Zhou X, Hou J, Zhou S (2024) In situ sprayed therapeutic hydrogel for oxygen actuated Janus regulation of postsurgical tumor recurrence/metastasis and wound healing. Nat Commun 15:814. https://doi.org/10.1038/s41467-024-45072-x
Cheng B, Fu X, Sheng Z (2004) Effect of exogenous basic fibroblast growth factor on proliferation and migration of endothelial cells of partial thickness scald in rats. Zhongguo Xiu Fu Chong Jian Wai Ke Za Zhi 18(3):200–204. Chinese
Cheng CH, Hong WJ, Li CN, Huang YH, Tsai JH, Huang CY et al (2025) Colostrum-derived exosomal lactoferrin promotes skin fibroblast regeneration by suppressing inflammatory responses. Curr Issues Mol Biol 47(7):549. https://doi.org/10.3390/cimb47070549
Ching YH, Sutton TL, Pierpont YN, Robson MC, Payne WG (2011) The use of growth factors and other humoral agents to accelerate and enhance burn wound healing. Eplasty 11:e41
Choi JH, Jun JH, Kim JH, Sung HJ, Lee JH (2014) Synergistic effect of interleukin-6 and hyaluronic acid on cell migration and ERK activation in human keratinocytes. J Korean Med Sci 29(Suppl 3):S210–S216. https://doi.org/10.3346/jkms.2014.29.S3.S210
Cioce A, Cavani A, Cattani C, Scopelliti F (2024) Role of the skin immune system in wound healing. Cells 13(7):624. https://doi.org/10.3390/cells13070624
Czauderna M, Białek M, Molik E, Zaworski K (2021) The improved method for determination of orotic acid in milk by ultra-fast liquid chromatography with optimized photodiode array detection. Animals (Basel) 11(11):3196. https://doi.org/10.3390/ani11113196
Darby IA, Hewitson TD (2007) Fibroblast differentiation in wound healing and fibrosis. Int Rev Cytol 257:143–179. https://doi.org/10.1016/S0074-7696(07)57004-X
Di Persio CM, Zheng R, Kenney J, Van De Water L (2016) Integrin-mediated regulation of epidermal wound functions. Cell Tissue Res 365(3):467–482. https://doi.org/10.1007/s00441-016-2446-2
Di-Luoffo M, Ben-Meriem Z, Lefebvre P, Delarue M, Guillermet-Guibert J (2021) PI3K functions as a hub in mechanotransduction. Trends Biochem Sci 46(11):878–888. https://doi.org/10.1016/j.tibs.2021.05.005
Dittmer J, Leyh B (2014) Paracrine effects of stem cells in wound healing and cancer progression (review). Int J Oncol 44(6):1789–1798. https://doi.org/10.3892/ijo.2014.2385
D'souza S, Addepalli V (2018) Preventive measures in oral cancer: An overview. Biomed Pharmacother 107:72–80. https://doi.org/10.1016/j.biopha.2018.07.114
Dupont S, Morsut L, Aragona M, Enzo E, Giulitti S, Cordenonsi M, Zanconato F, Le Digabel J, Forcato M, Bicciato S, Elvassore N, Piccolo S (2011) Role of YAP/TAZ in mechanotransduction. Nature 474:179–183. https://doi.org/10.1038/nature10137
Ebaid H, Abdel-salam B, Hassan I, Al-Tamimi J, Metwalli A, Alhazza I (2015) Camel milk peptide improves wound healing in diabetic rats by orchestrating the redox status and immune response. Lipids Health Dis 14:132. https://doi.org/10.1186/s12944-015-0136-9
El-Baky NA, Elkhawaga MA, Abdelkhalek ES, Sharaf MM, Redwan EM, Kholef HR (2021) De novo expression and antibacterial potential of four lactoferricin peptides in cell-free protein synthesis system. Biotechnol Rep 29:e00583. https://doi.org/10.1016/j.btre.2020.e00583
Emanuelli T, Burgeiro A, Carvalho E (2016) Effects of insulin on the skin: possible healing benefits for diabetic foot ulcers. Arch Dermatol Res 308(10):677–694. https://doi.org/10.1007/s00403-016-1686-z

Fan W, Yang X, Hu X, Huang R, Shi H, Liu G (2024) A novel conductive microtubule hydrogel for electrical stimulation of chronic wounds based on biological electrical wires. J Nanobiotechnology 22(1):258. https://doi.org/10.1186/s12951-024-02524-2
Farach-Carson MC, Wu D, França CM (2024) Proteoglycans in Mechanobiology of tissues and organs: Normal functions and Mechanopathology. Proteoglycan Res 2(2):e21. https://doi.org/10.1002/pgr2.21
Farahani PE, Lemke SB, Dine E, Uribe G, Toettcher JE, Nelson CM (2021) Substratum stiffness regulates erk signaling dynamics through receptor-level control. Cell Rep 37(13):110181. https://doi.org/10.1016/j.celrep.2021.110181
Flis Z, Molik E (2021) Importance of bioactive substances in sheep's Milk in human health. Int J Mol Sci 22(9):4364. https://doi.org/10.3390/ijms22094364
Flis Z, Szczecina J, Molik E (2022) The role of sheep's milk bioactive substances in the prevention of metabolic and viral diseases. J Anim Feed Sci 31(3):211–216. https://doi.org/10.22358/jafs/151020/2022
Flis Z, Szatkowski P, Pielichowska K, Molik E (2023) The potential of sheep or camel Milk constituents to contribute to novel dressings for diabetic wounds. Int J Mol Sci 24(24):17551. https://doi.org/10.3390/ijms242417551
Flis Z, Ptak A, Szatkowski P, Czauderna M, Molik E (2025) Bioactive compounds in sheep milk: the therapeutic potential of CLA, lactoferrin, and orotic acid in skin cancer treatment. J Anim Feed Sci 22(34):42. https://doi.org/10.3390/cimb47060456
Geneidy AK, Abdelnaby MA, Habib DA, Elbedaiwy HM, Shoueir KR (2025) Green synthesis of a lactoferrin-infused silver nanoparticle gel for enhanced wound healing. Sci Rep 15(1):15243. https://doi.org/10.1038/s41598-025-94450-y
Gracey E, Burssens A, Cambré I, Schett G, Lories R, McInnes IB, Asahara H, Elewaut D (2020) Tendon and ligament mechanical loading in the pathogenesis of inflammatory arthritis. Nat Rev Rheumatol 16:193–207. https://doi.org/10.1038/s41584-019-0364-x
Groeber F, Holeiter M, Hampel M, Hinderer S, Schenke-Layland K (2011) Skin tissue engineering—in vivo and in vitro applications. Adv Drug Deliv Rev 63:352–366
Häkkinen L, Koivisto L, Heino J, Larjava H (2015) Cell and molecular biology of wound healing. In: Vishwakarma A, Sharpe P, Shi S, Ramalingam M (eds) Stem cell biology and tissue engineering in dental sciences. Elsevier, Amsterdam, pp 165–188. https://doi.org/10.1016/B978-0-12-397157-9.00010-9
Hallan SS, Ferrara F, Cortesi R, Sguizzato M (2025) Potential of the Nano-encapsulation of antioxidant molecules in wound healing applications: An innovative strategy to enhance the bio-profile. Molecules 30(3):641. https://doi.org/10.3390/molecules30030641
Hassan ME, Ismail M (2019) Wound healing in the urinary tract. In: Aboumarzouk OM (ed) Blandy's urology, 3rd edn. Hoboken, Wiley-Blackwell, pp 19–26. https://doi.org/10.1002/9781118863343.ch6
Hocking AM, Gibran NS (2010) Mesenchymal stem cells: paracrine signaling and differentiation during cutaneous wound repair. Exp Cell Res 316(14):2213–2219. https://doi.org/10.1016/j.yexcr.2010.05.009
Hrynyk M, Neufeld RJ (2014) Insulin and wound healing. Burns J Int Soc Burn Inj 40(8):1433–1446. https://doi.org/10.1016/j.burns.2014.03.020
Hsu YH, Lin YF, Chen CH, Chiu YJ, Chiu HW (2017) Far infrared promotes wound healing through activation of Notch1 signaling. J Mol Med (Berl) 95(11):1203–1213. https://doi.org/10.1007/s00109-017-1580-y
Hu X, Yu W, Sun H, Wang X, Han C (2012) Epidermal cells delivered for cutaneous wound healing. J Dermatolog Treat 23(3):224–237. https://doi.org/10.3109/09546634.2010.495741
Hu X, Zhang H, Li X, Li Y, Chen Z (2020) Activation of mTORC1 in fibroblasts accelerates wound healing and induces fibrosis in mice. Wound Repair Regen 28(1):6–15. https://doi.org/10.1111/wrr.12759
Husain A, Khadka A, Ehrlicher A, Saint-Geniez M, Krishnan R (2022) Substrate stiffening promotes VEGF-A functions via the PI3K/Akt/mTOR pathway. Biochem Biophys Res Commun 586:27–33. https://doi.org/10.1016/j.bbrc.2021.11.030

Hwang DM, Kundu JK, Shin JW, Lee JC, Lee HJ, Surh YJ (2007) Cis-9,trans-11-conjugated linoleic acid down-regulates phorbol ester-induced NF-kappaB activation and subsequent COX-2 expression in hairless mouse skin by targeting IkappaB kinase and PI3K-Akt. Carcinogenesis 28(2):363–371. https://doi.org/10.1093/carcin/bgl151

Ishfaq B, Khan IU, Khalid SH, Asghar S (2023) Design and evaluation of sodium alginate based hydrogel dressings containing Betula utilis extract for cutaneous wound healing. Front Bioeng Biotechnol 11:1042077. https://doi.org/10.3389/fbioe.2023.1042077

Ito M, Liu Y, Yang Z, Nguyen J, Liang F, Morris RJ, Cotsarelis G (2005) Stem cells in the hair follicle bulge contribute to wound repair but not to homeostasis of the epidermis. Nat Med 11:1351. https://doi.org/10.1038/nm1328

Ito M, Yang Z, Andl T, Cui C, Kim N, Millar SE, Cotsarelis G (2007) Wnt-dependent de novo hair follicle regeneration in adult mouse skin after wounding. Nature 447:316–320. https://doi.org/10.1038/nature05766

Iyer K, Chen Z, Ganapa T, Wu BM, Tawil B, Linsley CS (2018) Keratinocyte migration in a three-dimensional in vitro wound healing model co-cultured with fibroblasts. Tissue Eng Regen Med 15(6):721–733. https://doi.org/10.1007/s13770-018-0145-7

Izadi A, Khedmat L, Mojtahedi SY (2019) Nutritional and therapeutic perspectives of camel milk and its protein hydrolysates: a review on versatile biofunctional properties. J Funct Foods 60:103441. https://doi.org/10.1016/j.jff.2019.103441

Jastrzębska E, Wadas E, Daszkiewicz T, Pietrzak-Fiećko R (2017) Nutritional value and health-promoting properties of mare's milk – a review. Czeh J Anim Sci 62(12):511–518. https://doi.org/10.17221/61/2016-CJAS

Jia X, Dou Z, Zhang Y, Li F, Xing B, Hu Z, Li X, Liu Z, Yang W, Liu Z (2023) Smart responsive and controlled release hydrogels for chronic wound treatment. Pharmaceutics. 15(12):2735. https://doi.org/10.3390/pharmaceutics15122735

JL HD (2019) Conjugated linoleic acid effects on cancer, obesity, and atherosclerosis: a review of pre-clinical and human trials with current perspectives. Nutrients 11(2):370. https://doi.org/10.3390/nu11020370

Johnson BZ, Stevenson AW, Prêle CM, Fear MW, Wood FM (2020) The role of IL-6 in skin fibrosis and cutaneous wound healing. Biomedicine 8:101. https://doi.org/10.3390/biomedicines8050101

Kalavathy G, Baskar G, Phoebe MGL, Muthaszeer AM, Subashini S, Chang SK (2025) Formulation of topical ointment using seaweed extracts of ulva lactuca and codium tomentosum and their physicochemical, antioxidant, anti-inflammatory, antibacterial and wound healing properties. Waste Biomass Valor:1–12. https://doi.org/10.1007/s12649-025-03052-2

Kananivand M, Nouri F, Yousefi MH, Pajouhi A, Ghorbani H, Afkhami H, Razavi ZS (2025) Mesenchymal stem cells and their exosomes: a novel approach to skin regeneration via signaling pathways activation. J Mol Histol 56(3):123–135. https://doi.org/10.1007/s10735-025-10394-7

Kanazawa S, Fujiwara T, Matsuzaki S, Shingaki K, Taniguchi M, Miyata S, Tohyama M, Sakai Y, Yano K, Hosokawa K, Kubo T (2010) bFGF regulates PI3-kinase-Rac1-JNK pathway and promotes fibroblast migration in wound healing. PLoS One 5(8):e12228. https://doi.org/10.1371/journal.pone.0012228

Karatas F, Aydin S, Kaygusuzoglu E, Yildiz H, Erulas FA, Ozkan Y (2008) Ghrelin and orotic acid increased in subclinical mastitis. Arch Physiol Biochem 114(3):178–182. https://doi.org/10.1080/13813450802180940

Kasowanjete P, Kumar SSD, Houreld NN (2024) A review of photobiomodulation on PI3K/AKT/mTOR in wound healing. J Photochem Photobiol 19:100215. https://doi.org/10.1016/j.jpap.2023.100215

Katoh K (2025) Integrin and its associated proteins as a mediator for Mechano-signal transduction. Biomolecules 15:166. https://doi.org/10.3390/biom15020166

Kell DB, Heyden EL, Pretorius E (2020) The biology of lactoferrin, an iron-binding protein that can help defend against viruses and bacteria. Front Immunol 11:86–99. https://doi.org/10.1016/j.smallrumres.2019.07.009

Khattab MA, Nabeh OA, Adel S, Abdelazeem M, Naser MM, Matter LM (2025) Topical insulin improves postoperative wound healing in controlled diabetic patients through regulating the expression of E-cadherin and Ki67: an open-label randomized controlled-trial. Future J Pharm Sci 11(1):80. https://doi.org/10.1186/s43094-025-00830-6

Kim EK, Tucker DF, Yun SJ, Do KH, Kim MS, Kim JH, Kim CD, Birnbaum MJ, Bae SS (2008) Linker region of Akt1/protein kinase bα mediates platelet-derived growth factor-induced translocation and cell migration. Cell Signal 20(11):2030–2037. https://doi.org/10.1016/j.cellsig.2008.07.012

Kim M, Kim J, Shin YK, Kim KY (2020) Gentisic acid stimulates keratinocyte proliferation through ERK1/2 phosphorylation. Int J Med Sci 17(5):626–631. https://doi.org/10.7150/ijms.36484

Kim J, Lee YI, Mun S, Jeong J, Lee DG, Kim M, Jo H, Lee S, Han K, Lee JH (2023) Efficacy and safety of epidermidibacterium keratini EPI-7 derived postbiotics in skin aging: a prospective clinical study. Int J Mol Sci 24(5):4634. https://doi.org/10.3390/ijms24054634

Kiwanuka E, Junker J, Eriksson E (2012) Harnessing growth factors to influence wound healing. Clin Plast Surg 39(3):239–248. https://doi.org/10.1016/j.cps.2012.04.003

Klein EA, Yin L, Kothapalli D, Castagnino P, Byfield FJ, Xu T, Levental I, Hawthorne E, Janmey PA, Assoian RK (2009) Cell-cycle control by physiological matrix elasticity and in vivo tissue stiffening. Curr Biol 19(18):1511–1518. https://doi.org/10.1016/j.cub.2009.07.069

Ko UH, Choi J, Choung J, Moon S, Shin JH (2019) Physicochemically tuned Myofibroblasts for wound healing strategy. Sci Rep 9(1):16070. https://doi.org/10.1038/s41598-019-52523-9

Koivisto L, Heino J, Häkkinen L, Larjava H (2014) Integrins in wound healing. Adv Wound Care (New Rochelle). 3(12):762–783. https://doi.org/10.1089/wound.2013.0436

Kondaveeti SB, Tripathi AK, Kumar D, Kumar A, Singh TG, Mahmood S, Ghosal K, Awasthi A, Kumar M (2025) Signalling pathways triggering therapeutic marine algae-derived natural polysaccharides for effective wound healing: a recent review. Carbohydr Polym 369:124253. https://doi.org/10.1016/j.carbpol.2025.124253

Krejner A, Litwiniuk M, Grzela T (2016) Matrix metalloproteinases in the wound microenvironment: therapeutic perspectives. Chronic Wound Care Manag Res 3:29–39. https://doi.org/10.2147/CWCMR.S84725

Krishnaswamy VR, Korrapati PS (2014) Role of dermatopontin in re-epithelialization: implications on keratinocyte migration and proliferation. Sci Rep 4:7385. https://doi.org/10.1038/srep07385

Kuehlmann B, Bonham CA, Zucal I, Prantl L, Gurtner GC (2020) Mechanotransduction in wound healing and fibrosis. J Clin Med 9(5):1423. https://doi.org/10.3390/jcm9051423

Kumar A, Boriek AM (2003) Mechanical stress activates the nuclear factor-kappaB pathway in skeletal muscle fibers: a possible role in Duchenne muscular dystrophy. FASEB J 17(3):386–396. https://doi.org/10.1096/fj.02-0542com

Kumar Srivastava A, Khare P, Kumar Nagar H, Raghuwanshi N, Srivastava R (2016) Hydroxyproline: a potential biochemical marker and its role in the pathogenesis of different diseases. Curr Protein Pept Sci 17(6):596–602. https://doi.org/10.2174/1389203717666151201192247

Kural MH, Billiar KL (2013) Regulating tension in three-dimensional culture environments. Exp Cell Res 319(16):2447–2459. https://doi.org/10.1016/j.yexcr.2013.06.019

Lacina L, Kolář M, Pfeiferová L, Gál P, Smetana K Jr (2024) Wound healing: insights into autoimmunity, ageing, and cancer ecosystems through inflammation and IL-6 modulation. Front Immunol 15:1403570. https://doi.org/10.3389/fimmu.2024.1403570

Lam M, Calvo F (2019) Regulation of mechanotransduction: emerging roles for septins. Cytoskeleton (Hoboken) 76(1):115–122. https://doi.org/10.1002/cm.21485

León Malaga FN, Espinoza MA, Munive-Degregori A, Barja-Ore J, Mauricio-Vilchez C, Mayta-Tovalino F (2023) Uses and applications of Lactoferrin in preclinical studies for successful bone regeneration process: a literature review. J Int Oral Health 15(4):337. https://doi.org/10.4103/jioh.jioh_221_22

Lewis CJ, Mardaryev A, Sharpe D, Botchkareva N (2015) Inhibition of bone morphogenetic protein signalling promotes wound healing in a human ex vivo model. Eur J Plast Surg 38(2):151–158. https://doi.org/10.1007/s00238-014-1031-8

Li J, Chen J, Kirsner R (2007) Pathophysiology of acute wound healing. Clin Dermatol 25:9–18. https://doi.org/10.1016/j.clindermatol.2006.09.007

Li M, Luan F, Zhao Y, Hao H, Liu J, Dong L, Fu X, Han W (2017) Mesenchymal stem cell-conditioned medium accelerates wound healing with fewer scars. Int Wound J 14(1):64–73. https://doi.org/10.1111/iwj.12551

Li C, An Y, Sun Y, Yang F, Xu Q, Wang Z (2022) Adipose mesenchymal stem cell-derived exosomes promote wound healing through the WNT/β-catenin signaling pathway in dermal fibroblasts. Stem Cell Rev Rep 18(6):2059–2073. https://doi.org/10.1007/s12015-022-10378-0

Li W, Yu Y, Huang R, Wang X, Lai P, Chen K, Shang L, Zhao Y (2023) Multi bioinspired functional conductive hydrogel patches for wound healing management. Adv Sci (Weinh). 10(25):e2301479. https://doi.org/10.1002/advs.202301479

Li N, Lu X, Yang Y, Ning S, Tian Y, Zhou M, Wang Z, Wang L, Zang J (2024) Calcium peroxide-based hydrogel patch with sustainable oxygenation for diabetic wound healing. Adv Healthcare Mater 13(16):2303314. https://doi.org/10.1002/adhm.202303314

Liang X, Ding L, Ma J, Li J, Cao L, Liu H, Teng M, Li Z, Peng Y, Chen H, Zheng Y, Cheng H, Liu G (2024) Enhanced mechanical strength and sustained drug release in carrier free silver coordinated anthraquinone natural antibacterial anti inflammatory hydrogel for infectious wound healing. Adv Healthcare Mater. 13(23):2400841. https://doi.org/10.1002/adhm.202400841

Lin Y, Zhang Y, Cai X, He H, Yang C, Ban J, Guo B (2024) Design and self assembly of peptide copolymer conjugates into nanoparticle hydrogel for wound healing in diabetes. Int J Nanomedicine 19:2487–2506. https://doi.org/10.2147/IJN.S452915

Lin Y, Song S, Guo H (2025) Recent advances in synergistic effect of lactoferrin. Curr Opin Food Sci 63:101285. https://doi.org/10.1111/j.1472-765X.2006.02076.x

Liu F, Lagares D, Choi KM, Stopfer L, Marinković A, Vrbanac V, Probst CK, Hiemer SE, Sisson TH, Horowitz JC, Rosas IO, Fredenburgh LE, Feghali-Bostwick CA, Varelas X, Tager AM, Tschumperlin DJ (2015) Mechanosignaling through YAP and TAZ drives fibroblast activation and fibrosis. Am J Physiol Lung Cell Mol Physiol 308(4):L344–L357. https://doi.org/10.1152/ajplung.00300.2014

Löffler M, Carrey EA, Zameitat E (2015) Orotic acid, more than just an intermediate of pyrimidine *de novo* synthesis. J Genet Genomics 42(5):207–219. https://doi.org/10.1016/j.jgg.2015.04.001

Lordan R, Walsh AM, Crispie F, Finnegan L, Cotter PD, Zabetakis I (2019) The effect of ovine milk fermentation on the antithrombotic properties of polar lipids. J Funct Foods 54:289–300. https://doi.org/10.1016/j.jff.2019.01.029

MacDonald HB (2000) Conjugated linoleic acid and disease prevention: a review of current knowledge. J Am Coll Nutr 19:111S–118S. https://doi.org/10.1080/07315724.2000.10718082

Mansour N, Shadeed W, Hamood S (2015) Clinical study on the effect of the sheep milk fat globules on deep second degree burns on pigs. Assiut Vet Med J 61(147):1–14. https://doi.org/10.21608/avmj.2015.170258

Marconi GD, Fonticoli L, Rajan TS, Lanuti P, Della Rocca Y, Pierdomenico SD, Trubiani O, Pizzicannella J, Diomede F (2021) Transforming growth factor-beta1 and human gingival fibroblast-to-myofibroblast differentiation: molecular and morphological modifications. Front Physiol 12:676512. https://doi.org/10.3389/fphys.2021.676512

Marynowicz W, Borski N, Flis Z, Ptak A, Molik E (2023) Orotic acid induces apoptotic death in ovarian adult granulosa tumour cells and increases mitochondrial activity in normal ovarian granulosa cells. Reprod Biol 23(3):100790. https://doi.org/10.1016/j.repbio.2023.100790

Mascharak S, HE dJ-P, Davitt MF, Guardino NJ, Gurtner GC, Wan DC, Longaker MT (2022) Modulating Cellular Responses to Mechanical Forces to Promote Wound Regeneration. Adv Wound Care (New Rochelle) 11(9):479–495. https://doi.org/10.1089/wound.2021.0040

Mascharak S, Griffin M, Talbott HE, Guo JL, Parker J, Morgan AG, Valencia C, Kuhnert MM, Li DJ, Liang NE, Kratofil RM, Daccache JA, Sidhu I, Davitt MF, Guardino N, Lu JM, Abbas DB,

Deleon NMD, Lavin CV, Adem S, Khan A, Chen K, Henn D, Spielman A, Cotterell A, Akras D, Downer M Jr, Tevlin R, Lorenz HP, Gurtner GC, Januszyk M, Naik S, Wan DC, Longaker MT (2025) Inhibiting mechanotransduction prevents scarring and yields regeneration in a large animal model. Sci Transl Med 17(809):eadt6387. https://doi.org/10.1126/scitranslmed.adt6387
Mayya C, Kharbhanda S, Haque A, Bhatia D (2021) Mechanisms of collective cell migration in wound healing: physiology and disease. Wound Heal Res Curr Trends Future Dir 1:75–92. https://doi.org/10.1007/978-981-33-6044-0_5
Miller B, Sewell-Loftin MK (2022) Mechanoregulation of vascular endothelial growth factor receptor 2 in angiogenesis. Front Cardiovasc Med 8:804934. https://doi.org/10.3389/fcvm.2021.804934
Mkrtschjan MA, Gaikwad SB, Kappenman KJ, Solís C, Dommaraju S, Le LV, Desai TA, Russell B (2018) Lipid signaling affects primary fibroblast collective migration and anchorage in response to stiffness and microtopography. J Cell Physiol 233(4):3672–3683. https://doi.org/10.1002/jcp.26236
Moatsou G, Sakkas L (2019) Sheep milk components: focus on nutritional advantages and biofunctional potential. Small Rumin Res 180:86–99
Molik E, Stańko K, Flis Z (2021) Prozdrowotne znaczenie substancji bioaktywnych mleka owczego. [The health benefits of bioactive substances in sheep's milk]. Prz Hod 89(3):18–21
Molik E, Szatkowski P, Kotowicz G, Flis Z, Pustkowiak H, Zapletal P (2024) Comparison of the content of bioactive substances in goat milk and goat milk lyophilizate. Anim Sci Genet 20(4):39–48. https://doi.org/10.5604/01.3001.0054.9892
Nam S, Gupta VK, Lee HP, Lee JY, Wisdom KM, Varma S, Flaum EM, Davis C, West RB, Chaudhuri O (2019) Cell cycle progression in confining microenvironments is regulated by a growth-responsive TRPV4-PI3K/Akt-p27Kip1 signaling axis. Sci Adv 5(8):eaaw6171. https://doi.org/10.1126/sciadv.aaw6171
Nath M, Vats M, Roy P (2013) Tri- and diorganotin (IV) complexes of biologically important orotic acid: synthesis, spectroscopic studies, in vitro anti-cancer, DNA fragmentation, enzyme assays and in vivo anti-inflammatory activities. Eur J Med Chem 59:310–321. https://doi.org/10.1016/j.ejmech.2012.11.023
Nong NTP, Hsu JL (2021) Characteristics of food protein-derived antidiabetic bioactive peptides: a literature update. Int J Mol Sci 22(17):9508. https://doi.org/10.3390/ijms22179508
Pang X, He X, Qiu Z, Zhang H, Xie R, Liu Z, Gu Y, Zhao N, Xiang Q, Cui Y (2023) Targeting integrin pathways: mechanisms and advances in therapy. Signal Transduct Target Ther 8(1):1. https://doi.org/10.1038/s41392-022-01259-6
Pardoux É, Boturyn D, Roupioz Y (2020) Antimicrobial peptides as probes in biosensors detecting whole bacteria: a review. Molecules 5(8):1998. https://doi.org/10.3390/molecules25081998
Parfenyuk EV, Dolinina ES, Kraev AS (2024) Synthesis and study of organo-modified silica based hydrogels: rheological properties and drug release kinetics. J Biomed Mater Res B Appl Biomater 112(6):e35418. https://doi.org/10.1002/jbm.b.35418
Park NY, Valacchi G, Lim Y (2010) Effect of dietary conjugated linoleic acid supplementation on early inflammatory responses during cutaneous wound healing. Mediat Inflamm 2010:342328. https://doi.org/10.1155/2010/342328
Park JS, Chu JS, Tsou AD, Diop R, Tang Z, Wang A, Li S (2011) The effect of matrix stiffness on the differentiation of mesenchymal stem cells in response to TGF-β. Biomaterials 32(16):3921–3930. https://doi.org/10.1016/j.biomaterials.2011.02.019
Park J, Kim TY, Kim Y, An S, Kim KS, Kang M, Kim SA, Kim J, Lee J, Cho SW, Seo J (2023) A mechanically resilient and tissue conformable hydrogel with hemostatic and antibacterial capabilities for wound care. Adv Sci (Weinh) 10(30):e2303651. https://doi.org/10.1002/advs.202303651
Patel JC, Shukla M, Shukla M (2025) Cellular and molecular interactions in CNS injury: the role of immune cells and inflammatory responses in damage and repair. Cells 14:918. https://doi.org/10.3390/cells14120918
Patilas C, Varsamos I, Galanis A, Vavourakis M, Zachariou D, Marougklianis V, Kolovos I, Tsalimas G, Karampinas P, Kaspiris A, Vlamis J, Pneumaticos S (2024) The role of Interleukin-

10 in the pathogenesis and treatment of a spinal cord injury. Diagnostics (Basel) 14(2):151. https://doi.org/10.3390/diagnostics14020151
Plikus MV, Wang X, Sinha S, Forte E, Thompson SM, Herzog EL, Driskell RR, Rosenthal N, Biernaskie J, Horsley V (2021) Fibroblasts: origins, definitions, and functions in health and disease. Cell 184(15):3852–3872. https://doi.org/10.1016/j.cell.2021.06.024
Porel P, Kaur M, Sharma V, Aran KR (2025) Understanding molecular mechanism of diabetic wound healing: addressing recent advancements in therapeutic managements. J Diabetes Metab Disord 24(1):76
Qin S, Ricotta V, Simon M, Clark RA, Rafailovich MH (2015) Continual cell deformation induced via attachment to oriented fibers enhances fibroblast cell migration. PLoS One 10(3):e0119094. https://doi.org/10.1371/journal.pone.0119094
Rafiq S, Huma N, Pasha I, Sameen A, Mukhtar O, Khan MI (2016) Chemical composition, nitrogen fractions and amino acids profile of milk from different animal species. Asian Australas J Anim Sci 29(7):1022–1028. https://doi.org/10.5713/ajas.15.0452
Ragazzini S, Scocozza F, Bernava G, Auricchio F, Colombo GI, Barbuto M, Conti M, Pesce M, Garoffolo G (2022) Mechanosensor YAP cooperates with TGF-β1 signaling to promote myofibroblast activation and matrix stiffening in a 3D model of human cardiac fibrosis. Acta Biomater 152:300–312. https://doi.org/10.1016/j.actbio.2022.08.063
Raina N, Rani R, Gupta M (2021) Angiogenesis: aspects in wound healing. In: Chatterjee S (ed) Endothelial signaling in vascular dysfunction and disease. Academic Press, Cambridge, pp 77–90
Raman N, Imran SAM, Ahmad Amin Noordin KB, Zaman WSWK, Nordin F (2022) Mechanotransduction in mesenchymal stem cells (MSCs) differentiation: a review. Int J Mol Sci 23(9):4580. https://doi.org/10.3390/ijms23094580
Ramirez-GarciaLuna JL, Rangel-Berridi K, Bergeron A, Kolosovas-Machuca ES, Wang SC, Berry GK, Martinez-Jimenez MA (2023) Local insulin improves wound healing: a systematic review and Bayesian network meta-analysis. Plast Reconstr Surg 52(6):1114e. https://doi.org/10.1097/PRS.0000000000010432
Raziyeva K, Kim Y, Zharkinbekov Z, Kassymbek K, Jimi S, Saparov A (2021) Immunology of acute and chronic wound healing. Biomolecules. 11:700. https://doi.org/10.3390/biom11050700
Rithika R, Vishali A, Malathi R (2025) Synthesis of collagen nanoparticles from eggshell membrane and its applications with special emphasis on wound healing and water purification. J Stress Physiol Biochem 21(3):47–64
Romero LO, Massey AE, Mata-Daboin AD, Sierra-Valdez FJ, Chauhan SC, Cordero-Morales JF, Vásquez V (2019) Dietary fatty acids fine-tune Piezo1 mechanical response. Nat Commun 10(1):1200. https://doi.org/10.1038/s41467-019-09055-7
Rousselle P, Braye F, Dayan G (2019) Re-epithelialization of adult skin wounds: cellular mechanisms and therapeutic strategies. Adv Drug Deliv Rev 146:344–365. https://doi.org/10.1016/j.addr.2018.06.019
Schilrreff P, Alexiev U (2022) Chronic inflammation in non-healing skin wounds and promising natural bioactive compounds treatment. Int J Mol Sci 23(9):4928. https://doi.org/10.3390/ijms23094928
Sée V, Rajala NK, Spiller DG, White MR (2004) Calcium-dependent regulation of the cell cycle via a novel MAPK–NF-kappaB pathway in Swiss 3T3 cells. J Cell Biol 166(5):661–672. https://doi.org/10.1083/jcb.200402136
Seetharaman S, Vianay B, Roca V, Farrugia AJ, De Pascalis C, Boëda B, Dingli F, Loew D, Vassilopoulos S, Bershadsky A, Théry M, Etienne-Manneville S (2022) Microtubules tune mechanosensitive cell responses. Nat Mater 21(3):366–377. https://doi.org/10.1038/s41563-021-01108-x
Shaw TJ, Martin P (2016) Wound repair: a showcase for cell plasticity and migration. Curr Opin Cell Biol 42:29–37. https://doi.org/10.1016/j.ceb.2016.04.001
Shibani M, Ringseis R, Alkazali M, Kerfakh O, Eder K (2011) Concentrations of conjugated linoleic acids in milk and tissues from single-humped Arabian camel (Camelus dromedaries)

kept under intensive standardized management. Afr J Agric Res 6(15):3470–3474. https://doi.org/10.5897/AJAR11.581

Shukla SK, Sharma AK, Gupta V, Yashavarddhan MH (2019) Pharmacological control of inflammation in wound healing. J Tissue Viability 28(4):218–222. https://doi.org/10.1016/j.jtv.2019.09.002

Sivaraj D, Chen K, Chattopadhyay A, Henn D, Wu W, Noishiki C, Magbual NJ, Mittal S, Mermin Bunnell AM, Bonham CA, Trotsyuk AA, Barrera JA, Padmanabhan J, Januszyk M, Gurtner GC (2021) Hydrogel scaffolds to deliver cell therapies for wound healing. Front Bioeng Biotechnol 9:660145. https://doi.org/10.3389/fbioe.2021.660145

Skutek M, van Griensven M, Zeichen J, Brauer N, Bosch U (2001) Cyclic mechanical stretching enhances secretion of interleukin 6 in human tendon fibroblasts. Knee Surg Sports Traumatol Arthrosc 9:322–326. https://doi.org/10.1007/s001670100217

Srinivasan A, Fults ML, Supronowicz P, Esquivel R, Zamilpa R (2018) Mesenchymal stem cell-derived products for tissue repair and regeneration. In: A roadmap to nonhematopoietic stem cell-based therapeutics: from the bench to the clinic, pp 419–441. https://doi.org/10.1016/B978-0-12-811920-4.00017-3

Sriramulu S, Banerjee A, Jothimani G, Pathak S (2020) Conditioned medium from the human umbilical cord-mesenchymal stem cells stimulate the proliferation of human keratinocytes. J Basic Clin Physiol Pharmacol 32(2):51–56. https://doi.org/10.1515/jbcpp-2019-0283

Stamm A, Reimers K, Strauß S, Vogt P, Scheper T, Pepelanova I (2016) In vitro wound healing assays—state of the art. BioNanoMaterials 17:1–10. https://doi.org/10.1515/bnm-2016-0002

Steward AJ, Kelly DJ (2015) Mechanical regulation of mesenchymal stem cell differentiation. J Anat 227(6):717–731. https://doi.org/10.1111/joa.12243

Su S, Begum S, Ezratty EJ (2020) An IFT20 mechanotrafficking axis is required for integrin recycling, focal adhesion dynamics, and polarized cell migration. Mol Biol Cell 31(17):1917–1930. https://doi.org/10.1091/mbc.E20-04-0232

Szatkowski P, Flis Z, Ptak A, Molik E (2024a) Biomateriały hydrożelowe możliwe do stosowania w leczeniu trudno gojących się ran. [hydrogel biomaterials that can be used in the treatment of difficult-to-heal wounds.]. Rocz Nauk Zoot 2051:2,135–2,145. https://doi.org/10.58146/th7n-ep73

Szatkowski P, Flis Z, Ptak A, Molik E (2024b) Hydrogel dressing biomaterial enriched with Vitamin C: synthesis and characterization. Int J Mol Sci 25(19):10565. https://doi.org/10.3390/ijms251910565

Takahashi M, Umehara Y, Yue H, Trujillo-Paez JV, Peng G, Nguyen HLT, Ikutama R, Okumura K, Ogawa H, Ikeda S, Niyonsaba F (2021) The antimicrobial peptide human β-Defensin-3 accelerates wound healing by promoting angiogenesis, cell migration, and proliferation through the FGFR/JAK2/STAT3 signaling pathway. Front Immunol 12:712781. https://doi.org/10.3389/fimmu.2021.712781

Takayama Y, Aoki R (2012) Roles of lactoferrin on skin wound healing. Biochem Cell Biol 90(3):293–301. https://doi.org/10.1139/o2012-016

Talati N, Kamato D, Piva TJ, Little PJ, Osman N (2018) Thrombin promotes PAI-1 expression and migration in keratinocytes via ERK dependent smad linker region phosphorylation. Cell Signal 47:37–43. https://doi.org/10.1016/j.cellsig.2018.03.009

Tang L, Cao X, Li X, Ding H (2021) Topical application with conjugated linoleic acid ameliorates 2, 4-dinitrofluorobenzene-induced atopic dermatitis-like lesions in BALB/c mice. Exp Dermatol 30(2):237–248. https://doi.org/10.1111/exd.14242

Thangavel P, Ramachandran B, Kannan R, Muthuvijayan V (2017) Biomimetic hydrogel loaded with silk and l-proline for tissue engineering and wound healing applications. J Biomed Mater Res B Appl Biomater 105(6):1401–1408. https://doi.org/10.1002/jbm.b.33675

Tong J, Wang Z (2017) Analysis of epidermal growth factor receptor-induced cell motility by wound healing assay. Methods Mol Biol 1652:159–163. https://doi.org/10.1007/978-1-4939-7219-7_12

Tschumperlin DJ, Liu F, Tager AM (2013) Biomechanical regulation of mesenchymal cell function. Curr Opin Rheumatol 25(1):92–100. https://doi.org/10.1097/BOR.0b013e32835b13cd

Valero C, Javierre E, García-Aznar JM, Gómez-Benito MJ (2014) A cell-regulatory mechanism involving feedback between contraction and tissue formation guides wound healing progression. PLoS One 9(3):e92774. https://doi.org/10.1371/journal.pone.0092774

Van De Water L, Varney S, Tomasek JJ (2013) Mechanoregulation of the myofibroblast in wound contraction, scarring, and fibrosis: opportunities for new therapeutic intervention. Adv Wound Care (New Rochelle) 2:122–141. https://doi.org/10.1089/wound.2012.0393

Veerman D, Prakash J (2023) Integrins: key targets in tissue fibrosis and tumor stroma. Biology of Extracellular Matrix 2023:99–133. https://doi.org/10.1007/978-3-031-23781-2_3

Vij N, Sharma A, Thakkar M, Sinha S, Mohan RR (2008) PDGF-driven proliferation, migration, and IL8 chemokine secretion in human corneal fibroblasts involve JAK2-STAT3 signaling pathway. Mol Vis 14:1020–1027

Vogel R, Brünig H, Häussler L (2015) Properties of polypropylene fibers using the green chemical orotic acid as nucleating agent. Adv Mater Res 4:207–214. https://doi.org/10.12989/amr.2015.4.4.207

Wager LJ, Leavesley DI (2015) MicroRNA regulation of epithelial-to-mesenchymal transition during re-epithelialisation: assessing an open wound. Wound Pract Res 23(3):132–142

Wang Y, Jones PJ (2004) Dietary conjugated linoleic acid and body composition123. Am J Clin Nutr 79(6):1153S–1158S. https://doi.org/10.1093/ajcn/79.6.1153S

Wang X, Zhu Y, Sun C, Wang T, Shen Y, Cai W, Sun J, Chi L, Wang H, Song N, Niu C, Shen J, Cong W, Zhu Z, Xuan Y, Li X, Jin L (2017) Feedback activation of basic fibroblast growth factor signaling via the Wnt/β-catenin pathway in skin fibroblasts. Front Pharmacol 8:32. https://doi.org/10.3389/fphar.2017.00032

Wang Z, Qi F, Luo H, Xu G, Wang D (2022) Inflammatory microenvironment of skin wounds. Front Immunol 13:789274. https://doi.org/10.3389/fimmu.2022.789274

Wang B, Pang M, Song Y, Wang H, Qi P, Bai S, Lei X, Wei S, Zong Z, Lin S, Zhang X, Cen X, Wang X, Yang Y, Li Y, Wang Y, Xu H, Huang L, Tortorella M, Cheng B, Lee Y, Qin D, Li G (2022a) Human fetal mesenchymal stem cells secretome promotes scarless diabetic wound healing through heat-shock protein family. Bioeng Transl Med 8(1):e10354. https://doi.org/10.1002/btm2.10354

Wang B, Zhao J, Lu W, Ma Y, Wang X, An X, Fan Z (2022b) The preparation of lactoferrin/magnesium silicate lithium injectable hydrogel and application in promoting wound healing. Int J Biol Macromol 220:1501–1511. https://doi.org/10.1016/j.ijbiomac.2022.09.126

Wang Z, Cao K, Yan D, Ge Y, Li R, Liu Y, Ma T, Sun X (2023) A study of the role of multiple layer-by-layer assembled bionic extracellular matrix in promoting wound healing via activation of the Wnt signaling pathway. J Biomed Mater Res B Appl Biomater 111(6):1023–1034. https://doi.org/10.1002/jbm.b.35222

Wang L, Lu X, Wang Y, Sun L, Fan X, Wang X, Bai J (2025) Research Progress on the application of novel wound healing dressings in different stages of wound healing. Pharmaceutics 17:976. https://doi.org/10.3390/pharmaceutics17080976

West T, Chunduru J, Murahari E (2017) Orotic acid: why it is important to understand its role in metabolism. Biochem Physiol open. Access 06:e157. https://doi.org/10.4172/2168-9652.1000e157

Wiesner J, Vilcinskas A (2010) Antimicrobial peptides: the ancient arm of the human immune system. Virulence 1(5):440–464. https://doi.org/10.4161/viru.1.5.12983

Wilkinson HN, Hardman MJ (2020) Wound healing: cellular mechanisms and pathological outcomes. Open Biol 10:200223. https://doi.org/10.1098/rsob.200223

Won KJ, Lee R, Choi SH, Kim JH, Hwang SH, Nah SY (2023) Gintonin-induced wound-healing-related responses involve epidermal-growth-factor-like effects in keratinocytes. Int J Mol Sci 24(18):14094. https://doi.org/10.3390/ijms241814094

Wong VW, Paterno J, Sorkin M, Glotzbach JP, Levi K, Januszyk M, Rustad KC, Longaker MT, Gurtner GC (2011) Mechanical force prolongs acute inflammation via T-cell-dependent pathways during scar formation. FASEB J 25(12):4498–4510. https://doi.org/10.1096/fj.10-178087

Wu Y, Chen L, Scott PG, Tredget EE (2007) Mesenchymal stem cells enhance wound healing through differentiation and angiogenesis. Stem Cells 25(10):2648–2659. https://doi.org/10.1634/stemcells.2007-0226

Wu F, Song L, Gong Y, Wang Y, Li H, Zhang S (2023) Antimicrobial peptide Pt5-1c promotes keratinocyte migration and proliferation via EGFR-mediated Akt/MAPK/STAT3 pathways. ChemistrySelect 8:e202203707. https://doi.org/10.1002/slct.202203707

Xu S, Zhao L, Li Y, Gu X, Liu Z, Han X, Li W, Ma W (2024) Activating the healing process: three-dimensional culture of stem cells in Matrigel for tissue repair. BMC Biotechnol 24(1):36. https://doi.org/10.1186/s12896-024-00862-5

Yao WD, Zhou JN, Tang C, Zhang JL, Chen ZY, Li Y, Gong XJ, Qu MY, Zeng Q, Jia YL, Wang HY, Fan T, Ren J, Guo LL, Xi JF, Pei XT, Han Y, Yue W (2024) Hydrogel microneedle patches loaded with stem cell mitochondria enriched microvesicles boost the chronic wound healing. ACS Nano 18(39):26733–26750. https://doi.org/10.1021/acsnano.4c06921

Yasmin H, Ramesh RPG, Joseph AM, Kishore U (2025) Fibroblast heterogeneity and its role in generating protective immunity in the secondary lymphoid organs. Front Immunol 16:1519789. https://doi.org/10.3389/fimmu.2025.1519789

Yin J, Zhang S, Yang C, Wang Y, Shi B, Zheng Q, Zeng N, Huang H (2022) Mechanotransduction in skin wound healing and scar formation: potential therapeutic targets for controlling hypertrophic scarring. Front Immunol 17(13):1028410. https://doi.org/10.3389/fimmu.2022.1028410

Yin D, Zhang H, Yang S-M, Yin R, Zhang W (2023) An image analysis tool for cell migration assay with a large sample size. Proc SPIE 12566:500–504. https://doi.org/10.1117/12.2667479

Yu Lin MO, Sampath D, Bosykh DA, Wang C, Wang X, Subramaniam T, Han W, Hong W, Chakraborty S (2025) YAP/TAZ drive Agrin-matrix metalloproteinase 12-mediated diabetic skin wound healing. J Invest Dermatol 145:155–170.e2. https://doi.org/10.1016/j.jid.2024.05.005

Yuan R, Fang Z, Liu F, He X, Du S, Zhang N, Zeng Q, Wei Y, Wu Y, Tao L (2024) Ferrocene based antioxidant self healing hydrogel via the Biginelli reaction for wound healing. ACS Macro Lett 13(5):475–482. https://doi.org/10.1021/acsmacrolett.4c00063

Yue Q, Song Y, Liu Z, Zhang L, Yang L, Li J (2022) Receptor for advanced glycation end products (RAGE): a pivotal hub in immune diseases. Molecules 27(15):4922. https://doi.org/10.3390/molecules27154922

Zebda N, Dubrovskyi O, Birukov KG (2012) Focal adhesion kinase regulation of mechanotransduction and its impact on endothelial cell functions. Microvasc Res 83(1):71–81. https://doi.org/10.1016/j.mvr.2011.06.007

Zhang Z, Lv L (2016) Effect of local insulin injection on wound vascularization in patients with diabetic foot ulcer. Exp Ther Med 11(2):397–402. https://doi.org/10.3892/etm.2015.2917

Zhang M, Liu NY, Wang XE, Chen YH, Li QL, Lu KR, Sun L, Jia Q, Zhang L, Zhang L (2011) Activin B promotes epithelial wound healing in vivo through RhoA-JNK signaling pathway. PLoS One 6(9):e25143. https://doi.org/10.1371/journal.pone.0025143

Zhang YM, Zhang ZQ, Liu YY, Zhou X, Shi XH, Jiang Q, Fan DL, Cao C (2015) Requirement of Gαi1/3-Gab1 signaling complex for keratinocyte growth factor-induced PI3K-AKT-mTORC1 activation. J Invest Dermatol 135(1):181–191. https://doi.org/10.1038/jid.2014.326

Zhang E, Gao B, Yang L, Wu X, Wang Z (2016) Notoginsenoside Ft1 promotes fibroblast proliferation via PI3K/Akt/mTOR signaling pathway and benefits wound healing in genetically diabetic mice. J Pharmacol Exp Ther 356:324–332. https://doi.org/10.1124/jpet.115.229369

Zhang X, Wei P, Yang Z, Liu Y, Yang K, Cheng Y, Yao H, Zhang Z (2022) Current Progress and outlook of Nano-based hydrogel dressings for wound healing. Pharmaceutics 15(1):68. https://doi.org/10.3390/pharmaceutics15010068

Zhang T, Guo L, Li R, Shao J, Lu L, Yang P, Zhao A, Liu Y (2023) Ellagic acid cyclodextrin inclusion complex loaded thiol ene hydrogel with antioxidant, antibacterial, and anti inflammatory properties for wound healing. ACS Appl Mater Interfaces 15(4):4959–4972. https://doi.org/10.1021/acsami.2c20229

Zhang L, Liu J, Wang Y, Wei M, Liu X, Jiang Y, Wang X, Zhu Z, Niu C, Liu S, Cui J, Chu T, Lu W, Zhang X, An X, Song Y (2024a) Mechanisms by which sheep milk consumption ameliorates insulin resistance in high-fat diet-fed mice. Food Res Int Ott Ont 179:114021. https://doi.org/10.1016/j.foodres.2024.114021

Zhang X, Liang Y, Huang S, Guo B (2024b) Chitosan based self healing hydrogel dressing for wound healing. Adv Colloid Interface Sci 332:103267. https://doi.org/10.1016/j.cis.2024.103267

Zhang X, Wu Y, Gong H, Xiong Y, Chen Y, Li L, Zhi B, Lv S, Peng T, Zhang H (2024c) A multifunctional herb derived glycopeptide hydrogel for chronic wound healing. Small 20(36):2400516. https://doi.org/10.1002/smll.202400516

Zhong X, Wang H, Jian X (2013) Insulin augments mechanical strain-induced ERK activation and cyclooxygenase-2 expression in MG63 cells through integrins. Exp Ther Med 7(1):295–299. https://doi.org/10.3892/etm.2013.1394

Zhou S, Xie M, Su J, Cai B, Li J, Zhang K (2023) New insights into balancing wound healing and scarless skin repair. J Tissue Eng 14:20417314231185848. https://doi.org/10.1177/20417314231185848

Zhou X, Yu X, You T, Zhao B, Dong L, Huang C, Zhou X, Xing M, Qian W, Luo G (2024) 3D printing based hydrogel dressings for wound healing. Adv Sci (Weinh) 11(47):e2404580. https://doi.org/10.1002/advs.202404580

Zhu G, Zhu X, Fan Q, Wan X (2011) Raman spectra of amino acids and their aqueous solutions. Spectrochim Acta A Mol Biomol Spectrosc 78(3):1187–1195. https://doi.org/10.1016/j.saa.2010.12.079

Zuo Z, Yang Z, Zhao J, Han L, Yang J, Wang Y, Zhang Z, Zhou D (2025) Recombinant human fibronectin mediates macrophage polarization via NF-κB/TGF-β1 pathway to enhance fibroblast proliferation. Biol Cell 117(9):e70025. https://doi.org/10.1111/boc.70025

GPSR Compliance

The European Union's (EU) General Product Safety Regulation (GPSR) is a set of rules that requires consumer products to be safe and our obligations to ensure this.

If you have any concerns about our products, you can contact us on ProductSafety@springernature.com

In case Publisher is established outside the EU, the EU authorized representative is:

Springer Nature Customer Service Center GmbH
Europaplatz 3
69115 Heidelberg, Germany

Batch number: 10755222

Printed by Printforce, the Netherlands